National Center for Construction Education and Research

Electrical Level One

Annotated Instructor's Guide

Prentice
Hall

Upper Saddle River, New Jersey
Columbus, Ohio

This information is general in nature and intended for training purposes only. Actual performance of activities described in this manual requires compliance with all applicable operating, service, maintenance, and safety procedures under the direction of qualified personnel. References in this manual to patented or proprietary devices do not constitute a recommendation of their use.

Prentice Hall

nccer

10 9 8 7 6 5 4 3 2
ISBN 0-13-018901-4

Preface

This volume was developed by the National Center for Construction Education and Research (NCCER) in response to the training needs of the construction and maintenance industries. It is one of many in the NCCER's standardized craft training program. The program, covering more than 30 craft areas and including all major construction skills, was developed over a period of years by industry and education specialists. Sixteen of the largest construction and maintenance firms in the U.S. committed financial and human resources to the teams that wrote the curricula and planned the nationally accredited training process. These materials are industry-proven and consist of competency-based textbooks and instructor's guides.

The NCCER is a nonprofit educational entity affiliated with the University of Florida and supported by the following industry and craft associations:

PARTNERING ASSOCIATIONS

- American Fire Sprinkler Association
- American Society for Training & Development
- American Welding Society
- Associated Builders & Contractors, Inc.
- Association for Career and Technical Education
- Associated General Contractors of America
- The Business Roundtable
- Carolinas AGC, Inc.
- Carolinas Electrical Contractors Association
- Construction Industry Institute
- Design-Build Institute of America
- Merit Contractors Association of Canada

- Metal Building Manufacturers Association
- National Association of Minority Contractors
- National Association of Women in Construction
- National Insulation Association
- National Ready Mixed Concrete Association
- National Utility Contractors Association
- National Vocational Technical Honor Society
- North American Crane Bureau
- Painting & Decorating Contractors of America
- Portland Cement Association
- SkillsUSA-VICA
- Steel Erectors Association of America
- Texas Gulf Coast Chapter ABC
- U.S. Army Corps of Engineers
- University of Florida
- Women Construction Owners & Executives, USA

Some of the features of the NCCER's standardized craft training program include:

- A proven record of success over many years of use by industry companies.
- National standardization providing portability of learned job skills and educational credits that will be of tremendous value to trainees.
- Recognition: upon successful completion of training with an accredited sponsor, trainees receive an industry-recognized certificate and transcript from the NCCER.
- Compliance with Apprenticeship, Training, Employer, and Labor Services (ATELS) requirements (formerly BAT) for related classroom training (CFR 29:29).
- Well-illustrated, up-to-date, and practical information.

COMMEMORATE 2000 WITH COLOR!

To kick off the new millennium, the NCCER and Prentice Hall will begin publishing select text-books in color. *Electrical Level One* is one of the first crafts to incorporate a brand-new design and lay-out. In addition to full-color photos and illustrations, also look for the following new pedagogical features designed to augment the technical material and make the book even more user friendly:

Inside Track provides a head start for those entering the field by presenting tricks of the trade from master electricians in a variety of disciplines.

Think About It uses "What If" questions to help trainees apply theory to real-world experiences and put ideas into action.

What's wrong with this picture? includes photos of actual code violations for identification and encourages trainees to approach each installation with a critical eye.

Case Histories emphasize the importance of safety by citing examples of the costly (and often devastating) consequences of ignoring NEC or OSHA regulations.

Profiles in Success share the apprenticeship and career experiences of various industry leaders, each of whom offers a unique perspective on how to succeed in the industry.

We're excited to offer you these new features and hope they help to foster a more rewarding learning experience. As always, your feedback is welcome! Visit the NCCER at www.NCCER.org or e-mail us at info@NCCER.org.

ANNOTATED INSTRUCTOR'S GUIDES—DESIGNED TO MAKE YOUR JOB EASIER!

We understand that the course material you choose may have a profound effect on your success. The best way we can help to ensure your success is to provide you with the finest tools we have available. Beginning in 2000, NCCER and Prentice Hall will be releasing select titles as Annotated Instructor's Guides. Each Annotated Instructor's Guide is actually the Trainee Module enhanced with specific directions to the instructor, space for the instructor's notes, suggestions for session break-outs, a comprehensive Materials and Equipment list, and teaching tips to coincide with the performance examinations, laboratories, and demonstrations.

The Electrical Level One Annotated Instructor's Guide is packaged with an accompanying test booklet (which includes the performance tests, written module exams, and answer keys) and the Instructor's version of the Trainee Workbook. Please note that the Transparency Masters package and Test Scrambling Software are sold separately for an additional charge.

For more information on the Annotated Instructor's Guides or to find out which titles are planned for annotations in the future, please contact your Prentice Hall Sales Specialist or NCCER Customer Service at info@nccer.org.

Acknowledgments

This curriculum was revised as a result of the farsightedness and leadership of the following organizations:

Bay Harbour Electric
Beacon Electric Company
Carolinas Electrical
   Contractors Association
Circle Electric, Inc.
Edgewel Electric, Inc.
Hamilton Electric

Pelican Chapter ABC
TIC, The Industrial Company
Tri-City Electric
Washington CITC
Watson Electrical Construction
Wisconsin ABC

This curriculum would not exist were it not for the dedication and unselfish energy of those volunteers who served on the Authoring Team. A sincere thanks is extended to:

Jerry Bass
Clarence "Ed" Cockrell
Gary Edgington
Tim Ely
Al Hamilton
E.L. Jarrell
L.J. LeBlanc

Jim Mitchem
Robert Mueller
Christine Porter
Mike Powers
Rob Soileau
Joe Sullivan

A final note: This book is the result of a collaborative effort involving the production, editorial, and development staff at Prentice-Hall, Inc. and the National Center for Construction Education and Research. Thanks to all of the dedicated people involved in the many stages of this project.

Contents

Electrical Safety

26101-00

MODULE OVERVIEW

This course introduces the electrical trainee to the safety rules and regulations for electricians, including the necessary precautions for avoiding various job site hazards.

PREREQUISITES

Please refer to the Course Map in the Trainee Module. Prior to training with this module, it is recommended that the trainee shall have successfully completed the following modules:

Core Curriculum

LEARNING OBJECTIVES

Upon completion of this module, the trainee will be able to:

1. Demonstrate safe working procedures in a construction environment.
2. Explain the purpose of OSHA and how it promotes safety on the job.
3. Identify electrical hazards and how to avoid or minimize them in the workplace.
4. Explain safety issues concerning lockout/tagout procedures, personal protection using assured grounding and isolation programs, confined space entry, respiratory protection, and fall protection systems.

PERFORMANCE OBJECTIVES

Under supervision of the instructor, the trainee should be able to:

1. Perform a visual inspection and an air test on rubber gloves.
2. Perform a hazard assessment of a job such as replacing the lights in your classroom.
   - Discuss the work to be performed and the hazards involved.
   - Locate the closest phone to the work site and ensure that the local emergency telephone numbers are either posted at the phone or known by you and your partner(s).
   - Plan an escape route from the location in the event of an accident.

NCCER STANDARDIZED CRAFT TRAINING PROGRAM

The National Center for Construction Education and Research (NCCER) provides a standardized national program of accredited craft training. Key features of the program include instructor certification, competency-based training, and performance testing. The program provides trainees, instructors, and companies with a standard form of recognition through a National Craft Training Registry. The program is described in full in the *Guidelines for Accreditation*, published by the NCCER. For more information on standardized craft training, contact the NCCER at P.O. Box 141104, Gainesville, FL 32614-1104, 352-334-0911, visit our Web site at www.NCCER.org, or e-mail info@NCCER.org.

HOW TO USE THIS ANNOTATED INSTRUCTOR'S GUIDE

Each page presents two sections of information. The larger section displays each page exactly as it appears in the Trainee Module. The narrow column ties suggested trainee and instructor actions to each page and provides icons to call your attention to material, safety, audiovisual, or testing requirements. The bottom of each page includes space for your notes.

If you see the Teaching Tip icon, that means there is a teaching tip associated with this section. Also refer to the suggested teaching tips at the end of the module.

SAFETY CONSIDERATIONS

Ensure that the trainees are equipped with appropriate personal protective equipment. Emphasize that the basic safety habits established in apprenticeship are crucial to a long and healthy career as an electrician.

PREPARATION

Before teaching this module, you should review the Module Outline, Learning and Performance Objectives, and the Materials and Equipment List. Be sure to allow ample time to prepare your own training or lesson plan and gather all required equipment and materials.

MATERIALS AND EQUIPMENT LIST

Materials:

Transparencies

Markers/chalk

Lockout/tagout devices and labels

Copy of the latest edition of the *National Electrical Code*®

OSHA Electrical Safety Guidelines (pocket guide)

Module Examinations*

Performance Profile Sheets*

Equipment:

Overhead projector and screen

Whiteboard/chalkboard

Various types of personal protective and safety equipment, including rubber gloves, insulating blankets, lockout/tagout devices, hot sticks, fuse pullers, shorting probes, safety glasses, and face shields

Access to an eye wash station

*Located in the Test Booklet packaged with this Annotated Instructor's Guide.

ADDITIONAL RESOURCES

This module is intended to present thorough resources for task training. The following reference works are suggested for both instructors and motivated trainees interested in further study. These are optional materials for continued education rather than for task training.

29 CFR Parts 1900 – 1910, Standards for General Industry. Occupational Safety and Health Administration, US Department of Labor.

29 CFR Part 1926, Standards for the Construction Industry. Occupational Safety and Health Administration, US Department of Labor.

National Electrical Code Handbook, Latest Edition. Quincy, MA: National Fire Protection Association.

National Electrical Safety Code, Latest Edition. Quincy, MA: National Fire Protection Association.

NOTES

The designations "National Electrical Code," "NE Code," and "NEC," where used in this document, refer to the *National Electrical Code*®, which is a registered trademark of the National Fire Protection Association, Quincy, MA. All National Electrical Code (NEC) references in this module refer to the 1999 edition of the NEC.

If you feel that additional math instruction would be helpful, Prentice Hall offers a basic math textbook entitled *Fundamentals of Electrical and Mechanical Mathematics.* It covers the basic math requirements for electrical trainees and may be ordered by contacting Prentice Hall Customer Service at 1-800-922-0579.

TEACHING TIME FOR THIS MODULE

An outline for use in developing your lesson plan is presented below. Note that each Roman numeral in the outline equates to one session of instruction. Each session has a suggested time period of 2½ hours. This includes 10 minutes at the beginning of each session for administrative tasks and one 10-minute break during the session. Approximately 12½ hours are suggested to cover *Electrical Safety*. You will need to adjust the time required for hands-on activity and testing based on your class size and resources.

Topic	Planned Time
Session I. Introduction to Electrical Hazards; Laboratory	
A. Electrical Shock	_____
1. Body Resistance	_____
2. Burns	_____
B. Protective Equipment	_____
1. Rubber Protective Equipment	_____
2. Protective Apparel	_____
3. Personal Clothing	_____
4. Hot Sticks	_____
5. Fuse Pullers	_____
6. Shorting Probes	_____
7. Eye and Face Protection	_____
C. Laboratory	_____
Under instructor supervision, have the trainees practice examining rubber gloves using the visual inspection and air tests.	_____
D. Verifying Deenergized Circuits	_____
E. Basic Safety Precautions	_____
Session II. OSHA Standards	
A. Section-by-Section Review of OSHA Standards 1910 and 1926	_____
1. Design Safety Standards for Electrical Systems	_____
2. Electric Utilization Systems	_____
3. General Requirements	_____
4. Wiring Design and Protection	_____
5. Wiring Methods, Components, and Equipment for General Use	_____
6. Specific Purpose Equipment and Installations	_____
7. Hazardous (Classified) Locations	_____
8. Special Systems	_____
9. Scope	_____
10. Training	_____
11. Selection and Use of Work Practices	_____
12. Use of Equipment	_____
13. Safeguards for Personnel Protection	_____
B. OSHA Safety Topics	_____
1. Safety Philosophy and General Safety Precautions	_____
2. Electrical Regulations and Lockout/Tagout Rule	_____
3. Other OSHA Regulations	_____

Session III. Ladders and Scaffolds

    A. Ladders   _____

       1. Straight Ladders   _____

       2. Extension Ladders   _____

       3. Step Ladders   _____

    B. Scaffolds   _____

Session IV. General Construction Safety Topics; Laboratory;
**   Confined Space Entry; First Aid**

    A. Lifts, Hoists, and Cranes   _____

    B. Lifting   _____

    C. Laboratory   _____

       Under instructor supervision, have the trainees practice proper
       lifting procedures.   _____

    D. Basic Tool Safety   _____

    E. Confined Space Entry Procedures   _____

       1. General Guidelines   _____

       2. Confined Space Hazard Review   _____

       3. Entry and Work Procedures   _____

    F. First Aid   _____

Session V. Hazardous Materials; Laboratory; Fall Protection;
**   Module Examination and Performance Testing**

    A. Solvents and Toxic Vapors   _____

       1. Precautions When Using Solvents   _____

       2. Respiratory Protection   _____

    B. Asbestos   _____

       1. Monitoring   _____

       2. Regulated Areas   _____

       3. Methods of Compliance   _____

    C. Batteries   _____

       1. Acids   _____

       2. Eye Wash Stations   _____

    D. Laboratory   _____

       Under instructor supervision, have the trainees practice using an eye
       wash station.   _____

    E. PCBs _____

    F. Fall Protection   _____

       1. Fall Protection Procedures   _____

       2. Types of Fall Protection Systems   _____

    G. Module Examination   _____

       1. Trainees must score 70% or higher to receive recognition from the NCCER.

       2. Record the testing results on Craft Training Report Form 200 and submit
          the results to the Training Program Sponsor.

    H. Performance Testing   _____

       1. Trainees must perform each task to the satisfaction of the instructor to
          receive recognition from the NCCER.

       2. Record the testing results on Craft Training Report Form 200 and submit
          the results to the Training Program Sponsor.

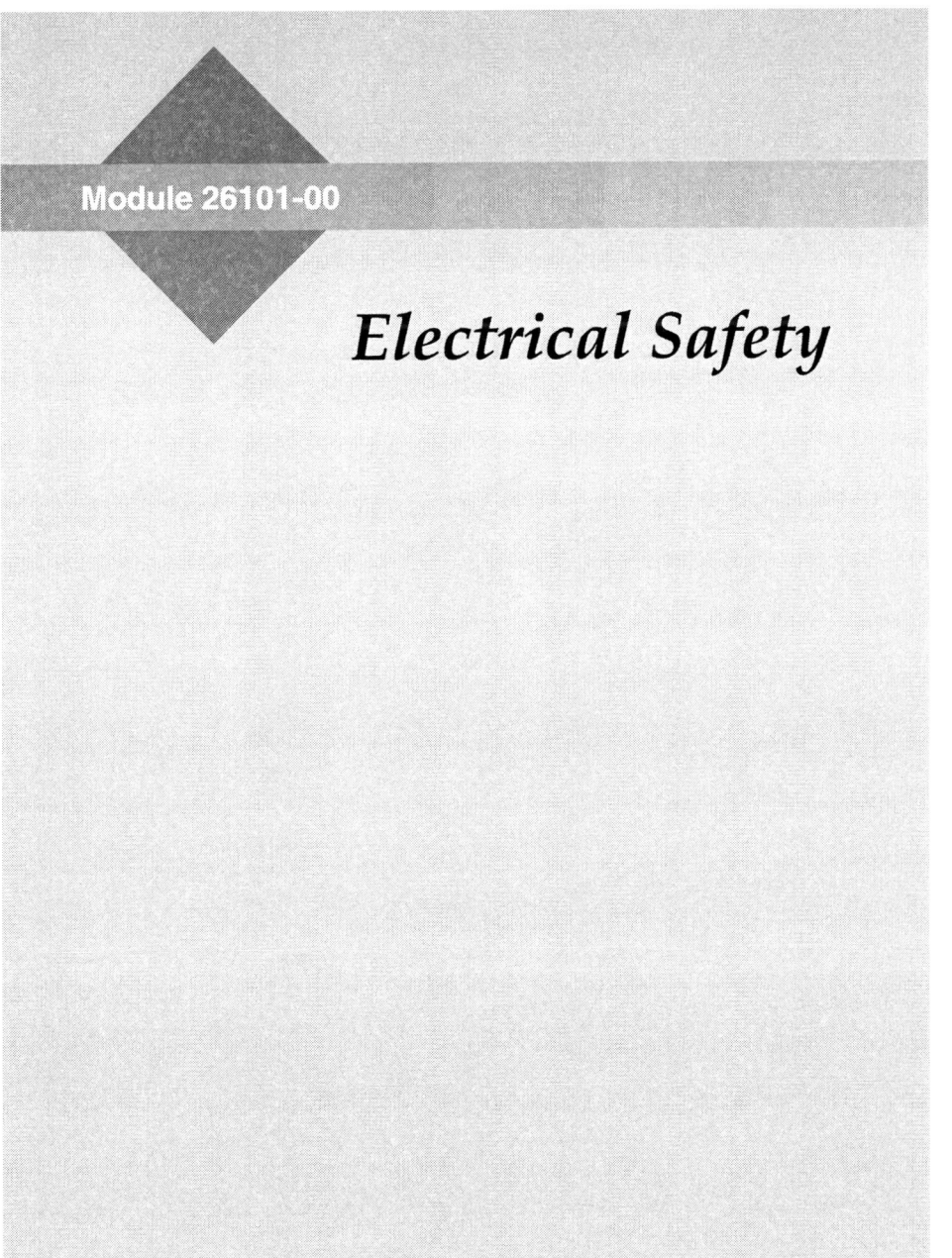

Module 26101-00

Electrical Safety

Instructor's Notes:

ELECTRICAL LEVEL ONE—INSTRUCTOR'S GUIDE MODULE 26101

Course Map

This course map shows all of the task modules in the first level of the Electrical curricula. The suggested training order begins at the bottom and proceeds up. Skill levels increase as a trainee advances on the course map. The training order may be adjusted by the local Training Program Sponsor.

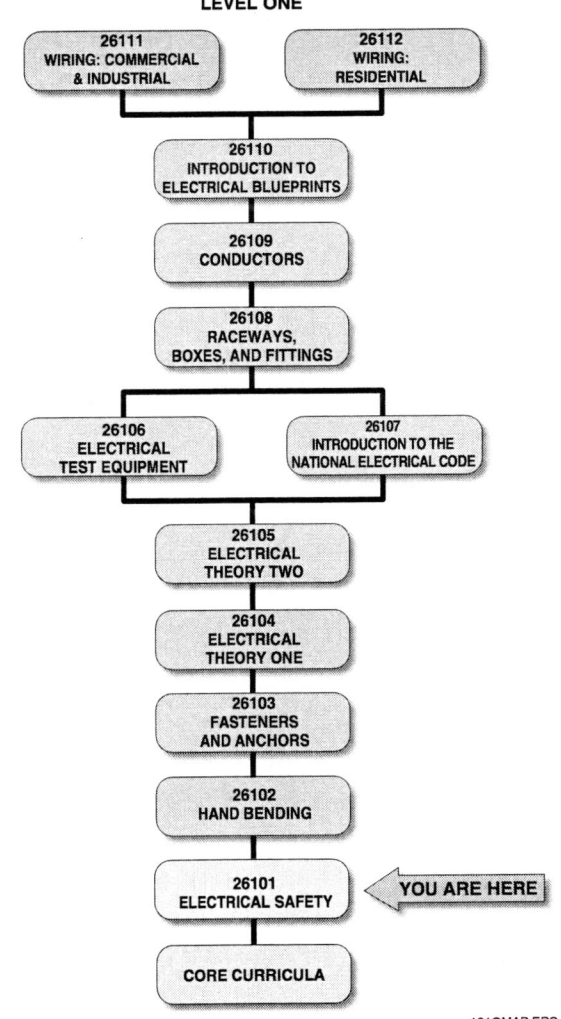

LEVEL ONE

- 26111 WIRING: COMMERCIAL & INDUSTRIAL
- 26112 WIRING: RESIDENTIAL
- 26110 INTRODUCTION TO ELECTRICAL BLUEPRINTS
- 26109 CONDUCTORS
- 26108 RACEWAYS, BOXES, AND FITTINGS
- 26106 ELECTRICAL TEST EQUIPMENT
- 26107 INTRODUCTION TO THE NATIONAL ELECTRICAL CODE
- 26105 ELECTRICAL THEORY TWO
- 26104 ELECTRICAL THEORY ONE
- 26103 FASTENERS AND ANCHORS
- 26102 HAND BENDING
- 26101 ELECTRICAL SAFETY ◄ YOU ARE HERE
- CORE CURRICULA

101CMAP.EPS

Assign reading of Module 26101.

1–iii

Instructor's Notes:

_____

_____

_____

_____

_____

_____

_____

_____

Figures

Tables

Materials

Ensure you have everything required to teach the course. Check the Materials and Equipment List at the front of this Instructor's Guide.

ELECTRICAL LEVEL ONE—INSTRUCTOR'S GUIDE MODULE 26101

Electrical Safety

Objectives

Upon completion of this module, the trainee will be able to:

1. Demonstrate safe working procedures in a construction environment.
2. Explain the purpose of OSHA and how it promotes safety on the job.
3. Identify electrical hazards and how to avoid or minimize them in the workplace.
4. Explain safety issues concerning lockout/tagout procedures, personal protection using assured grounding and isolation programs, confined space entry, respiratory protection, and fall protection systems.

Prerequisites

Successful completion of the following Task Modules is recommended before beginning study of this Task Module: Core Curricula.

Required Trainee Materials

1. Trainee Task Module
2. Copy of the latest edition of the *National Electrical Code*
3. *OSHA Electrical Safety Guidelines* (pocket guide)
4. Appropriate Personal Protective Equipment

1.0.0 ◆ INTRODUCTION

As an electrician, you will be exposed to many potentially hazardous conditions that will exist on the job site. No training manual, set of rules and regulations, or listing of hazards can make working conditions completely safe. However, it is possible for an electrician to work a full career without serious accident or injury. To reach this goal, you need to be aware of potential hazards and stay constantly alert to these hazards. You must take the proper precautions and practice the basic rules of safety. You must be safety-conscious at all times. Safety should become a habit. Keeping a safe attitude on the job will go a long way in reducing the number and severity of accidents. Remember that your safety is up to you.

As an apprentice electrician, you need to be especially careful. You should only work under the direction of experienced personnel who are familiar with the various job site hazards and the means of avoiding them.

The most life-threatening hazards on a construction site are:

- Falls when you are working in high places
- Electrocution caused by coming into contact with live electrical circuits
- The possibility of being crushed by falling materials or equipment

Note: The designations "National Electrical Code," "NE Code," and "NEC," where used in this document, refer to the *National Electrical Code*®, which is a registered trademark of the National Fire Protection Association, Quincy, MA. *All National Electrical Code (NEC) references in this module refer to the 1999 edition of the NEC.*

Show Transparency 1, Course Objectives.

Show Transparency 2, Performance Profile Tasks.

Discuss the main types of hazards found on construction sites.

1–1

Electrical Safety in the Workplace

Each year in the U.S., there are approximately 20,000 electricity-related accidents at home and in the workplace. In a recent year, these accidents resulted in 700 deaths. Electrical accidents are the third leading cause of death in the workplace.

- The possibility of being struck by flying objects or moving equipment/vehicles such as trucks, forklifts, and construction equipment

Other hazards include cuts, burns, back sprains, and getting chemicals or objects in your eyes. Most injuries, both those that are life-threatening and those that are less severe, are preventable if the proper precautions are taken.

What's wrong with this picture?

MIKE POWERS   101PO101.EPS

2.0.0 ◆ ELECTRICAL SHOCK

Electricity can be described as a potential that results in the movement of electrons in a conductor. This movement of electrons is called *electrical current*. Some substances, such as silver, copper, steel, and aluminum, are excellent conductors. The human body is also a conductor. The conductivity of the human body greatly increases when the skin is wet or moistened with perspiration.

Electrical current flows along the path of least resistance to return to its source. The source return point is called the *neutral* or *ground* of a circuit. If the human body contacts an electrically energized point and is also in contact with the ground or another point in the circuit, the human body becomes a path for the current to return to its source. *Table 1* shows the effects of current passing through the human body. One mA is one milliamp, or one one-thousandth of an ampere.

A primary cause of death from electrical shock is when the heart's rhythm is overcome by an electrical current. Normally, the heart's operation uses a very low-level electrical signal to cause the heart

Table 1 Current Level Effects on the Human Body

Current Value	Typical Effects
Less than 1mA	No sensation.
1 to 20mA	Sensation of shock, possibly painful. May lose some muscular control between 10 and 20mA.
20 to 50mA	Painful shock, severe muscular contractions, breathing difficulties.
50 to 200mA	Same symptoms as above, only more severe, up to 100mA. Between 100 and 200mA, ventricular fibrillation may occur. Typically results in almost immediate death unless special medical equipment and treatment are available.
Over 200mA	Severe burns and muscular contractions. The chest muscles contract and stop the heart for the duration of the shock.

ELECTRICAL LEVEL ONE—TRAINEE MODULE 26101

Teaching Tip

Discuss "What's wrong with this picture?"

Instructor's Notes:

_____

_____

_____

_____

_____

_____

_____

_____

_____

ELECTRICAL LEVEL ONE—INSTRUCTOR'S GUIDE MODULE 26101

THINK ABOUT IT

Severity of Shock

In Table 1, how many milliamps separate a mild shock from a potentially fatal one? What is the fractional equivalent of this in amps? How many amps are drawn by a 60W light bulb?

Teaching Tip

Discuss the "Think About It."

to contract and pump blood. When an abnormal electrical signal, such as current from an electrical shock, reaches the heart, the low-level heartbeat signals are overcome. The heart begins twitching in an irregular manner and goes out of rhythm with the pulse. This twitching is called fibrillation. Unless the normal heartbeat rhythm is restored using special defibrillation equipment (paddles), the individual will die. No known case of heart fibrillation has ever been corrected without the use of defibrillation equipment by a qualified medical practitioner. Other effects of electrical shock may include immediate heart stoppage and burns. In addition, the body's reaction to the shock can cause a fall or other accident. Delayed internal problems can also result.

2.1.0 The Effect of Current

The amount of current measured in amperes that passes through a body determines the outcome of an electrical shock. The higher the voltage, the greater the chance for a fatal shock. In a one-year study in California, the following results were observed by the State Division of Industry Safety:

- Thirty percent of all electrical accidents were caused by contact with conductors. Of these accidents, 66% involved low-voltage conductors (those carrying 600 volts [V] or less).

Note

Electric shocks or burns are a major cause of accidents in our industry. According to the Bureau of Labor Statistics, electrical shock is the leading cause of death in the electrical industry.

- Portable, electrically operated hand tools made up the second largest number of injuries (15%). Almost 70% of these injuries happened when the frame or case of the tool became energized. These injuries could have been prevented by following proper safety practices, using grounded or double-insulated tools, and using ground fault circuit interrupter (GFCI) protection.

In one ten-year study, investigators found 9,765 electrical injuries that occurred in accidents. Over 18% of these injuries involved contact with voltage levels of over 600 volts. A little more than 13% of these high-voltage injuries resulted in death. These high-voltage totals included limited-amperage contacts, which are often found on electronic equipment. When tools or equipment touch high-voltage overhead lines, the chance that a resulting injury will be fatal climbs to 28%. Of the low-voltage injuries, 1.4% were fatal.

CAUTION

High voltage, defined as 600 volts or more, is almost ten times as likely to kill as low voltage. However, on the job you spend most of your time working on or near lower voltages. Due to the frequency of contact, most electrocution deaths actually occur at low voltages. Attitude about the harmlessness of lower voltages undoubtedly contributes to this statistic.

These statistics have been included to help you gain respect for the environment where you work and to stress how important safe working habits really are.

Classroom

Explain that terms shown in bold (blue) are defined in the Glossary at the back of this module.

Safety

Emphasize that shock is the leading cause of death in the electrical industry.

INSIDE TRACK

Dangers of Electricity

Never underestimate the power of electricity. For example, the current through a 25W light bulb is more than enough to kill you.

Electrocution

Why can a bird perch safely on an electric wire? Squirrels are a common cause of shorts at substations; why does a squirrel get electrocuted when a bird does not?

- HAND TO HAND 1000Ω
- 120 VOLT
- FORMULA: $I = E/R$
- $120/1000 = 0.120$ AMPS OR 120 MILLIAMPS

101F01.EPS

Figure 1 ◆ Body resistance.

2.1.1 Body Resistance

Electricity travels in closed circuits, and its normal route is through a conductor. Shock occurs when the body becomes part of the electric circuit (*Figure 1*). The current must enter the body at one point and leave at another. Shock normally occurs in one of three ways: the person must come in contact with both wires of the electric circuit; one wire of the electric circuit and the ground; or a metallic part that has become hot by being in contact with an energized wire while the person is also in contact with the ground.

To fully understand the harm done by electrical shock, we need to understand something about the physiology of certain body parts: the skin, the heart, and muscles.

- Skin covers the body and is made up of three layers. The most important layer, as far as electric shock is concerned, is the outer layer of dead cells referred to as the *horny layer*. This layer is composed mostly of a protein called *keratin*, and it is the keratin that provides the largest percentage of the body's electrical resistance. When it is dry, the outer layer of skin may have a resistance of several thousand ohms, but when it is moist, there is a radical drop in resistance, as is also the case if there is a cut or abrasion that pierces the horny layer. The amount of resistance provided by the skin will vary widely from individual to individual. A worker with a thick horny layer will have a much higher resis-

tance than a child. The resistance will also vary widely at different parts of the body. For instance, the worker with high-resistance hands may have low-resistance skin on the back of his calf. The skin, like any insulator, has a breakdown voltage at which it ceases to act as a resistor and is simply punctured, leaving only the lower-resistance body tissue to impede the flow of current in the body. The breakdown voltage will vary with the individual, but is in the area of 600V. Since most industrial power distribution systems operate at 480V or higher, technicians working at these levels need to have special awareness of the shock potential.

- The heart is the pump that sends life-sustaining blood to all parts of the body. The blood flow is caused by the contractions of the heart muscle, which is controlled by electrical impulses. The electrical impulses are delivered by an intricate system of nerve tissue with built-in timing mechanisms, which make the chambers of the heart contract at exactly the right time. An outside electric current of as little as 75 milliamperes can upset the rhythmic, coordinated beating of the heart by disturbing the nerve impulses. When this happens, the heart is said to be in *fibrillation*, and the pumping action stops. Death will occur quickly if the normal beat is not restored. Remarkable as it may seem, what is needed to defibrillate the heart is a shock of an even higher intensity.
- The other muscles of the body are also controlled by electrical impulses delivered by nerves. Electric shock can cause loss of muscular control, resulting in the inability to let go of an electrical conductor. Electric shock can also cause injuries of an indirect nature in which involuntary muscle reaction from the electric shock can cause bruises, fractures, and even deaths resulting from collisions or falls.

Severity of shock—The severity of shock received when a person becomes a part of an electric circuit is affected by three primary factors: the amount of current flowing through the body (measured in amperes), the path of the current through the

ELECTRICAL LEVEL ONE—TRAINEE MODULE 26101

Instructor's Notes:

_____

_____

_____

_____

_____

_____

_____

_____

_____

body, and the length of time the body is in the circuit. Other factors that may affect the severity of the shock are the frequency of the current, the phase of the heart cycle when shock occurs, and the general health of the person prior to the shock. Effects can range from a barely perceptible tingle to immediate cardiac arrest. Although there are no absolute limits, or even known values that show the exact injury at any given amperage range, *Table 1* lists the general effects of electric current on the body for different current levels. As this table illustrates, a difference of only 100 milliamperes exists between a current that is barely perceptible and one that can kill.

A severe shock can cause considerably more damage to the body than is visible. For example, a person may suffer internal hemorrhages and destruction of tissues, nerves, and muscle. In addition, shock is often only the beginning in a chain of events. The final injury may well be from a fall, cuts, burns, or broken bones.

2.1.2 Burns

The most common shock-related injury is a burn. Burns suffered in electrical accidents may be of three types: electrical burns, arc burns, and thermal contact burns.

- Electrical burns are the result of electric current flowing through the tissues or bones. Tissue damage is caused by the heat generated by the current flow through the body. An electrical burn is one of the most serious injuries you can receive, and should be given immediate attention. Since the most severe burning is likely to be internal, what may appear at first to be a small surface wound could, in fact, be an indication of severe internal burns.
- Arc burns make up a substantial portion of the injuries from electrical malfunctions. The electric arc between metals can be up to 35,000°F, which is about four times hotter than the surface of the sun. Workers several feet from the source of the arc can receive severe or fatal burns. Since most electrical safety guidelines recommend safe working distances based on shock considerations, workers can be following these guidelines and still be at risk from arc. Electric arcs can occur due to poor electrical contact or failed insulation. Electrical arcing is caused by the passage of substantial amounts of current through the vaporized terminal material (usually metal or carbon).

> **CAUTION**
>
> Since the heat of the arc is dependent on the short circuit current available at the arcing point, arcs generated on 480V systems can be just as dangerous as those generated at 13,000V.

- The third type of burn is a thermal contact burn. It is caused by contact with objects thrown during the blast associated with an electric arc. This blast comes from the pressure developed by the near-instantaneous heating of the air surrounding the arc, and from the expansion of the metal as it is vaporized. (Copper expands by a factor in excess of 65,000 times in boiling.) These pressures can be great enough to hurl people, switchgear, and cabinets considerable distances. Another hazard associated with the blast is the hurling of molten metal droplets, which can also cause thermal contact burns and associated damage. A possible beneficial side effect of the blast is that it could hurl a nearby person away from the arc, thereby reducing the effect of arc burns.

3.0.0 ◆ REDUCING YOUR RISK

There are many things that can be done to greatly reduce the chance of receiving an electrical shock. Always comply with your company's safety policy and all applicable rules and regulations, including job site rules. In addition, the Occupational Safety and Health Administration (OSHA) publishes the *Code of Federal Regulations (CFR)*. CFR Part 1910

Bodily Harm

What factors affect the amount of damage to the body during an electric shock?

Discuss the effects of electrical, arc, and thermal contact burns.

Discuss the "Think About It."

Discuss the "Case History."

Arc Blast

Last summer, an electrician in Louisville, KY was cleaning a high-voltage switch cabinet. He removed the padlock securing the switch enclosure and opened the door. He used a voltage meter to verify 0V on the three load phases at the rear of the cabinet. However, he did not test all potentially energized parts within the cabinet, and some components were still energized. He was wearing standard work boots and safety glasses, but was not wearing protective rubber gloves or an arc suit. As he used a paintbrush to clean the switch, an arc blast occurred, which lasted approximately one-sixth of a second. The electrician was knocked down by the blast. He suffered third-degree burns and required skin grafts on his arms and hands. The investigation determined that the blast was caused by debris, such as a cobweb, falling across the open switch.

The Bottom Line: Always wear the appropriate personal protective equipment and test all components for voltage before working in or around electrical devices.

covers the OSHA standards for general industry and CFR Part 1926 covers the OSHA standards for the construction industry.

Do not approach any electrical conductors closer than indicated in *Table 2* unless you are sure they are deenergized and your company has designated you as a qualified individual. Also, the values given in the table are *minimum* safe clearance distances; if you already have standard distances established, these are provided only as supplemental information. These distances are listed in CFR 1910.333/1926.416.

Discuss approach distances.

3.1.0 Protective Equipment

You should also become familiar with common personal protective equipment. In particular, know the voltage rating of each piece of equipment. Rubber gloves are used to prevent the skin from coming into contact with energized circuits. A separate leather cover protects the rubber glove from punc-

Discuss the OSHA requirements for personal protective equipment.

Table 2 Approach Distances for Qualified Employees— Alternating Current

Voltage Range (Phase-to-Phase)	Minimum Approach Distance
300V and less	Avoid contact
Over 300V, not over 750V	1 ft. 0 in. (30.5 cm)
Over 750V, not over 2kV	1 ft. 6 in. (46 cm)
Over 2kV, not over 15kV	2 ft. 0 in. (61 cm)
Over 15kV, not over 37kV	3 ft. 0 in. (91 cm)
Over 37kV, not over 87.5kV	3 ft. 6 in. (107 cm)
Over 87.5kV, not over 121kV	4 ft. 0 in. (122 cm)
Over 121kV, not over 140kV	4 ft. 6 in. (137 cm)

tures and other damage (see *Figure 2*). OSHA addresses the use of protective equipment, apparel, and tools in CFR 1910.335(a). This article is divided into two sections: *Personal Protective Equipment* and *General Protective Equipment and Tools*.

The first section, *Personal Protective Equipment*, includes the following requirements:

- Employees working in areas where there are potential electrical hazards shall be provided with, and shall use, electrical protective equipment that is appropriate for the specific parts of the body to be protected and for the work to be performed.
- Protective equipment shall be maintained in a safe, reliable condition and shall be periodically

VERONICA WESTFALL   101F02.EPS

Figure 2 ◆ Rubber gloves and leather protectors.

ELECTRICAL LEVEL ONE—TRAINEE MODULE 26101

Instructor's Notes:

_____

_____

_____

_____

_____

_____

_____

_____

_____

ELECTRICAL LEVEL ONE—INSTRUCTOR'S GUIDE MODULE 26101

inspected or tested, as required by CFR 1910.137/1926.95.

- If the insulating capability of protective equipment may be subject to damage during use, the insulating material shall be protected.
- Employees shall wear nonconductive head protection wherever there is a danger of head injury from electric shock or burns due to contact with exposed energized parts.
- Employees shall wear protective equipment for the eyes and face wherever there is danger of injury to the eyes or face from electric arcs or flashes or from flying objects resulting from an electrical explosion.

The second section, *General Protective Equipment and Tools,* includes the following requirements:

- When working near exposed energized conductors or circuit parts, each employee shall use insulated tools or handling equipment if the tools or handling equipment might make contact with such conductors or parts. If the insulating capability of insulated tools or handling equipment is subject to damage, the insulating material shall be protected.
- Fuse handling equipment, insulated for the circuit voltage, shall be used to remove or install fuses when the fuse terminals are energized.
- Ropes and handlines used near exposed energized parts shall be nonconductive.
- Protective shields, protective barriers, or insulating materials shall be used to protect each employee from shock, burns, or other electrically related injuries while that employee is working near exposed energized parts that might be accidentally contacted or where dangerous electric heating or arcing might occur. When normally enclosed live parts are exposed for maintenance or repair, they shall be guarded to protect unqualified persons from contact with the live parts.

The types of electrical safety equipment, protective apparel, and protective tools available for use are quite varied. We will discuss the most common types of safety equipment. These include:

- Rubber protective equipment, including gloves and blankets
- Protective apparel
- Personal clothing
- Hot sticks
- Fuse pullers
- Shorting probes
- Eye and face protection

3.1.1 1910.335(a)(1)/1926.951(a)
Rubber Protective Equipment

At some point during the performance of their duties, all electrical workers will be exposed to energized circuits or equipment. Two of the most important articles of protection for electrical workers are insulated rubber gloves and rubber blankets, which must be matched to the voltage rating for the circuit or equipment. Rubber protective equipment is designed for the protection of the user. If it fails during use, a serious injury could occur.

Rubber protective equipment is available in two types. Type 1 designates rubber protective equipment that is manufactured of natural or synthetic rubber that is properly vulcanized, and Type 2 designates equipment that is ozone resistant, made from any elastomer or combination of elastomeric compounds. Ozone is a form of oxygen that is produced from electricity and is present in the air surrounding a conductor under high voltages. Normally, ozone is found at voltages of 10kV and higher, such as those found in electric utility transmission and distribution systems. Type 1 protective equipment can be damaged by *corona cutting,* which is the cutting action of ozone on natural rubber when it is under mechanical stress. Type 1 rubber protective equipment can also be damaged by ultraviolet rays. However, it is very important that the rubber protective equipment in use today be made of natural rubber or Type 1 equipment. Type 2 rubber protective equipment is very stiff and is not as easily worn as Type 1 equipment.

Various classes—The American National Standards Institute (ANSI) and the American Society for Testing of Materials (ASTM) have designated a specific classification system for rubber protective equipment. The voltage ratings are as follows:

- Class 0   1,000V
- Class 1   7,500V
- Class 2   17,500V
- Class 3   26,500V
- Class 4   36,000V

Inspection of protective equipment—Before rubber protective equipment can be worn by personnel in the field, all equipment must have a current test date stenciled on the equipment, and it must be inspected by the user. Insulating gloves must be tested each day by the user before they can be used. They must also be tested during the day if their insulating value is ever in question. Because rubber protective equipment is going to be used

Discuss the various classes of rubber protective equipment.

for personal protection and serious injury could result from its misuse or failure, it is important that an adequate safety factor be provided between the voltage on which it is to be used and the voltage at which it was tested.

All rubber protective equipment must be marked with the appropriate voltage rating and last inspection date. The markings that are required to be on rubber protective equipment must be applied in a manner that will not interfere with the protection that is provided by the equipment.

 WARNING!
Never work on anything energized without direct instruction from your employer!

Gloves—Both high- and low-voltage rubber gloves are of the gauntlet type and are available in various sizes. To get the best possible protection and service life, here are a few general rules that apply whenever they are used in electrical work:

- Always wear leather protectors over your gloves. Any direct contact with sharp or pointed objects may cut, snag, or puncture the gloves and take away the protection you are depending on. Leather protectors are required by the National Fire Protection Association's Standard 70-E if the insulating capabilities of the gloves are subject to damage. (The standards of the National Fire Protection Association [NFPA] are incorporated into the OSHA standards.)

- Always wear rubber gloves right side out (serial number and size to the outside). Turning gloves inside out places a stress on the preformed rubber.

- Always keep the gauntlets up. Rolling them down sacrifices a valuable area of protection.

- Always inspect and field check gloves before using them. Always check the inside for any debris. The inspection of gloves is covered in more detail later in this section.

- Use light amounts of talcum powder or cotton liners with the rubber gloves. This gives the user more comfort, and it also helps to absorb some of the perspiration that can damage the gloves over years of use.

- Wash the rubber gloves in lukewarm, clean, fresh water after each use. Dry the gloves inside and out prior to returning to storage. Never use any type of cleaning solution on the gloves.

- Once the gloves have been properly cleaned, inspected, and tested, they must be properly stored. They should be stored in a cool, dry, dark place that is free from ozone, chemicals, oils, solvents, or other materials that could damage the gloves. Such storage should not be in the vicinity of hot pipes or direct sunlight. Both gloves and sleeves should be stored in their natural shape and kept in a bag or box inside their protectors. They should be stored undistorted, right side out, and unfolded.

- Gloves can be damaged by many different chemicals, especially petroleum-based products such as oils, gasoline, hydraulic fluid inhibitors, hand creams, pastes, and salves. If contact is made with these or other petroleum-based products, the contaminant should be wiped off immediately. If any signs of physical damage or chemical deterioration are found (e.g., swelling, softness, hardening, stickiness, ozone deterioration, or sun checking), the protective equipment must not be used.

- Never wear watches or rings while wearing rubber gloves; this can cause damage from the inside out and defeats the purpose of using rubber gloves. Never wear anything conductive.

- Rubber gloves must be tested every six months by a certified testing laboratory. Always check the inspection date before using gloves.

- Use rubber gloves only for their intended purpose, not for handling chemicals or other work. This also applies to the leather protectors.

Before rubber gloves are used, a visual inspection and an air test should be made. This should be done prior to use and as many times during the day as you feel necessary. To perform a visual inspection, stretch a small area of the glove, checking to see that no defects exist, such as:

Classroom

Discuss the proper use and maintenance of rubber gloves.

Instructor's Notes:

_____

_____

_____

_____

_____

_____

_____

_____

_____

101F03.EPS

101F04.EPS

101F05.EPS

101F06.EPS

Figure 3 ◆ Inspection.   *Figure 4* ◆ Trapping air.   *Figure 5* ◆ Inflated glove.   *Figure 6* ◆ Inspecting glove.

Show Transparencies 4 through 7 (Figures 3 through 6).

- Embedded foreign material
- Deep scratches
- Pinholes or punctures
- Snags or cuts

Gloves and sleeves can be inspected by rolling the outside and inside of the protective equipment between the hands. This can be done by squeezing together the inside of the gloves or sleeves to bend the outside area and create enough stress to the inside surface to expose any cracks, cuts, or other defects. When the entire surface has been checked in this manner, the equipment is then turned inside out, and the procedure is repeated. It is very important not to leave the rubber protective equipment inside out.

Remember, any damage at all reduces the insulating ability of the rubber glove. Look for signs of deterioration from age, such as hardening and slight cracking. Also, if the glove has been exposed to petroleum products, it should be considered suspect because deterioration can be caused by such exposure. Gloves that are found to be defective must be turned in for disposal. Never leave a damaged glove lying around; someone may think it is a good glove and not perform an inspection prior to using it.

After visually inspecting the glove, other defects may be observed by applying the air test.

Step 1 Stretch the glove and look for any defects, as shown in *Figure 3*.

Step 2 Twirl the glove around quickly or roll it down from the glove gauntlet to trap air inside, as shown in *Figure 4*.

Step 3 Trap the air by squeezing the gauntlet with one hand. Use the other hand to squeeze the palm, fingers, and thumb to check for weaknesses and defects. See *Figures 5* and *6*.

Step 4 Hold the glove up to your ear to try to detect any escaping air.

Step 5 If the glove does not pass this inspection, it must be turned in for disposal.

CAUTION
Never use compressed gas for the air test as this can damage the glove.

Insulating blankets—An insulating blanket is a versatile cover-up device best suited for the protection of maintenance technicians against accidental contact with energized electrical equipment.

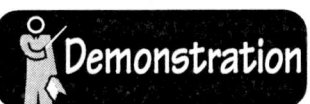
Demonstrate how to inspect rubber gloves using the visual inspection and air tests.

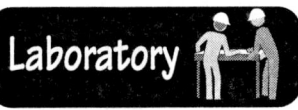
Have the trainees practice inspecting rubber gloves.

Discuss the use of insulating blankets.

ELECTRICAL SAFETY—TRAINEE MODULE 26101   1–9

These blankets are designed and manufactured to provide insulating quality and flexibility for use in covering. Insulating blankets are designed only for covering equipment and should not be used on the floor. (Special rubber mats are available for floor use.) Use caution when installing on sharp edges or covering pointed objects.

Blankets must be inspected yearly and should be checked before each use. To check rubber blankets, place the blanket on a flat surface and roll the blanket from one corner to the opposite corner. If there are any irregularities in the rubber, this method will expose them. After the blanket has been rolled from each corner, it should then be turned over and the procedure repeated.

Insulating blankets are cleaned in the same manner as rubber gloves. Once the protective equipment has been properly cleaned, inspected, and tested, it must be properly stored. It should be stored in a cool, dry, dark place that is free from ozone, chemicals, oils, solvents, or other materials that could damage the equipment. Such storage should not be in the vicinity of hot pipes or direct sunlight. Blankets may be stored rolled in containers that are designed for this use; the inside diameter of the roll should be at least two inches.

3.1.2 Protective Apparel

Besides rubber gloves, there are other types of special application protective apparel, such as fire suits, face shields, and rubber sleeves.

Manufacturing plants should have other types of special application protective equipment available for use, such as high-voltage sleeves, high-voltage boots, nonconductive protective helmets, nonconductive eyewear and face protection, and switchboard blankets.

All equipment should be inspected before use and during use, as necessary. The equipment used and the extent of the precautions taken depend on each individual situation; however, it is better to

be overprotected than underprotected when you are trying to prevent electrocution.

When working with high voltages, flash suits may be required in some applications. Some plants require them to be worn for all switching and rack-in or rack-out operations.

Face shields should also be worn during all switching operations where arcs are a possibility. The thin plastic type of face shield should be avoided because it will melt when exposed to the extremely high temperatures of an electrical arc.

Rubber sleeves are another type of protective apparel that should be worn during switching operations and breaker racking. Sleeves must be inspected yearly.

3.1.3 Personal Clothing

Any individual who will perform work in an electrical environment or in plant substations should dress accordingly. Avoid wearing synthetic-fiber clothing; these types of materials will melt when exposed to high temperatures and will actually increase the severity of a burn. Wear cotton clothing, fiberglass-toe boots or shoes, and hard hats. Use hearing protection where needed.

3.1.4 Hot Sticks

Hot sticks are insulated tools designed for the manual operation of disconnecting switches, fuse removal and insertion, and the application and removal of temporary grounds.

A hot stick is made up of two parts, the head or *hood* and the insulating rod. The head can be made of metal or hardened plastic, while the insulating section may be wood, plastic, laminated wood, or other effective insulating materials. There are also telescoping sticks available.

Most plants have hot sticks available for different purposes. Select a stick of the correct type and size for the application.

To the extent possible, demonstrate the use of and allow the trainees to examine various types of protective apparel.

Discuss the "Think About It."

Dressing for Safety

How could minor flaws in clothing cause harm? What about metal components, such as the rivets in jeans, or synthetic materials, such as polyester? How are these dangerous? How does protective apparel prevent accidents?

ELECTRICAL LEVEL ONE—TRAINEE MODULE 26101

Instructor's Notes:

ELECTRICAL LEVEL ONE—INSTRUCTOR'S GUIDE MODULE 26101

Storage of hot sticks is important. They should be hung up vertically on a wall to prevent any damage. They should also be stored away from direct sunlight and prevented from being exposed to petroleum products. The preferred method of storage is to place the stick in a long section of capped pipe.

3.1.5 Fuse Pullers

Use the plastic or fiberglass style of fuse puller for removing and installing low-voltage cartridge fuses. All fuse pulling and replacement operations must be done using fuse pullers.

The best type of fuse puller is one that has a spread guard installed. This prevents the puller from opening if resistance is met when installing fuses.

3.1.6 Shorting Probes

Before working on deenergized circuits that have capacitors installed, you must discharge the ca-pacitors using a safety shorting probe. When using a shorting probe, first connect the test clip to a good ground to make contact. If necessary, scrape the paint from the metal surface. Then, hold the shorting probe by the handle and touch the probe end of the shorting rod to the points to be shorted. The probe end can be hooked over the part or terminal to provide a constant connection to ground. Never touch any metal part of the shorting probe while grounding circuits or components. Whenever possible, especially when working on or near any deenergized high-voltage circuits, shorting probes should be connected and then left attached to the deenergized portion of any circuit for the duration of the work. This action serves as an extra safety precaution against any accidental application of voltage to the circuit.

3.1.7 Eye and Face Protection

NFPA 70-E requires that protective equipment for the eyes and face shall be used whenever there is danger of injury to the eyes or face from electrical arcs or flashes, or from flying or falling objects resulting from an electrical explosion.

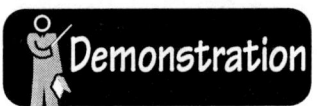

To the extent possible, demonstrate the use of and allow the trainees to examine hot sticks, fuse pullers, shorting probes, safety glasses, and face shields.

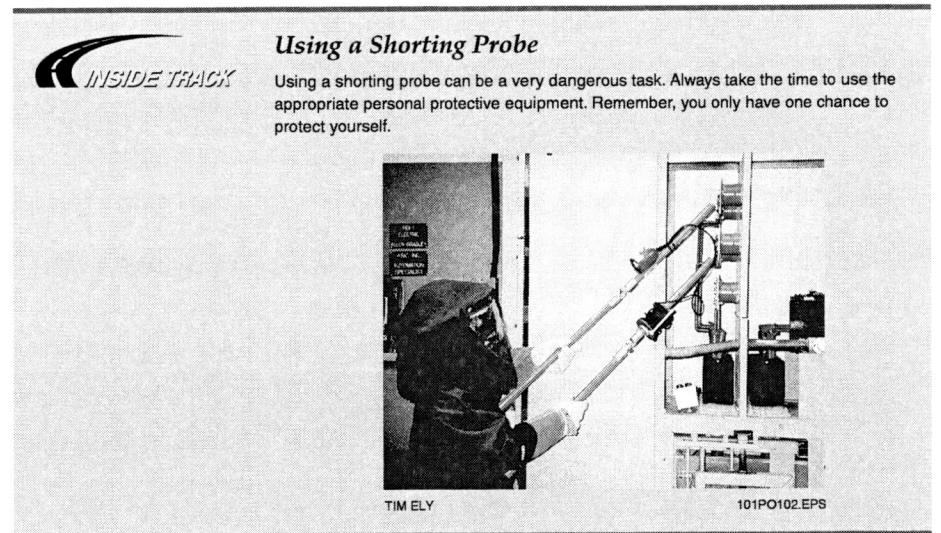

INSIDE TRACK

Using a Shorting Probe

Using a shorting probe can be a very dangerous task. Always take the time to use the appropriate personal protective equipment. Remember, you only have one chance to protect yourself.

TIM ELY   101PO102.EPS

Explain how to verify that a circuit has been deenergized.

3.2.0 Verify That Circuits Are Deenergized

You should always assume that all the circuits are energized until you have verified that the circuit is deenergized. Follow these steps to verify that a circuit is deenergized:

Step 1 Ensure that the circuit is properly tagged and locked out (OSHA 1910.333/1926.417).

Step 2 Verify the test instrument operation on a known source.

Step 3 Using the test instrument, check the circuit to be deenergized. The voltage should be zero.

Step 4 Verify the test instrument operation, once again on a known power source.

3.3.0 Other Precautions

There are several other precautions you can take to help make your job safer. For example:

- Always remove all jewelry (e.g., rings, watches, bracelets, and necklaces) before working on electrical equipment. Most jewelry is made of conductive material and wearing it can result in a shock, as well as other injuries if the jewelry gets caught in moving components.
- When working on energized equipment, it is safer to work in pairs. In doing so, if one of the workers experiences a harmful electrical shock, the other worker can quickly deenergize the circuit and call for help.
- Plan each job before you begin it. Make sure you understand exactly what it is you are going to do. If you are not sure, ask your supervisor.
- You will need to look over the appropriate prints and drawings to locate isolation devices and potential hazards. Never defeat safety interlocks. Remember to plan your escape route before starting work. Know where the nearest phone is and the emergency number to dial for assistance.
- If you realize that the work will go beyond the scope of what was planned, stop and get instructions from your supervisor before continuing. Do not attempt to plan as you go.
- It is critical that you stay alert. Work places are dynamic, and situations relative to safety are always changing. If you leave the work area to pick up material, take a break, or have lunch, reevaluate your surroundings when you return. Remember, plan ahead.

Review the safety checklist.

4.0.0 ◆ OSHA

The purpose of the Occupational Safety and Health Administration (OSHA) is "to assure safe and healthful working conditions for working men and women." OSHA is authorized to enforce standards and assist and encourage the states in their efforts to ensure safe and healthful working conditions. OSHA assists states by providing for research, information, education, and training in the field of occupational safety and health.

The law that established OSHA specifies the duties of both the employer and employee with respect to safety. Some of the key requirements are outlined below. This list does not include everything, nor does it override the procedures called for by your employer.

- Employers shall provide a place of employment free from recognized hazards likely to cause death or serious injury.
- Employers shall comply with the standards of the act.
- Employers shall be subject to fines and other penalties for violation of those standards.

 WARNING!
OSHA states that employees have a duty to follow the safety rules laid down by the employer. Additionally, some states can reduce the amount of benefits paid to an injured employee if that employee was not following known, established safety rules. Your company may also terminate you if you violate an established safety rule.

4.1.0 Safety Standards

The OSHA standards are split into several sections. As discussed earlier, the two that affect you the most are CFR 1926, construction specific, and CFR 1910, which is the standard for general industry. Either or both may apply depending on where you are working and what you are doing. If a job site condition is covered in the 1926 book, then that standard takes precedent. However, if a more stringent requirement is listed in the 1910 standard, it should also be met. An excellent example is the current difference in the two standards on confined spaces; if someone gets hurt or killed, the decision to use the less stringent 1926 standard could be called into question. OSHA's

Instructor's Notes:

_____
_____
_____
_____
_____
_____
_____
_____
_____

General Duty Clause states that an employer should have known all recognized hazards and removed the hazard or protected the employee.

To protect workers from the occupational injuries and fatalities caused by electrical hazards, OSHA has issued a set of design safety standards for electrical utilization systems. These standards are 1926.400–449 and 1910.302–308. OSHA also recognizes the *National Electrical Code (NEC)* for certain installations. The OSHA electrical standards for construction are included in *Appendix A* in the back of this module.

Note

OSHA does *not* recognize the current edition of the NEC; it generally takes several years for that to occur.

The CFR 1910 standard must be followed whenever the construction standard CFR 1926 does not address an issue that is covered by CFR 1910, or for a pre-existing installation. If the CFR 1910 standard is more stringent than CFR 1926, then the more stringent standard should be followed. OSHA does not update their standards in a timely manner, and as such, there are often differences in similar sections of the two standards.

Safety should always be first, and the more protective work rules should always be chosen.

4.1.1 1910.302–308/1926.402–408 Design Safety Standards for Electrical Systems

This section contains design safety regulations for all the electrical equipment and installations used to provide power and light to employee workplaces. The articles listed are outlined in the following sections.

4.1.2 1910.302/1926.402 Electric Utilization Systems

This article identifies the scope of the standard. Listings are included to show which electrical installations and equipment are covered under the standard, and which installations and equipment are not covered under the standard. Furthermore, certain sections of the standard apply only to utilization equipment installed after March 15, 1972, and some apply only to equipment installed after April 16, 1981. Article 1910.302 (1926.402) addresses these oddities and provides guidance to clarify them.

4.1.3 1910.303/1926.403 General Requirements

This article covers topics that mostly concern equipment installation clearances, identification,

Safety on the Job Site

Uncovered openings present several hazards:

- Workers may trip over them.
- If they are large enough, workers may actually fall through them.
- If there is a work area below, tools or other objects may fall through them, causing serious injury to workers below.

Did you know that OSHA could cite your company for working around the open pipe in the photo shown here, even if you weren't responsible for creating it? If you are working near uncovered openings, you have only two options:

- Cover the opening properly (either you may do it or you may have the responsible contractor cover it).

or:

- Stop work and leave the job site until the opening has been covered.

MIKE POWERS   101PO103.EPS

Show and have the trainees examine the NEC and OSHA books. Emphasize that the more stringent regulation should always be applied.

Review the applicable sections of CFR 1910/1926.

Emphasize a contractor's responsibility for maintaining a safe job site.

Continue review of applicable sections of CFR 1910/1926.

and examination. Some of the major subjects addressed in this article are:

- Equipment installation examinations
- Splicing
- Marking
- Identification of disconnecting means
- Workspace around electrical equipment

4.1.4   1910.304/1926.404
Wiring Design and Protection

This article covers the application, identification, and protection requirements of grounding conductors, outside conductors, service conductors, and equipment enclosures. Some of the major topics discussed are:

- Grounded conductors
- Outside conductors
- Service conductors
- Overcurrent protection
- System grounding requirements

4.1.5   1910.305/1926.405
Wiring Methods, Components, and Equipment for General Use

In general, this article addresses the wiring method requirements of raceways, cable trays, pull and junction boxes, switches, and switchboards; the application requirements of temporary wiring installations; the equipment and conductor requirements for general wiring; and the protection requirements of motors, transformers, capacitors, and storage batteries. Some of the major topics are:

- Wiring methods
- Cabinets, boxes, and fittings
- Switches
- Switchboards and panelboards
- Enclosures for damp or wet locations
- Conductors for general wiring
- Flexible cords and cables
- Portable cables
- Equipment for general use

4.1.6   1910.306/1926.406
Specific Purpose Equipment and Installations

This article addresses the requirements of special equipment and installations not covered in other articles. Some of the major types of equipment and installations found in this article are:

- Electric signs and outline lighting
- Cranes and hoists

- Elevators, dumbwaiters, escalators, and moving walks
- Electric welders
- Data processing systems
- X-ray equipment
- Induction and dielectric heating equipment
- Electrolytic cells
- Electrically driven or controlled irrigation machines
- Swimming pools, fountains, and similar installations

4.1.7   1910.307/1926.407
Hazardous (Classified) Locations

This article covers the requirements for electric equipment and wiring in locations that are classified because they contain: (1) flammable vapors, liquids, and/or gases, or combustible dust or fibers; and (2) the likelihood that a flammable or combustible concentration or quantity is present. Some of the major topics covered in this article are:

- Scope
- Electrical installations in hazardous locations
- Conduit
- Equipment in Division 2 locations

The OSHA standard for hazardous (classified) locations is included in *Appendix A* at the back of this module.

4.1.8   1910.308/1926.408 Special Systems

This article covers the wiring methods, grounding, protection, identification, and other general requirements of special systems not covered in other articles. Some of the major subtopics found in this article are:

- Systems over 600 volts nominal
- Emergency power systems
- Class 1, 2, and 3 remote control, signaling, and power-limited circuits
- Fire-protective signaling systems
- Communications systems

4.1.9   1910.331/1926.416 Scope

This article serves as an overview of the following articles and also provides a summary of the installations that this standard allows qualified and unqualified persons to work on or near, as well as the installations that this standard does *not* cover.

1–14   ELECTRICAL LEVEL ONE—TRAINEE MODULE 26101

Instructor's Notes:

_____

_____

_____

_____

_____

_____

_____

_____

_____

4.1.10 1910.332 Training

The training requirements contained in this article apply to employees who face a risk of electric shock. Some of the topics that appear in this article are:

- Content of training
- Additional requirements for unqualified persons
- Additional requirements for qualified persons
- Type of training

4.1.11 1910.333/1926.416–417
Selection and Use of Work Practices

This article covers the implementation of safety-related work practices necessary to prevent electrical shock and other related injuries to the employee. Some of the major topics addressed in this article are listed below:

- General
- Working on or near exposed deenergized parts
- Working on or near exposed energized parts

4.1.12 1910.334/1926.431
Use of Equipment

This article was added to reinforce the regulations pertaining to portable electrical equipment, test equipment, and load break switches. Major topics include:

- Portable electric equipment
- Electric power and lighting circuits
- Test instruments and equipment

4.1.13 1910.335/1926.416
Safeguards for Personnel Protection

This article covers the personnel protection requirements for employees in the vicinity of electrical hazards. It addresses regulations that protect personnel working on equipment as well as personnel working nearby. Some of the major topics are:

- Use of protective equipment
- Alerting techniques

Now that background topics have been covered and an overview of the OSHA electrical safety standards has been provided, it is time to move on to topics related directly to safety. As we discuss these topics, we will continually refer to the OSHA standards to identify the requirements that govern them.

OSHA 1926 Subpart K also addresses electrical safety requirements that are necessary for the practical safeguarding of employees involved in construction work.

4.2.0 Safety Philosophy and General Safety Precautions

The most important piece of safety equipment required when performing work in an electrical environment is common sense. All areas of electrical safety precautions and practices draw upon common sense and attention to detail. One of the most dangerous conditions in an electrical work area is a poor attitude toward safety.

 WARNING!
Only qualified individuals may work on electrical equipment. Your employer will determine who is qualified. Remember, your employer's safety rules must always be followed.

As stated in CFR 1910.333(a)/1926.403, safety-related work practices shall be employed to prevent electric shock or other injuries resulting from either direct or indirect electrical contact when work is performed near or on equipment or circuits that are or may be energized. The specific safety-related work practices shall be consistent with the nature and extent of the associated electrical hazards. The following are considered some of the basic and necessary attitudes and electrical safety precautions that lay the groundwork for a proper safety program. Before going on any electrical work assignment, these safety precautions should be reviewed and adhered to.

- *All work on electrical equipment should be done with circuits deenergized and cleared or grounded*—It is obvious that working on energized equipment is much more dangerous than working on equipment that is deenergized. Work on energized electrical equipment should be avoided if at all possible. CFR 1910.333(a)(1)/1926.403 states that live parts to which an employee may be exposed shall be deenergized before the employee works on or near them, unless the employer can demonstrate that deenergizing introduces additional or increased hazards or is not possible because of equipment design or operational limitations. Live parts that operate at less than 50 volts to ground need not be deenergized if there will be no increased exposure to electrical burns or to explosion due to electric arcs.

Define the term *qualified individual*. Explain that it involves having the proper training and experience to complete a certain task.

Review OSHA general safety precautions.

Continue review of OSHA general safety precautions.

- *All conductors, buses, and connections should be considered energized until proven otherwise*—As stated in 1910.333(b)(1)/1926.417, conductors and parts of electrical equipment that have not been locked out or tagged out in accordance with this section should be considered energized. Routine operation of the circuit breakers and disconnect switches contained in a power distribution system can be hazardous if not approached in the right manner. Several basic precautions that can be observed in switchgear operations are:
  - Wear proper clothing made of 100% cotton or fire-resistant fabric.
  - Eye, face, and head protection should be worn. Turn your head away whenever closing devices.
  - Whenever operating circuit breakers in low-voltage or medium-voltage systems, always stand off to the side of the unit.
  - Always try to operate disconnect switches and circuit breakers under a no-load condition.
  - Never intentionally force an interlock on a system or circuit breaker.
  - Always verify what you are closing a device into; you could violate a lockout or close into a hard fault.

Often, a circuit breaker or disconnect switch is used for providing lockout on an electrical system. To ensure that a lockout is not violated, perform the following procedures when using the device as a lockout point:

- Breakers must always be locked out and tagged as discussed previously whenever you are working on a circuit that is tied to an energized breaker. Breakers capable of being opened and racked out to the disconnected position should have this done. Afterward, approved safety locks must be installed. The breaker may be removed from its cubicle completely to prevent unexpected mishaps. Always follow the standard rack-out and removal procedures that were supplied with the switchgear. Once removed, a sign must be hung on the breaker identifying its use as a lockout point, and approved safety locks must be installed when the breaker is used for isolation. In addition, the closing springs should be discharged.
- Some of the circuit breakers used are equipped with keyed interlocks for protection during operation. These locks are generally called *kirklocks* and are relied upon to ensure proper sequence of operation only. These are not to be used for the purpose of locking out a circuit or system. Where disconnects are installed for use

in isolation, they should never be opened under load. When opening a disconnect manually, it should be done quickly with a positive force. Again, lockouts should be used when the disconnects are open.
- Whenever performing switching or fuse replacements, always use the protective equipment necessary to ensure personnel safety. *Never* make the assumption that because things have gone fine the last 999 times, they will not go wrong this time. Always prepare yourself for the worst case accident when performing switching.
- Whenever reenergizing circuits following maintenance or removal of a faulted component, extreme care should be used. Always verify that the equipment is in a condition to be reenergized safely. All connections should be insulated and all covers should be installed. Have all personnel stand clear of the area for the initial reenergization. *Never* assume everything is in perfect condition. Verify the conditions!

The following procedure is provided as a guideline for ensuring that equipment and systems will not be damaged by reclosing low-voltage circuit breakers into faults. If a low-voltage circuit breaker has opened for no apparent reason, perform the following:

Step 1 Verify that the equipment being supplied is not physically damaged and shows no obvious signs of overheating or fire.

Step 2 Make all appropriate tests to locate any faults.

Step 3 Reclose the feeder breaker. Stand off to the side when closing the breaker.

Step 4 If the circuit breaker trips again, do not attempt to reclose the breaker. In a plant environment, Electrical Engineering should be notified, and the cause of the trip should be isolated and repaired.

The same general procedure should be followed for fuse replacement, with the exception of transformer fuses. If a transformer fuse blows, the transformer and feeder cabling should be inspected and tested before reenergizing. A blown fuse to a transformer is very significant because it normally indicates an internal fault. Transformer failures are catastrophic in nature and can be extremely dangerous. If applicable, contact the in-plant Electrical Engineering Department prior to commencing any effort to reenergize a transformer.

Power must always be removed from a circuit when removing and installing fuses. The air break

ELECTRICAL LEVEL ONE—TRAINEE MODULE 26101

Instructor's Notes:

_____

_____

_____

_____

_____

_____

_____

_____

_____

ELECTRICAL LEVEL ONE—INSTRUCTOR'S GUIDE MODULE 26101

THINK ABOUT IT

Working on Energized Systems

Some electricians commonly work on energized systems because they think it's too much trouble to turn off the power. What practices have you seen around your home or workplace that could be deadly?

disconnects (or quick disconnects) provided on the upstream side of a large transformer must be opened prior to removing the transformer's fuses. Otherwise, severe arcing will occur as the fuse is removed. This arcing can result in personnel injury and equipment damage.

To replace fuses servicing circuits below 600 volts:

- Secure power to fuses or ensure all downstream loads have been disconnected.
- Always use a positive force to remove and install fuses.

When replacing fuses servicing systems above 600 volts:

- Open and lock out the disconnect switches.
- Unlock the fuse compartment.
- Verify that the fuses are deenergized.
- Attach the fuse removal hot stick to the fuse and remove it.

4.3.0 Electrical Regulations

OSHA has certain regulations that apply to job site electrical safety. These regulations include:

- All electrical work shall be in compliance with the latest NEC and OSHA standards.

> **Note**
> OSHA may not recognize the current edition of the NEC, which can sometimes cause problems; however, OSHA typically will *not* cite for any differences.

- The noncurrent-carrying metal parts of fixed, portable, and plug-connected equipment shall be grounded. It is best to choose grounded tools. However, portable tools and appliances protected by an approved system of double insulation need not be grounded. *Figure 7* shows an example of a double-insulated/ungrounded tool.
- Extension cords shall be the three-wire type, shall be protected from damage, and shall not be fastened with staples or hung in a manner that could cause damage to the outer jacket or

NOTE: COLORED AREAS SHOW INSULATING MATERIAL.

101F07.EPS

Figure 7 ◆ Double-insulated electric drill.

ELECTRICAL SAFETY—TRAINEE MODULE 26101

1–17

Discuss the "Case History."

Ask the trainees whether they have ever put themselves at risk by working without GFCI protection. Continue to review the OSHA safety guidelines.

insulation. Never run an extension cord through a doorway or window that can pinch the cord. Also, never allow vehicles or equipment to drive over cords.

- Exposed lamps in temporary lights shall be guarded to prevent accidental contact, except where lamps are deeply recessed in the reflector. Temporary lights shall not be suspended, except in accordance with their listed labeling.
- Receptacles for attachment plugs shall be of an approved type and properly installed. Installation of the receptacle will be in accordance with the listing and labeling for each receptacle and shall be GFCI-protected if the setting is a temporarily wired construction site. If permanent receptacles are used with extension cords, then you must use GFCI protection.
- Each disconnecting means for motors and appliances and each service feeder or branch circuit at the point where it originates shall be legibly marked to indicate its purpose and voltage.
- Flexible cords shall be used in continuous lengths (no splices) and shall be of a type listed in *NEC Table 400-4*.
- Personnel safety ground fault protection for temporary construction is required. There are two methods for accomplishing this: an assured grounding program (this method is still used but should be used in conjunction with a GFCI program) or ground fault protection receptacles or breakers. Each employer will set the standard and method to be used. *Figure 8* shows a typical ground-fault circuit interrupter. The OSHA standard for ground fault protection on construction sites is included in *Appendix A* at the back of this module.

Refer the trainees to Appendix A.

4.3.1 OSHA Lockout/Tagout Rule

OSHA released the 29 CFR 1926 lockout/tagout rule in December 1991. This rule covers the specific procedure to be followed for the "servicing and maintenance of machines and equipment in which the unexpected energization or startup of the machines or equipment, or releases of stored energy, could cause injury to employees. This standard establishes minimum performance requirements for the control of such hazardous energy."

The purpose of the OSHA procedure is to ensure that equipment is isolated from all potentially hazardous energy (e.g., electrical, mechanical, hydraulic, chemical, or thermal), and tagged and locked out before employees perform any servicing or maintenance activities in which the unexpected energization, startup, or release of stored energy could cause injury. All employees shall be instructed in the lockout/tagout procedure.

 CAUTION

Although 99% of your work may be electrical, be aware that you may also need to lock out mechanical equipment.

VERONICA WESTFALL   101F08.EPS

Figure 8 ◆ Typical GFCI.

Review the OSHA lockout/tagout rule.

ELECTRICAL LEVEL ONE—TRAINEE MODULE 26101

Instructor's Notes:

_____

_____

_____

_____

_____

_____

_____

_____

_____

Teaching Tip

Discuss the "Think About It."

Classroom

Continue review of lockout/tagout procedure.

> **GFCIs**
>
> THINK ABOUT IT
>
> Explain how GFCIs protect people. Where should a GFCI be installed in the circuit to be most effective?

The following is an example of a lockout/tagout procedure. Make sure to use the procedure that is specific to your employer or job site.

> **WARNING!**
>
> This procedure is provided for your information only. The OSHA procedure provides only the minimum requirements for lockouts/tagouts. Consult the lockout/tagout procedure for your company and the plant or job site at which you are working. Remember that your life could depend on the lockout/tagout procedure. It is critical that you use the correct procedure for your site.

I. Introduction
  A. This lockout/tagout procedure has been established for the protection of personnel from potential exposure to hazardous energy sources during construction, installation, service, and maintenance of electrical energy systems.
  B. This procedure applies to and must be followed by all personnel who may be potentially exposed to the unexpected startup or release of hazardous energy (e.g., electrical, mechanical, pneumatic, hydraulic, chemical, or thermal).

Exception: This procedure does not apply to process and/or utility equipment or systems with cord and plug power supply systems when the cord and plug are the only source of hazardous energy, are removed from the source, and remain under the exclusive control of the authorized employee.

Exception: This procedure does not apply to troubleshooting (diagnostic) procedures and installation of electrical equipment and systems when the energy source cannot be deenergized because continuity of service is essential or shutdown of the system is impractical. Additional personal protective equipment for such work is required and the safe work practices identified for this work must be followed.

II. *Definitions*
  • *Affected employee*—Any person working on or near equipment or machinery when maintenance or installation tasks are being performed by others during lockout/tagout conditions.
  • *Appointed authorized employee*—Any person appointed by the job site supervisor to coordinate and maintain the security of a group lockout/tagout condition.
  • *Authorized employee*—Any person authorized by the job site supervisor to use lockout/tagout procedures while working on electrical equipment.
  • *Authorized supervisor*—The assigned job site supervisor who is in charge of coordination or procedures and maintenance of security of all lockout/tagout operations at the job site.
  • *Energy isolation device*—An approved electrical disconnect switch capable of accepting approved lockout/tagout hardware for the purpose of isolating and securing a hazardous electrical source in an open or safe position.
  • *Lockout/tagout hardware*—A combination of padlocks, danger tags, and other devices designed to attach to and secure electrical isolation devices.

III. *Training*
  A. Each authorized supervisor, authorized employee, and appointed authorized employee shall receive initial and as-needed user-level training in lockout/tagout procedures.
  B. Training is to include recognition of hazardous energy sources, the type and magnitude of energy sources in the workplace, and the procedures for energy isolation and control.
  C. Retraining will be conducted on an as-needed basis whenever lockout/tagout procedures are changed or there is evidence that procedures are not being followed properly.

Continue review of lockout/tagout procedure.

Show Transparency 9 (Figure 10).

IV. *Protective Equipment and Hardware*
  A. Lockout/tagout devices shall be used exclusively for controlling hazardous electrical energy sources.
  B. All padlocks must be numbered and assigned to one employee only.
  C. No duplicate or master keys will be made available to anyone except the site supervisor.
  D. A current list with the lock number and authorized employee's name must be maintained by the site supervisor.
  E. Danger tags must be of the standard white, red, and black *DANGER—DO NOT OPER-ATE* design and shall include the authorized employee's name, the date, and the appropriate network company (use permanent markers).
  F. Danger tags must be used in conjunction with padlocks, as shown in *Figure 9*.

V. *Procedures*
  A. Preparation for lockout/tagout:
    1. Check the procedures to ensure that no changes have been made since you last used a lockout/tagout.
    2. Identify all authorized and affected employees involved with the pending lockout/tagout.

B. Sequence for lockout/tagout:
  1. Notify all authorized and affected personnel that a lockout/tagout is to be used and explain the reason why.
  2. Shut down the equipment or system using the normal OFF or STOP procedures.
  3. Lock out energy sources and test disconnects to be sure they cannot be moved to the ON position and open the control cutout switch. If there is no cutout switch, block the magnet in the switch open position before working on electrically operated equipment/apparatus such as motors, relays, etc. Remove the control wire.
  4. Lock and tag the required switches in the open position. Each authorized employee must affix a separate lock and tag. An example is shown in *Figure 10*.

VERONICA WESTFALL   101F09.EPS

Figure 9 ◆ Lockout/tagout device.

101F10.EPS

Figure 10 ◆ Multiple lockout/tagout device.

ELECTRICAL LEVEL ONE—TRAINEE MODULE 26101

Instructor's Notes:

_____

_____

_____

_____

_____

_____

_____

_____

_____

ELECTRICAL LEVEL ONE—INSTRUCTOR'S GUIDE MODULE 26101

5. Dissipate any stored energy by attaching the equipment or system to ground.
6. Verify that the test equipment is functional via a known power source.
7. Confirm that all switches are in the open position and use test equipment to verify that all parts are deenergized.
8. If it is necessary to temporarily leave the area, upon returning, retest to ensure that the equipment or system is still deenergized.

C. Restoration of energy:
1. Confirm that all personnel and tools, including shorting probes, are accounted for and removed from the equipment or system.
2. Completely reassemble and secure the equipment or system.
3. Replace and/or reactivate all safety controls.
4. Remove locks and tags from isolation switches. Authorized employees must remove their own locks and tags.
5. Notify all affected personnel that the lockout/tagout has ended and the equipment or system is energized.
6. Operate or close isolation switches to restore energy.

VI. Emergency Removal Authorization
A. In the event a lockout/tagout device is left secured, and the authorized employee is absent, or the key is lost, the authorized supervisor can remove the lockout/tagout device.
B. The authorized employee must be informed that the lockout/tagout device has been removed.
C. Written verification of the action taken, including informing the authorized employee of the removal, must be recorded in the job journal.

What's wrong with this picture?

MIKE POWERS   101PO104.EPS

Lockout/Tagout Dilemma

Last year in Georgia, electricians found energized switches after the lockout of a circuit panel in an older system that had been upgraded several times. The existing wiring did not match the current site drawings. A subsequent investigation found many such situations in older facilities.

The Bottom Line: Never rely solely on drawings. It is mandatory that the circuit be tested after lockout to verify that it is deenergized.

Discuss "What's wrong with this picture?"

Review the "Case History."

Discuss the "Think About It."

Emphasize the importance of assuming that all circuits are energized until proven otherwise.

Lockout/Tagout—Who Does It and When?

What situations are likely to require lockout/tagout? Who is responsible for performing the lockout/tagout? When would more than one person be responsible?

4.4.0 Other OSHA Regulations

There are other OSHA regulations that you need to be aware of on the job site. For example:

- OSHA requires the posting of hard hat areas. Be alert to those areas and always wear your hard hat properly, with the bill in front. Hard hats should be worn whenever overhead hazards exist, or there is the risk of exposure to electric shock or burns.
- You should wear safety shoes on all job sites. Keep them in good condition.
- Do not wear clothing with exposed metal zippers, buttons, or other metal fasteners. Avoid wearing loose-fitting or torn clothing.
- Protect your eyes. Your eyesight is threatened by many activities on the job site. Always wear safety glasses with full side shields. In addition, the job may also require protective equipment such as face shields or goggles.

4.4.1 Testing for Voltage

OSHA also requires that you inspect or test existing conditions before beginning work on electrical equipment or lines. Usually, you will use a voltmeter/sensor or voltage tester to do this. You should assume that all electrical equipment and lines are energized until you have determined that they are not. Do not proceed to work on or near energized parts until the operating voltage is determined.

After the electrical equipment to be worked on has been locked and tagged out, the equipment must be verified as deenergized before work can proceed. This section sets the requirements that must be met before any circuits or equipment can be considered deenergized. First, and most importantly, only qualified persons may verify that a circuit or piece of equipment is deenergized. Before approaching the equipment to be worked on, the qualified person shall operate the equipment's normal operating controls to check that the proper energy sources have been disconnected.

Upon opening a control enclosure, the qualified person shall note the presence of any components that may store electrical energy. Initially, these components should be avoided.

To verify that the lockout was adequate and the equipment is indeed deenergized, a qualified person must use appropriate test equipment to check for power, paying particular attention to induced voltages and unrelated feedback voltage.

Ensure that your testing equipment is working properly by performing the *live-dead-live* check before each use. To perform this test, first check your voltmeter on a known live voltage source. This known source must be in the same range as the electrical equipment you will be working on. Next, without changing scales on your voltmeter, check for the presence of power in the equipment you have locked out. Finally, to ensure that your voltmeter did not malfunction, check it again on the known live source. Performing this test will assure you that your voltage testing equipment is reliable.

In accordance with OSHA section 1910.333(b)(2)(iv)/1926.417(d)(4)(ii), if the circuit to be tested normally operates at more than 600 volts, the live-dead-live check must be performed.

Once it has been verified that power is not present, stored electrical energy that might endanger personnel must be released. A qualified person must use the proper devices to release the stored energy, such as using a shorting probe to discharge a capacitor.

5.0.0 ◆ LADDERS AND SCAFFOLDS

Ladders and scaffolds account for about half of the injuries from workplace electrocutions. The involuntary recoil that can occur when a person is shocked can cause the person to be thrown from a ladder or high place.

5.1.0 Ladders

Many job site accidents involve the misuse of ladders. Make sure to follow these general rules every time you use any ladder. Following these rules can prevent serious injury or even death.

ELECTRICAL LEVEL ONE—TRAINEE MODULE 26101

Instructor's Notes:

_____

_____

_____

_____

_____

_____

_____

_____

_____

ELECTRICAL LEVEL ONE—INSTRUCTOR'S GUIDE MODULE 26101

- Before using any ladder, inspect it. Look for loose or missing rungs, cleats, bolts, or screws, and check for cracked, broken, or badly worn rungs, cleats, or side rails.
- If you find a ladder in poor condition, do not use it. Report it and tag it for repair or disposal.
- Never modify a ladder by cutting it or weakening its parts.
- Do not set up ladders where they may be run into by others, such as in doorways or walkways. If it is absolutely necessary to set up a ladder in such a location, protect the ladder with barriers.
- Do not increase a ladder's reach by standing it on boxes, barrels, or anything other than a flat surface.
- Check your shoes for grease, oil, or mud before climbing a ladder. These materials could make you slip.
- Always face the ladder and hold on with both hands when climbing up or down.
- Never lean out from the ladder. Keep your belt buckle centered between the rails. If something is out of reach, get down and move the ladder.

 WARNING!
When performing electrical work, always use ladders made of nonconductive material.

5.1.1 Straight and Extension Ladders

There are some specific rules to follow when working with straight and extension ladders:

- Always place a straight ladder at the proper angle. The distance from the ladder feet to the base of the wall or support should be about one-fourth the working height of the ladder (see *Figure 11*).
- Secure straight ladders to prevent slipping. Use ladder shoes or hooks at the top and bottom. Another method is to secure a board to the floor against the ladder feet. For brief jobs, someone can hold the straight ladder.
- Side rails should extend above the top support point by at least 36 inches (see *Figure 11*).
- It takes two people to safely extend and raise an extension ladder. Extend the ladder only after it has been raised to an upright position.
- Never carry an extended ladder.
- Never use two ladders spliced together.
- Ladders should not be painted because paint can hide defects.

What's wrong with this picture?

MIKE POWERS   101PO105.EPS

5.1.2 Step Ladders

There are also a few specific rules to use with a step ladder:

- Always open the step ladder all the way and lock the spreaders to avoid collapsing the ladder accidentally.
- Use a step ladder that is high enough for the job so that you do not have to reach. Get someone to hold the ladder if it is more than 10 feet high.
- Never use a step ladder as a straight ladder.
- Never stand on or straddle the top two rungs of a step ladder.
- Ladders are not shelves.

 WARNING!
Do not leave tools or materials on a step ladder.

Sometimes you will need to move or remove protective equipment, guards, or guardrails to

Discuss "What's wrong with this picture?"

Review the safety rules for ladders.

Show Transparency 10 (Figure 11).

HEIGHT TO BEARING POINT "E"	HORIZ. DISTANCE RESTING POINT TO LADDER BASE "H"
9½'	2½'
13½'	3½'
17½'	4½'
21½'	5½'
25'	6½'
28'	7'
31'	8'

"H" SHOULD BE ¼ OF "E"

101F11.EPS

Figure 11 ◆ Straight ladder positioning.

complete a task using a ladder. Remember, always replace what you moved or removed before leaving the area.

5.2.0 Scaffolds

Working on scaffolds also involves being safe and alert to hazards. In general, keep scaffold platforms clear of unnecessary material or scrap. These can become deadly tripping hazards or falling objects. Carefully inspect each part of the scaffold as it is erected. Your life may depend on it! Makeshift scaffolds have caused many injuries and deaths on job sites. Use only scaffolding and planking materials designed and marked for their specific use. When working on a scaffold, follow the established specific requirements set by OSHA for the use of fall protection. When appropriate, wear an approved harness with a lanyard properly anchored to the structure.

> **Note**
> The following requirements represent a compilation of the more stringent requirements of both CFR 1910 and CFR 1926.

The following are some of the basic OSHA rules for working safely on scaffolds:

- Scaffolds must be erected on sound, rigid footing that can carry the maximum intended load.
- Guardrails and toe boards must be installed on the open sides and ends of platforms that are higher than six feet above the ground or floor.
- There must be a screen of ½-inch maximum openings between the toe board and the midrail where persons are required to work or pass under the scaffold.
- Scaffold planks must extend over their end supports not less than six inches nor more than 12 inches and must be properly blocked.
- If the scaffold does not have built-in ladders that meet the standard, then it must have an attached ladder access.
- All employees must be trained to erect, dismantle, and use scaffold(s).
- Unless it is impossible, fall protection must be worn while building or dismantling all scaffolding.
- Work platforms must be completely decked for use by employees.
- Your hard hat is the first line of protection from falling objects. Your hard hat, however, cannot protect your shoulders, arms, back, or feet from the danger of falling objects. The person working below depends on those working above. When you are working above the ground, be careful so that material, including your tools, cannot fall from your work site. Use trash containers or other similar means to keep debris from falling and never throw or sweep material from above.

Classroom

Review the safety guidelines for scaffolds.

Instructor's Notes:

_____

_____

_____

_____

_____

_____

_____

_____

Scaffolds and Electrical Hazards

Remember that scaffolds are excellent conductors of electricity. Recently, a maintenance crew needed to move a scaffold and although time was allocated in the work order to dismantle and rebuild the scaffold, the crew decided to push it instead. They did not follow OSHA recommendations for scaffold clearance and did not perform a job site survey. During the move, the five-tier scaffold contacted a 12,000V overhead power line. All four members of the crew were killed and the crew chief received serious injuries.

The Bottom Line: Never take shortcuts when it comes to your safety and the safety of others. Trained safety personnel should survey each job site prior to the start of work to assess potential hazards. Safe working distances should be maintained between scaffolding and power lines.

Review the "Case History."

What's wrong with this picture?

MIKE POWERS   101PO106.EPS

Discuss "What's wrong with this picture?"

- Always warn others of moving or approaching overhead loads.
- Never attempt to distract signal persons or operators of overhead equipment.
- Obey warning signs.
- Do not use equipment that you are not qualified to operate.

7.0.0 ◆ LIFTING

Back injuries cause many lost working hours every year. That is in addition to the misery felt by the person with the hurt back! Learn how to lift properly and size up the load. To lift, first stand close to the load. Then, squat down and keep your back straight. Get a firm grip on the load and keep the load close to your body. Lift by straightening your legs. Make sure that you lift with your legs and not your back. Do not be afraid to ask for help if you feel the load is too heavy. See *Figure 12* for an example of proper lifting.

8.0.0 ◆ BASIC TOOL SAFETY

When using any tools for the first time, read the operator's manual to learn the recommended safety precautions. If you are not certain about the operation of any tool, ask the advice of a more experienced worker. Before using a tool, you should know its function and how it works.

Always use the right tool for the job. Incorrectly using tools is one of the leading causes of job site injury. Using a hammer as a pry bar or a screwdriver as a chisel can cause damage to the tool and injure you in the process.

6.0.0 ◆ LIFTS, HOISTS, AND CRANES

On the job, you may be working in the operating area of lifts, hoists, or cranes. The following safety rules are for those who are working in the area with overhead equipment but are not directly involved in its operation.

- Stay alert and pay attention to the warning signals from operators.
- Never stand or walk under a load, regardless of whether it is moving or stationary.

Show Transparency 11 (Figure 12).

ELECTRICAL SAFETY—TRAINEE MODULE 26101

1–25

Figure 12 ◆ Proper lifting.

9.0.0 ◆ CONFINED SPACE ENTRY PROCEDURES

Occasionally, you may be required to do your work in a manhole or vault. If this is the case, there are some special safety considerations that you need to be aware of. For details on the subject of working in manholes and vaults, refer to 1910.146/1926.21(a)(6)(i) and (ii) and the *National Electrical Safety Code*. The general precautions are listed in the following paragraphs.

9.1.0 General Guidelines

A confined space includes (but is not limited to) any of the following: a manhole, boiler, tank,

1–26

ELECTRICAL LEVEL ONE—TRAINEE MODULE 26101

Instructor's Notes:

Lifting

trench (four feet or deeper), tunnel, hopper, bin, sewer, vat, pipeline, vault, pit, air duct, or vessel. A confined space is identified as follows:

- It has limited entry and exit.
- It is not intended for continued human occupancy.
- It has poor ventilation.
- It has the potential for entrapment/engulfment.
- It has the potential for accumulating a dangerous atmosphere.
- Entry into a confined space occurs when any part of the body crosses the plane of entry. No employee shall enter a confined space unless the employee has been trained in confined space entry procedures.
- All hazards must be eliminated or controlled before a confined space entry is made.
- All appropriate personal protective equipment shall be worn at all times during confined space entry and work. The minimum required equipment includes a hard hat, safety glasses, full body harness, and life line.
- Ladders used for entry must be secured.
- A rescue retrieval system must be in use when entering confined spaces and while working in permit-required confined spaces (discussed later). Each employee must be capable of being rescued by the retrieval system.
- Only no-entry rescues will be performed by company personnel. Entry rescues will be performed by trained rescue personnel identified on the entry permit.
- The area outside the confined space must be properly barricaded, and appropriate warning signs must be posted.

- Entry permits can be issued and signed by job site supervisors only. Permits must be kept at the confined space while work is being conducted. At the end of the shift, the entry permits must be made part of the job journal and retained for one year.

9.2.0 Confined Space Hazard Review

Before determining the proper procedure for confined space entry, a hazard review shall be performed. The hazard review shall include, but not be limited to, the following conditions:

- The past and current uses of the confined space
- The physical characteristics of the space including size, shape, air circulation, etc.
- Proximity of the space to other hazards
- Existing or potential hazards in the confined space, such as:
  - Atmospheric conditions (oxygen levels, flammable/explosive levels, and/or toxic levels)
  - Presence/potential for liquids
  - Presence/potential for particulates
- Potential for mechanical/electrical hazards in the confined space (including work to be done)

Once the hazard review is completed, the supervisor, in consultation with the project managers and/or safety manager, shall classify the confined space as one of the following:

- A nonpermit confined space
- A permit-required confined space controlled by ventilation
- A permit-required confined space

Review the guidelines for conducting a confined space hazard review.

Working in Tight Spaces

THINK ABOUT IT We routinely work in cramped quarters, from closets to ceiling spaces, without concern. What hazards should we be aware of? What hazards do we typically ignore?

Review the "Think About It."

Once the confined space has been properly classified, the appropriate entry and work procedures must be followed.

9.3.0 Entry and Work Procedures

Nonpermit spaces—A hazard review checklist must be completed before a confined space is designated as a *nonpermit space*. The checklist must be made part of the job journal, and a copy of the checklist must be sent to the safety office. A nonpermit confined space must meet the following criteria:

- There is no actual or potential atmospheric hazard.

> **Note**
> Using ventilation to clear the atmosphere does not meet this criterion.

- There are no actual or potential physical, electrical, or mechanical hazards capable of causing harm or death.

Documentation using the hazards checklist and entry permit forms, and verifying that the confined space is hazard-free, must be made available to employees and maintained at the confined space while work is conducted. If it is necessary to enter the space to verify that it is hazard-free or to eliminate hazards, entry must be made under the requirements of a permit-required space.

An employee may enter the confined space using the minimum fall protection of harness and anchored life line. Once in the space, the employee may disconnect the life line and reconnect it before exiting.

If the work being done creates a hazard, the space must be reclassified as a permit-required space. If any other atmospheric, physical, electrical, or mechanical hazards arise, the space is to be evacuated immediately and reclassified as a permit-required entry space.

Permit-required spaces controlled by ventilation—A hazard review checklist must be completed before a confined space is designated as a *permit-required space controlled by ventilation*. The checklist must be made part of the job journal, and a copy of the checklist must be sent to the safety office. A permit-required confined space controlled by ventilation must meet the following criteria:

- The only hazard in the confined space is an actual/potential atmospheric hazard.

- Continuous forced-air ventilation maintains a safe atmosphere (i.e., within the limits designated on the entry permit).
- Inspection and monitoring data are documented.
- No other physical, electrical, or mechanical hazard exists.

An entry permit must be issued and signed by the job site supervisor and be kept at the confined space while work is being conducted.

Atmospheric testing must be conducted before entry into the confined space and in the following order:

- Oxygen content
- Flammable gases and vapors
- Toxic contaminants

Unacceptable atmospheric conditions must be eliminated with forced air ventilation. If continuous forced air ventilation is required to maintain an acceptable atmosphere, employees may not enter until forced air ventilation has eliminated any hazardous atmosphere. Periodic atmospheric testing must be conducted during the work shift to ensure that the atmosphere remains clear. Periodic monitoring must be documented on the entry permit. If atmospheric conditions change, employees must exit the confined space immediately, and atmospheric conditions must be re-evaluated. Continuous communication must be maintained with the employees working in the confined space.

If hot work is to be performed, a hot work permit is required, and the hazard analysis must document that the hot work does not create additional hazards that are not controlled by ventilation only. Hot work is defined as any work that produces arcs, sparks, flames, heat, or other sources of ignition.

A rescue plan using trained rescue personnel must be in place prior to the start of work in the confined space. All employees should be aware of the rescue plan and how to activate it.

Permit-required confined spaces—A hazard review checklist must be completed before a confined space is designated as a *permit-required confined space*. The checklist must be made part of the job journal, and a copy must be sent to the safety office. A permit-required space meets the following criteria:

- There are actual/potential hazards, other than a hazardous atmosphere.
- Ventilation alone does not eliminate atmospheric hazards.
- Conditions in and around the confined space must be continually monitored.

Review entry permits and rescue plans.

Instructor's Notes:

_____

_____

_____

_____

_____

_____

_____

_____

_____

An entry permit must be issued and signed by the job site supervisor. The permit is to be kept at the confined space while work is being performed in the space.

Atmospheric testing must be conducted before entry into the confined space and in the following order:

- Oxygen content
- Flammable gases and contaminants
- Toxic contaminants

Unacceptable atmospheric conditions must be eliminated/controlled prior to employee entry. Methods of elimination may include isolation, purging, flushing, or ventilating. Continuous atmospheric monitoring must be conducted while employees are in the confined space. Triggering of a monitoring alarm means employees should evacuate the confined space immediately. Any other physical hazards must be eliminated or controlled by engineering and work practice controls before entry. Additional personal protective equipment should be used as a follow-up to the above methods. An attendant, whose job it is to monitor conditions in and around the confined space and to maintain contact with the employees in the space, must be stationed outside the confined space for the duration of entry operations.

If hot work is to be performed, a hot work permit is required, and the hazard analysis must document the additional hazards and precautions to be considered.

A rescue plan using trained rescue personnel must be in place before confined space entry. The attendant should be aware of the rescue plan and have the means to activate it.

10.0.0 ◆ FIRST AID

You should be prepared in case an accident does occur on the job site or anywhere else. First aid training that includes certification classes in CPR and artificial respiration could be the best insurance you and your fellow workers ever receive. Make sure that you know where first aid is available at your job site. Also, make sure you know the accident reporting procedure. Each job site should also have a first aid manual or booklet giving easy-to-find emergency treatment procedures for various types of injuries. Emergency first aid telephone numbers should be readily available to everyone on the job site. Refer to CFR 1910.151/1926.23 and 1926.50 for specific requirements.

11.0.0 ◆ SOLVENTS AND TOXIC VAPORS

The solvents that are used by electricians may give off vapors that are toxic enough to make people temporarily ill or even cause permanent injury. Many solvents are skin and eye irritants. Solvents can also be systemic poisons when they are swallowed or absorbed through the skin.

Solvents in spray or aerosol form are dangerous in another way. Small aerosol particles or solvent vapors mix with air to form a combustible mixture with oxygen. The slightest spark could cause an explosion in a confined area because the mix is perfect for fast ignition. There are procedures and methods for using, storing, and disposing of most solvents and chemicals. These procedures are normally found in the Material Safety Data Sheets (MSDSs) available at your facility.

An MSDS is required for all materials that could be hazardous to personnel or equipment. These sheets contain information on the material, such as the manufacturer and chemical makeup. As much information as possible is kept on the hazardous material to prevent a dangerous situation; or, in the event of a dangerous situation, the information is used to rectify the problem in as safe a manner as possible. See *Figure 13* for an example of procedures you may find on the job.

11.1.0 Precautions When Using Solvents

It is always best to use a nonflammable, nontoxic solvent whenever possible. However, any time solvents are used, it is essential that your work area be adequately ventilated and that you wear the appropriate personal protective equipment:

- A chemical face shield with chemical goggles should be used to protect the eyes and skin from sprays and splashes.
- A chemical apron should be worn to protect your body from sprays and splashes. Remember that some solvents are acid-based. If they come into contact with your clothes, solvents can eat through your clothes to your skin.
- A paper filter mask does not stop vapors; it is used only for nuisance dust. In situations where a paper mask does not supply adequate protection, chemical cartridge respirators might be needed. These respirators can stop many vapors if the correct cartridge is selected. In areas where ventilation is a serious problem, a self-contained breathing apparatus (SCBA) must be used.

Review the hazards involved with the use of solvents.

Show Transparency 12 (Figure 13).

Section VII — Precautions for Safe Handling and Use

Steps to Be Taken in Case Material is Released or Spilled

Isolate from oxidizers, heat, sparks, electric equipment, and open flames.

Waste Disposal Method

Recycle or incinerate observing local, state and federal health, safety and pollution laws.

Precautions to Be Taken in Handling and Storing

Store in a cool dry area. Observe label cautions and instructions.

Other Precautions

SEE ATTACHMENT PARA #3

Section VIII — Control Measures

Respiratory Protection *(Specify Type)*

Suitable for use with organic solvents

| Ventilation | Local Exhaust   preferable | Special   none |
| Mechanical *(General)*   acceptable | Other   none |

Protective Gloves   recommended
(must not dissolve in solvents)   Eye Protection   goggles

Other Protective Clothing or Equipment   none

Work/Hygenic Practices   Use with adequate ventilation. Observe label cautions.

101F13.TIF

Figure 13 ◆ Portion of an MSDS.

- Make sure that you have been given a full medical evaluation and that you are properly trained in using respirators at your site.

11.2.0 Respiratory Protection

Protection against high concentrations of dust, mist, fumes, vapors, gases, and/or oxygen deficiency is provided by appropriate respirators.

Appropriate respiratory protective devices should be used for the hazardous material involved and the extent and nature of the work performed.

An air-purifying respirator is, as its name implies, a respirator that removes contaminants from air inhaled by the wearer. The respirators may be divided into the following types: particulate-removing (mechanical filter), gas- and vapor-removing (chemical filter), and a combination of particulate-removing and gas- and vapor-removing.

Particulate-removing respirators are designed to protect the wearer against the inhalation of particulate matter in the ambient atmosphere. They may be designed to protect against a single type of particulate, such as pneumoconiosis-producing and nuisance dust, toxic dust, metal fumes or mist, or against various combinations of these types.

Gas- and vapor-removing respirators are designed to protect the wearer against the inhalation of gases or vapors in the ambient atmosphere. They are designated as gas masks, chemical cartridge respirators (nonemergency gas respirators), and self-rescue respirators. They may be designed to protect against a single gas such as chlorine; a single type of gas, such as acid gases; or a combination of types of gases, such as acid gases and organic vapors.

If you are required to use a respiratory protective device, you must be evaluated by a physician to ensure that you are physically fit to use a respirator. You must then be fitted and thoroughly instructed in the respirator's use.

Any employee whose job entails having to wear a respirator must keep his face free of facial hair in the seal area.

Respiratory protective equipment must be inspected regularly and maintained in good condition. Respiratory equipment must be properly

Classroom

List the various types of respiratory protection and explain the purpose of each type. Emphasize the importance of proper training and adherence to company safety policies.

ELECTRICAL LEVEL ONE—TRAINEE MODULE 26101

Instructor's Notes:

ELECTRICAL LEVEL ONE—INSTRUCTOR'S GUIDE MODULE 26101

Altered Respiratory Equipment

A self-employed man applied a solvent-based coating to the inside of a tank. Instead of wearing the proper respirator, he used nonstandard air supply hoses and altered the face mask. All joints and the exhalation slots were sealed with tape. He collapsed and was not discovered for several hours.

The Bottom Line: Never alter or improvise safety equipment.

Review the "Case History." Emphasize the dangers involved in altering any type of equipment.

WARNING!
Do not use any respirator unless you have been fitted for it and thoroughly understand its use. As with all safety rules, follow your employer's respiratory program and policies.

---

cleaned on a regular basis and stored in a sanitary, dustproof container.

12.0.0 ◆ ASBESTOS

Asbestos is a mineral-based material that is resistant to heat and corrosive chemicals. Depending on the chemical composition, asbestos fibers may range in texture from coarse to silky. The properties that make asbestos fibers so valuable to industry are its high tensile strength, flexibility, heat and chemical resistance, and good frictional properties.

Asbestos fibers enter the body by inhalation of airborne particles or by ingestion and can become embedded in the tissues of the respiratory or digestive systems. Years of exposure to asbestos can cause numerous disabling or fatal diseases. Among these diseases are asbestosis, an emphysema-like condition; lung cancer; mesothelioma, a cancerous tumor that spreads rapidly in the cells of membranes covering the lungs and body organs; and gastrointestinal cancer.

12.1.0 Monitoring

Employers who have a workplace or work operation covered by OSHA 3096 (*Asbestos Standard for the Construction Industry*) must perform initial monitoring to determine the airborne concentrations of asbestos to which employees may be exposed. If employers can demonstrate that employee exposures are below the action level and/or excursion limit by means of objective or historical data, initial monitoring is not required.

If initial monitoring indicates that employee exposures are below the action level and/or excursion limit, then periodic monitoring is not required. Within regulated areas, the employer must conduct daily monitoring unless all workers are equipped with supplied-air respirators operated in the positive-pressure mode. If daily monitoring by statistically reliable measurements indicates that employee exposures are below the action level and/or excursion limit, then no further monitoring is required for those employees whose exposures are represented by such monitoring. Employees must be given the chance to observe monitoring, and affected employees must be notified as soon as possible following the employer's receipt of the results.

12.2.0 Regulated Areas

The employer must establish a regulated area where airborne concentrations of asbestos exceed or can reasonably be expected to exceed the locally determined exposure limit, or when certain types of construction work are performed, such as cutting asbestos-cement sheets and removing asbestos-containing floor tiles. Only authorized personnel may enter regulated areas. All persons entering a regulated area must be supplied with an appropriate respirator. No smoking, eating, drinking, or applying cosmetics is permitted in regulated areas. Warning signs must be displayed at each regulated area and must be posted at all approaches to regulated areas. These signs must bear the following information:

DANGER
ASBESTOS
CANCER AND LUNG DISEASE HAZARD
AUTHORIZED PERSONNEL ONLY
RESPIRATORS AND PROTECTIVE CLOTHING
ARE REQUIRED IN THIS AREA

Discuss the hazards involved with handling asbestos.

Emphasize that even though contact with asbestos may not present any immediate discomfort, the long-term effects can be deadly.

Discuss the explosion and acid hazards involved when working with batteries.

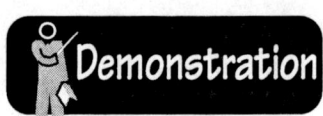

If available, demonstrate the proper use of an eye wash station.

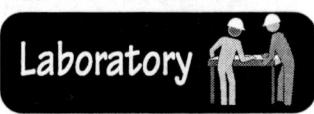

Have each trainee practice operating the wash station.

Where feasible, the employer shall establish negative-pressure enclosures before commencing asbestos removal, demolition, and renovation operations. The setup and monitoring requirements for negative-pressure enclosures are as follows:

- A competent person shall be designated to set up the enclosure and ensure its integrity and supervise employee activity within the enclosure.
- Exemptions are given for small-scale, short-duration maintenance or renovation operations.
- The employer shall conduct daily monitoring of the exposure of each employee who is assigned to work within a regulated area. Short-term monitoring is required whenever asbestos concentrations will not be uniform throughout the workday and where high concentrations of asbestos may reasonably be expected to be released or created in excess of the local limit.

In addition, warning labels must be affixed on all asbestos products and to all containers of asbestos products, including waste containers, that may be in the workplace. The label must include the following information:

DANGER

CONTAINS ASBESTOS FIBERS
AVOID CREATING DUST
CANCER AND LUNG DISEASE HAZARD

12.3.0 Methods of Compliance

To the extent feasible, engineering and work practice controls must be used to reduce employee exposure to within the permissible exposure limit (PEL). The employer must use one or more of the following control methods to achieve compliance:

- Local exhaust ventilation equipped with high-efficiency particulate air (HEPA) filter dust collection systems
- General ventilation systems
- Vacuum cleaners equipped with HEPA filters
- Enclosure or isolation of asbestos dust-producing processes
- Use of wet methods, wetting agents, or removal encapsulants during asbestos handling, mixing, removal, cutting, application, and cleanup
- Prompt disposal of asbestos-containing wastes in leak-tight containers

Prohibited work practices include the following:

- The use of high-speed abrasive disc saws that are not equipped with appropriate engineering controls
- The use of compressed air to remove asbestos-containing materials, unless the compressed air is used in conjunction with an enclosed ventilation system

Where engineering and work practice controls have been instituted but are insufficient to reduce employee exposure to a level that is at or below the PEL, respiratory protection must be used to supplement these controls.

13.0.0 ◆ BATTERIES

Working around wet cell batteries can be dangerous if the proper precautions are not taken. Batteries often give off hydrogen gas as a byproduct. When hydrogen mixes with air, the mixture can be explosive in the proper concentration. For this reason, smoking is strictly prohibited in battery rooms, and only insulated tools should be used. Proper ventilation also reduces the chance of explosion in battery areas. Follow your company's procedures for working near batteries. Also, ensure that your company's procedures are followed for lifting heavy batteries.

13.1.0 Acids

Batteries also contain acid, which will eat away human skin and many other materials. Personal protective equipment for battery work typically includes chemical aprons, sleeves, gloves, face shields, and goggles to prevent acid from contacting skin and eyes. Follow your site procedures for dealing with spills of these materials. Also, know the location of first aid when working with these chemicals.

13.2.0 Wash Stations

Because of the chance that battery acid may contact someone's eyes or skin, wash stations are located near battery rooms. Do not connect or disconnect batteries without proper supervision. Everyone who works in the area should know where the nearest wash station is and how to use it. Battery acid should be flushed from the skin and eyes with large amounts of water or with a neutralizing solution.

14.0.0 ◆ PCBs

Polychlorinated biphenyls (PCBs) are chemicals that were marketed under various trade names as

Instructor's Notes:

CAUTION

If you come in contact with battery acid, report it immediately to your supervisor.

a liquid insulator/cooler in older transformers. In addition to being used in older transformers, PCBs are also found in some large capacitors and in the small ballast transformers used in street lighting and ordinary fluorescent light fixtures. Disposal of these materials is regulated by the EPA and must be done through a regulated disposal company; use extreme caution and follow your facility procedures.

WARNING!

Do not come into contact with PCBs. They present a variety of serious health risks, including lung damage and cancer.

15.0.0 ◆ FALL PROTECTION

15.1.0 Fall Protection Procedures

Fall protection must be used when employees are on a walking or working surface that is six feet or more above a lower level and has an unprotected edge or side. The areas covered include, but are not limited to:

- Finished and unfinished floors or mezzanines
- Temporary or permanent walkways/ramps
- Finished or unfinished roof areas
- Elevator shafts and hoist-ways
- Floor, roof, or walkway holes
- Working six feet or more above dangerous equipment

Exception: If the dangerous equipment is unguarded, fall protection must be used at all heights regardless of the fall distance.

Note

Walking/working surfaces do not include ladders, scaffolds, vehicles, or trailers. Also, an unprotected edge or side is an edge/side where there is no guardrail system at least 39 inches high.

Fall protection is not required during inspection, investigation, or assessment of job site conditions before or after construction work.

These fall protection guidelines do not apply to the following areas. Fall protection for these areas is located in the subparts cited in parentheses.

- Cranes and derricks (1926 subpart N/1910 subpart N)
- Scaffolding (1926 subpart L/1910 subpart D)
- Electrical power transmission and distribution (1926 subpart V/1910 subpart R)
- Stairways and ladders (1926 subpart X/1910 subpart D)
- Excavations (1926 subpart P)

What's wrong with this picture?

MIKE POWERS   101PO107.EPS

Discuss PCB hazards.

Discuss fall protection systems.

Discuss "What's wrong with this picture?" Note the lack of hard hat and fall protection.

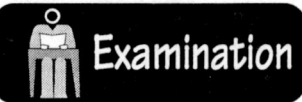

Fall protection must be selected in order of preference as listed below. Selection of a lower-level system (e.g., safety nets) must be based only on feasibility of protection. The list includes, but is not limited to, the following:

- Guardrail systems and hole covers
- Personal fall arrest systems
- Safety nets

These fall protection procedures are designed to warn, isolate, restrict, or protect workers from a potential fall hazard.

15.2.0 Types of Fall Protection Systems

The type of system selected shall depend on the fall hazards associated with the work to be performed. First, a hazard analysis shall be conducted by the job site supervisor prior to the start of work. Based on the hazard analysis, the job site supervisor and project manager, in consultation with the safety manager, will select the appropriate fall protection system. All employees will be instructed in the use of the fall protection system before starting work.

Summary

Safety must be your concern at all times so that you do not become either the victim of an accident or the cause of one. Safety requirements and safe work practices are provided by OSHA and your employer. It is essential that you adhere to all safety requirements and follow your employer's safe work practices and procedures. Also, you must be able to identify the potential safety hazards of your job site. The consequences of unsafe job site conduct can often be expensive, painful, or even deadly. Report any unsafe act or condition immediately to your supervisor. You should also report all work-related accidents, injuries, and illnesses to your supervisor immediately. Remember, proper construction techniques, common sense, and a good safety attitude will help to prevent accidents, injuries, and fatalities.

Review Questions

1. The most life-threatening hazards on a construction site are _____.
   a. falls
   b. electrocution
   c. being crushed or struck by falling or flying objects
   d. All of the above.

2. If a person's heart begins to fibrillate due to an electrical shock, the solution is to _____.
   a. leave the person alone until the fibrillation stops
   b. administer heart massage
   c. use the Heimlich maneuver
   d. have a qualified person use emergency defibrillation equipment

3. The majority of injuries due to electrical shock are caused by _____.
   a. electrically operated hand tools
   b. contact with low-voltage conductors
   c. contact with high-voltage conductors
   d. lightning

4. Class 0 rubber gloves are used when working with voltages less than _____.
   a. 1,000 volts
   b. 7,500 volts
   c. 17,500 volts
   d. 26,500 volts

5. An important use of a hot stick is to _____.
   a. keep cattle moving
   b. keep your hands warm
   c. replace fuses
   d. test circuits to see if they are live

6. Which of these statements correctly describes a double-insulated power tool?
   a. There is twice as much insulation on the power cord.
   b. It can safely be used in place of a grounded tool.
   c. It is made entirely of plastic or other nonconducting material.
   d. The entire tool is covered in rubber.

Instructor's Notes:

_____

_____

_____

_____

_____

_____

_____

_____

7. Which of the following applies in a lockout/tagout procedure?
   a. Only the supervisor can install lockout/tagout devices.
   b. If several employees are involved, the lockout/tagout equipment is applied only by the first employee to arrive at the disconnect.
   c. Lockout/tagout devices applied by one employee can be removed by another employee as long as it can be verified that the first employee has left for the day.
   d. Lockout/tagout devices are installed by every authorized employee involved in the work.

8. What is the proper distance from the feet of a straight ladder to the wall?
   a. one-fourth the working height of the ladder
   b. one-half the height of the ladder
   c. three feet
   d. one-fourth of the square root of the height of the ladder

9. What are the minimum and maximum distances (in inches) that a scaffold plank can extend beyond its end support?
   a. 4, 8
   b. 6, 10
   c. 6, 12
   d. 8, 12

10. Which of these conditions applies to a permit-required confined space, but not to a permit-required space controlled by ventilation?
   a. A hazard review checklist must be completed.
   b. An attendant, whose job is to monitor the space, must be stationed outside the space.
   c. Unacceptable atmospheric conditions must be eliminated.
   d. Atmospheric testing must be conducted.

Michael J. Powers, Tri-City Electrical Contractors, Inc.

How did you choose a career in the electrical field?
My father was an electrician and after I "burned out" with a career in fast-food management, I decided to choose a completely different field.

Tell us about your apprenticeship experience.
It was excellent! I worked under several very knowledgeable electricians and had a pretty good selection of teachers. Over my four-year apprenticeship, I was able to work on a variety of jobs, from Photomats to kennels to colleges.

What positions have you held and how did they help you to get where you are now?
I have been an electrical apprentice, a licensed electrician, a job site superintendent, a master electrician, and am currently a corporate safety and training director. The knowledge I acquired in electrical theory in apprenticeship school and preparing for my licensing exams, as well as the practical on-the-job experience in my twenty years in the electrical field, were wonderful training for my current position.

I also serve on the authoring team for NCCER's Electrical Curricula, which has provided me with not only the opportunity to share what I have learned, but is also a great way to keep current in other areas by meeting with electricians from a variety of disciplines (we have commercial, residential, and industrial electricians on the team, as well as instructors).

What would you say was the single greatest factor that contributed to your success?
A company that recognized and rewarded competent, hard workers and provided them with the support and guidance to allow them to develop and succeed in the industry.

What does your current job entail?
I am responsible for safe work practices and procedures through the job site management team at Tri-City Electrical Contractors, Inc., which currently has a workforce of 1,300. I also assist in developing, delivering, and administering the training program, from apprenticeship to in-house to outsourced training.

What advice do you have for trainees?
Training in all its aspects is the key to your success and advancement in the industry. Any time you are given a training opportunity, take it, even if it might not appear relevant at the time. Eventually, all knowledge can be applied to some situation. Most importantly, have fun! The construction industry is composed of good people. I firmly believe that construction workers, as a group, are much more honest and direct than any other comparable group. Wait—did I say comparable group? That's a misstatement—there is no comparable group. Construction workers build America!

Instructor's Notes:

_____
_____
_____
_____
_____
_____
_____
_____
_____

Trade Terms Introduced in This Module

Double-insulated/ungrounded tool: An electrical tool that is constructed so that the case is insulated from electrical energy. The case is made of a nonconductive material.

Fibrillation: Very rapid irregular contractions of the muscle fibers of the heart that result in the heartbeat and pulse going out of rhythm with each other.

Grounded tool: An electrical tool with a three-prong plug at the end of its power cord or some other means to ensure that stray current travels to ground without passing through the body of the user. The ground plug is bonded to the conductive frame of the tool.

Ground fault circuit interrupter (GFCI): A protective device that functions to deenergize a circuit or portion thereof within an established period of time when a current to ground exceeds some predetermined value that is less than that required to operate the overcurrent protective device of the supply circuit.

Polychlorinated biphenyls (PCBs): Toxic chemicals that may be contained in liquids used to cool certain types of large transformers and capacitors.

Summary of OSHA Construction Standards

Instructor's Notes:

ELECTRICAL STANDARDS FOR CONSTRUCTION

INTRODUCTION

Electricity has long been recognized as a serious workplace hazard, exposing employees to such dangers as electric shock, electrocution, fires, and explosions.

Experts in electrical safety have traditionally looked toward the widely used *National Electrical Code* (NEC) for help in the practical safeguarding of persons from these hazards. The Occupational Safety and Health Administration (OSHA) recognized the important role of the NEC in defining basic requirements for safety in electrical installations by including the entire 1971 NEC by reference in Subpart K of 29 *Code of Federal Regulations* Part 1926 (Construction Safety and Health Standards).

In a final rule dated July 11, 1986, OSHA updated, simplified, and clarified Subpart K, 29 CFR 1926. The revisions serve these objectives:

- NEC requirements that directly affect employees in construction workplaces have been placed in the text of the OSHA standard, eliminating the need for the NEC to be incorporated by reference.

- Certain requirements that supplemented the NEC have been integrated in the new format.

- Performance language is utilized and superfluous specifications omitted and changes in technology accommodated.

In addition, the standard is easier for employers and employees to use and understand. Also, the OSHA revision of the electrical standards has been made more flexible, eliminating the need for constant revision to keep pace with the NEC, which is revised every three years.

101APX01.TIF

SUBPART K

The NEC provisions directly related to employee safety are included in the body of the standard itself - making it unnecessary to continue the adoption by reference of the NEC. Subpart K is divided into four major groups plus a general definitions section:

- Installation Safety Requirements
  [29 CFR 1926.402 - 1926.415]

- Safety-Related Work Practices
  [29 CFR 1926.416 - 1926.430]

- Safety-Related Maintenance and Environmental Considerations
  [29 CFR 1926.431 - 1926.440]

- Safety Requirements for Special Equipment
  [29 CFR 1926.441 - 1926.448]

- Definitions
  [29 CFR 1926.449]

101APX02.TIF

Instructor's Notes:

I. INSTALLATION SAFETY REQUIREMENTS

Part I of the standard is very comprehensive. Only some of the major topics and brief summaries of these requirements are included in this discussion.

Sections 29 CFR 1926.402 through 1926.408 contain installation safety requirements for electrical equipment and installations used to provide electric power and light at the jobsite. These sections apply to installations, both temporary and permanent, used on the jobsite; but they *do not* apply to existing permanent installations that were in place before the construction activity commenced.

If an installation is made in accordance with the 1984 *National Electrical Code*, it will be considered to be in compliance with Sections 1926.403 through 1926.408, except for:

1926.404(b)(1)	Ground-fault protection for employees
1926.405(a)(2)(ii)(E)	Protection of lamps on temporary wiring
1926.405(a)(2)(ii)(F)	Suspension of temporary lights by cords
1926.405(a)(2)(ii)(G)	Portable lighting used in wet or conductive locations
1926.405(a)(2)(ii)(J)	Extension cord sets and flexible cords

101APX03.TIF

Approval

The electrical conductors and equipment used by the employer must be approved.

Examination, Installation, and Use of Equipment

The employer must ensure that electrical equipment is free from recognized hazards that are likely to cause death or serious physical harm to employees. Safety of equipment must be determined by the following:

- Suitability for installation and use in conformity with the provisions of the standard. Suitability of equipment for an identified purpose may be evidenced by a listing, by labeling, or by certification for that identified purpose.

- Mechanical strength and durability. For parts designed to enclose and protect other equipment, this includes the adequacy of the protection thus provided.

- Electrical insulation.

- Heating effects under conditions of use.

- Arcing effects.

- Classification by type, size, voltage, current capacity, and specific use.

- Other factors that contribute to the practical safeguarding of employees who use or are likely to come in contact with the equipment.

Guarding

Live parts of electric equipment operating at 50 volts or more must be guarded against accidental contact. Guarding of live parts must be accomplished as follows:

- Location in a cabinet, room, vault, or similar enclosure accessible only to qualified persons.

101APX04.TIF

Instructor's Notes:

_____

_____

_____

_____

_____

_____

_____

_____

_____

- Use of permanent, substantial partitions or screens to exclude unqualified persons.

- Location on a suitable balcony, gallery, or platform elevated and arranged to exclude unqualified persons.

- Elevation of eight feet or more above the floor.

Entrance to rooms and other guarded locations containing exposed live parts must be marked with conspicuous warning signs forbidding unqualified persons to enter.

Electric installations that are over 600 volts and that are open to unqualified persons must be made with metal-enclosed equipment or enclosed in a vault or area controlled by a lock. In addition, equipment must be marked with appropriate caution signs.

Overcurrent Protection

The following requirements apply to overcurrent protection of circuits rated 600 volts, nominal, or less.

- Conductors and equipment must be protected from overcurrent in accordance with their ability to safely conduct current and the conductors must have sufficient current-carrying capacity to carry the load.

- Overcurrent devices must not interrupt the continuity of the grounded conductor unless all conductors of the circuit are opened simultaneously, except for motor-running overload protection.

- Overcurrent devices must be readily accessible and not located where they could create an employee safety hazard by being exposed to physical damage or located in the vicinity of easily ignitable material.

- Fuses and circuit breakers must be so located or shielded that employees will not be burned or otherwise injured by their operation, e.g., arcing.

101APX05.TIF

Grounding of Equipment Connected by Cord and Plug

Exposed noncurrent-carrying metal parts of cord- and plug-connected equipment that may become energized must be grounded in the following situations:

- When in a hazardous (classified) location.

- When operated at over 150 volts to ground, except for guarded motors and metal frames of electrically heated appliances if the appliance frames are permanently and effectively insulated from ground.

- When one of the types of equipment listed below. But see Item 6 for exemption.

    1. Hand held motor-operated tools.

    2. Cord- and plug-connected equipment used in damp or wet locations or by employees standing on the ground or on metal floors or working inside metal tanks or boilers.

    3. Portable and mobile X-ray and associated equipment.

    4. Tools likely to be used in wet and/or conductive locations.

    5. Portable hand lamps.

    6. [Exemption] Tools likely to be used in wet and/or conductive locations need not be grounded if supplied through an isolating transformer with an ungrounded secondary of not over 50 volts. Listed or labeled portable tools and appliances protected by a system of double insulation, or its equivalent, need not be grounded. If such a system is employed, the equipment must be distinctively marked to indicate that the tool or appliance uses a system of double insulation.

101APX06.TIF

Instructor's Notes:

_____

_____

_____

_____

_____

_____

_____

_____

_____

II. SAFETY-RELATED WORK PRACTICES

Protection of Employees

The employer must not permit an employee to work near any part of an electric power circuit that the employee could contact in the course of work, unless the employee is protected against shock by de-energizing the circuit and grounding it or by guarding it effectively by insulation or other means.

Where the exact location of underground electric power lines is unknown, employees using jack hammers or hand tools that may contact a line must be provided with insulated protective gloves.

Even before work is begun, the employer must determine by inquiry, observation, or instruments where any part of an exposed or concealed energized electric power circuit is located. This is necessary because a person, tool or machine could come into physical or electrical contact with the electric power circuit.

The employer is required to advise employees of the location of such lines, the hazards involved, and protective measures to be taken as well as to post and maintain proper warning signs.

Passageways and Open Spaces

The employer must provide barriers or other means of guarding to ensure that workspace for electrical equipment will not be used as a passageway during the time when energized parts of electrical equipment are exposed. Walkways and similar working spaces must be kept clear of electric cords. Other standards cover load ratings, fuses, cords, and cables.

Lockout and Tagging of Circuits

Tags must be placed on controls that are to be deactivated during the course of work on energized or de-energized equipment or circuits. Equipment or circuits that are de-energized must be rendered inoperative and have tags attached at all points where such equipment or circuits can be energized. 101APX07.TIF

III. SAFETY-RELATED MAINTENANCE AND ENVIRONMENTAL CONSIDERATIONS

Maintenance of Equipment

The employer must ensure that all wiring components and utilization equipment in hazardous locations are maintained in a dust-tight, dust-ignition-proof, or explosion-proof condition without loose or missing screws, gaskets, threaded connections, seals, or other impairments to a tight condition.

Environmental Deterioration of Equipment

Unless identified for use in the operating environment, no conductors or equipment can be located:

- In damp or wet locations.

- Where exposed to gases, fumes, vapors, liquids, or other agents having a deteriorating effect on the conductors or equipment.

- Where exposed to excessive temperatures.

Control equipment, utilization equipment, and busways approved for use in dry locations only must be protected against damage from the weather during building construction.

For protection against corrosion, metal raceways, cable armor, boxes, cable sheathing, cabinets, elbows, couplings, fittings, supports, and support hardware must be of materials appropriate for the environment in which they are installed.

101APX08.TIF

Instructor's Notes:

_____

_____

_____

_____

_____

_____

_____

_____

_____

IV. SAFETY REQUIREMENTS FOR SPECIAL EQUIPMENT

Batteries

Batteries of the unsealed type must be located in enclosures with outside vents or in well-ventilated rooms arranged to prevent the escape of fumes, gases, or electrolyte spray into other areas. Other provisions include the following:

Ventilation--to ensure diffusion of the gases from the battery and to prevent the accumulation of an explosive mixture.

Racks and trays--treated to make them resistant to the electrolyte.

Floors--acid-resistant construction unless protected from acid accumulations.

Face shields, aprons, and rubber gloves--for workers handling acids or batteries.

Facilities for quick drenching of the eyes and body--within 25 feet (7.62 m) of battery handling areas.

Facilities--for flushing and neutralizing spilled electrolytes and for fire protection.

Battery Charging

Battery charging installations must be located in areas designated for that purpose. When batteries are being charged, vent caps must be maintained in functioning condition and kept in place to avoid electrolyte spray. Also, charging apparatus must be protected from damage by trucks.

101APX09.TIF

HAZARDOUS (CLASSIFIED) LOCATIONS

The *National Electrical Code* (NEC) defines hazardous locations as those areas "where fire or explosion hazards may exist due to flammable gases or vapors, flammable liquids, combustible dust, or ignitable fibers or flyings."

A substantial part of the NEC is devoted to the discussion of hazardous locations. That's because electrical equipment can become a source of ignition in these volatile areas. Articles 500 through 504, and 510 through 517 provide classification and installation standards for the use of electrical equipment in these locations. The writers of the NEC developed a short-hand method of describing areas classified as hazardous locations. One of the purposes of this discussion is to explain this classification system. Hazardous locations are classified in three ways by the *National Electrical Code*: TYPE, CONDITION, and NATURE.

Hazardous Location Types

Class I Locations

According to the NEC, there are three types of hazardous locations. The first type of hazard is one which is created by the presence of <u>flammable gases or vapors</u> in the air, such as natural gas or gasoline vapor. When these materials are found in the atmosphere, a potential for explosion exists, which could be ignited if an electrical or other source of ignition is present. The Code writers have referred to this first type of hazard as <u>Class I</u>. So, a *Class I Hazardous Location* is one in which *flammable gases or vapors* may be present in the air in sufficient quantities to be explosive or ignitable. Some typical Class I locations are:

- Petroleum refineries, and gasoline storage and dispensing areas;

- Dry cleaning plants where vapors from cleaning fluids can be present;

- Spray finishing areas;

- Aircraft hangars and fuel servicing areas; and

101APX10.TIF

Instructor's Notes:

_____

_____

_____

_____

_____

_____

_____

_____

- Utility gas plants, and operations involving storage and handling of liquified petroleum gas or natural gas.

All of these are Class I . . . gas or vapor . . . hazardous locations. All require special Class I hazardous location equipment.

Class II Locations
The second type of hazard listed by the *National Electrical Code* are those areas made hazardous by the presence of combustible dust. These are referred to in the Code as "Class II Locations." Finely pulverized material, suspended in the atmosphere, can cause as powerful an explosion as one occurring at a petroleum refinery. Some typical Class II locations are:

- Grain elevators;

- Flour and feed mills;

- Plants that manufacture, use or store magnesium or aluminum powders;

- Producers of plastics, medicines and fireworks;

- Producers of starch or candies;

- Spice-grinding plants, sugar plants and cocoa plants; and

- Coal preparation plants and other carbon handling or processing areas.

Class III Locations
Class III hazardous locations, according to the NEC, are areas where there are easily-ignitable fibers or flyings present, due to the types of materials being handled, stored, or processed. The fibers and flyings are not likely to be suspended in the air, but can collect around machinery or on lighting fixtures and where heat, a spark or hot metal can ignite them. Some typical Class III locations are:

101APX11.TIF

- Textile mills, cotton gins;

- Cotton seed mills, flax processing plants; and

- Plants that shape, pulverize or cut wood and create sawdust or flyings.

Hazardous Location Conditions

In addition to the types of hazardous locations, the *National Electrical Code* also concerns itself with the kinds of conditions under which these hazards are present. The Code specifies that hazardous material may exist in several different kinds of conditions which, for simplicity, can be described as, first, normal conditions, and, second, abnormal conditions.

In the normal condition, the hazard would be expected to be present in everyday production operations or during frequent repair and maintenance activity.

When the hazardous material is expected to be confined within closed containers or closed systems and will be present only through accidental rupture, breakage or unusual faulty operation, the situation could be called "abnormal."

The Code writers have designated these two kinds of conditions very simply, as Division 1 - normal and Division 2 - abnormal. Class I, Class II and Class III hazardous locations can be either Division 1 or Division 2.

Good examples of Class I, Division 1 locations would be the areas near open dome loading facilities or adjacent to relief valves in a petroleum refinery, because the hazardous material would be present during normal plant operations.

Closed storage drums containing flammable liquids in an inside storage room would not normally allow the hazardous vapors to escape into the atmosphere. But, what happens if one of the containers is leaking? You've got a Division 2 - abnormal - condition . . . a Class I, Division 2 hazardous location.

101APX12.TIF

Instructor's Notes:

_____

_____

_____

_____

_____

_____

_____

_____

_____

So far we've covered the three types of hazardous locations:

Class I - gas or vapor
Class II - dust, and
Class III - fibers and flyings.

And secondly, kinds of conditions:

Division 1 - normal conditions, and
Division 2 - abnormal conditions.

Now let's move on to a discussion of the <u>nature</u> of hazardous substances.

Nature of Hazardous Substances

The gases and vapors of Class I locations are broken into four groups by the Code: A, B, C, and D. These materials are grouped according to the ignition temperature of the substance, its explosion pressure, and other flammable characteristics.

The only substance in Group A is acetylene. Acetylene makes up only a very small percentage of hazardous locations. Consequently, little equipment is available for this type of location. Acetylene is a gas with extremely high explosion pressures.

Group B is another relatively small segment of classified areas. This group includes hydrogen and other materials with similar characteristics. If you follow certain specific restrictions in the Code, some of these Group B locations, other than hydrogen, can actually be satisfied with Group C and Group D equipment.

Group C and Group D are by far the most usual Class I groups. They comprise the greatest percentage of all Class I hazardous locations. Found in Group D are many of the most common flammable substances such as butane, gasoline, natural gas and propane.

101APX13.TIF

In Class II - dust locations - we find the hazardous materials in Groups E, F, and G. These groups are classified according to the ignition temperature and the conductivity of the hazardous substance. Conductivity is an important consideration in Class II locations, especially with metal dusts.

Metal dusts are categorized in the Code as Group E. Included here are aluminum and magnesium dusts and other metal dusts of similar nature.

Group F atmospheres contain such materials as carbon black, charcoal dust, coal and coke dust.

In Group G we have grain dusts, flour, starch, cocoa, and similar types of materials.

Review

Let's quickly review. Hazardous locations are classified in three ways by the *National Electrical Code*: TYPE, CONDITION, and NATURE.

There are three types of hazardous conditions: Class I - gas and vapor, Class II - dust, and Class III - fibers and flyings.

There are two kinds of hazardous conditions: Division 1 - normal, and Division 2 - abnormal.

And finally, there is the nature of the hazardous substance . . . where we find Groups A, B, C, and D in Class I locations, and, in Class II locations: Groups E, F, and G.

Let's illustrate our Code "translation" with an example. How would we classify a storage area where LP gas is contained in closed tanks? LP gas is a Class I substance (gas or vapor). It's Division 2 because it would only be in the atmosphere if an accidental rupture or leakage occurred, and it is Group D material.

The table below summarizes the various hazardous (classified) locations.

101APX14.TIF

Instructor's Notes:

_____

_____

_____

_____

_____

_____

_____

_____

SUMMARY OF CLASS I, II, III HAZARDOUS LOCATIONS			
CLASSES	GROUPS	DIVISIONS	
		1	2
I Gases, vapors, and liquids (Art. 501)	A: Acetylene B: Hydrogen, etc. C: Ether, etc. D: Hydrocarbons, fuels, solvents, etc.	Normally explosive and hazardous	Not normally present in an explosive concentration (but may accidentally exist)
II Dusts (Art. 502)	E: Metal dusts (conductive,* and explosive) F: Carbon dusts (some are conductive,* and all are explosive) G: Flour, starch, grain, combustible plastic or chemical dust (explosive)	Ignitable quantities of dust normally are or may be in suspension, or conductive dust may be present	Dust not normally suspended in an ignitable concentration (but may accidentally exist). Dust layers are present.
III Fibers and flyings (Art. 503)	Textiles, wood-working, etc. (easily ignitable, but not likely to be explosive)	Handled or used in manufacturing	Stored or handled in storage (exclusive of manufacturing)

* NOTE: Electrically conductive dusts are dusts with a resistivity less than 10^5 ohm-centimeter.

101APX15.TIF

Hazardous Location Equipment

Sources of Ignition

Now that we've completed our Code translation, we're ready to move to the next part of our discussion - hazardous location equipment. To do this, let's first take a look at the ways in which electrical equipment can become a source of ignition. There are three of them.

Arcs and sparks produced by the normal operation of equipment, like motor starters, contactors, and switches, can ignite a hazardous location atmosphere.

The high temperatures of some heat-producing equipment, such as lamps and lighting fixtures, can ignite flammable atmospheres if they exceed the ignition temperature of the hazardous material. The *National Electrical Code* requires special marking of heat - producing equipment with temperatures above 100°C (212°F).

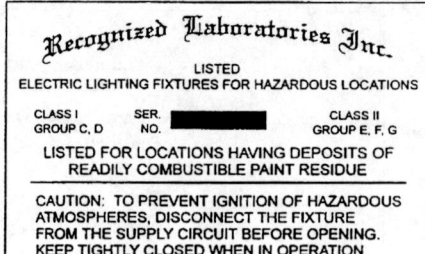

Electrical equipment failure is another way an explosion could be set off. A burn out of a lamp socket or shorting of a terminal could spark a real disaster in a hazardous location.

Equipment Design and Construction

Now let's get down to specific hardware and how it is designed and constructed to be suitable for hazardous locations . . . starting with those designed for Class I . . . gas or vapor . . . applications.

The first requirement for a Class I enclosure is strength. The enclosure must be strong enough to contain an explosion within. The walls must be thick enough to withstand the internal strain. It has to be explosion-proof in case gas or vapors get inside. Secondly, it must function at a temperature below the ignition temperature

101APX16.TIF

Instructor's Notes:

of the surrounding atmosphere.

The equipment must also provide a way for the burning gases to escape from the device as they expand during an internal explosion; but, only after they have been cooled off and their flames "quenched." This escape route for the exploding gases is provided through several types of <u>flame paths</u>.

One type is the <u>ground surface</u> flame path. Here the surfaces are ground, mated, and held to a tolerance of 15 ten-thousandths of an inch. This permits gases to escape, but only after they've been sufficiently <u>cooled</u>, so they won't ignite the volatile surrounding atmosphere.

Another kind of flame path is the <u>threaded</u> flame path. After an explosion, the gas travels out the threaded joint . . . but as it does, it cools off.

Exploded gases may also escape around the shafts of operators used in the enclosure. But, here again, close tolerances are used to quench the burning gas.

Examples of two flame paths are shown below.

101APX17.TIF

FLAME PATHS

Hot gases escape through openings
designed into threaded joint

OPENINGS DESIGNED INTO THREADED JOINT

Explosion-Proof
Enclosure

Accurately Machined Ground-Joint
of Flange, Maximum Gap 0.0015 in.

Cool Gases

Threaded Boss

Minimum Flame Path
Length is 3/8 Inch

Hot Gases

OPENINGS DESIGNED INTO GROUND JOINT

101APX18.TIF

Instructor's Notes:

You can see how important it is to make certain that all flame paths are protected during installation and maintenance, and even during handling, shipping, and storage of explosion-proof material. Even slight damage to a flame path can permit burning gases to escape, igniting the surrounding atmosphere. Also, all cover bolts must be installed for the same reason. A single missing bolt could allow the release of flaming gases.

In designing equipment for Class I, Division 1 locations, it is assumed that the hazardous gases or vapors will be present and eventually seep into the enclosure, so there is a very real chance for an internal explosion to occur.

In the case of Class II, however, the assumptions are different and so the design is different. In Class II, the explosive dust is kept away from equipment housed within the enclosure so that no internal explosion can take place and there is no longer any need for heavy explosion-containing construction, or flame paths. This difference explains why Class I, Division 1 equipment can be called explosion-proof, and Class II equipment is called dust-ignition proof. Class II equipment has a different set of requirements:

1. It must seal out the dust.

2. It must operate below the ignition temperature of the hazardous substance.

3. It must allow for a <u>dust blanket</u>. That is, the build-up of dust collecting on top of the device that can cause it to run "hot" and ignite the surrounding atmosphere.

For Class III equipment, there is very little difference in the design from Class II. Class III equipment must minimize entrance of fibers and flyings; prevent the escape of sparks, burning material or hot metal particles resulting from failure of equipment; and operate at a temperature that will prevent the ignition of fibers accumulated on the equipment.

101APX19.TIF

There are many enclosures, devices, and fixtures suitable for all three classes. This simply means that it meets the specifications for each individual type. A Class I device which could contain an explosion of a specified gas would also have to prevent dust from entering the enclosure to be suitable for Class II. The close tolerance of the flame path which cools the burning gases is also close enough to exclude explosive dust so that a gasket would not be needed.

Proper installation of hazardous location equipment calls for the use of seals. Special fittings are required to keep hot gases from traveling through the conduit system igniting other areas if an internal explosion occurs in a Class I device. They are also needed in certain situations to keep flammable dusts from entering dust-ignition-proof enclosures through the conduit. As shown in the figure below, when arcs and sparks cause ignition of flammable gases and vapors, the equipment contains the explosion and vents only cool gases into the surrounding hazardous area.

101APX20.TIF

Instructor's Notes:

_____

_____

_____

_____

_____

_____

_____

_____

_____

Hazardous Location Equipment Seals

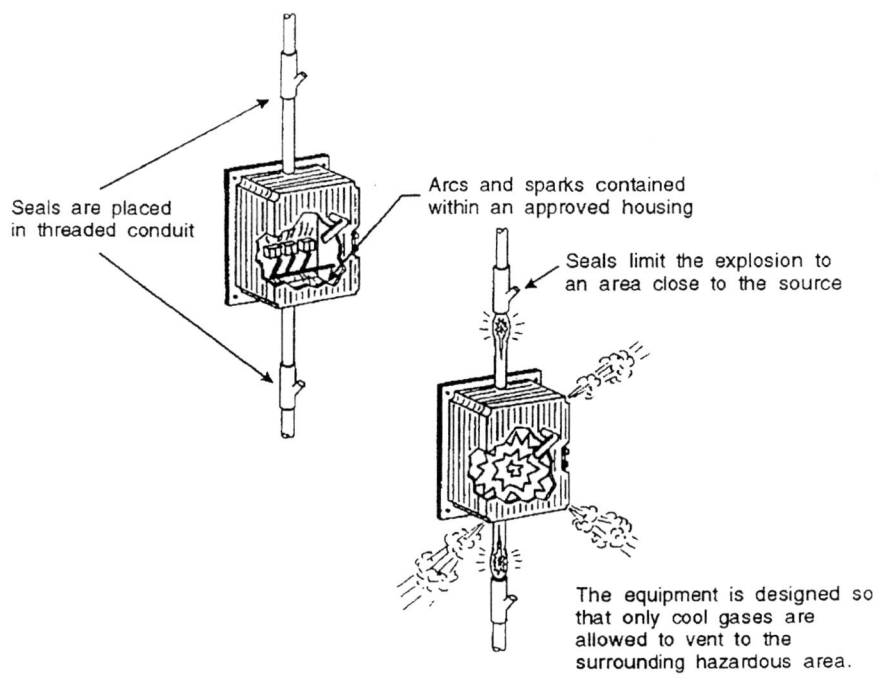

Seals are placed
in threaded conduit

Arcs and sparks contained
within an approved housing

Seals limit the explosion to
an area close to the source

The equipment is designed so
that only cool gases are
allowed to vent to the
surrounding hazardous area.

101APX21.TIF

Sealing fittings are designed to be filled with a chemical compound after the wires have been pulled. As the compound hardens, it seals passageways for dusts and gases. As shown in the figure below, in each conduit run entering an enclosure for switches, circuit breakers, fuses, relays, resistors, or other apparatus which may produce arcs, sparks, or high temperatures within Class I locations, conduit seals shall be placed as close as practicable and in no case more than 18 inches (457 mm) from such enclosures. Again, consult the Code for specific rules for the use of seals.

Sealing compound to prevent passage of flames or vapor through the conduit

101APX22.TIF

Instructor's Notes:

_____

_____

_____

_____

_____

_____

_____

_____

_____

Rigorous standards for hazardous location equipment have been set. Nationally Recognized Testing Laboratories conduct actual explosion tests under laboratory conditions. For each Class I enclosure they experiment with different mixtures of gas and air . . . from very lean mixtures (a small percentage of gas) to very rich mixtures (a high percentage of gas) until they find the one that creates the greatest explosion pressure. To pass inspection, the equipment must not only prevent the ignition of the surrounding atmosphere, but also be able to withstand a hydrostatic test where oil is pumped into the enclosure at high pressure to test the limits of its strength. The device will not pass unless it can resist rupture at four times the maximum pressure found in the explosion tests. For example, if explosion testing shows a maximum pressure for a junction box of 250 pounds per square inch (psi), to get approval, the box must be able to withstand 1,000 psi of hydrostatic pressure - FOUR TIMES the maximum anticipated pressure of 250 psi.

101APX23.TIF

Summary

Regardless of the cause of a hazardous location, it is necessary that every precaution be taken to guard against ignition of the atmosphere. Electrical equipment can be a potential source of ignition through one of three ways:

1. Arcs and sparks
2. High temperatures
3. Electrical equipment failure

Hazardous location equipment is designed and constructed to eliminate the potential for ignition of the atmosphere.

The *National Electrical Code* is the "Bible" of the Electrical Industry, and the primary source of reference for hazardous locations. The NEC is also the basis for OSHA standard 1926.407, Hazardous (Classified) Locations. There are several OSHA standards that require the installation of electrical wiring and equipment in hazardous (classified) locations according to the requirements of Subpart K, Electrical. The NEC should be consulted as a supplement to the OSHA standards for additional background information concerning hazardous locations.

101APX24.TIF

Instructor's Notes:

_____

_____

_____

_____

_____

_____

_____

_____

GROUND-FAULT PROTECTION ON CONSTRUCTION SITES

INSULATION AND GROUNDING

Insulation and grounding are two recognized means of preventing injury during electrical equipment operation. Conductor insulation may be provided by placing nonconductive material such as plastic around the conductor. Grounding may be achieved through the use of a direct connection to a known ground such as a metal cold water pipe.

Consider, for example, the metal housing or enclosure around a motor or the metal box in which electrical switches, circuit breakers, and controls are placed. Such enclosures protect the equipment from dirt and moisture and prevent accidental contact with exposed wiring. However, there is a hazard associated with housings and enclosures. A malfunction within the equipment--such as deteriorated insulation--may create an electrical shock hazard. Many metal enclosures are connected to a ground to eliminate the hazard. If a "hot" wire contacts a grounded enclosure, a ground fault results which normally will trip a circuit breaker or blow a fuse. Metal enclosures and containers are usually grounded by connecting them with a wire going to ground. This wire is called an equipment grounding conductor. Most portable electric tools and appliances are grounded by this means. There is one disadvantage to grounding: a break in the grounding system may occur without the user's knowledge.

Insulation may be damaged by hard usage on the job or simply by aging. If this damage causes the conductors to become exposed, the hazards of shocks, burns, and fire will exist. Double insulation may be used as additional protection on the live parts of a tool, but double insulation does not provide protection against defective cords and plugs or against heavy moisture conditions.

The use of a ground-fault circuit interrupter (GFCI) is one method used to overcome grounding and insulation deficiencies.

101APX25.TIF

WHAT IS A GFCI?

The ground-fault circuit interrupter (GFCI) is a fast-acting circuit breaker which senses small imbalances in the circuit caused by current leakage to ground and, in a fraction of a second, shuts off the electricity. The GFCI continually matches the amount of current going to an electrical device against the amount of current returning from the device along the electrical path. Whenever the amount "going" differs from the amount "returning" by approximately 5 milliamps, the GFCI interrupts the electric power within as little as 1/40 of a second. (See diagram.)

Instructor's Notes:

_____

_____

_____

_____

_____

_____

_____

_____

_____

Ground-Fault Circuit Interrupter

GFCI monitors the difference in current flowing into the "hot" and out to the grounded neutral conductors. The difference (1/2 ampere in this case) will flow back through any available path, such as the equipment grounding conductor, and through a person holding the tool, if the person is in contact with a grounded object.

101APX27.TIF

However, the GFCI will not protect the employee from line-to-line contact hazards (such as a person holding two "hot" wires or a hot and a neutral wire in each hand). It does provide protection against the most common form of electrical shock hazard--the ground fault. It also provides protection against fires, overheating, and destruction of insulation on wiring.

WHAT ARE THE HAZARDS?

With the wide use of portable tools on construction sites, the use of flexible cords often becomes necessary. Hazards are created when cords, cord connectors, receptacles, and cord- and plug-connected equipment are improperly used and maintained.

Generally, flexible cords are more vulnerable to damage than is fixed wiring. Flexible cords must be connected to devices and to fittings so as to prevent tension at joints and terminal screws. Because a cord is exposed, flexible, and unsecured, joints and terminals become more vulnerable. Flexible cord conductors are finely stranded for flexibility, but the strands of one conductor may loosen from under terminal screws and touch another conductor, especially if the cord is subjected to stress or strain.

A flexible cord may be damaged by activities on the job, by door or window edges, by staples or fastenings, by abrasion from adjacent materials, or simply by aging. If the electrical conductors become exposed, there is a danger of shocks, burns, or fire. A frequent hazard on a construction site is a cord assembly with improperly connected terminals.

When a cord connector is wet, hazardous leakage can occur to the equipment grounding conductor and to humans who pick up that connector if they also provide a path to ground. Such leakage is not limited to the face of the connector but also develops at any wetted portion of it.

When the leakage current of tools is below 1 ampere, and the grounding conductor has a low resistance, no shock should be perceived. However, should the resistance

101APX28.TIF

Instructor's Notes:

_____

_____

_____

_____

_____

_____

_____

_____

_____

of the equipment grounding conductor increase, the current through the body also will increase. Thus, if the resistance of the equipment grounding conductor is significantly greater than 1 ohm, tools with even small leakages become hazardous.

PREVENTING AND ELIMINATING HAZARDS

GFCIs can be used successfully to reduce electrical hazards on construction sites. Tripping of GFCIs--interruption of current flow--is sometimes caused by wet connectors and tools. It is good practice to limit exposure of connectors and tools to excessive moisture by using watertight or sealable connectors. Providing more GFCIs or shorter circuits can prevent tripping caused by the cumulative leakage from several tools or by leakages from extremely long circuits.

EMPLOYER'S RESPONSIBILITY

OSHA ground-fault protection rules and regulations have been determined necessary and appropriate for employee safety and health. Therefore, it is the employer's responsibility to provide either: (a) ground-fault circuit interrupters on construction sites for receptacle outlets in use and not part of the permanent wiring of the building or structure; or (b) a scheduled and recorded assured equipment grounding conductor program on construction sites, covering all cord sets, receptacles which are not part of the permanent wiring of the building or structure, and equipment connected by cord and plug which are available for use or used by employees.

GROUND-FAULT CIRCUIT INTERRUPTERS

The employer is required to provide approved ground-fault circuit interrupters for all 120-volt, single-phase, 15- and 20-ampere receptacle outlets on construction sites which are not a part of the permanent wiring of the building or structure and which are in use by employees. Receptacles on the ends of extension cords are not part of the permanent wiring and, therefore, must be protected by GFCIs whether or not the extension cord is plugged into permanent wiring. These GFCIs monitor the current-to-the-load for leakage to ground. When this leakage exceeds 5 mA ± 1 mA,

101APX29.TIF

the GFCI interrupts the current. They are rated to trip quickly enough to prevent electrocution. This protection is required in addition to, not as a substitute for, the grounding requirements of OSHA safety and health rules and regulations, 29 CFR 1926. The requirements which employers must meet, if they choose the GFCI option, are stated in 29 CFR 1926.404(b)(1)(ii). (See appendix.)

ASSURED EQUIPMENT GROUNDING CONDUCTOR PROGRAM

The assured equipment grounding conductor program covers all cord sets, receptacles which are not a part of the permanent wiring of the building or structure, and equipment connected by cord and plug which are available for use or used by employees. The requirements which the program must meet are stated in 29 CFR 1926.404(b)(1)(iii), but employers may provide additional tests or procedures. (See appendix.) OSHA requires that a written description of the employer's assured equipment grounding conductor program, including the specific procedures adopted, be kept at the jobsite. This program should outline the employer's specific procedures for the required equipment inspections, tests, and test schedule.

The required tests must be recorded, and the record maintained until replaced by a more current record. The written program description and the recorded tests must be made available, at the jobsite, to OSHA and to any affected employee upon request. The employer is required to designate one or more **competent persons** to implement the program.

Electrical equipment noted in the assured equipment grounding conductor program must be visually inspected for damage or defects before each day's use. Any damaged or defective equipment must not be used by the employee until repaired.

Two tests are required by OSHA. One is a continuity test to ensure that the equipment grounding conductor is electrically continuous. It must be performed on all cord sets, receptacles which are not part of the permanent wiring of the building or structure, and on cord- and plug-connected equipment which is required to be grounded. This test may be performed using a simple continuity tester, such as a

101APX30.TIF

ELECTRICAL LEVEL ONE—TRAINEE MODULE 26101

Instructor's Notes:

_____

_____

_____

_____

_____

_____

_____

_____

lamp and battery, a bell and battery, an ohmmeter, or a receptacle tester.

The other test must be performed on receptacles and plugs to ensure that the equipment grounding conductor is connected to its proper terminal. This test can be performed with the same equipment used in the first test.

These tests are required before first use, after any repairs, after damage is suspected to have occurred, and at 3-month intervals. Cord sets and receptacles which are essentially fixed and not exposed to damage must be tested at regular intervals not to exceed 6 months. Any equipment which fails to pass the required tests shall not be made available or used by employees.

SUMMARY

This discussion provides information to help guide employers and employees in protecting themselves against 120-volt electrical hazards on the construction site, through the use of ground-fault circuit interrupters or through an assured equipment grounding conductor program.

When planning your program, remember to use the OSHA rules and regulations as a guide to ensure employee safety and health. Following these rules and regulations will help reduce the number of injuries and accidents from electrical hazards. Work disruptions should be minor, and the necessary inspections and maintenance should require little time.

An effective safety and health program requires the cooperation of both the employer and employees.

If you need additional information planning your program, contact the OSHA office nearest you.

101APX31.TIF

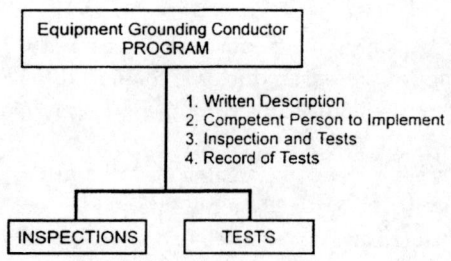

Visual inspection of following:

1. cord sets

2. cap, plug and receptacle of cord sets

3. equipment connected by cord and plug

Exceptions:

• receptacles and cord sets which are fixed and not exposed to damage

Frequency of Inspections:

• before each day's use

Conduct tests for:

1. continuity of equipment grounding conductor

2. proper terminal connection of equipment grounding conductor

Frequency of Tests:

• before first use

• after repair, and before placing back in service

• before use, after suspected damage

• every 3 months, except that cord sets and receptacles that are fixed and not exposed to damage must be tested at regular intervals not to exceed 6 months.

101APX32.TII

Instructor's Notes:

_____

_____

_____

_____

_____

_____

_____

_____

APPENDIX

Construction Safety and Health Regulations Part 1926 Subpart K (Partial)

§1926.404 Wiring design and protection.

(b) Branch circuits--(1) Ground-fault protection--(i) General.
The employer shall use either ground-fault circuit interrupters as specified in paragraph (b)(l)(ii) of this section or an assured equipment grounding conductor program as specified in paragraph (b)(l)(iii) of this section to protect employees on construction sites. These requirements are in addition to any other requirements for equipment grounding conductors.

(ii) Ground-fault circuit interrupters. All 120-volt, single-phase, 15- and 20-ampere receptacle outlets on construction sites, which are not a part of the permanent wiring of the building or structure and which are in use by employees, shall have approved ground-fault circuit interrupters for personnel protection. Receptacles on a two-wire, single-phase portable or vehicle-mounted generator rated not more than 5kW, where the circuit conductors of the generator are insulated from the generator frame and all other grounded surfaces, need not be protected with ground-fault circuit interrupters.

(iii) Assured equipment grounding conductor program. The employer shall establish and implement an assured equipment grounding conductor program on construction sites covering cord sets, receptacles which are not a part of the building or structure, and equipment connected by cord and plug which are available for use or used by employees. This program shall comply with the following minimum requirements:

(A) A written description of the program, including the specific procedures adopted by the employer, shall be available at the jobsite for inspection and copying by the Assistant Secretary and any affected employee.

(B) The employer shall designate one or more competent persons [as defined in §1926.32(f)] to implement the program.

(C) Each cord set, attachment cap, plug and receptacle of cord sets, and any equipment connected by cord and plug, except cord sets and receptacles which are fixed and not exposed to damage, shall be visually inspected before each day's use for external defects, such as deformed or missing pins or insulation damage, and for indications of possible internal damage. Equipment found damaged or defective shall not be used until repaired.

(D) The following tests shall be performed on all cord sets, receptacles which are not a part of the permanent wiring of the building or structure, and cord- and plug-connected equipment required to be grounded:

(1) All equipment grounding conductors shall be tested for continuity and shall be electrically continuous.

(2) Each receptacle and attachment cap or plug shall be tested for correct attachment of the equipment grounding conductor. The equipment grounding conductor shall be connected to its proper terminal.

(E) All required tests shall be performed:

(1) Before first use;

(2) Before equipment is returned to service following any repairs;

(3) Before equipment is used after any incident which can be reasonably suspected to have caused damage (for example, when a cord set is run over); and

(4) At intervals not to exceed 3 months, except that cord sets and receptacles which are fixed and not exposed to damage shall be tested at intervals not exceeding 6 months.

101APX34.TIF

Instructor's Notes:

_____

_____

_____

_____

_____

_____

_____

_____

_____

(F) The employer shall not make available or permit the use by employees of any equipment which has not met the requirements of this paragraph (b)(l)(iii) of this section.

(G) Tests performed as required in this paragraph shall be recorded. This test record shall identify each receptacle, cord set, and cord- and plug-connected equipment that passed the test and shall indicate the last date it was tested or the interval for which it was tested. This record shall be kept by means of logs, color coding, or other effective means and shall be maintained until replaced by a more current record. The record shall be made available on the jobsite for inspection by the Assistant Secretary and any affected employee.

101APX35.TIF

Answers to Review Questions

Answer	Section
1. d	1.0.0
2. d	2.0.0
3. b	2.1.0
4. a	3.1.1
5. c	3.1.4
6. b	4.3.0
7. d	4.3.1
8. a	5.1.1
9. c	5.2.0
10. b	9.3.0

Instructor's Notes:

_____

_____

_____

_____

_____

_____

_____

_____

_____

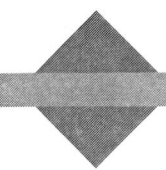

Additional Resources

For advanced study of topics covered in this task module, the following books are suggested:

29 CFR Parts 1900–1910, Standards for General Industry. Occupational Safety and Health Administration, U.S. Department of Labor.

29 CFR Part 1926, Standards for the Construction Industry. Occupational Safety and Health Administration, U.S. Department of Labor.

National Electrical Code Handbook, Latest Edition. Quincy, MA: National Fire Protection Association.

National Electrical Safety Code, Latest Edition. Quincy, MA: National Fire Protection Association.

NCCER CRAFT TRAINING USER UPDATES

The NCCER makes every effort to keep these textbooks up-to-date and free of technical errors. We appreciate your help in this process. If you have an idea for improving this textbook, or if you find an error, a typographical mistake, or an inaccuracy in the NCCER's Craft Training textbooks, please write us, using this form or a photocopy. Be sure to include the exact module number, page number, a detailed description, and the correction, if applicable. Your input will be brought to the attention of the Technical Review Committee. Thank you for your assistance.

Instructors – If you found that additional materials were necessary in order to teach this module effectively, please let us know so that we may include them in the Equipment/Materials list in the Instructor's Guide.

Write: Curriculum Revision and Development Department
National Center for Construction Education and Research
P.O. Box 141104, Gainesville, FL 32614-1104

Fax: 352-334-0932

E-mail: curriculum@nccer.org

Craft _____   Module Name _____

Copyright Date _____   Module Number _____   Page Number(s) _____

Description _____

_____

_____

_____

(Optional) Correction _____

_____

_____

(Optional) Your Name and Address _____

_____

_____

TEACHING TIPS

Section 1.0.0   *What's wrong with this picture?*

Have the trainees call out as many violations as they can find. These include improper lifting, reaching, ladder use, sloppy work area, etc. If desired, divide the class into teams and see which team comes up with the most hazards.

Section 2.0.0   *Think About It – Severity of Shock*

Note that less than 200mA (⅕A) separates a mild shock from a fatal shock. A 60W light bulb draws ½A at 120V.

Section 2.1.0   *Think About It – Electrocution*

Explain that the bird is safe because it does not provide a path to ground. The squirrel does provide a path to ground when its body contacts both the pole and the wire as it attempts to get up onto the wire.

Section 2.1.1   *Think About It – Bodily Harm*

Explain that the amount of damage depends on the amount of current, length of contact, effectiveness of ground, path of the current through the body, and moisture present on the body, as well as the thickness of the skin and any broken areas on the skin.

Section 3.1.2   *Think About It – Dressing for Safety*

Explain that minor clothing flaws such as tears may become caught in machinery or expose the skin to toxic substances. Rivets are conductive, while polyester has a tendency to melt into the skin under intense heat. Protective apparel is designed to place a nonconductive shield between the body and electrical and/or chemical sources of harm.

Section 4.2.0   *Think About It – Working on Energized Systems*

Sample answers might include using metal utensils in a toaster in the home or using tools without GFCI protection in the home or on the job.

Section 4.3.1   *Think About It – GFCIs*

Explain that GFCIs respond to very small current imbalances, thereby protecting personnel from higher ground fault currents. GFCIs should be installed at the source.

Note: The NEC has specific requirements regarding the placement of GFCIs in areas such as kitchens and bathrooms. See *NEC Article 210.*

Section 4.3.1   *What's wrong with this picture?*

Explain how dangerous it is to trust your life to a piece of electrical tape. Emphasize that a proper lockout/tagout provides the electrician with the security of a physical lock to prevent unauthorized persons from reenergizing the circuit.

Section 4.3.1   *Case History – Lockout/Tagout Dilemma*

Ask the trainees to apply this scenario to their own homes. In how many cases does the existing wiring match what would have been shown on the original plans?

Section 4.3.1 *Think About It – Lockout/Tagout—Who Does It and When?*

Note that whenever the power is removed, a lockout/tagout procedure should be followed. Each individual is responsible for protecting himself/herself, but the responsibility of coordination generally falls to the supervisor at the site.

Section 5.1.1 *What's wrong with this picture?*

Note that the ladder is facing the wrong way for proper access and that it does not extend 36" beyond the top support. Ask the trainees to explain why this is dangerous.

Section 5.2.0 *What's wrong with this picture?*

Note the dangerous construction of this scaffold. Have the trainees spot violations by comparing this photo with the scaffold guidelines on the previous page.

Section 9.2.0 *Think About It – Working in Tight Spaces*

Ask the trainees to list some of the hazards of working in a small area, such as a crawl space. They should be concerned with the physical hazards, such as ventilation (particularly when exposed to toxic vapors or when working in hot weather); contact with protruding nails or fiberglass insulation; and proper work clearances. They should also be aware of potential electrical hazards, such as exposed wiring.

Section 15.1.0 *Think About It – Putting It All Together*

Ask the trainees to identify some of the key points of this module. Emphasize that a professional approach to safety involves recognizing hazards, acquiring the proper training to minimize hazards, and accepting the responsibility for protecting both yourself and others.

Hand Bending

26102-00

MODULE OVERVIEW

This course introduces the electrical trainee to the methods and procedures used in cutting, bending, and reaming conduit.

PREREQUISITES

Please refer to the Course Map in the Trainee Module. Prior to training with this module, it is recommended that the trainee shall have successfully completed the following modules:

Core Curriculum; Electrical Level One, Module 26101

LEARNING OBJECTIVES

Upon completion of this module, the trainee will be able to:

1. Identify the methods of hand bending conduit.
2. Identify the various methods used to install conduit.
3. Use math formulas to determine conduit bends.
4. Make 90° bends, back-to-back bends, offsets, kicks, and saddle bends using a hand bender.
5. Cut, ream, and thread conduit.

PERFORMANCE OBJECTIVES

Under supervision of the instructor, the trainee should be able to:

1. Given a piece of EMT, complete the following using a hand bender, hacksaw, and reaming tool. No couplings are allowed, and there should be no kinks in the pipe.
   - Offset
   - Saddle
   - 90° bend

NCCER STANDARDIZED CRAFT TRAINING PROGRAM

The National Center for Construction Education and Research (NCCER) provides a standardized national program of accredited craft training. Key features of the program include instructor certification, competency-based training, and performance testing. The program provides trainees, instructors, and companies with a standard form of recognition through a National Craft Training Registry. The program is described in full in the *Guidelines for Accreditation*, published by the NCCER. For more information on standardized craft training, contact the NCCER at P.O. Box 141104, Gainesville, FL 32614-1104, 352-334-0911, visit our Web site at www.NCCER.org, or e-mail info@NCCER.org.

HOW TO USE THIS ANNOTATED INSTRUCTOR'S GUIDE

Each page presents two sections of information. The larger section displays each page exactly as it appears in the Trainee Module. The narrow column ties suggested trainee and instructor actions to each page and provides icons to call your attention to material, safety, audiovisual, or testing requirements. The bottom of each page includes space for your notes.

If you see the Teaching Tip icon, that means there is a teaching tip associated with this section.

SAFETY CONSIDERATIONS

Ensure that the trainees are equipped with appropriate personal protective equipment. Stress the importance of avoiding contact with sharp edges when handling conduit.

PREPARATION

Before teaching this module, you should review the Module Outline, Learning and Performance Objectives, and the Materials and Equipment List. Be sure to allow ample time to prepare your own training or lesson plan and gather all required equipment and materials.

MATERIALS AND EQUIPMENT LIST

Materials:

Transparencies

Markers/chalk

Pencils/scratch paper

Various pieces of conduit

One piece of EMT per trainee

PVC pieces

PVC cement

Cutting oil

Shop towels

No. 10 or No. 12 solid wire

Copy of the latest edition of the *National Electrical Code®*

OSHA Electrical Safety Guidelines (pocket guide)

Module Examinations*

Performance Profile Sheets*

Equipment:

Overhead projector and screen

Whiteboard/chalkboard

Appropriate personal protective equipment

Tape measure

Hickey

Hand bender

Calculator

Hacksaw

Pipe vise

Pipe cutter

Reamer

Hand-operated threader

Sandbox or drip pan

Torpedo level

*Located in the Test Booklet packaged with this Annotated Instructor's Guide.

ADDITIONAL RESOURCES

This module is intended to present thorough resources for task training. The following reference works are suggested for both instructors and motivated trainees interested in further study. These are optional materials for continued education rather than for task training.

Benfield Conduit Bending Manual, Latest Edition. New York, NY: McGraw-Hill Publishing Company.

National Electrical Code Handbook, Latest Edition. Quincy, MA: National Fire Protection Association.

Tom Henry's Conduit Bending Package (includes video, book, and bending chart). Winter Park, FL: Code Electrical Classes, Inc.

NOTES

The designations "National Electrical Code," "NE Code," and "NEC," where used in this document, refer to the *National Electrical Code®*, which is a registered trademark of the National Fire Protection Association, Quincy, MA. All National Electrical Code (NEC) references in this module refer to the 1999 edition of the NEC.

If you feel that additional math instruction would be helpful, Prentice Hall offers a basic math textbook entitled *Fundamentals of Electrical and Mechanical Mathematics*. It covers the basic math requirements for electrical trainees and may be ordered by contacting Prentice Hall Customer Service at 1-800-922-0579.

TEACHING TIME FOR THIS MODULE

An outline for use in developing your lesson plan is presented below. Note that each Roman numeral in the outline equates to one session of instruction. Each session has a suggested time period of 2½ hours. This includes 10 minutes at the beginning of each session for administrative tasks and one 10-minute break during the session. Approximately 7½ hours are suggested to cover *Hand Bending*. You will need to adjust the time required for hands-on activity and testing based on your class size and resources.

Topic	Planned Time
Session I. Introduction to Hand Bending; Laboratory; Gain; Back-to-Back Bends	
A. Introduction to Hand Bending	_____
1. Equipment	_____
2. Geometry Required to Make a Bend	_____
3. Making a 90° Bend	_____
B. Laboratory	_____
Under instructor supervision, have the trainees practice making 90° bends.	_____
C. Gain	_____
D. Back-to-Back 90° Bends	_____
Session II. Offset Bends; Laboratory; Saddle Bends; Laboratory	
A. Making an Offset	_____
B. Laboratory	_____
Under instructor supervision, have the trainees practice making offset bends.	_____
C. Parallel Offsets	_____
D. Saddle Bends	_____
E. Laboratory	_____
Under instructor supervision, have the trainees practice making saddle bends.	_____
F. Four-Bend Saddles	_____
Session III. Cutting and Reaming Conduit; Laboratory; Threading Conduit; Cutting and Joining PVC Conduit; Module Examination and Performance Testing	
A. Cutting and Reaming Conduit	_____
1. Using a Hacksaw	_____
2. Using a Pipe Cutter	_____
3. Reaming Conduit	_____
B. Laboratory	_____
Under instructor supervision, have the trainees practice cutting and reaming conduit.	_____
C. Threading Conduit	_____
D. Cutting and Joining PVC Conduit	_____

F. Module Examination   _____

    1. Trainees must score 70% or higher to receive recognition from the NCCER.

    2. Record the testing results on Craft Training Report Form 200 and submit the results to the Training Program Sponsor.

G. Performance Testing   _____

    1. Trainees must perform each task to the satisfaction of the instructor to receive recognition from the NCCER.

    2. Record the testing results on Craft Training Report Form 200 and submit the results to the Training Program Sponsor.

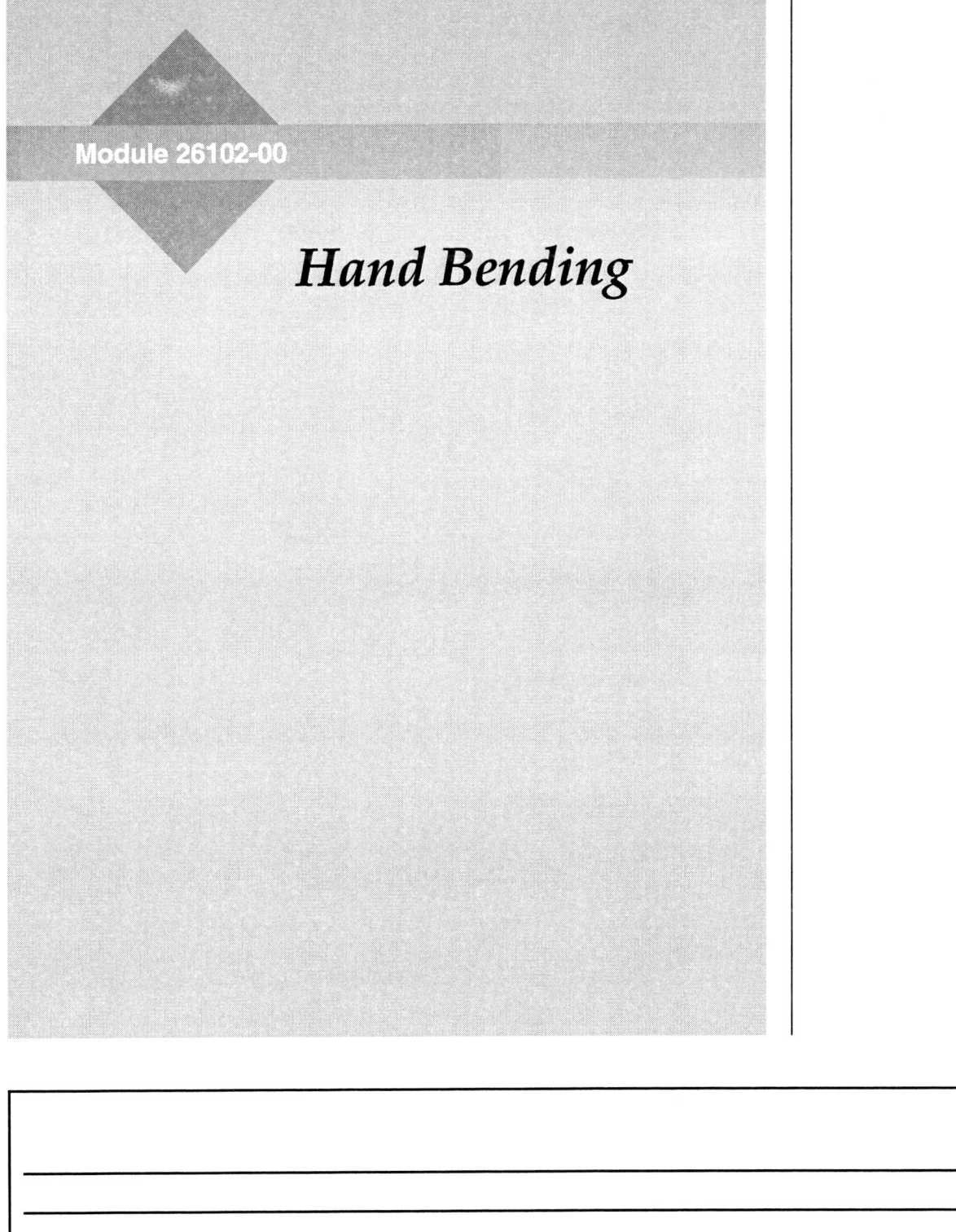

Module 26102-00

Hand Bending

Course Map

This course map shows all of the task modules in the first level of the Electrical curricula. The suggested training order begins at the bottom and proceeds up. Skill levels increase as a trainee advances on the course map. The training order may be adjusted by the local Training Program Sponsor.

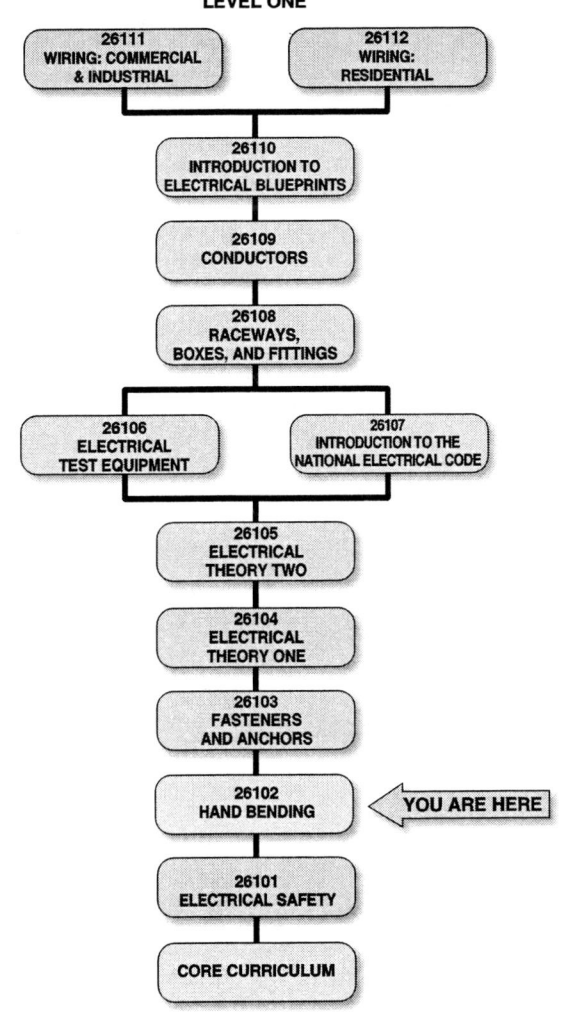

LEVEL ONE

26111
WIRING: COMMERCIAL & INDUSTRIAL

26112
WIRING: RESIDENTIAL

26110
INTRODUCTION TO ELECTRICAL BLUEPRINTS

26109
CONDUCTORS

26108
RACEWAYS, BOXES, AND FITTINGS

26106
ELECTRICAL TEST EQUIPMENT

26107
INTRODUCTION TO THE NATIONAL ELECTRICAL CODE

26105
ELECTRICAL THEORY TWO

26104
ELECTRICAL THEORY ONE

26103
FASTENERS AND ANCHORS

26102
HAND BENDING ◁ YOU ARE HERE

26101
ELECTRICAL SAFETY

CORE CURRICULUM

102CMAP.EPS

Homework

Assign reading of Module 26102.

2–iii

Figures

Tables

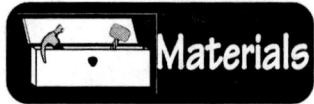

Ensure you have everything required to teach the course. Check the Materials and Equipment List at the front of this Instructor's Guide.

Instructor's Notes:

_____

_____

_____

_____

_____

_____

_____

_____

_____

Hand Bending

Objectives

Upon completion of this module, the trainee will be able to:

1. Identify the methods of hand bending conduit.
2. Identify the various methods used to install conduit.
3. Use math formulas to determine conduit bends.
4. Make 90° bends, back-to-back bends, offsets, kicks, and saddle bends using a hand bender.
5. Cut, ream, and thread conduit.

Prerequisites

Successful completion of the following Task Modules is recommended before beginning study of this Task Module: Core Curricula; Electrical Level One, Module 26101.

Required Trainee Materials

1. Trainee Task Module
2. Copy of the latest edition of the *National Electrical Code*
3. *OSHA Electrical Safety Guidelines* (pocket guide)
4. Appropriate Personal Protective Equipment

1.0.0 ◆ INTRODUCTION

The art of conduit bending is dependent upon the skills of the electrician and requires a working knowledge of basic terms and proven procedures.

Practice, knowledge, and training will help you gain the skills necessary for proper conduit bending and installation. You will be able to practice conduit bending in the lab and in the field under the supervision of experienced coworkers. In this module, the techniques for using hand-operated and step conduit benders such as the hand bender and the hickey will be covered. The process of hand bending conduit, and cutting, reaming, and threading conduit will also be explained.

2.0.0 ◆ HAND BENDING EQUIPMENT

Figure 1 shows a hand bender. Hand benders are convenient to use on the job because they are portable and no electrical power is required. Hand benders have a shape that supports the walls of the conduit being bent.

These benders are used to make various bends in smaller-size conduit (½" to 1¼"). Most hand benders are sized to bend rigid conduit and electrical metallic tubing (EMT) of corresponding sizes. For example, a single hand bender can bend either ¾" EMT or ½" rigid conduit. The next larger size of hand bender will bend either 1" EMT or ¾" rigid conduit. This is because the corresponding sizes of conduit have nearly equal outside diameters.

The first step in making a good bend is familiarizing yourself with the bender. The manufacturer of the bender will typically provide documentation indicating starting points, distance between offsets, gains, and other important values associated with

Note: The designations "National Electrical Code," "NE Code," and "NEC," where used in this document, refer to the National Electrical Code®, which is a registered trademark of the National Fire Protection Association, Quincy, MA. All National Electrical Code (NEC) references in this module refer to the 1999 edition of the NEC.

Show Transparency 1, Course Objectives.

Show Transparency 2, Performance Profile Tasks.

Discuss hand benders and their applications.

Explain that terms shown in bold (blue) are defined in the Glossary at the back of this module.

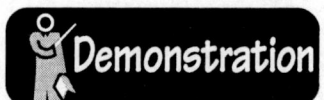

Demonstrate how to mock up a conduit run using a piece of solid wire.

Point out the NEC requirements pertaining to conduit bends.

Explain the difference between a hand bender and a hickey.

Show Transparency 3 (Figure 2).

Working with Conduit

Unprotected electrical cable is susceptible to moisture and physical damage; therefore, protect the wiring with conduit.

Bending Conduit

A good way to practice bending conduit is to use a piece of No. 10 or No. 12 solid wire and bend it to resemble the bends you need. This gives you some perspective on how to bend the conduit and it will also help you to anticipate any problems with the bends.

VERONICA WESTFALL   102F01.EPS

Figure 1 ◆ Hand benders.

that particular bender. There is no substitute for taking the time to review this information. It will make the job go faster and result in better bends.

CAUTION

When making bends, be sure you have a firm grip on the handle to avoid slippage and possible injury.

When performing a bend, it is important to keep the conduit on a stable, firm, flat surface for the entire duration of the bend. Hand benders are designed to have force applied using one foot and the hands. See *Figure 2*. It is important to use constant foot pressure as well as force on the handle to achieve uniform bends. Allowing the conduit to rise up or performing the bend on soft ground can result in distorting the conduit outside the bender.

102F02.EPS

Figure 2 ◆ Pushing down on the bender to complete the bend.

Note

Bends should be made in accordance with the guidelines of *NEC Article 345* (intermediate metal conduit or IMC), *Article 346* (rigid metal conduit), *Article 347* (rigid nonmetallic conduit such as polyvinyl chloride or PVC), or *Article 348* (electrical metallic tubing).

A hickey should not be confused with a hand bender. The hickey, which is used for rigid conduit only, functions quite differently. See *Figure 3*.

When you use a hickey to bend conduit, you are forming the bend as well as the radius. When using a hickey, be careful not to flatten or kink the conduit. Hickeys should only be used with rigid conduit because very little support is given to the walls of the conduit being bent. A hickey is a seg-

ELECTRICAL LEVEL ONE—TRAINEE MODULE 26102

Instructor's Notes:

_____

_____

_____

_____

_____

_____

_____

_____

_____

ELECTRICAL LEVEL ONE—INSTRUCTOR'S GUIDE MODULE 26102

VERONICA WESTFALL   102F03.EPS

Figure 3 ◆ Hickeys.

VERONICA WESTFALL   102F04.EPS

Figure 4 ◆ Typical PVC heating units.

ment bending device. First, a small bend of about 10° is made. Then, the hickey is moved to a new position and another small bend is made. This process is continued until the bend is completed. A hickey can be used for conduit stub-ups in slabs and decks.

PVC conduit is bent using a heating unit (*Figure 4*). The PVC must be rotated regularly while it is in the heater so that it heats evenly. Once heated, the PVC is removed, and the bending is performed by hand. Some units use an electric heating element, while others use liquid propane (LP). After bending, a damp sponge or cloth is often used so that the PVC sets up faster.

When bending PVC that is 2" or larger in diameter, there is a risk of wrinkling or flattening the bend. A plug set eliminates this problem (*Figure 5*). A plug is inserted into each end of the piece of PVC

 CAUTION

Avoid contact with the case of the heating unit; it can become very hot and cause burns. Also, to avoid a fire hazard, ensure that the unit is cool before storage. If using an LP unit, keep a fire extinguisher nearby.

being bent. Then, a hand pump is used to pressurize the conduit before bending it. The pressure is about 3 to 5 psi.

 Note

The plugs must remain in place until the pipe is cooled and set.

 Classroom

Describe how to bend PVC using a heating unit.

Safety

Discuss the precautions that must be taken when using PVC heating units.

VERONICA WESTFALL   102F05.EPS

Figure 5 ◆ Typical plug set.

What's wrong with this picture?

VERONICA WESTFALL   102PO202.EPS

2.1.0 Geometry Required to Make a Bend

Bending conduit requires that you use some basic geometry. You may already be familiar with most of the concepts needed; however, here is a review of the concepts directly related to this task. A right triangle is defined as any triangle with a 90° angle. The side directly opposite the 90° angle is called the *hypotenuse,* and the side on which the triangle sits is the *base.* The vertical side is called the *height.* On the job, you will apply the relationships in a right triangle when making an offset bend. The offset forms the hypotenuse of a right triangle (*Figure 6*).

> **Note**
> There are reference tables for sizing offset bends based on these relationships (see *Appendix A*).

A circle is defined as a closed curved line whose points are all the same distance from its center. The distance from the center point to the edge of the circle is called the *radius.* The length from one edge of the circle to the other edge through the center point is the *diameter.* The distance around the circle is called the *circumference.* A circle can be divided into four equal quadrants. Each quadrant accounts for 90°, making a total of 360°. When you make a 90° bend, you will use ¼ of a circle, or one quadrant. Concentric circles are circles that have a common center but different radii. The concept of concentric circles can be applied to concentric bends in conduit. The angle of each bend is 90°.

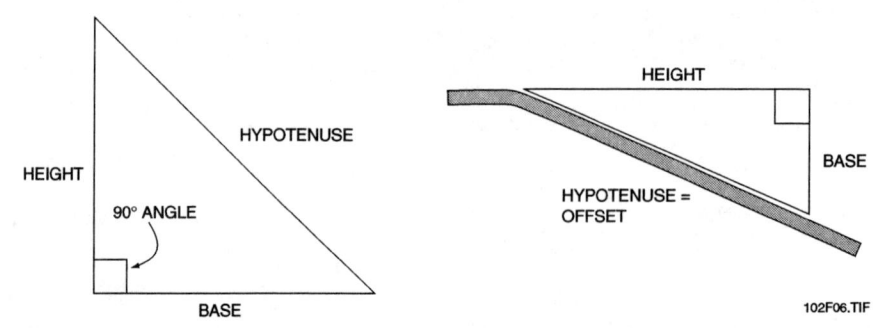

102F06.TIF

Figure 6 ◆ Right triangle and offset bend.

ELECTRICAL LEVEL ONE—TRAINEE MODULE 26102

Instructor's Notes:

_____

_____

_____

_____

_____

_____

_____

_____

_____

Practical Bending

When making offset bends of 45° to step conduit up to another level, square floor tiles make a convenient grid to gauge the distance and angles.

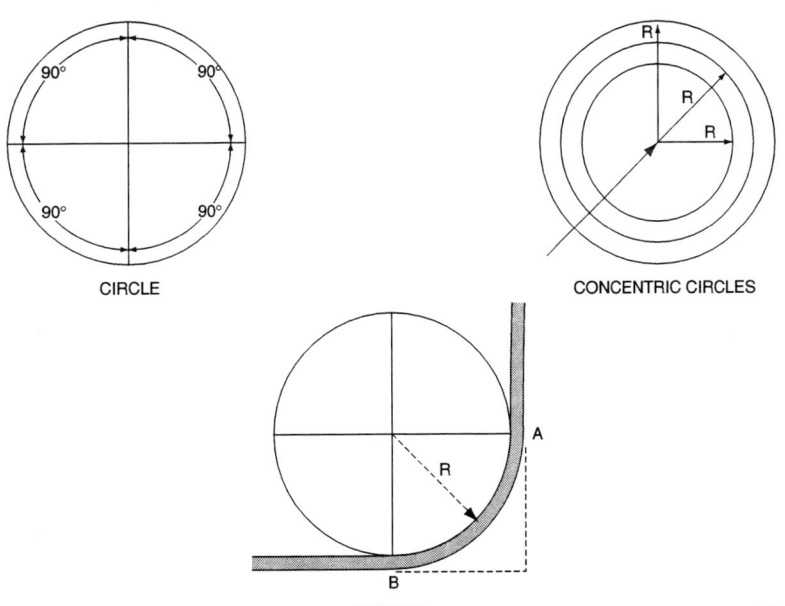

CIRCLE

CONCENTRIC CIRCLES

90° BEND

102F07.EPS

Figure 7 ◆ Circles and 90° bends.

Such bends have the same center point, but the radius of each is different. See *Figure 7*.

To calculate the circumference of a circle, use the following formula:

$$C = \pi \times D \text{ or } C = \pi D$$

In this formula, C = circumference, D = diameter, and π = 3.14. Another way of stating the formula for circumference is C = 2πR, where R equals the radius or ½ the diameter.

To figure the arc of a quadrant use:

$$\text{Length of arc} = (.25)\ 2\pi R = 1.57R$$

For this formula, the arc of a quadrant equals ¼ the circumference of the circle or 1.57 times the radius.

A bending radius table is included in *Appendix B*.

2.2.0 Making a 90° Bend

The 90° stub bend is probably the most basic bend of all. The stub bend is used much of the time, regardless of the type of conduit being installed. Before beginning to make the bend, you need to know two measurements:

- Desired rise or stub-up
- Take-up distance of the bender

The desired rise is the height of the stub-up. The *take-up* is the amount of conduit the bender will use to form the bend. Take-up distances are usually listed in the manufacturer's instruction manual. Typical bender take-up distances are shown in *Table 1*.

Show Transparency 5 (Figure 7).

Explain how to calculate circumference and arc length.

Refer the trainees to *Appendix B* in the back of this module.

Table 1 Typical Bender Take-Up Distances

EMT	Rigid/IMC	Take-Up
½"	—	5"
¾"	½"	6"
1"	¾"	8"
1¼"	1"	11"

Once you have determined the take-up, subtract it from the stub-up height. Mark that distance on the conduit (all the way around) at that distance from the end. The mark will indicate the point at which you will begin to bend the conduit. Line up the starting point on the conduit with the starting point on the bender. Most benders have a mark, like an arrow, to indicate this point. *Figure 8* shows the take-up required to achieve an 18" stub-up on a piece of ½" EMT.

Once you have lined up the bender, use one foot to hold the conduit steady. Keep your heel on the floor for balance. Apply pressure on the bend-er foot pedal with your other foot. Make sure you hold the bender handle level, as far up as possible, to get maximum leverage. Then, bend the conduit in one smooth motion, pulling as steadily as possible. Avoid overstretching.

> **Note**
> When bending conduit using the take-up method, always place the bender on the conduit and make the bend facing the hook of the conduit from which the measurements were taken.

After finishing the bend, check to make sure you have the correct angle and measurement. Use the following steps to check a 90° bend:

Step 1 With the back of the bend on the floor, measure to the end of the conduit stub-up to make sure it is the right length.

Matching Bends in Parallel Runs

INSIDE TRACK

Suppose you are running 1" rigid conduit along with a 2" rigid conduit in a rack and you come to a 90° bend. If you used a 1" shoe, the radius would not match that of the 2" conduit bend. To match the 2" 90° bend, take your 1" conduit and put it in the 2" shoe of your bender, then bend as usual. This 1" 90° bend will now have the same radius as the 2" 90° bend. This trick will only work on rigid conduit. If done with EMT, it will flatten the pipe.

Figure 8 ◆ Bending an 18-inch stub-up.

ELECTRICAL LEVEL ONE—TRAINEE MODULE 26102

Show Transparency 6 (Figure 8).

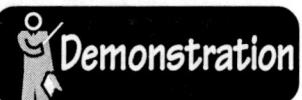

Demonstrate how to make and check a 90° one-shot bend.

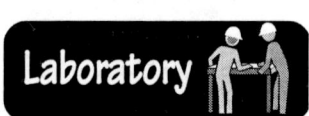

Have the trainees practice making and checking a 90° one-shot bend.

Instructor's Notes:

Take-Up Method

When bending conduit using the take-up method, always place the bender on the conduit and make the bend facing the end of the conduit from which the measurements were taken. It helps to make a narrow mark with a soft lead pencil or marker completely around the conduit to ensure straight bends.

Checking Vertical Rise

Use a torpedo level to check for plumb on a vertical rise.

VERONICA WESTFALL   102PO203.EPS

Demonstrate how to check for plumb using a torpedo level.

Step 2 Check the 90° angle of the bend with a square or at the angle formed by the floor and a wall. A torpedo level may also be used.

Note

If you overbend a conduit slightly past the desired angle, you can use the bender to bend the conduit back to the correct angle.

The above procedure will produce a 90° *one-shot bend*. That means that it took a single bend to form the conduit bend. A *segment bend* is any bend that is formed by a series of bends of a few degrees each, rather than a single one-shot bend.

A shot is actually one bend in a segment bend. Segment or sweep bends must conform to the provisions of the NEC.

2.3.0 Gain

The gain is the distance saved by the arc of a 90° bend. Knowing the gain can help you to precut, ream, and prethread both ends of the conduit before you bend it. This will make your work go more quickly because it is easier to work with conduit while it is straight. *Figure 9* shows that the overall developed length of a piece of conduit with a 90° bend is less than the sum of the horizontal and vertical distances when measured square to the corner. This is shown by the following equation:

$$\text{Developed length} = (A + B) - \text{gain}$$

Show Transparency 7 (Figure 9).

Discuss the "Think About It." Explain that smooth bends help to prevent conductor damage.

Discuss the "Think About It." Explain that take-up represents the length needed while gain is the distance saved.

Show Transparency 8 (Figure 10A).

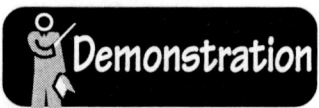

Demonstrate how to make a back-to-back bend.

Smooth Bends

Why are smooth bends so important?

GAIN = DISTANCE SAVED

Rigid Conduit Size	NEC Radius	90° Gain
1/2"	4"	2-5/8"
3/4"	5"	3-1/4"
1"	6"	4"
1-1/4"	8"	5-5/8"

TYPICAL GAIN TABLE

102F09.TIF

Figure 9 ◆ Gain.

Gain

What is the difference between the gain and the take-up of a bend?

An example of a manufacturer's gain table is also shown in *Figure 9*. These tables are used to determine the gain for a certain size conduit.

2.4.0 Back-to-Back 90° Bends

A back-to-back bend consists of two 90° bends made on the same piece of conduit and placed back-to-back (*Figure 10*).

To make a back-to-back bend, make the first bend (labeled *X* in *Figure 10*) in the usual manner. To make the second bend, measure the required distance between the bends from the back of the first bend. This distance is labeled *L* in the figure. Reverse the bender on the conduit, as shown in *Figure 10*. Place the bender's back-to-back indicating mark at point *Y* on the conduit. Note that outside measurements from point *X* to point *Y* are used. Holding the bender in the reverse position and properly aligned, apply foot pressure and complete the second bend.

2.5.0 Making an Offset

Many situations require that the conduit be bent so that it can pass over objects such as beams and other conduits, or enter meter cabinets and junction boxes. Bends used for this purpose are called *offsets (kicks)*. To produce an offset, two equal bends of less than 90° are required, a specified distance apart, as shown in *Figure 11*.

Offsets are a trade-off between space and the effort it will take to pull the wire. The larger the degree of bend, the harder it will be to pull the wire. The smaller the degree of bend, the easier it will be to pull the wire. Use the shallowest degree of bend that will still allow the conduit to bypass the obstruction and fit in the given space.

When conduit is offset, some of the conduit length is used. If the offset is made into the area, an allowance must be made for this shrinkage. If the offset angle is away from the obstruction, the shrinkage can be ignored. *Table 2* shows the

---

ELECTRICAL LEVEL ONE—TRAINEE MODULE 26102

Instructor's Notes:

_____

_____

_____

_____

_____

_____

_____

_____

_____

VERONICA WESTFALL   102F10B.EPS

Figure 10 ◆ Back-to-back bends.

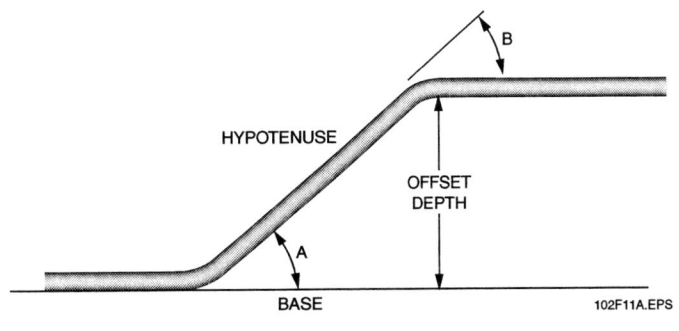

BASE   102F11A.EPS

Figure 11 ◆ Offsets.

VERONICA WESTFALL 102F11B.EPS

**Show Transparency 9
(Figure 11A).**

Demonstrate how to make an offset bend.

Have the trainees practice making offset bends.

Show Transparencies 10 through 12 (Figures 12 through 14).

Discuss the "Think About It."

6" Offset:
Distance between bends = 12"
Shrink = 1½"

10" Offset:
Distance between bends = 14"
Shrink = 3¾"

amount of shrinkage per inch of rise for common offset angles.

The formula for figuring the distance between bends is as follows:

$$\text{Distance between bends} = \text{depth of offset} \times \text{multiplier}$$

The distance between the offset bends can generally be found in the manufacturer's documentation for the bender. *Table 3* shows the distance between bends for the most common offset angles.

Calculations related to offsets are derived from the branch of mathematics known as *trigonometry*, which deals with triangles. The multipliers shown in *Table 2* represent the *cosecant* (COS) of the related offset angle. The multiplier is determined by

Table 2 Shrinkage Calculation

Offset Angle	Multiplier	Shrinkage (per inch of rise)
10° × 10°	6	¹⁄₁₆"
22½° × 22½°	2.6	³⁄₁₆"
30° × 30°	2	¼"
45° × 45°	1.4	⅜"
60° × 60°	1.2	½"

dividing the depth of the offset by the hypotenuse of the triangle created by the offset (*Figure 11*).

Basic trigonometry (trig) functions are briefly covered in *Appendix A*. As you will see in the next section, the *tangent* (TAN) of the offset angle is also used in calculating parallel offsets. Understanding trig functions will help you understand how offsets are determined. If you have a scientific calculator and understand these functions, you can calculate offset angles when you know the dimensions of the triangle created by the offset and the obstacle.

2.6.0 Parallel Offsets

Often, multiple pieces of conduit must be bent around a common obstruction. In this case, parallel offsets are made. Since the bends are laid out along a common radius, an adjustment must be made to ensure that the ends do not come out uneven, as shown in *Figure 12*.

The center of the first bend of the innermost conduit is found first, as shown in *Figure 13*. Each successive conduit must have its centerline moved farther away from the end of the pipe, as shown in *Figure 14*. The amount to add is calculated as follows:

$$\text{Amount added} = \text{center-to-center spacing} \times \text{tangent (TAN) of ½ offset angle}$$

Table 3 Common Offset Factors (in Inches)

Offset Depth	22½° Between Bends	22½° Shrinkage	30° Between Bends	30° Shrinkage	45° Between Bends	45° Shrinkage	60° Between Bends	60° Shrinkage
2	5¼	⅜	—	—	—	—	—	—
3	7¾	⁹⁄₁₆	6	¾	—	—	—	—
4	10½	¾	8	1	—	—	—	—
5	13	1³⁄₁₆	10	1¼	7	1⅞	—	—
6	15½	1⅛	12	1½	8½	2¼	7¼	3
7	18¼	1⁵⁄₁₆	14	1¾	9¾	2⅝	8⅜	3½
8	20¾	1½	16	2	11¼	3	9⅝	4
9	23½	1¾	18	2¼	12½	3⅜	10⅞	4½
10	26	1⅞	20	2½	14	3¾	12	5

Calculating Shrinkage

You're making a 30° by 30° offset to clear a 6" obstruction. What will be the distance between bends? What will be the developed length shrink? Make the same calculations for a 10" offset with 45° bends.

Instructor's Notes:

_____

_____

_____

_____

_____

_____

_____

_____

_____

Equal Angles

Why is it important that the angles be identical when making an offset bend?

Discuss the "Think About It." Explain that uniform bends are necessary to ensure that the run remains parallel.

Tangents can be found using the trig tables provided in *Appendix A.*

For example, *Figure 15* shows three pipes laid out as parallel and offset. The angle of the offset is 30°. The center-to-center spacing is 3". The start of the innermost pipe's first bend is 12".

The starting point of the second pipe will be:

$$12" + [\text{center-to-center spacing} \times \text{TAN (}\tfrac{1}{2}\text{ offset angle)}]$$
$$12" + (3" \times \text{TAN } 15°) = 12" + (3" \times .2679)$$
$$= 12" + .8037"$$

This is approximately 12¹³⁄₁₆".

The starting point for the outermost pipe is:

$$12¹³⁄₁₆" + ¹³⁄₁₆" = 13⅝"$$

 Audiovisual

Show Transparency 13 (Figure 15).

2.7.0 Saddle Bends

A saddle bend is used to go around obstructions. *Figure 16* illustrates an example of a saddle bend that is required to clear a pipe obstruction. Making a saddle bend will cause the center of the saddle to shorten ³⁄₁₆" for every inch of saddle depth (see *Table 4*). For example, if the pipe diameter is 2 inches, this would cause a ⅜" shortening of the conduit on each side of the bend.

When making saddle bends, the following steps should apply:

Step 1 Locate the center mark A on the conduit by using the size of the obstruction (i.e., pipe diameter) and calculate the shrink rate

Show Transparency 14 (Figure 16).

Figure 12 ◆ Incorrect parallel offsets.

 Demonstration

Demonstrate how to make a saddle bend.

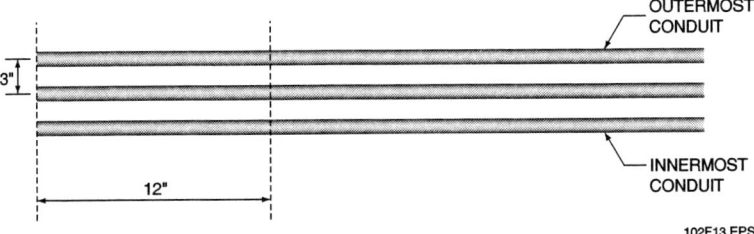

OUTERMOST CONDUIT

3"

12"

INNERMOST CONDUIT

102F13.EPS

Figure 13 ◆ Center of first bend.

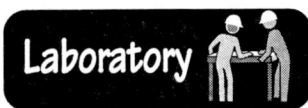 Laboratory

Have the trainees practice making saddle bends.

OUTERMOST
CONDUIT

INNERMOST
CONDUIT

3"

12"

12¹³/₁₆"

13⁵/₈"

102F14.EPS

Figure 14 ◆ Successive centerlines.

3"

12"

12¹³/₁₆"

13⁵/₈"

30°

102F15.EPS

Figure 15 ◆ Parallel offset pipes.

Discuss the "Think About It." Starting point of second pipe = 13⅝".

Calculating Parallel Offsets

You're making parallel offsets of 45° and the lengths of conduit are spaced 4" center to center. If the offset starts 12" down the pipe, what is the starting point for the bend on the second pipe?

BEFORE — 15³/₈"

PLACE CENTER MARK AHEAD ³/₈"

AFTER — 15"

³/₈" CENTER SHRINK

TOTAL
SHRINKAGE = ³/₄"

2" DIAMETER

B (22¹/₂°)   A (45°)   C (22¹/₂°)

102F16.EPS

Figure 16 ◆ Saddle measurement.

Instructor's Notes:

Table 4 Shrinkage Chart for Saddle Bends with a 45° Center Bend and Two 22½° Bends

Obstruction Depth	Shrinkage Amount (Move Center Mark Forward)	Make Outside Marks from *New* Center Mark
1	³⁄₁₆"	2½"
2	⅜"	5"
3	⁹⁄₁₆"	7½"
4	¾"	10"
5	¹⁵⁄₁₆"	12½"
6	1⅛"	15"
For each additional inch, add	³⁄₁₆"	2½"

Step 2 Locate marks B and C on the conduit by measuring 2½" for every 1" of saddle depth *from* the A mark (i.e., for the saddle depth of 2 inches, the B mark would be 5" before the A mark and the C mark would be 5" after the A mark). See *Figure 17*.

Step 3 Refer to *Figure 18* and make a 45° bend at point A, make a 22½° bend at point B, and make a 22½° bend at point C. (Be sure to check the manufacturer's specifications.)

2.8.0 Four-Bend Saddles

Four-bend saddles can be difficult. The reason is that four bends must be aligned exactly on the same plane. Extra time spent laying it out and performing the bends will pay off in not having to scrap the whole piece and start over.

Figure 19 illustrates that the four-bend saddle is really two offsets formed back-to-back. Working left to right, the procedure for forming this saddle is as follows:

Step 1 Determine the height of the offset.

Step 2 Determine the correct spacing for the first offset and mark the conduit.

Step 3 Bend the first offset.

of the obstruction (for example, if the pipe diameter is 2 inches, ⅜" of conduit will be lost on each side of the bend for a total shrinkage of ¾"). This figure will be added to the measurement from the end of the conduit to the centerline of the obstruction (for example, if the distance measured from the conduit end and the obstruction centerline was 15", the distance to A would be 15⅜").

Figure 17 ◆ Measurement locations.

Figure 18 ◆ Location of bends.

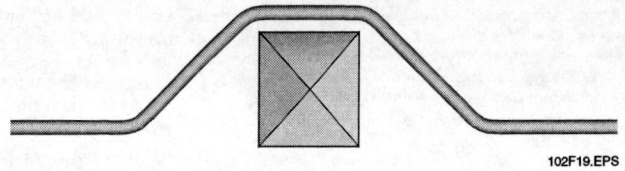

102F19.EPS

Figure 19 ◆ Typical four-bend saddle.

Step 4 Mark the start point for the second offset at the trailing edge of the obstruction.

Step 5 Mark the spacing for the second offset.

Step 6 Bend the second offset.

Using *Figure 20* as an example, a four-bend saddle using ½" EMT is laid out as follows:

- Height of the box = 6"
- Width of the box = 8"
- Distance to the obstruction = 36"

Two 30° offsets will be used to form the saddle. It is created as follows:

Step 1 See *Figure 21*. Working from left to right, calculate the start point for the first bend. The distance to the obstruction is 36", the offset is 6", and the 30° multiplier from *Table 2* is 2.00:

Distance to the obstruction − (offset × constant for the angle) + shrinkage = distance to the first bend

36" − (6" × 2.00) + 1½" = 25½"

Step 2 Determine where the second bend will end to ensure the conduit clears the obstruction. See *Figure 22*.

Distance to the first bend + distance to second bend + shrinkage = total length of the first offset

25½" + 12" + 1½" = 39"

Step 3 Determine the start point of the second offset. The width of the box is 8"; therefore, the start point of the second offset should be 8" beyond the end of the first offset:

8" + 39" = 47"

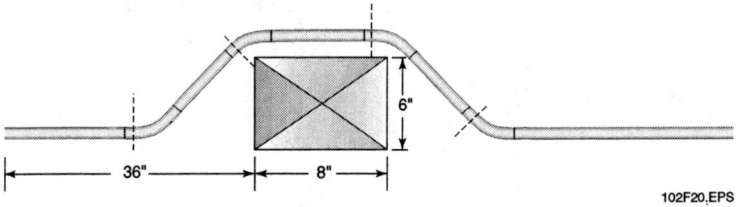

6"

36"   8"

102F20,EPS

Figure 20 ◆ Four-bend saddle.

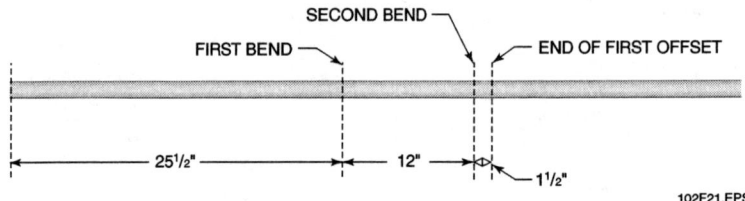

SECOND BEND

FIRST BEND   END OF FIRST OFFSET

25¹/₂"   12"   1¹/₂"

102F21.EPS

Figure 21 ◆ Four-bend saddle measurements.

Instructor's Notes:

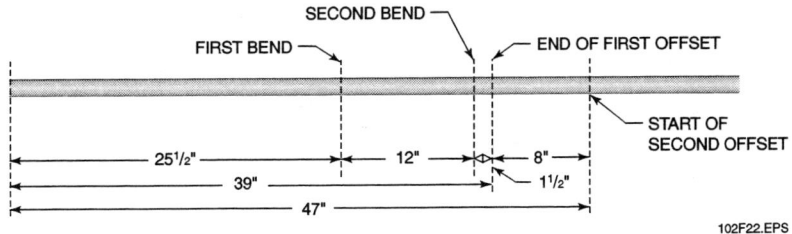

Figure 22 ◆ Bend and offset measurements.

Planning Bends

The more bends you make, the more difficult it is to pull the wires through the conduit. Therefore, plan your bends in advance, avoid sharp bends, and make as few bends as possible. The NEC allows the bends in a single run of conduit to total 360°; however, 360° is not as much as you might think. For example, if you bend the conduit 90° for two corners of a room, with two 45° offsets where the conduit connects to a panelboard and junction box, you've used up your 360°.

Using Your Bender Head to Secure Conduit

To secure conduit while cutting, insert the conduit into one of the holes in the bender head, brace your knee against the bender to secure it, then proceed to cut the conduit.

Step 4 Determine the spacing for the second off-set. Since the first and second offsets have the same rise and angle, the distance between bends will be the same, or 12".

3.0.0 ◆ CUTTING, REAMING, AND THREADING CONDUIT

Rigid conduit, IMC, and EMT are available in standard 10-foot lengths. When installing conduit, it is cut to fit the job requirements.

3.1.0 Hacksaw Method of Cutting Conduit

Conduit is normally cut using a hacksaw. To cut conduit with a hacksaw:

Step 1 Inspect the blade of the hacksaw and replace it, if needed. A blade with 18, 24, or 32 cutting teeth per inch is recommended for conduit. Use a higher tooth count for EMT and a lower tooth count for rigid conduit and IMC. If the blade needs to be replaced, point the teeth toward the front of the saw when installing the new blade.

Step 2 Secure the conduit in a pipe vise.

Step 3 Rest the middle of the hacksaw blade on the conduit where the cut is to be made. Position the saw so the end of the blade is pointing slightly down and the handle is pointing slightly up. Push forward gently until the cut is started. Make even strokes until the cut is finished.

CAUTION

To avoid bruising your knuckles on the newly cut pipe, use gentle strokes for the final cut.

Emphasize the importance of careful planning to ensure proper bends.

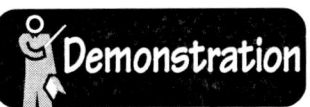

Demonstrate how to use a hacksaw to cut conduit.

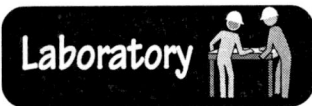

Have the trainees practice cutting conduit using a hacksaw.

Step 4 Check the cut. The end of the conduit should be straight and smooth. *Figure 23* shows correct and incorrect cuts. Ream the conduit.

3.2.0 Pipe Cutter Method

A pipe cutter can also be used to cut rigid IMC conduit. To use a pipe cutter:

Step 1 Secure the conduit in a pipe vise and mark a place for the cut.

Step 2 Open the cutter and place it over the conduit with the cutter wheel on the mark.

Step 3 Tighten the cutter by rotating the screw handle.

CAUTION

Do not overtighten the cutter. Overtightening can break the cutter wheel and distort the wall of the conduit.

Step 4 Rotate the cutter counterclockwise to start the cut. *Figure 24* shows the proper way to rotate the cutter.

Step 5 Tighten the cutter handle ¼ turn for each full turn around the conduit. Again, make sure that you do not overtighten it.

INCORRECT   CORRECT

102F23.EPS

Figure 23 ◆ Conduit ends after cutting.

Using Tape

Use a piece of tape as a guide for marking your cutting lines around the conduit. This ensures a straight cut.

102F24A.EPS   VERONICA WESTFALL   102F24B.EPS

Figure 24 ◆ Cutter rotation.

Instructor's Notes:

_____

_____

_____

_____

_____

_____

_____

_____

_____

Step 6 Add a few drops of cutting oil to the groove and continue cutting. Avoid skin contact with the oil.

Step 7 When the cut is almost finished, stop cutting and snap the conduit to finish the cut. This reduces the ridge that can be formed on the inside of the conduit.

Step 8 Clean the conduit and cutter with a shop towel rag.

Step 9 Ream the conduit.

3.3.0 Reaming Conduit

When the conduit is cut, the inside edge is sharp. This edge will damage the insulation of the wire when it is pulled through. To avoid this damage, the inside edge must be smoothed or *reamed* using a reamer (*Figure 25*).

To ream the inside edge of a piece of conduit using a hand reamer, proceed as follows:

VERONICA WESTFALL   102F25.EPS

Figure 25 ◆ Rigid conduit reamer.

Step 1 Place the conduit in a pipe vise.

Step 2 Insert the reamer tip in the conduit.

Step 3 Apply light forward pressure and start rotating the reamer. *Figure 26* shows the

102F26A.EPS

VERONICA WESTFALL   102F26B.EPS

Figure 26 ◆ Reamer rotation.

Show Transparency 23 (Figure 26A).

Demonstrate how to ream conduit.

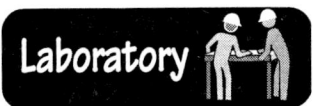

Have the trainees practice reaming conduit.

Safety

Emphasize that burrs are sharp and can cause cuts if gloves are not worn.

Demonstration

Demonstrate the use of a ratchet threader.

proper way to rotate the reamer. It should be rotated using a downward motion. The reamer can be damaged if you rotate it in the wrong direction. The reamer should bite as soon as you apply the proper pressure.

Step 4 Remove the reamer by pulling back on it while continuing to rotate it. Check the progress and then reinsert the reamer. Rotate the reamer until the inside edge is smooth. You should stop when all burrs have been removed.

 Note
If a conduit reamer is not available, use a half-round file (the tang of the file must have a handle attached). EMT may be reamed using the nose of diagonal cutters or small hand reamers.

3.4.0 Threading Conduit

After conduit is cut and reamed, it is usually threaded so it can be properly joined. Only rigid conduit and IMC have walls thick enough for threading.

The tool used to cut threads in conduit is called a *die*. Conduit dies are made to cut a taper of ¾ inch per foot. The number of threads per inch varies from 8 to 18, depending upon the diameter of the conduit. A thread gauge is used to measure how many threads per inch are cut.

The threading dies are contained in a die head. The die head can be used with a hand-operated ratchet threader (*Figure 27*) or with a portable power drive.

To thread conduit using a hand-operated threader, proceed as follows:

VERONICA WESTFALL   102F27.EPS

Figure 27 ◆ Hand-operated ratchet threader.

Step 1 Insert the conduit in a pipe vise. Make sure the vise is fastened to a strong surface. Place supports, if necessary, to help secure the conduit.

Step 2 Determine the correct die and head. Inspect the die for damage such as broken teeth. Never use a damaged die.

Step 3 Insert the die securely in the head. Make sure the proper die is in the appropriately numbered slot on the head.

Step 4 Determine the correct thread length to cut for the conduit size used (match the manufacturer's thread length).

Step 5 Lubricate the die with cutting oil at the beginning and throughout the threading operation. Avoid skin contact with the oil.

Step 6 Cut threads to the proper length. Make sure that the conduit enters the tapered side of the die. Apply pressure and start turning the head. You should back off the head each ¼ turn to clear away chips.

Step 7 Remove the die when the proper cut is made. Threads should be cut only to the length of the die. Overcutting will leave the threads exposed to corrosion.

Step 8 Inspect the threads to make sure they are clean, sharp, and properly made. Use a thread gauge to measure the threads. The finished end should allow for a wrench-tight fit with one or two threads exposed.

 Note
The conduit should be reamed again after threading to remove any burrs and edges. Cutting oil must be swabbed from the inside and outside of the conduit. Use a sandbox or drip pan under the threader to collect drips and shavings.

Die heads can also be used with portable power drives. You will follow the same steps when using a portable power drive. Threading machines are often used on larger conduit and where frequent threading is required. Threading machines hold and rotate the conduit while the die is fed onto the conduit for cutting. When using a threading machine, make sure you secure the legs properly and follow the manufacturer's instructions.

2–18   ELECTRICAL LEVEL ONE—TRAINEE MODULE 26102

Instructor's Notes:

_____

_____

_____

_____

_____

_____

_____

_____

_____

3.5.0 Cutting and Joining PVC Conduit

PVC conduit may be easily cut with a fine-tooth handsaw. To ensure square cuts, a miter box or similar device is recommended for cutting 2" and larger PVC. You can deburr the cut ends using a pocket knife. Smaller diameter PVC conduit, up to 1½", may be cut using a PVC cutter.

Use the following steps to join PVC conduit sections or attachments to plastic boxes:

Step 1 Wipe all the contacting surfaces clean and dry.

Step 2 Apply a coat of cement (a brush or aerosol can is recommended) to the male end to be attached.

Step 3 Press the conduit and fitting together and rotate about a half-turn to evenly distribute the cement.

Forming PVC in the field requires a special tool called a *hot box* or other specialized methods. PVC may not be threaded when it is used for electrical applications.

 Note
Cementing the PVC must be done quickly. The aerosol spray cans of cement or the cement/brush combination are usually provided by the PVC manufacturer. Make sure you use the recommended cement.

 CAUTION
Solvents and cements used with PVC are hazardous. Wear gloves and follow the product instructions.

Summary

You must choose a conduit bender to suit the kind of conduit being installed and the type of bend to be made. Some knowledge of the geometry of

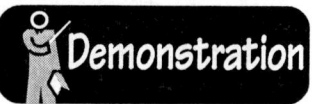 **Demonstration**

Demonstrate how to cut and join PVC conduit.

 Safety

Emphasize the precautions required when working with PVC solvents and cements.

 INSIDE TRACK

Oiling the Threader
For smoother operation, oil the threader often while threading the conduit.

VERONICA WESTFALL   102PO204.EPS

INSIDE TRACK

Threading Conduit
The key to threading conduit is to start with a square cut. If you don't get it right the first time, the fitting won't thread properly.

Demonstration

Demonstrate how to cut a piece of PVC using a nylon string.

Classroom

Discuss the "Think About It." Emphasize that poor bending techniques can result in pulling problems, damaged conductors, and a sloppy appearance.

Have the trainees complete the Review Questions and go over the answers prior to administering the Module Examination.

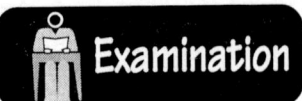

Examination

Administer the Module Examination and Performance Test. Be sure to record the results on Craft Training Report Form 200 and submit the results to the Training Program Sponsor.

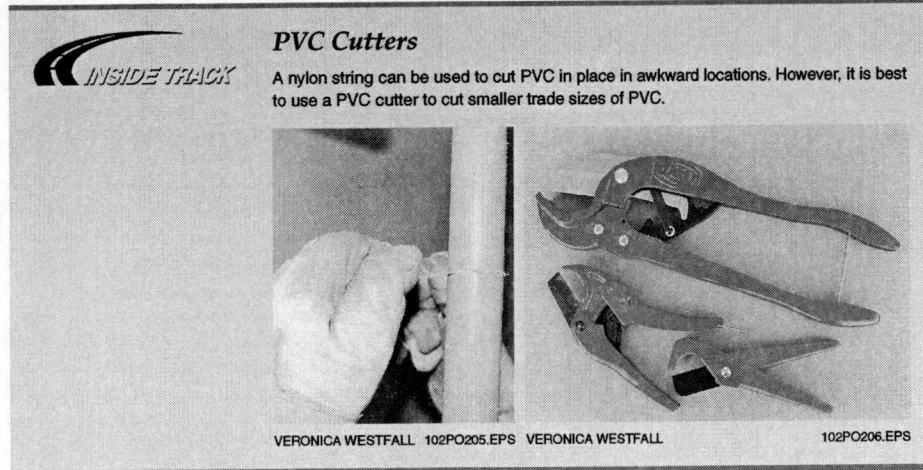

INSIDE TRACK

PVC Cutters

A nylon string can be used to cut PVC in place in awkward locations. However, it is best to use a PVC cutter to cut smaller trade sizes of PVC.

VERONICA WESTFALL 102PO205.EPS VERONICA WESTFALL   102PO206.EPS

THINK ABOUT IT

Putting It All Together

This module has stressed the precision necessary for creating accurate and uniform bends. Why is this important? What practical problems can result from sloppy or inaccurate bends?

right triangles and circles needs to be mastered to make the necessary calculations. You must be able to calculate, lay out, and perform bending operations on a single run of conduit and also on two or more parallel runs of conduit. At times, data tables for the figures may be consulted for the calculations. All work must conform to the requirements of the NEC.

Review Questions

1. The field bending of PVC requires a _____.
   a. hickey
   b. heating unit
   c. segmented bender
   d. one-shot bender

2. After bending PVC, the bend can be set by using _____.
   a. a damp sponge or cloth
   b. dry ice
   c. ice-cold water
   d. a blow dryer

3. A hickey is used for bending _____.
   a. rigid conduit
   b. EMT and IMC
   c. PVC conduit
   d. rigid conduit, EMT, IMC, and PVC conduit

4. A plug set is typically used to prevent _____ PVC when bending it.
   a. overpressurizing
   b. flattening
   c. corroding
   d. cutting

5. What is the key to accurate bending with a hand bender?
   a. Correct size and length of handle
   b. Constant foot pressure on the back piece
   c. Using only the correct brand of bender
   d. Correct inverting of the conduit bender

ELECTRICAL LEVEL ONE—TRAINEE MODULE 26102

Instructor's Notes:

_____
_____
_____
_____
_____
_____
_____
_____
_____
_____

ELECTRICAL LEVEL ONE—INSTRUCTOR'S GUIDE MODULE 26102

6. In a right triangle, the side directly opposite the 90° angle is called the _____.
   a. right side
   b. hypotenuse
   c. altitude
   d. base

7. The formula for calculating the circumference of a circle is _____.
   a. $\pi \times R^2$
   b. $2\pi \times R^2$
   c. $\pi \times D$
   d. $2\pi \times D$

8. Prior to making a 90° bend, what two measurements must be known?
   a. Length of conduit and size of conduit
   b. Desired rise and length of conduit
   c. Size of bender and size of conduit
   d. Stub-up distance and take-up distance

9. A back-to-back bend is _____.
   a. a two-shot 90° bend
   b. two 90° bends made back-to-back
   c. an offset with four bends
   d. a segmented bend

10. To ensure the conduit enters straight into the junction box, a(n) _____ may be required.
    a. back-to-back bend
    b. saddle bend
    c. offset
    d. take-up

11. To prevent the ends of the conduit from being staggered, what additional information must be used when making parallel offset bends?
    a. Center-to-center spacing and tangent of ½ the offset angle
    b. Length of conduit and size of conduit
    c. Stub-up distance and take-up distance
    d. Offset angle and length of conduit

12. When making a saddle bend, the center of the saddle will cause the conduit to shrink _____ for every inch of saddle depth.
    a. ⅜"
    b. ³⁄₁₆"
    c. ¾"
    d. ³⁄₃₂"

13. When using a pipe cutter, always rotate the cutter _____ to start the cut.
    a. in a clockwise direction
    b. with the grain
    c. in a counterclockwise direction
    d. against the grain

14. You would use _____ to smooth the sharp inside edge of metal conduit after it has been cut.
    a. a flat file
    b. rough sandpaper
    c. a reamer
    d. a pocket knife

15. What tool is used to cut threads in rigid conduit or IMC?
    a. A thread gauge
    b. A cutter
    c. A tap
    d. A die

Timothy Ely, Beacon Electric Company

Tim Ely is a man who believes in giving something back to the industry that nurtured his successful career. Despite working in a demanding executive position, he serves on many industry committees and was instrumental in the development of the NCCER Electrical Program.

What made you decide to become an electrician?
During my last two years of high school, I worked for a do-it-all construction company. We laid concrete, installed roofs, hung drywall, installed plumbing, and did electrical work. I liked the electrical work the best.

How did you learn the trade?
I learned through on-the-job training, hard work, and studying on my own. I had good teachers who were patient with me and took the time to help me succeed.

What kinds of jobs did you hold on the way to your current position?
I started out wiring houses and did that for the first two years. Then I switched over to commercial and industrial work, and worked as an apprentice in that area for two more years before becoming a journeyman. From there, I served as a lead electrician, then foreman, then finally city superintendent, before being promoted to my current job as general superintendent.

What factor or factors have contributed most to your success?
Hard work helps a lot. I also try to bring a positive attitude to work with me every day. My family and friends have supported me throughout my career.

What does a general superintendent do in your company?
In my job, I have responsibility for all the job sites, as well as the warehouse and service trucks. I also have responsibility for employee hiring, safety training, job planning and scheduling, quality control, and licensing. I personally hold 26 different state and city licenses, and I firmly believe that getting the training to obtain your licenses and then doing the in-service training to keep your licenses current are important factors in an electrician's success. For example, an electrical contractor can bid on jobs in a wide geographical area. Electricians working for that contractor can work on projects in different cities, even different states. Every place you go will require you to have a valid license.

What advice would you give to someone entering the electrical trade?
Work hard, treat people with respect, and keep an open mind. Be careful how you deal with people. Someone you offend today may wind up being your boss or a potential customer tomorrow.

Instructor's Notes:

_____

_____

_____

_____

_____

_____

_____

_____

_____

Trade Terms Introduced in This Module

90° bend: A bend that changes the direction of the conduit by 90°.

Back-to-back bend: Any bend formed by two 90° bends with a straight section of conduit between the bends.

Concentric bends: Making 90° bends in two or more parallel runs of conduit and increasing the radius of each conduit from the inside of the run toward the outside.

Developed length: The actual length of the conduit that will be bent.

Gain: Because a conduit bends in a radius and not at right angles, the length of conduit needed for a bend will not equal the total determined length. Gain is the distance saved by the arc of a 90° bend.

Offsets: An offset (kick) is two bends placed in a piece of conduit to change elevation to go over or under obstructions or for proper entry into boxes, cabinets, etc.

Rise: The length of the bent section of conduit measured from the bottom, centerline, or top of the straight section to the end of the bent section.

Segment bend: A large bend formed by multiple short bends or *shots*.

Stub-up: Another name for the rise in a section of conduit. Also, a term used for conduit penetrating a slab or the ground.

Using Trigonometry to Determine Offset Angles and Multipliers

You do not have to be a mathematician to use trigonometry. Understanding the basic trig functions and how to use them can help you calculate unknown distances or angles. Assume that the right triangle below represents a conduit offset. If you know the length of one side and the angle, you can calculate the length of the other sides, or if you know the length of any two of the sides of the triangle, you can then find the offset angle using one or more of these trig functions. You can use a trig table such as that shown on the following pages or a scientific calculator to determine the offset angle. For example, if the cosecant of angle A is 2.6, the trig table tells you that the offset angle is 22½°.

$$\text{SINE } a = \frac{O}{H}$$

$$\text{COSINE } a = \frac{A}{H}$$

$$\text{TANGENT } a = \frac{O}{A}$$

$$\text{COSECANT } a = \frac{H}{O}$$

$$\text{SECANT } a = \frac{H}{A}$$

$$\text{COTANGENT } a = \frac{A}{O}$$

Fractions represent the ratios of the lengths of the sides.

O = 30

$$\frac{1}{\text{Sine}} = \text{Multiplier}$$

Sine 30 = 0.5

b = 90° - a

Hypotenuse

Side **O**pposite

Side **A**djacent

To determine the multiplier for the distance between bends in an offset:

1. Determine the angle of the offset: 30°

2. Find the sine of the angle: 0.5

3. Find the inverse (reciprocal) of the sine: $\frac{1}{0.5}$ = 2. This is also listed in trig tables as the cosecant of the angle.

4. This number multiplied by the height of the offset gives the hypotenuse of the triangle, which is equal to the distance between bends.

102APX01.EPS

Instructor's Notes:

Angle	Sine	Cosine	Tangent	Cotangent	Cosecant
1°	.0175	.9998	.0175	57.3	57.3065
2°	.0349	.9994	.0349	28.6	28.6532
3°	.0523	.9986	.0524	19.1	19.1058
4°	.0698	.9976	.0699	14.3	14.3348
5°	.0872	.9962	.0875	11.4	11.4731
6°	.1045	.9945	.1051	9.51	9.5666
7°	.1219	.9925	.1228	8.14	8.2054
8°	.1392	.9903	.1405	7.12	7.1854
9°	.1564	.9877	.1584	6.31	6.3926
10°	.1736	.9848	.1763	5.67	5.7587
11°	.1908	.9816	.1944	5.14	5.2408
12°	.2079	.9781	.2126	4.70	4.8097
13°	.2250	.9744	.2309	4.33	4.4454
14°	.2419	.9703	.2493	4.01	4.1335
15°	.2588	.9659	.2679	3.73	3.8636
16°	.2756	.9613	.2867	3.49	3.5915
17°	.2924	.9563	.3057	3.27	3.4203
18°	.3090	.9511	.3249	3.08	3.2360
19°	.3256	.9455	.3443	2.90	3.0715
20°	.3420	.9397	.3640	2.75	2.9238
21°	.3584	.9336	.3839	2.61	2.7904
22°	.3744	.9272	.4040	2.48	2.6694
23°	.3907	.9205	.4245	2.36	2.5593
24°	.4067	.9135	.4452	2.25	2.4585
25°	.4226	.9063	.4663	2.14	2.3661
26°	.4384	.8988	.4877	2.05	2.2811
27°	.4540	.8910	.5095	1.96	2.2026
28°	.4695	.8829	.5317	1.88	2.1300
29°	.4848	.8746	.5543	1.80	2.0626
30°	.5000	.8660	.5774	1.73	2.0000
31°	.5150	.8572	.6009	1.66	1.9415
32°	.5299	.8480	.6249	1.60	1.8870
33°	.5446	.8387	.6494	1.54	1.8360
34°	.5592	.8290	.6745	1.48	1.7883
35°	.5736	.8192	.7002	1.43	1.7434
36°	.5878	.8090	.7265	1.38	1.7012
37°	.6018	.7986	.7536	1.33	1.6616
38°	.6157	.7880	.7813	1.28	1.6242
39°	.6293	.7771	.8098	1.23	1.5890
40°	.6428	.7660	.8391	1.19	1.5557
41°	.6561	.7547	.8693	1.15	1.5242
42°	.6691	.7431	.9004	1.11	1.4944
43°	.6820	.7314	.9325	1.07	1.4662
44°	.6947	.7193	.9657	1.04	1.4395
45°	.7071	.7071	1.0000	1.00	1.4142

102APX02.TIF

TRIGONOMETRY TABLE (continued)

Angle	Sine	Cosine	Tangent	Cotangent	Cosecant
46°	.7193	.6947	1.035	.966	1.4395
47°	.7314	.6820	1.0724	.933	1.3673
48°	.7431	.6691	1.1106	.900	1.3456
49°	.7547	.6561	1.1504	.869	1.3250
50°	.7660	.6428	1.1918	.839	1.3054
51°	.7771	.6293	1.2349	.810	1.2867
52°	.7880	.6157	1.2799	.781	1.2690
53°	.7986	.6018	1.3270	.754	1.2521
54°	.8090	.5878	1.3764	.727	1.2360
55°	.8192	.5736	1.4281	.700	1.2207
56°	.8290	.5592	1.4826	.675	1.2062
57°	.8387	.5446	1.5399	.649	1.1923
58°	.8480	.5299	1.6003	.625	1.1791
59°	.8572	.5150	1.6643	.601	1.1666
60°	.8660	.5000	1.7321	.577	1.1547
61°	.8746	.4848	1.8040	.554	1.1433
62°	.8829	.4695	1.8807	.532	1.1325
63°	.8910	.4540	1.9626	.510	1.1223
64°	.8988	.4384	2.0503	.488	1.1126
65°	.9063	.4226	2.1445	.466	1.1033
66°	.9135	.4067	2.2460	.445	1.0946
67°	.9205	.3907	2.3559	.424	1.0863
68°	.9272	.3746	2.4751	.404	1.0785
69°	.9336	.3584	2.6051	.384	1.0711
70°	.9397	.3420	2.7475	.364	1.0641
71°	.9455	.3256	2.9042	.344	1.0576
72°	.9511	.3090	3.0777	.325	1.0514
73°	.9563	.2924	3.2709	.306	1.0456
74°	.9613	.2756	3.4874	.287	1.0402
75°	.9659	.2588	3.7321	.268	1.0352
76°	.9703	.2419	4.0108	.249	1.0306
77°	.9744	.2250	4.3315	.231	1.0263
78°	.9781	.2079	4.7046	.213	1.0223
79°	.9816	.1908	5.1446	.194	1.0187
80°	.9848	.1736	5.6713	.176	1.0154
81°	.9877	.1564	6.3138	.158	1.0124
82°	.9903	.1392	7.1154	.141	1.0098
83°	.9925	.1219	8.1443	.123	1.0075
84°	.9945	.1045	9.5144	.105	1.0055
85°	.9962	.0872	11.4300	.088	1.0038
86°	.9976	.0698	14.3010	.070	1.0024
87°	.9986	.0523	19.0810	.052	1.0013
88°	.9994	.0349	28.6360	.035	1.0006
89°	.9998	.0175	57.2900	.018	1.0001
90°	1.0000	.0000	00	.000	1.0000

102APX03.TIF

Instructor's Notes:

_____

_____

_____

_____

_____

_____

_____

_____

_____

BENDING RADIUS TABLE

RADIUS (INCHES)	RADIUS INCREMENTS (INCHES)									
	0	1	2	3	4	5	6	7	8	9
0	0.00	1.57	3.14	4.71	6.28	7.85	.942	10.99	12.56	14.13
10	15.70	17.27	18.84	20.41	21.98	23.85	25.12	26.69	28.26	29.83
20	31.40	32.97	34.54	36.11	37.68	39.25	40.82	42.39	43.96	45.83
30	47.10	48.67	50.24	51.81	53.38	54.95	56.52	58.09	59.66	61.23
40	62.80	64.37	65.94	67.50	69.03	70.65	72.22	73.79	75.36	76.93
50	87.50	80.07	81.64	83.21	84.78	86.35	87.92	89.49	91.06	92.63
60	94.20	95.77	97.34	98.91	100.48	102.05	103.62	105.19	106.76	108.33
70	109.90	111.47	113.04	114.61	116.18	117.75	119.32	120.89	122.46	124.03
80	125.60	127.17	128.74	130.31	131.88	133.45	135.02	136.59	138.16	139.73
90	141.30	142.87	144.44	146.01	147.58	149.15	150.72			

Developed length for following angles use fraction of 90° chart.

For	15°	22-1/2°	30°	45°	60°	67-1/2°	75°	90°
Take	1/6	1/4	1/3	1/2	2/3	3/4	5/6	See Chart

For any other degrees: Developed length = .01744 x radius x degrees.

102APX04.TIF

Answers to Review Questions

Answer	Section
1. b	2.0.0
2. a	2.0.0
3. a	2.0.0
4. b	2.0.0
5. b	2.0.0
6. b	2.1.0
7. c	2.1.0
8. d	2.2.0
9. b	2.4.0
10. c	2.5.0
11. a	2.6.0
12. b	2.7.0
13. c	3.2.0
14. c	3.3.0
15. d	3.4.0

Instructor's Notes:

_____

_____

_____

_____

_____

_____

_____

_____

_____

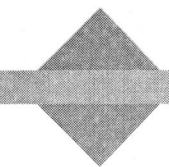

Additional Resources

For advanced study of topics covered in this task module, the following books are suggested:

Benfield Conduit Bending Manual, Latest Edition. New York: McGraw-Hill Publishing Company.

National Electrical Code Handbook, Latest Edition. Quincy, MA: National Fire Protection Association.

Tom Henry's Conduit Bending Package (includes video, book, and bending chart). Winter Park, FL: Code Electrical Classes, Inc.

NCCER CRAFT TRAINING USER UPDATES

The NCCER makes every effort to keep these textbooks up-to-date and free of technical errors. We appreciate your help in this process. If you have an idea for improving this textbook, or if you find an error, a typographical mistake, or an inaccuracy in the NCCER's Craft Training textbooks, please write us, using this form or a photocopy. Be sure to include the exact module number, page number, a detailed description, and the correction, if applicable. Your input will be brought to the attention of the Technical Review Committee. Thank you for your assistance.

Instructors – If you found that additional materials were necessary in order to teach this module effectively, please let us know so that we may include them in the Equipment/Materials list in the Instructor's Guide.

Write: Curriculum Revision and Development Department
National Center for Construction Education and Research
P.O. Box 141104, Gainesville, FL 32614-1104

Fax: 352-334-0932

E-mail: curriculum@nccer.org

Craft _____ Module Name _____

Copyright Date _____ Module Number _____ Page Number(s) _____

Description _____

_____

_____

_____

_____

(Optional) Correction _____

_____

_____

(Optional) Your Name and Address _____

_____

_____

_____

Fasteners and Anchors

26103-00

MODULE OVERVIEW

This course introduces the electrical trainee to the applications and installation procedures for various types of fasteners and anchors.

PREREQUISITES

Please refer to the Course Map in the Trainee Module. Prior to training with this module, it is recommended that the trainee shall have successfully completed the following modules:

Core Curriculum; Electrical Level One, Modules 26101 and 26102

LEARNING OBJECTIVES

Upon completion of this module, the trainee will be able to:

1. Identify and explain the use of threaded fasteners.
2. Identify and explain the use of non-threaded fasteners.
3. Identify and explain the use of anchors.
4. Demonstrate the correct applications for fasteners and anchors.
5. Install fasteners and anchors.

PERFORMANCE OBJECTIVES

Under supervision of the instructor, the trainee should be able to:

1. Install selected threaded fasteners.
2. Install blind (pop) rivets.
3. Install selected screws.
4. Install selected anchors.
5. Install selected toggle bolts.

NCCER STANDARDIZED CRAFT TRAINING PROGRAM

The National Center for Construction Education and Research (NCCER) provides a standardized national program of accredited craft training. Key features of the program include instructor certification, competency-based training, and performance testing. The program provides trainees, instructors, and companies with a standard form of recognition through a National Craft Training Registry. The program is described in full in the *Guidelines for Accreditation*, published by the NCCER. For more information on standardized craft training, contact the NCCER at P.O. Box 141104, Gainesville, FL 32614-1104, 352-334-0911, visit our Web site at www.NCCER.org, or e-mail info@NCCER.org.

HOW TO USE THIS ANNOTATED INSTRUCTOR'S GUIDE

Each page presents two sections of information. The larger section displays each page exactly as it appears in the Trainee Module. The narrow column ties suggested trainee and instructor actions to each page and provides icons to call your attention to material, safety, audiovisual, or testing requirements. The bottom of each page includes space for your notes.

If you see the Teaching Tip icon, that means there is a teaching tip associated with this section. Also refer to the suggested teaching tips at the end of the module.

SAFETY CONSIDERATIONS

Ensure that the trainees are equipped with appropriate personal protective equipment. Stress the importance of following all safety precautions and procedures when working with the tools associated with the installation of fasteners and anchors. Emphasize the importance of safety goggles/glasses.

PREPARATION

Before teaching this module, you should review the Module Outline, Learning and Performance Objectives, and the Materials and Equipment List. Be sure to allow ample time to prepare your own training or lesson plan and gather all required equipment and materials.

MATERIALS AND EQUIPMENT LIST

Materials:

Transparencies

Markers/chalk

Assorted threaded fasteners, including:

    Bolts

    Cap screws

    Studs

    Machine screws

    Nuts

    Washers

Assorted non-threaded fasteners, including:

    Retainers

    Keys

    Pins

    Blind (pop) rivets and rivet tool

    Tie wraps

Assorted special threaded fasteners, including:

    Eye bolts

    Anchor bolts

    J-bolts

Assorted screws, including:

    Wood screws

    Lag screws and shields

    Concrete/masonry screws

    Thread-forming (sheet metal) and thread-cutting screws

    Deck screws

    Drywall screws

Hammer-driven tools and related pin and stud fasteners

Powder-actuated tool, powder charges, and related pin and stud fasteners

Assorted mechanical anchors and associated anchor setting tool(s), including:

    Wedge

    Stud

    Sleeve

    One-piece

    Hammer-driven

    Drop-in

    Expansion shields

    Lead (caulk-in)

    Screw (fiber, lead, and plastic)

    Self-drilling

    Toggle bolts

    Sleeve-type

    Wallboard

    Metal drive-in

Concrete, masonry, wood

Module Examinations*

Performance Profile Sheets*

Equipment:

Overhead projector and screen

Whiteboard/chalkboard

Appropriate personal protective equipment

Mechanic's tool box with an assortment of wrenches, screwdrivers, etc.

Torque wrenches

Drills/drivers and assorted drill bits

*Located in the Test Booklet packaged with this Annotated Instructor's Guide.

ADDITIONAL RESOURCES

This module is intended to present thorough resources for task training. The following reference works are suggested for both instructors and motivated trainees interested in further study. These are optional materials for continued education rather than for task training.

American Electrician's Handbook, Latest Edition. New York, NY: McGraw-Hill.

Carpentry, Latest Edition. New York, NY: Delmar Publishers.

The Sheet Metal Toolbox Manual, Latest Edition. Upper Saddle River, NJ: Prentice Hall.

NOTES

The designations "National Electrical Code," "NE Code," and "NEC," where used in this document, refer to the *National Electrical Code*®, which is a registered trademark of the National Fire Protection Association, Quincy, MA. All National Electrical Code (NEC) references in this module refer to the 1999 edition of the NEC.

If you feel that additional math instruction would be helpful, Prentice Hall offers a basic math textbook entitled *Fundamentals of Electrical and Mechanical Mathematics.* It covers the basic math requirements for electrical trainees and may be ordered by contacting Prentice Hall Customer Service at 1-800-922-0579.

TEACHING TIME FOR THIS MODULE

An outline for use in developing your lesson plan is presented below. Note that each Roman numeral in the outline equates to one session of instruction. Each session has a suggested time period of 2-1/2 hours. This includes 10 minutes at the beginning of each session for administrative tasks and one 10-minute break during the session. Approximately 5 hours are suggested to cover *Fasteners and Anchors.* You will need to adjust the time required for hands-on activity and testing based on your class size and resources.

Topic	Planned Time
Session I. Threaded Fasteners; Non-Threaded Fasteners; Special Threaded Fasteners; Screws	
A. Threaded Fasteners	_____
1. Thread Standards	_____
a. Thread Series	_____
b. Thread Classes	_____
c. Thread Identification	_____
d. Grade Markings	_____
2. Bolt and Screw Types	_____
a. Machine Screws	_____
b. Machine Bolts	_____
c. Cap Screws	_____
d. Set Screws	_____
e. Stud Bolts	_____
3. Nuts	_____
a. Jam Nuts	_____
b. Castellated, Slotted, and Self-Locking Nuts	_____
c. Acorn Nuts	_____
d. Wing Nuts	_____
4. Washers	_____
a. Lock Washers	_____
b. Flat and Fender Washers	_____

    5. Installing Fasteners   _____
      a. Torque Tightening   _____
      b. Installing Threaded Fasteners   _____

B. Laboratory   _____

Under instructor supervision, have the trainees practice installing selected threaded fasteners.

C. Non-Threaded Fasteners   _____
    1. Retainer Fasteners   _____
    2. Keys   _____
    3. Pin Fasteners   _____
      a. Dowel Pins   _____
      b. Taper and Spring Pins   _____
      c. Cotter Pins   _____
    4. Blind/Pop Rivets   _____
    5. Tie Wraps   _____

D. Laboratory   _____

Under instructor supervision, have the trainees practice installing blind (pop) rivets.   _____

E. Special Threaded Fasteners   _____
    1. Eye Bolts   _____
    2. Anchor Bolts   _____
    3. J-Bolts   _____

F. Screws   _____
    1. Wood Screws   _____
    2. Lag Screws and Shields   _____
    3. Concrete/Masonry Screws   _____
    4. Thread-Forming and Thread-Cutting Screws   _____
    5. Deck Screws   _____
    6. Drywall Screws   _____
    7. Drive Screws   _____

G. Laboratory   _____

Under instructor supervision, have the trainees practice installing selected screws.

Session II. Hammer-Driven Pins and Studs; Powder-Actuated Tools and Fasteners; Mechanical Anchors; Epoxy Anchoring Systems; Module Examination and Performance Testing

A. Hammer-Driven Pins and Studs   _____

B. Powder-Actuated Tools and Fasteners   _____

C. Mechanical Anchors   _____

   1. One-Step Anchors   _____

      a. Wedge Anchors   _____

      b. Stud Bolt Anchors   _____

      c. Sleeve Anchors   _____

      d. One-Piece Anchors   _____

      e. Hammer-Set Anchors   _____

   2. Bolt Anchors   _____

      a. Drop-In Anchors   _____

      b. Single- and Double-Expansion Anchors   _____

      c. Lead (Caulk-In) Anchors   _____

   3. Screw Anchors   _____

   4. Self-Drilling Anchors   _____

   5. Guidelines for Drilling Anchor Holes in Hardened Concrete or Masonry   _____

   6. Hollow-Wall Anchors   _____

      a. Toggle Bolts   _____

      b. Sleeve-Type Wall Anchors   _____

      c. Wallboard Anchors   _____

      d. Metal Drive-In Anchors   _____

D. Laboratory   _____

Under instructor supervision, have the trainees practice installing selected anchors and toggle bolts.

E. Epoxy Anchoring Systems   _____

F. Module Examination   _____

   1. Trainees must score 70% or higher to receive recognition from the NCCER.

   2. Record the testing results on Craft Training Report Form 200 and submit the results to the Training Program Sponsor.

G. Performance Testing   _____

   1. Trainees must perform each task to the satisfaction of the instructor to receive recognition from the NCCER.

   2. Record the testing results on Craft Training Report Form 200 and submit the results to the Training Program Sponsor.

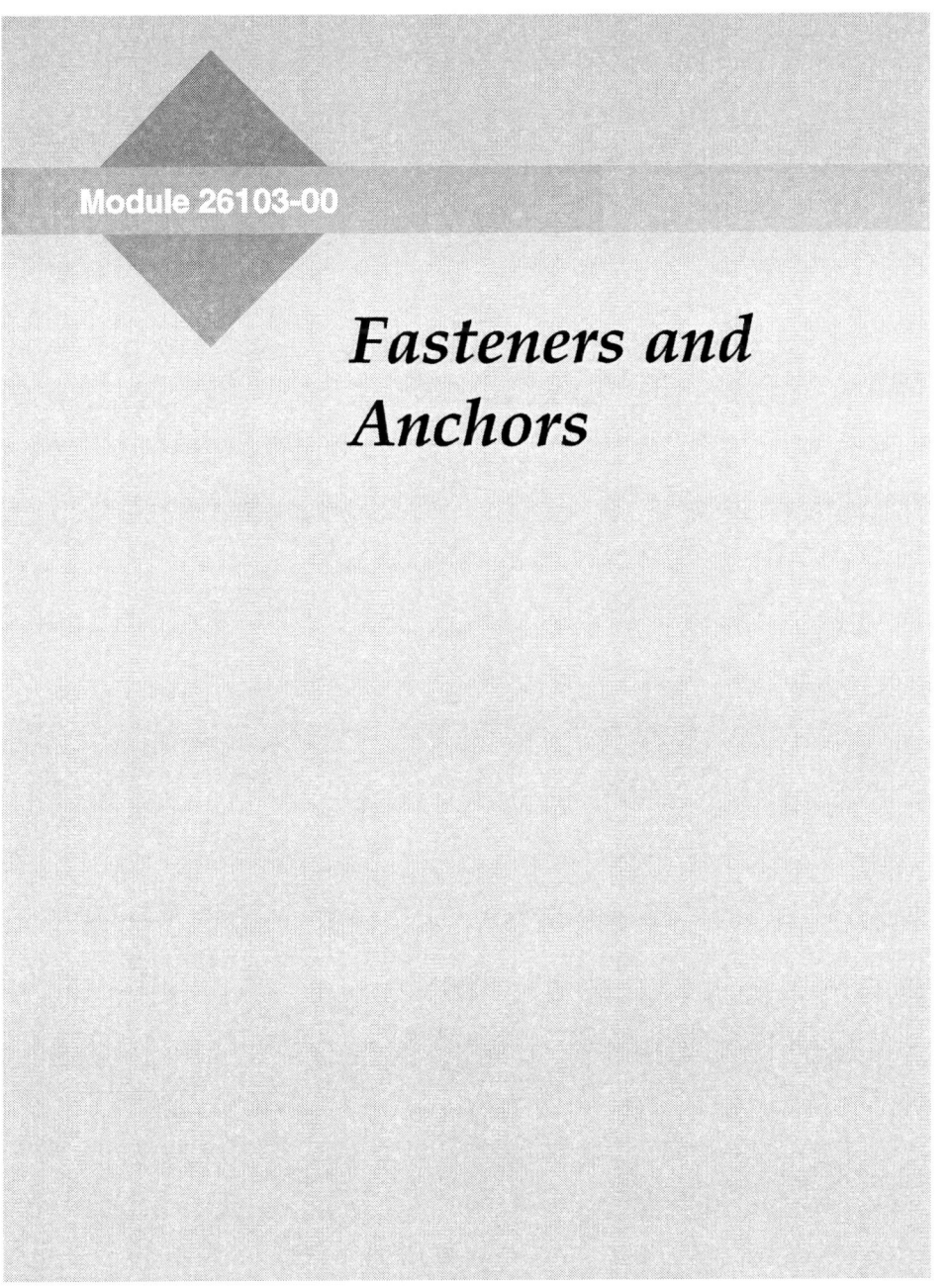

Module 26103-00

Fasteners and Anchors

Course Map

This course map shows all of the task modules in the first level of the Electrical curricula. The suggested training order begins at the bottom and proceeds up. Skill levels increase as a trainee advances on the course map. The training order may be adjusted by the local Training Program Sponsor.

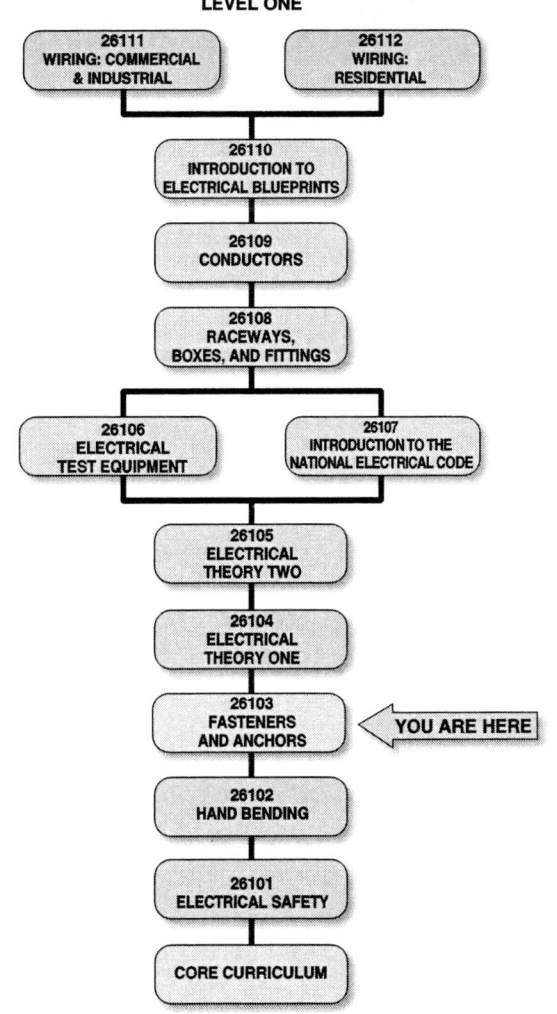

LEVEL ONE

26111
WIRING: COMMERCIAL & INDUSTRIAL

26112
WIRING: RESIDENTIAL

26110
INTRODUCTION TO ELECTRICAL BLUEPRINTS

26109
CONDUCTORS

26108
RACEWAYS, BOXES, AND FITTINGS

26106
ELECTRICAL TEST EQUIPMENT

26107
INTRODUCTION TO THE NATIONAL ELECTRICAL CODE

26105
ELECTRICAL THEORY TWO

26104
ELECTRICAL THEORY ONE

26103
FASTENERS AND ANCHORS ◁ YOU ARE HERE

26102
HAND BENDING

26101
ELECTRICAL SAFETY

CORE CURRICULUM

103CMAP.EPS

Homework

Assign reading of Module 26103.

3–iii

ELECTRICAL LEVEL ONE—INSTRUCTOR'S GUIDE MODULE 26103

Instructor's Notes:

_____

_____

_____

_____

_____

_____

_____

_____

_____

Figures

Ensure you have everything required to teach the course. Check the Materials and Equipment List at the front of this Instructor's Guide.

Instructor's Notes:

_____

_____

_____

_____

_____

_____

_____

_____

_____

Fasteners and Anchors

Objectives

Upon completion of this module, the trainee will be able to:

1. Identify and explain the use of threaded fasteners.
2. Identify and explain the use of non-threaded fasteners.
3. Identify and explain the use of anchors.
4. Demonstrate the correct applications for fasteners and anchors.
5. Install fasteners and anchors.

Prerequisites

Successful completion of the following Task Modules is recommended before beginning study of this Task Module: Core Curricula; Electrical Level One, Modules 26101 and 26102.

Required Trainee Materials

1. Trainee Task Module
2. Copy of the latest edition of the *National Electrical Code*
3. Appropriate Personal Protective Equipment

1.0.0 ◆ INTRODUCTION

Fasteners are used to assemble and install many different types of equipment, parts, and materials. Fasteners include screws, bolts, nuts, pins, clamps, retainers, tie wraps, rivets, and keys. Fasteners are used extensively in the electrical craft. You need to be familiar with the many different types of fasteners in order to identify, select, and properly install the correct fastener for a specific application.

The two primary categories of fasteners are:

- Threaded fasteners
- Non-threaded fasteners

Within each of these two categories, there are numerous different types and sizes of fasteners. Each type of fastener is designed for a specific application. The kind of fastener used for a job may be listed in the project specifications, or you may have to select an appropriate fastener.

Failure of fasteners can result in a number of different problems. To perform quality electrical work, it is important to use the correct type and size of fastener for the particular job. It is equally important that the fastener be installed properly.

In this module, you will be introduced to fasteners commonly used in electrical work.

2.0.0 ◆ THREADED FASTENERS

Threaded fasteners are the most commonly used type of fastener. Many threaded fasteners are assembled with nuts and washers. The following sections describe standard threads used on threaded fasteners, as well as different types of bolts, screws, nuts, and washers. *Figure 1* shows several types of threaded fasteners.

Note: The designations "National Electrical Code," "NE Code," and "NEC," where used in this document, refer to the *National Electrical Code®*, which is a registered trademark of the National Fire Protection Association, Quincy, MA. *All National Electrical Code (NEC) references in this module refer to the 1999 edition of the NEC.*

3–1

Show Transparency 1, Course Objectives.

Show Transparency 2, Performance Profile Tasks.

Discuss threaded fasteners and their applications.

Show Transparency 3 (Figure 1).

What's wrong with this picture?

VERONICA WESTFALL   103PO301.EPS

2.1.0 Thread Standards

There are many different types of threads used for manufacturing fasteners. The different types of threads are designed to be used for different jobs. Threads used on fasteners are manufactured to industry-established standards for uniformity. The most common thread standard is the Unified standard, sometimes referred to as the *American standard.* Unified standards are used to establish thread series and classes.

2.1.1 Thread Series

Unified standards are established for three series of threads, depending on the number of threads per inch for a certain diameter of fastener. These three series are:

- *Unified National Coarse (UNC) thread*—Used for bolts, screws, nuts, and other general purposes. Fasteners with UNC threads are commonly used for rapid assembly or disassembly of parts and where corrosion or slight damage may occur.
- *Unified National Fine (UNF) thread*—Used for bolts, screws, nuts, and other applications where a finer thread for a tighter fit is desired.
- *Unified National Extra Fine (UNEF) thread*—Used on thin-walled tubes, nuts, ferrules, and couplings.

2.1.2 Thread Classes

The Unified standards also establish thread classes. Classes 1A, 2A, and 3A apply to external

HEX HEAD BOLT   HEX HEAD CAP SCREW

SQUARE HEAD BOLT   HEX SOCKET HEAD CAP SCREW

CONTINUOUS THREAD STUD   DOUBLE-END STUD

103F01.EPS

Figure 1 ◆ Threaded fasteners.

Instructor's Notes:

_____

_____

_____

_____

_____

_____

_____

_____

_____

threads only. Classes 1B, 2B, and 3B apply to internal threads only. Thread classes are distinguished from each other by the amounts of tolerance provided. Classes 3A and 3B provide a minimum clearance and classes 1A and 1B provide a maximum clearance.

Classes 2A and 2B are the most commonly used. Classes 3A and 3B are used when close tolerances are needed. Classes 1A and 1B are used where quick and easy assembly is needed and a large tolerance is acceptable.

2.1.3 Thread Identification

Thread identification is done using a standard method. *Figure 2* shows how screw threads are designated for a common fastener.

- *Nominal size*—The nominal size is the approximate diameter of the fastener.

- *Number of threads per inch (TPI)*—The TPI is standard for all diameters.
- *Thread series symbol*—The Unified standard thread type (UNC, UNF, or UNEF).
- *Thread class symbol*—The closeness of fit between the bolt threads and nut threads.
- *Left-hand thread symbol*—Specified by the symbol LH. Unless threads are specified with the LH symbol, the threads are right-hand threads.

2.1.4 Grade Markings

Special markings on the head of a bolt or screw can be used to determine the quality of the fastener. The Society of Automotive Engineers (SAE) and the American Society for Testing of Materials (ASTM) have developed the standards for these markings. These grade or line markings for steel bolts and screws are shown in *Figure 3*.

Discuss thread identification and grade markings.

Show Transparencies 4 and 5 (Figures 2 and 3).

Thread Classes

External and internal thread classes differ widely with regard to tolerances. Why would you use Classes 3A or 3B for high-grade commercial products such as precision tools and machines? What household products would warrant the use of Classes 1A and 1B?

Discuss the "Think About It." Precision tools require the close tolerances provided by Classes 3A/3B, while Classes 1A/1B provide for quick assembly. Household applications vary widely.

3/4 - 10 - UNC - 2A - LH

NOMINAL SIZE (DIAMETER)
NO. OF THREADS PER INCH
THREAD SERIES SYMBOL
THREAD CLASS SYMBOL
LEFT-HAND THREAD (NO DESIGNATION IF R.H. THREAD)

103F02.EPS

Figure 2 ◆ Screw thread designations.

Screw Diameters

A screw's diameter and TPI for a given project depend on what condition(s)?

Discuss the "Think About It." The screw diameter and TPI depend on the strength required and the materials to be joined.

Display a variety of bolts and screws and demonstrate how to identify them using ASTM and SAE grade markings.

ASTM AND SAE GRADE MARKINGS FOR
STEEL BOLTS & SCREWS

GRADE MARKING	SPECIFICATION	MATERIAL
	SAE-GRADE 0	STEEL
	SAE-GRADE 1 ASTM-A 307	LOW CARBON STEEL
	SAE-GRADE 2	LOW CARBON STEEL
	SAE-GRADE 3	MEDIUM CARBON STEEL, COLD WORKED
A 449	SAE-GRADE 5	MEDIUM CARBON STEEL, QUENCHED AND TEMPERED
	ASTM-A 449	
A 325	ASTM-A 325	MEDIUM CARBON STEEL, QUENCHED AND TEMPERED
BB	ASTM-A 354 GRADE BB	LOW ALLOY STEEL, QUENCHED AND TEMPERED
BC	ASTM-A 354 GRADE BC	LOW ALLOY STEEL, QUENCHED AND TEMPERED
	SAE-GRADE 7	MEDIUM CARBON ALLOY STEEL, QUENCHED AND TEMPERED ROLL THREADED AFTER HEAT TREATMENT
	SAE-GRADE 8	MEDIUM CARBON ALLOY STEEL, QUENCHED AND TEMPERED
	ASTM-A 354 GRADE BD	ALLOY STEEL, QUENCHED AND TEMPERED
A 490	ASTM-A 490	ALLOY STEEL, QUENCHED AND TEMPERED

ASTM SPECIFICATIONS
A 307 - LOW CARBON STEEL EXTERNALLY AND INTERNALLY THREADED STANDARD FASTENERS.
A 325 - HIGH STRENGTH STEEL BOLTS FOR STRUCTURAL STEEL JOINTS, INCLUDING SUITABLE
      NUTS AND PLAIN HARDENED WASHERS.
A 449 - QUENCHED AND TEMPERED STEEL BOLTS AND STUDS.
A 354 - QUENCHED AND TEMPERED ALLOY STEEL BOLTS AND STUDS WITH SUITABLE NUTS.
A 490 - HIGH STRENGTH ALLOY STEEL BOLTS FOR STRUCTURAL STEEL JOINTS, INCLUDING
      SUITABLE NUTS AND PLAIN HARDENED WASHERS.
SAE SPECIFICATION
J 429 - MECHANICAL AND QUALITY REQUIREMENTS FOR THREADED FASTENERS.

103F03.EPS

Figure 3 ◆ Grade markings for steel bolts and screws.

Generally, the higher-quality steel fasteners have a greater number of marks on the head. If the head is unmarked, the fastener is usually considered to be made of mild steel (having low carbon content).

2.2.0 Bolt and Screw Types

Bolts and screws are made in many different sizes and shapes and from a variety of materials. They are usually identified by the head type or other

ELECTRICAL LEVEL ONE—TRAINEE MODULE 26103

Instructor's Notes:

_____

_____

_____

_____

_____

_____

_____

_____

_____

ELECTRICAL LEVEL ONE—INSTRUCTOR'S GUIDE MODULE 26103

Figure 4 ◆ Machine screws.

special characteristics. The following sections describe several different types of bolts and screws.

2.2.1 Machine Screws

Machine screws are used for general assembly work. They come in a variety of types with slotted or recessed heads. Machine screws are generally available in diameters ranging from 0 (0.060") to ½" (0.500"). The length of machine screws typically varies from ⅛" to 3". Machine screws are also manufactured in metric sizes. *Figure 4* shows different types of machine screws.

As shown, the heads of machine screws are made in different shapes and with slots made to fit various kinds of manual and power tool screwdrivers. Flat-head screws are used in a countersunk hole and tightened so that the head is flush with the surface. Oval-head screws are also used in a countersunk hole in applications where a more decorative finish is desired. Pan and round-head screws are general-use fastening screws. Fillister, hex socket, and TORX® socket screws are typically used in confined space applications on machined assemblies that need a finished appearance. They are often installed in a recessed hole. Truss screws

Discuss the "Think About It."

Discuss machine screws and their applications.

Show Transparency 6 (Figure 4).

Display a variety of machine screws and demonstrate how to identify them by head type.

Discuss the "Think About It." Answers will vary, but the instructor should emphasize use in electrical applications.

are a low-profile screw generally used without a washer. To prevent damage when tightening and removing machine screws (regardless of head type), make sure to use a screwdriver or power tool bit with the proper tip to drive them.

2.2.2 Machine Bolts

Machine bolts are generally used to assemble parts where close tolerances are not required. Machine bolts have square or hexagonal heads and are generally available in diameters ranging from ¼" to 3". The length of machine bolts typically varies from ½" to 30". Nuts used with machine bolts are similar in shape to the bolt heads. The nuts are usually purchased at the same time as the bolts. *Figure 5* shows two different types of machine bolts.

2.2.3 Cap Screws

Cap screws are often used on high-quality assemblies requiring a finished appearance. The cap screw passes through a clearance hole in one of the assembly parts and screws into a threaded hole in the other part. The clamping action occurs by tightening the cap screw.

Cap screws are made to close tolerances and are provided with a machined or semi-finished bearing surface under the head. They are normally made in coarse and fine thread series and in diameters from ¼" to 2". Lengths may range from ⅜" to 10". Metric sizes are also available. *Figure 6* shows typical cap screws.

2.2.4 Set Screws

Heat-treated steel is normally used to make set screws. Common uses of set screws include preventing pulleys from slipping on shafts, holding collars in place on shafts, and holding shafts in place. The head style and point style are typically used to classify set screws. *Figure 7* shows several set screw heads and point styles.

2.2.5 Stud Bolts

Stud bolts (*Figure 8*) are headless bolts that are threaded over the entire length of the bolt or for a length on both ends of the bolt. One end of the

HEX HEAD BOLT

SQUARE HEAD BOLT

103F05.EPS

Figure 5 ◆ Machine bolts.

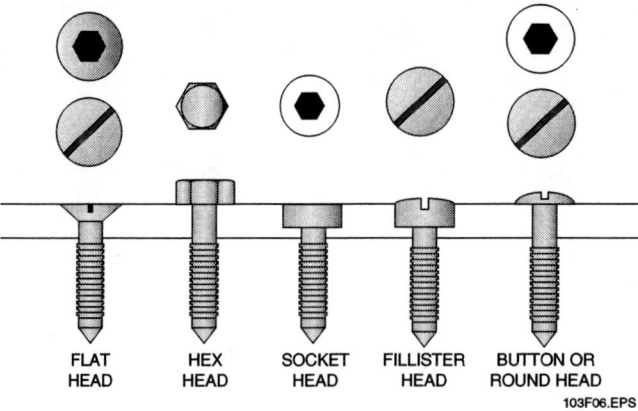

FLAT HEAD   HEX HEAD   SOCKET HEAD   FILLISTER HEAD   BUTTON OR ROUND HEAD

103F06.EPS

Figure 6 ◆ Cap screws.

ELECTRICAL LEVEL ONE—TRAINEE MODULE 26103

Instructor's Notes:

_____

_____

_____

_____

_____

_____

_____

_____

_____

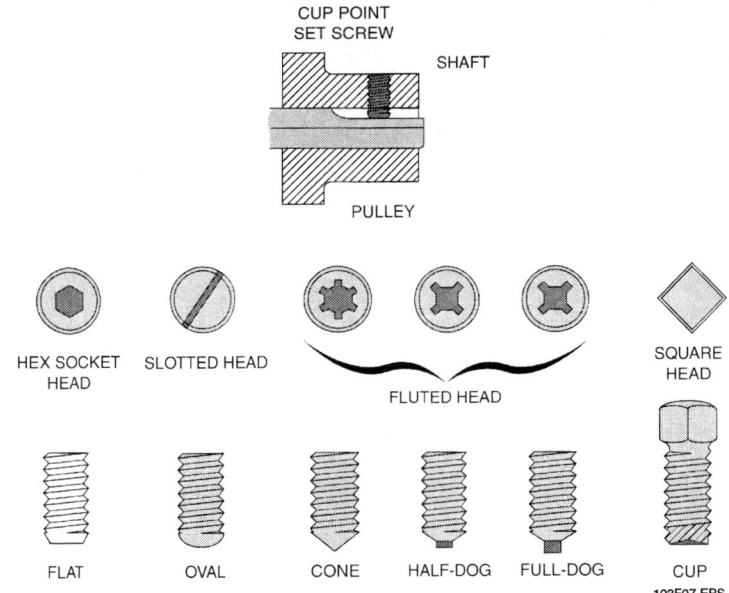

Figure 7 ◆ Set screws.

Figure 8 ◆ Stud bolt.

stud bolt is screwed into a tapped hole. The part to be clamped is placed over the remaining portion of the stud, and a nut and washer are screwed on to clamp the two parts together. Other stud bolts have machine-screw threads on one end and lag-screw threads on the other so that they can be screwed into wood.

Stud bolts are used for several purposes, including holding together inspection covers on equipment and bearing caps.

Discuss nut shapes and classifications.

Show Transparencies 11 and 12 (Figures 9 and 10).

Figure 9 ◆ Nuts.

2.3.0 Nuts

Most nuts used with threaded fasteners are hexagonal or square. They are usually used with bolts having the same shape head. *Figure 9* shows several different types of nuts that are used with threaded fasteners.

Nuts are typically classified as regular, semi-finished, or finished. The only machining done on regular nuts is to the threads. In addition to the threads, semi-finished nuts are also machined on the bearing face. Machining the bearing face makes a truer surface for fitting the washer. The only difference between semi-finished and finished nuts is that finished nuts are made to closer tolerances.

The standard machine screw nut has a regular finish. Regular and semi-finished nuts are shown in *Figure 10*.

Instructor's Notes:

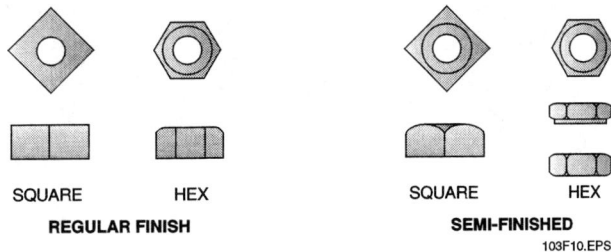

SQUARE   HEX   SQUARE   HEX
REGULAR FINISH   **SEMI-FINISHED**
103F10.EPS

Figure 10 ◆ Nut finishes.

103F11.TIF

Figure 11 ◆ Jam nut.

2.3.1 Jam Nuts

A jam nut is used to lock a standard nut in place. A jam nut is a thin nut installed on top of the standard nut. *Figure 11* shows an example of a jam nut installation. Note that a regular nut can also be used as a jam nut.

2.3.2 Castellated, Slotted, and Self-Locking Nuts

Castellated (castle) and slotted nuts are slotted across the flat part of the nut. They are used with specially manufactured bolts in applications where little or no loosening of the fastener can be tolerated. After the nut has been tightened, a cotter pin is fitted in through a hole in the bolt and one set of slots in the nut. The cotter pin keeps the nut from loosening under working conditions.

Self-locking nuts are also used in many applications where loosening of the fastener cannot be tolerated. Self-locking nuts are designed with nylon inserts, or they are deliberately deformed in such a manner so they cannot work loose. An advantage of self-locking nuts is that no hole in the bolt is needed. *Figure 12* shows typical castellated, slotted, and self-locking nuts.

2.3.3 Acorn Nuts

When appearance is important or exposed, sharp thread edges on the fastener must be avoided, acorn (cap) nuts are used. The acorn nut tightens on the bolt and covers the ends of the threads. The tightening capability of an acorn nut is limited by the depth of the nut. *Figure 13* shows a typical acorn nut.

2.3.4 Wing Nuts

Wing nuts are designed to allow rapid loosening and tightening of the fastener without the need for a wrench. They are used in applications where limited torque is required and where frequent adjustments and service are necessary. *Figure 14* shows a typical wing nut.

> **Note**
>
> Wing nuts should be used for applications where hand tightening is sufficient.

 Classroom

Discuss various types of nuts and their applications. Pass around examples for the trainees to examine.

 Audiovisual

Show Transparencies 13 through 16 (Figures 11 through 14).

CASTELLATED   SLOTTED   SELF-LOCKING

103F12.EPS

Figure 12 ◆ Castellated, slotted, and self-locking nuts.

TOP VIEW   SIDE VIEW

103F13.EPS

Figure 13 ◆ Acorn nut.

Note

The threads of the bolt or screw should have minimal clearance from the hole in the washer.

2.4.0 Washers

There are several different types and sizes of washers. They fit over a bolt or screw to provide an enlarged surface for bolt heads and nuts. Wash-

ers also serve to distribute the fastener load over a larger area and to prevent marring of the surfaces. Standard washers are made in light, medium, heavy-duty, and extra heavy-duty series. *Figure 15* shows different types of washers.

2.4.1 Lock Washers

Lock washers are designed to keep bolts or nuts from working loose. There are various types of lock washers for different applications.

- *Split-ring*—Commonly used with bolts and cap screws.
- *External*—Used for the greatest resistance.
- *Internal*—Used with small screws.
- *Internal-external*—Used for oversized mounting holes.
- *Countersunk*—Used with flat or oval-head screws.

2.4.2 Flat and Fender Washers

Flat washers are used under bolts or nuts to spread the load over a larger area and protect the surface. Common flat washers are made to fit bolt or screw

TOP VIEW   SIDE VIEW

103F14.EPS

Figure 14 ◆ Wing nut.

ELECTRICAL LEVEL ONE—TRAINEE MODULE 26103

Instructor's Notes:

_____
_____
_____
_____
_____
_____
_____
_____
_____

WEDGE   FENDER   FLAT
STANDARD WASHERS

SPLIT RING   EXTERNAL   INTERNAL   INTERNAL-   COUNTERSUNK
                                      EXTERNAL
LOCK WASHERS

103F15.EPS

Figure 15 ◆ Washers.

Carrying Fasteners

INSIDE TRACK   A nail apron is handy for carrying a large amount of fasteners.

---

sizes ranging from No. 6 up to 1" with outside diameters ranging from ⅜" to 2", respectively.

Fender washers are wide-surfaced washers made to bridge oversized holes or other wide clearances to keep bolts or nuts from pulling through the material being fastened. They are flat washers that have a larger diameter and surface area than regular washers. They may also be thinner than a regular washer. Fender washers are typically made to fit bolt or screw sizes ranging from ⁿ∕₁₆" to ½" with outside diameters ranging from ¾" to 2", respectively.

2.5.0 Installing Fasteners

Different types of fasteners require different installation techniques. However, all installations require knowing the proper installation methods, tightening sequence, and torque specifications for the type of fastener being used. Some bolts and nuts require that special safety wires or pins be installed to keep them from working loose.

Most fastener manufacturers provide charts that specify the size hole that should be drilled into the base material for use with each of their products (*Figure 16*). The charts typically show the proper size drill bit to use if it is necessary to first drill and

tap holes for use with machine bolts, screws, or other threaded fasteners. They also show the proper size drill to use for drilling pilot holes used with metal and wood screws. (Various kinds of screws are described in detail later in this module.)

2.5.1 Torque Tightening

To properly tighten a threaded fastener, two primary factors must be considered:

- The strength of the fastener material
- The degree to which the fastener is tightened

A torque wrench is used to control the degree of tightness. The torque wrench measures how much a fastener is being tightened. *Torque* is the turning force applied to the fastener. Torque is normally expressed in inch pounds (in. lbs.) or foot pounds (ft. lbs.). A one-pound force applied to a wrench that is 1 foot long exerts 1 foot pound, or 12 inch pounds, of torque. The torque reading is shown on the indicator on the torque wrench as the fastener is being tightened. *Figure 17* shows two types of torque wrenches.

Different types of bolts, nuts, and screws are torqued to different values depending on the

Explain how to determine pilot hole sizes.

Show Transparency 18 (Figure 16).

Discuss torque.

FASTENERS AND ANCHORS—TRAINEE MODULE 26103   3–11

FASTENERS AND ANCHORS—INSTRUCTOR'S GUIDE MODULE 26103   **3–11**

DRILL THIS SIZE HOLE		To Tap For This Size Bolt or Screw	For This Size Wood Screw Pilot In Hard Wood
Drill Size	Dec. Equiv.		
60	.0400		
59	.0410		
58	.0420		
57	.0430		
56	.0465	0 x 80	
3/64	.0469		
55	.0520		
54	.0550	1 x 56	No. 3
53	.0595	1 x 64-72	
1/16	.0625		
52	.0635		No. 4
51	.0670		
50	.0700	2 x 56-64	
49	.0730		No. 5
48	.0760		
5/64	.0781	3 x 48	No. 6
47	.0785		
46	.0810		
45	.0820	3 x 56	
44	.0860	4 x 36	No. 7
43	.0890	4 x 40	
42	.0935	4 x 48	
3/32	.0937		
41	.0960		
40	.0980	5 x 36	No. 8
39	.0995		
38	.1015	5 x 40	
37	.1040	5 x 44	No. 9
36	.1069		
7/64	.1094		
35	.1100	6 x 32	
34	.1110	6 x 36	
33	.1130	6 x 40	No. 10
32	.1160		
31	.1200		No. 11
1/8	.1250	7 x 36	
30	.1285	8 x 30	No. 12
29	.1360	8 x 32-36	
28	.1405	8 x 40	
27	.1440	9 x 30	
26	.1470	3/16 x 24	
25	.1495	10 x 24	No. 14
24	.1520		
23	.1540	10 x 28	
5/32	.1562		
22	.1570	10 x 30	
21	.1590	10 x 32	
20	.1610	3/16 x 32	
19	.1660		
18	.1695		No. 16
11/64	.1719		
17	.1730		
16	.1770	12 x 24	
15	.1800		
14	.1820	12 x 28	
13	.1850	12 x 32	No. 18
3/16	.1875		
12	.1890		

DRILL THIS SIZE HOLE		To Tap For This Size Bolt or Screw	For This Size Wood Screw Pilot In Hard Wood
Drill Size	Dec. Equiv.		
11	.1910		
10	.1935	15 x 20	
9	.1960		
8	.1990		
7	.2010	1/4 x 20	
13/64	.2031		
6	.2040		
5	.2055		
4	.2090	1/4 x 24	No. 20
3	.2130	1/4 x 28	
7/32	.2187	1/4 x 32	
2	.2210		
1	.2280		No. 24
A	.2340		
15/64	.2344		
B	.2380		
C	.2420		
D	.2460		
1/4	.2500		

DRILL THIS SIZE HOLE		To Tap For This Size Bolt or Screw
Drill Size	Dec. Equiv.	
E	.2500	
F	.2570	
G	.2610	5/16 x 18
17/64	.2656	
H	.2660	5/16 x 18
I	.2720	
J	.2770	5/16 x 24-32*
K	.2810	
9/32	.2812	5/16 x 24-32*
L	.2900	
M	.2950	
19/64	.2969	
N	.3020	
5/16	.3125	3/8* x 16-1/8* P
O	.3160	
P	.3230	
21/64	.3281	3/8 x 20-24
Q	.3332	
R	.3390	
11/32	.3437	
S	.3480	
T	.3580	
23/64	.3594	
U	.3680	
3/8	.3750	7/16 x 14
V	.3770	
W	.3860	
25/64	.3906	7/16 x 14
X	.3970	
Y	.4040	
13/32	.4062	
Z	.4130	
27/64	.4219	1/2 x 12-13
7/16	.4375	1/4* Pipe
29/64	.4531	1/2 x 20-24
15/32	.4687	1/2 x 27
31/64	.4844	9/16 x 12
1/2	.5000	

* All tap drill sizes are for 75% full thread except asterisked sizes which are 60% full thread.

103F16.EPS

Figure 16 ◆ Fastener hole guide chart.

Instructor's Notes:

DIAL TORQUE
TYPE

MICROMETER
TYPE

103F17.EPS

Figure 17 ◆ Torque wrenches.

Demonstration

Demonstrate the use of torque wrenches and screwdrivers.

application. Always check the project specifications and the manufacturer's manual to determine the proper torque for a particular type of fastener. *Figure 18* shows selected torque values for various graded steel bolts.

2.5.2 Installing Threaded Fasteners

The following general procedure can be used to install threaded fasteners in a variety of applications.

INSIDE TRACK

Torquing Screwdrivers

Torquing screwdrivers are calibrated in inch pounds.

VERONICA WESTFALL

103PO302.EPS

Torque Wrenches

THINK ABOUT IT

Think about situations in which the precise torque is important when installing a fastener. Do you commonly use an ordinary wrench in situations that dictate the use of a torque wrench?

VERONICA WESTFALL

103PO303.EPS

TORQUE IN FOOT POUNDS

FASTENER DIAMETER	THREADS PER INCH	MILD STEEL	STAINLESS STEEL 18-8	ALLOY STEEL
1/4	20	4	6	8
5/16	18	8	11	16
3/8	16	12	18	24
7/16	14	20	32	40
1/2	13	30	43	60
5/8	11	60	92	120
3/4	10	100	128	200
7/8	9	160	180	320
1	8	245	285	490

SUGGESTED TORQUE VALUES FOR GRADED STEEL BOLTS

GRADE		SAE 1 OR 2	SAE 5	SAE 6	SAE 8
TENSILE STRENGTH		64,000 PSI	105,000 PSI	130,000 PSI	150,000 PSI
GRADE MARK					
BOLT DIAMETER	THREADS PER INCH	FOOT POUNDS TORQUE			
1/4	20	5	7	10	10
5/16	18	9	14	19	22
3/8	16	15	25	34	37
7/16	14	24	40	55	60
1/2	13	37	60	85	92
9/16	12	53	88	120	132
5/8	11	74	120	169	180
3/4	10	120	200	280	296
7/8	9	190	302	440	473
1	8	282	466	660	714

103F18.EPS

Figure 18 ◆ Torque value chart.

Instructor's Notes:

_____

_____

_____

_____

_____

_____

_____

_____

_____

Note

When installing threaded fasteners for a specific job, make sure to check all installation requirements.

WARNING!

Follow all safety precautions.

Step 1 Select the proper bolts or screws for the job.

Step 2 Check for damaged or dirty internal and external threads.

Step 3 Clean the bolt or screw threads. Do not lubricate the threads if a torque wrench is to be used to tighten the nuts.

Step 4 Insert the bolts through the predrilled holes and tighten the nuts by hand. Or, insert the screws through the holes and start the threads by hand.

Note

Turn the nuts or screws several turns by hand and check for cross threading.

Step 5 Following the proper tightening sequence, tighten the bolts or screws snugly.

Step 6 Check the torque specification. Following the proper tightening sequence, tighten each bolt, nut, or screw several times approaching the specified torque. Tighten to the final torque specification.

Step 7 If required to keep the bolts or nuts from working loose, install jam nuts, cotter pins, or safety wire. *Figure 19* shows fasteners with a safety wire installed.

3.0.0 ◆ NON-THREADED FASTENERS

Non-threaded fasteners have many uses in the electrical field. Different types of non-threaded fasteners include retainers, keys, pins, clamps, washers, rivets, and tie wraps.

3.1.0 Retainer Fasteners

Retainer fasteners, also called *retaining rings,* are used for both internal and external applications. Some retaining rings are seated in grooves in the fastener. Other types of retainer fasteners are self-locking and do not require a groove. To easily remove internal and external retainer rings without damaging the ring or the fastener, special pliers are used. *Figure 20* shows several types of retainer fasteners.

3.2.0 Keys

To prevent a gear or pulley from rotating on a shaft, keys are inserted. Half of the key fits into a

Figure 19 ◆ Safety-wired fasteners.

Demonstrate the procedure for installing threaded fasteners.

Show Transparency 21 (Figure 19).

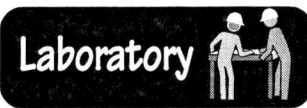

Have the trainees practice installing threaded fasteners.

Discuss retainer fasteners and their applications. Pass around examples for the trainees to examine.

Show Transparency 22 (Figure 20).

BASIC INTERNAL RING   BASIC EXTERNAL RING   INVERTED INTERNAL RING   INVERTED EXTERNAL RING

INTERNAL SELF-LOCKING RING   EXTERNAL SELF-LOCKING RING   TRIANGLE SELF-LOCKING RING

103F20.EPS

Figure 20 ◆ Retainer fasteners (rings).

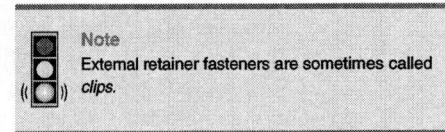
keyseat on the shaft. The other half fits into a key-way in the hub of the gear or pulley. The key fastens the two parts together, stopping the gear or pulley from turning on the shaft. *Figure 21* shows several types of keys and keyways and their uses.

Some different types of keys include:

• *Square key*—Usually one-quarter of the shaft diameter. It may be slightly tapered on the top for easier fitting.
• *Pratt and Whitney key*—Similar to the square key, but rounded at both ends. It fits into a keyseat of the same shape.
• *Gib head key*—Interchangeable with the square key. The head design allows easy removal from the assembly.
• *Woodruff key*—Semicircular shape that fits into a keyseat of the same shape. The top of the key fits into the keyway of the mating part.

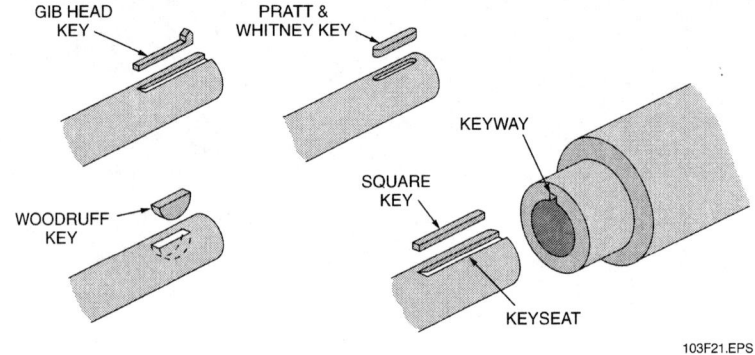

GIB HEAD KEY   PRATT & WHITNEY KEY

WOODRUFF KEY   SQUARE KEY   KEYWAY   KEYSEAT

103F21.EPS

Figure 21 ◆ Keys and keyways.

Classroom

Discuss various types of keys and their applications. Pass around examples for the trainees to examine.

Audiovisual

Show Transparency 23 (Figure 21).

Instructor's Notes:

_____

_____

_____

_____

_____

_____

_____

_____

_____

_____

3.3.0 Pin Fasteners

Pin fasteners come in several types and sizes. They have a variety of applications. Common uses of pin fasteners include holding moving parts together, aligning mating parts, fastening hinges, holding gears and pulleys on shafts, and securing slotted nuts. *Figure 22* shows several pin fasteners.

3.3.1 Dowel Pins

Dowel pins fit into holes to position mating parts. They may also support a portion of the load placed on the parts. *Figure 23* shows an application of dowel pins used to position mating parts.

3.3.2 Taper and Spring Pins

Taper and spring pins are used to fasten gears, pulleys, and collars to a shaft. *Figure 24* shows how taper and spring pins are used to attach a component to a shaft. The groove in a spring pin allows it to compress against the walls in a spring-like fashion.

HARDENED AND GROUND DOWEL PIN

TAPER PIN

CLEVIS PIN

COTTER PIN

FULL-LENGTH TAPER

HALF-LENGTH TAPER

FULL-LENGTH WITH PILOT

REVERSE TAPER

GROOVED PINS

103F22.EPS

Figure 22 ◆ Pin fasteners.

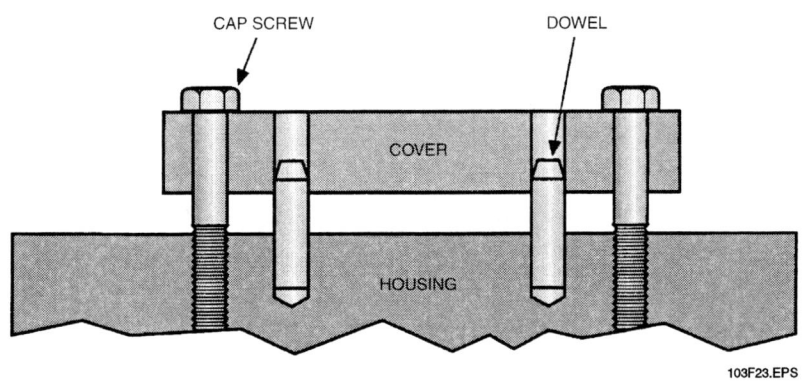

CAP SCREW

DOWEL

COVER

HOUSING

103F23.EPS

Figure 23 ◆ Dowel pins.

TAPER PIN

SHAFTS

GEARS

SPRING PIN

103F24.EPS

Figure 24 ◆ Taper and spring pins.

3.3.3 Cotter Pins

There are several different types of cotter pins used as a locking device for a variety of applications. Cotter pins are often inserted through a hole drilled crosswise through a shaft to prevent parts from slipping on or off the shaft. They are also used to keep slotted nuts from working loose. Standard cotter pins are general-use pins. When installed, the extended prong is normally bent back over the nut to provide the locking action. If it is ever removed, throw it away and replace it with a new one. The humped, cinch, and hitch-type cotter pins are self-locking pins. The humped and cinch type should also be thrown away and replaced with a new one if removed. The hitch pin, also called a *hair pin*, is a reusable pin made to

be installed and removed quickly. *Figure 25* shows several common types of cotter pins.

3.4.0 Blind/Pop Rivets

When only one side of a joint can be reached, blind rivets can be used to fasten the parts together. Some applications of blind rivets include fastening light to heavy gauge sheet metal, fiberglass, plastics, and belting. Blind rivets are made of a variety of materials and come in several sizes and lengths. They are installed using special riveting tools. *Figure 26* shows a typical blind rivet installation.

Blind rivets are installed through drilled or punched holes using a special blind (pop) rivet gun. *Figure 27* shows a typical rivet gun.

STANDARD PIN   HUMPED PIN   CINCH PIN   HITCH PIN

103F25.EPS

Figure 25 ◆ Cotter pins.

ELECTRICAL LEVEL ONE—TRAINEE MODULE 26103

Discuss blind/pop rivets and their applications. Pass around examples for the trainees to examine.

Show Transparencies 28 and 29 (Figures 26 and 27).

Instructor's Notes:

_____

_____

_____

_____

_____

_____

_____

_____

_____

RIVET TOOL

FULL MANDREL

BREAK MANDREL

BLIND RIVET

103F26.EPS

Figure 26 ◆ Blind rivet installation.

Use the following general procedure to install blind rivets.

 WARNING!
Follow all safety precautions. Make sure to wear proper eye and face protection when riveting.

Step 1 Select the correct length and diameter of blind rivet to be used.

Step 2 Select the appropriate drill bit for the size of rivet being used.

Step 3 Drill a hole through both parts being connected.

Rivet Guns

Rivet guns are designed for blind riveting (i.e., when you can reach only one side of the material).

SQUEEZE HANDLE

RIVET

SETTING TOOL

103F27.EPS

Figure 27 ◆ Rivet gun.

Using a Rivet Gun

Before you attach materials using a rivet gun and possibly do something you'll regret later, make a spacing template. Hold the template up to your material to ensure that all rivets fall into the correct locations, then secure your pieces. When using the template for actual riveting, hold it in position with an awl or drift pin inserted into your predrilled rivet holes. Note that rivets may pull through certain materials or when joining two pieces of flexible material together when the hole is slightly larger than required.

Figure 28 ◆ Joining parts.

Step 4 Inspect the rivet gun for any defects that might make it unsafe for use.

Step 5 Place the rivet mandrel into the proper size setting tool.

Step 6 Insert the rivet end into the predrilled hole.

Step 7 Install the rivet by squeezing the handle of the rivet gun, causing the jaws in the setting tool to grip the mandrel. The mandrel is pulled up, expanding the rivet until it breaks at the shear point. *Figure 28* shows the rivet and tool positioned for joining parts together.

Step 8 Inspect the rivet to make sure the pieces are firmly riveted together and that the rivet is properly installed. *Figure 29* shows a properly installed blind rivet.

3.5.0 Tie Wraps

A tie wrap is a one-piece, self-locking cable tie, usually made of nylon, that is used to fasten a bundle of wires and cables together. Tie wraps can be quickly installed either manually or using a special installation tool. Black tie wraps resist ultraviolet light and are recommended for outdoor use.

Tie wraps are made in standard, cable strap and clamp, and identification configurations (*Figure

Figure 29 ◆ Properly installed blind rivets.

30*). All types function to clamp bundled wires or cables together. In addition, the cable strap and clamp has a molded mounting hole in the head used to secure the tie with a rivet, screw, or bolt after the tie wrap has been installed around the wires or cable. Identification tie wraps have a large flat area provided for imprinting or writing cable identification information. There is also a releasable version available. It is a non-permanent tie used for bundling wires or cables that may require frequent additions or deletions. Cable ties are made in various lengths ranging from about 3"

Instructor's Notes:

_____

_____

_____

_____

_____

_____

_____

_____

_____

Figure 30 ◆ Tie wraps.

to 30", allowing them to be used for fastening wires and cables into bundles with diameters ranging from about ½" to 9", respectively. Tie wraps can also be attached to a variety of adhesive mounting bases made for that purpose.

4.0.0 ◆ SPECIAL THREADED FASTENERS

Special threaded fasteners consist of hardware manufactured in several shapes and sizes and designed to perform specific jobs. Certain types of nuts may be considered special threaded fasteners if they are designed especially for a particular application. In the electrical craft, special threaded fasteners are used on a number of different jobs.

The three types of special threaded fasteners described below are eye bolts, toggle bolts, and J-bolts.

4.1.0 Eye Bolts

Eye bolts get their name from the eye or loop at one end. The other end of an eye bolt is threaded. There are many types of eye bolts. The eye on some eye bolts is formed and welded, while the eye on other types is forged. Shoulder-forged eye bolts are commonly used as lifting devices and guides for wires, cables, and cords. *Figure 31* shows some typical eye bolts.

4.2.0 Anchor Bolts

An anchor bolt is used to fasten parts, machines, and equipment to concrete or masonry founda-

tions, floors, and walls. There are several types of anchor bolts designed for different applications. *Figure 32* shows a type of anchor bolt for use in wet concrete. If the concrete has already hardened, expansion anchor bolts are used. These are covered later in this module.

One common method used to install anchor bolts in wet concrete involves making a wooden template to locate the anchor bolts. The template positions the anchor bolts so that they correspond to those in the equipment to be fastened.

4.3.0 J-Bolts

J-bolts get their name from the curve on one end that gives them a J shape. The other end of a J-bolt is threaded. There are many types of J-bolts. Some J-bolts are used to hold tubing bundles and include a plastic jacket to protect the tubing. Others are used to attach equipment to existing grating. Most J-bolts used in tubing racks are attached using two nuts. The upper nut allows for adjustment. The tubing bundle is clamped firmly, but not flattened. Both nuts are tightened against the tube track for positive holding. *Figure 33* shows a typical J-bolt.

5.0.0 ◆ SCREWS

Screws are made in a variety of shapes and sizes for different fastening jobs. The finish or coating used on a screw determines whether it is for interior or exterior use, corrosion resistant, etc. Screws of all types have heads with different shapes and slots similar to those previously described for machine

Discuss eye bolts, anchor bolts, and J-bolts. Pass around examples for the trainees to examine.

Show Transparencies 33 through 35 (Figures 31 through 33).

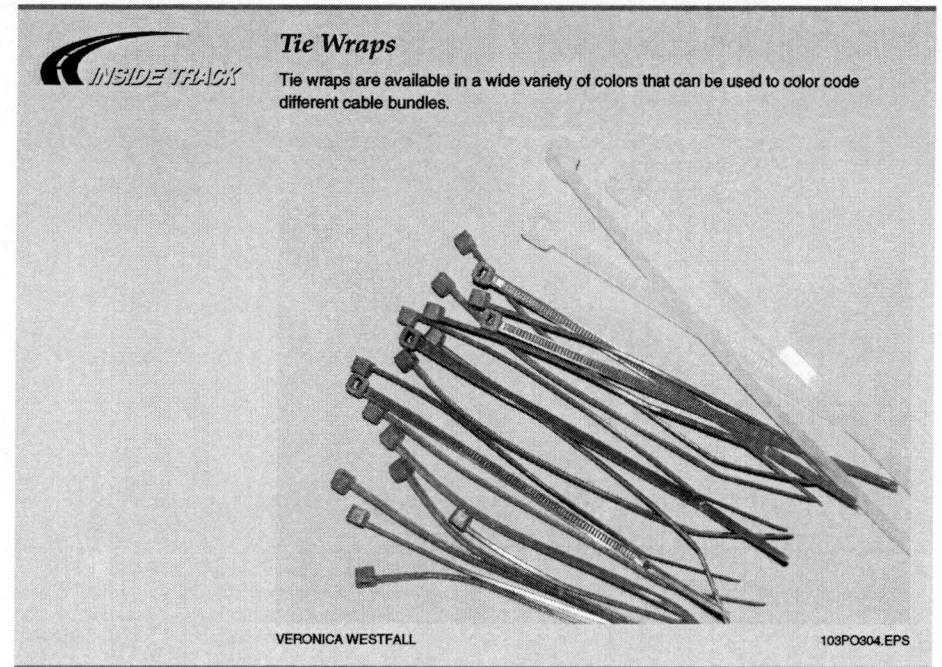

screws. Some have machine threads and are self-drilling. The size or diameter of a screw body or shank is given in gauge numbers ranging from No. 0 to No. 24, and in fractions of an inch for screws with diameters larger than ¼". The higher the gauge number, the larger the diameter of the

shank. Screw lengths range from ¼" to 6", measured from the tip to the part of the head that is flush to the surface when driven in. When choosing a screw for an application, you must consider the type and thickness of the materials to be fastened, the size of the screw, the material it is made of, the

PLAIN FORGED EYE SHOULDER FORGED EYE OPEN EYE CLOSED EYE

103F31.EPS

Figure 31 ◆ Eye bolts.

Instructor's Notes:

_____

_____

_____

_____

_____

_____

_____

_____

_____

shape of its head, and the type of driver. Because of the wide diversity in the types of screws and their application, always follow the manufacturer's recommendation to select the right screw for the job. To prevent damage to the screw head or the material being fastened, always use a screwdriver or power driver bit with the proper size and shape tip to fit the screw.

Some of the more common types of screws are:

- Wood screws
- Lag screws
- Masonry/concrete screws
- Thread-forming and thread-cutting screws

- Deck screws
- Drywall screws
- Drive screws

5.1.0 Wood Screws

Wood screws (*Figure 34*) are typically used to fasten boxes, panel enclosures, etc. to wood framing or structures where greater holding power is

Discuss wood screws and their applications. Pass around examples for the trainees to examine.

Show Transparency 36 (Figure 34).

103F32.EPS

Figure 32 ◆ Anchor bolt.

103F33.EPS

Figure 33 ◆ Typical J-bolt.

Screws

INSIDE TRACK

In most applications, either threaded or non-threaded fasteners could be used. However, threaded fasteners are sometimes preferred because they can usually be tightened and removed without damaging the surrounding material.

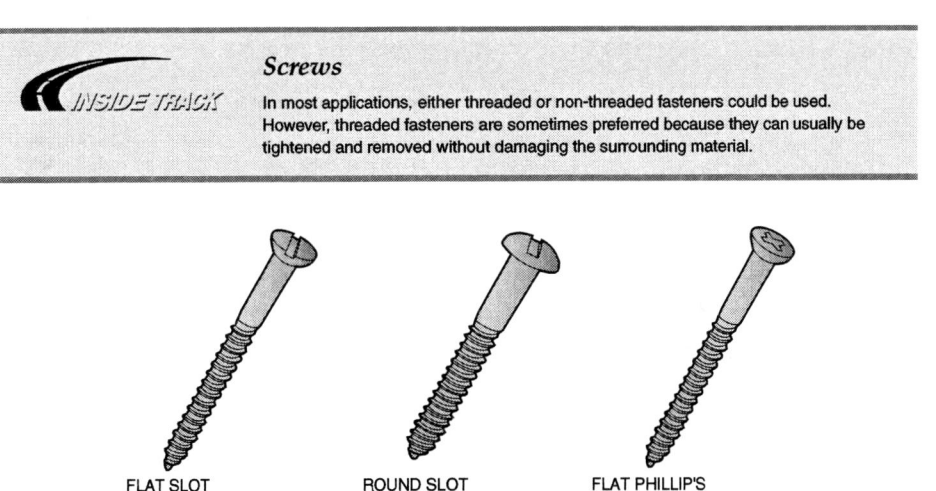

FLAT SLOT   ROUND SLOT   FLAT PHILLIP'S

103F34.EPS

Figure 34 ◆ Wood screws.

Driving Wood Screws

To maintain holding power, be careful not to drill your pilot hole too large. It's wise to drill a pilot hole deep enough to equal about two-thirds of the length of the threaded portion of the screw. Additionally, to lubricate screw threads, use soap, which makes the screw easier to drive.

Classroom

Discuss lag screws and their applications. Pass around examples for the trainees to examine.

Discuss concrete/masonry screws and their applications. Pass around examples for the trainees to examine.

Audiovisual

Show Transparency 37 (Figure 36).

needed than can be provided by nails. They are also used to fasten equipment to wood in applications where it may occasionally need to be unfastened and removed. Wood screws are commonly made in lengths from ¼" to 4", with shank gauge sizes ranging from 0 to 24. The shank size used is normally determined by the size hole provided in the box, panel, etc. to be fastened. When determining the length of a wood screw to use, a good rule of thumb is to select screws long enough to allow about ⅔ of the screw length to enter the piece of wood that is being gripped.

5.2.0 Lag Screws and Shields

Lag screws (*Figure 35*) or lag bolts are heavy-duty wood screws with square- or hex-shaped heads that provide greater holding power. Lag screws with diameters ranging between ¼" and ½" and lengths ranging from 1" to 6" are common. They are typically used to fasten heavy equipment to wood, but can also be used to fasten equipment to concrete when a lag shield is used.

WALTER JOHNSON   103F35.EPS

Figure 35 ◆ Lag screws and shields.

A lag shield is a lead tube that is split lengthwise but remains joined at one end. It is placed in a predrilled hole in the concrete. When a lag screw is screwed into the lag shield, the shield expands in the hole, firmly securing the lag screw. In hard masonry, short lag shields (typically 1" to 2" long) may be used to minimize drilling time. In soft or weak masonry, long lag shields (typically 1½" to 3" long) should be used to achieve maximum holding strength.

Make sure to use the proper length lag screw to achieve proper expansion. The length of the lag screw used should be equal to the thickness of the component being fastened plus the length of the lag shield. Also, drill the hole in the masonry to a depth approximately ½" longer than the shield being used. If the head of a lag screw rests directly on wood when installed, a flat washer should be placed under the head to prevent the head from digging into the wood as the lag screw is tightened down. Be sure to take the thickness of any washers used into account when selecting the length of the screw.

5.3.0 Concrete/Masonry Screws

Concrete/masonry screws (*Figure 36*), commonly called *self-threading anchors*, are used to fasten a device or fixture to concrete, block, or brick. No anchor is needed. To provide a matched tolerance anchor-

PHILLIP'S FLAT
HEAD

HEX WASHER
HEAD

103F36.EPS

Figure 36 ◆ Concrete screws.

Instructor's Notes:

_____

_____

_____

_____

_____

_____

_____

_____

ing system, the screws are installed using specially designed carbide drill bits and installation tools made for use with the screws. These tools are typically used with a standard rotary drill hammer. The installation tool, along with an appropriate drive socket or bit, is used to drive the screws directly into predrilled holes that have a diameter and depth specified by the screw manufacturer. When being driven into the concrete, the widely spaced threads on the screws cut into the walls of the hole to provide a tight friction fit. Most types of concrete/masonry screws can be removed and reinstalled to allow for shimming and leveling of the fastened device.

5.4.0 Thread-Forming and Thread-Cutting Screws

Thread-forming screws (*Figure 37*), commonly called *sheet metal screws,* are made of hard metal. They form a thread as they are driven into the work. This thread-forming action eliminates the need to tap a hole before installing the screw. To achieve proper holding, it is important to make sure to use the proper size bit when drilling pilot holes for thread-forming screws. The correct drill bit size

used for a specific size screw is usually marked on the box containing the screws. Some types of thread-forming screws also drill their own holes, eliminating drilling, punching, and aligning parts. Thread-forming screws are primarily used to fasten light-gauge metal parts together. They are made in the same diameters and lengths as wood screws.

Hardened steel thread-cutting metal screws with blunt points and fine threads (*Figure 38*) are used to join heavy-gauge metals, metals of different gauges, and nonferrous metals. They are also used to fasten sheet metal to building structural members. These screws are made of hardened steel that is harder than the metal being tapped.

STANDARD THREAD-
FORMING SCREW

SELF-DRILLING
SCREW

103F37.EPS

Figure 37 ◆ Thread-forming screws.

Discuss thread-forming and thread-cutting screws and their applications. Pass around examples for the trainees to examine.

Show Transparencies 38 and 39 (Figures 37 and 38).

Discuss the "Think About It." Self-drilling screws are commonly used to mount boxes to metal studs.

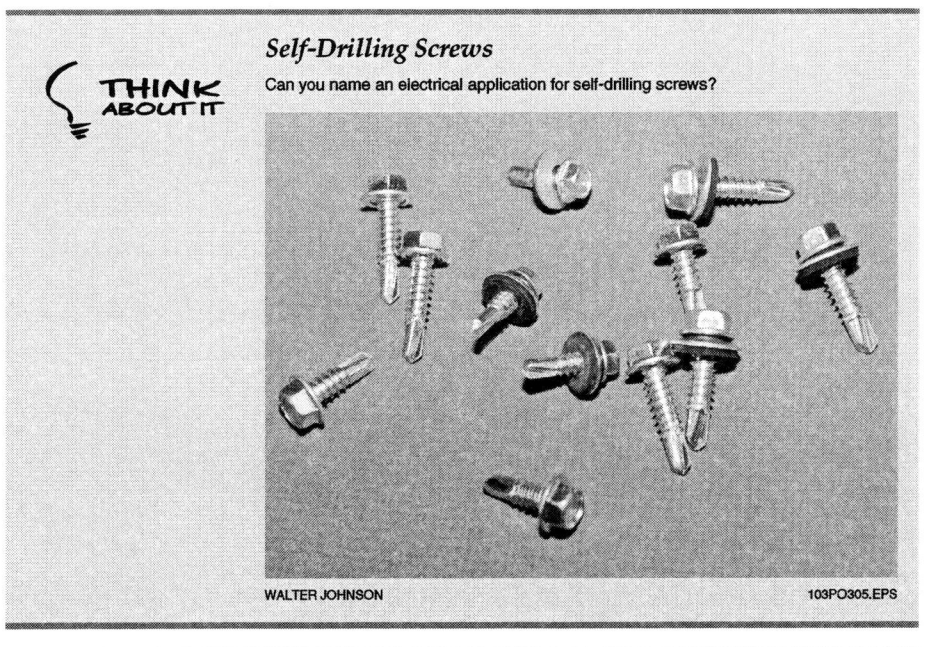

Self-Drilling Screws

THINK
ABOUT IT

Can you name an electrical application for self-drilling screws?

WALTER JOHNSON   103PO305.EPS

ROUND HEAD   PAN HEAD   HEX HEAD
                                 103F38.EPS

Figure 38 ◆ Thread-cutting screws.

They cut threads by removing and cutting a portion of the metal as they are driven into a pilot hole and through the material.

5.5.0 Deck Screws

Deck screws (*Figure 39*) are made in a wide variety of shapes and sizes for different indoor and outdoor applications. Some are made to fasten pressure-treated and other types of wood decking to wood framing. Self-drilling types are made to fasten wood decking to different gauges of metal support structures. Similarly, other self-drilling kinds are made to fasten metal decking and sheeting to different gauges and types of metal structural support members. Because of their wide diversity, it is important to follow the manufacturer's recommendations for selection of the proper screw for a particular application. Many manufacturers make a stand-up installation tool used for driving their deck screws. Use of this tool eliminates angle driving, underdriven or overdriven screws, screw wobble, etc. It also reduces operator fatigue.

5.6.0 Drywall Screws

Drywall screws (*Figure 40*) are thin, self-drilling screws with bugle-shaped heads. Depending on

the type of screw, it cuts through the wallboard and anchors itself into wood and/or metal studs, holding the wallboard tight to the stud. Coarse thread screws are normally used to fasten wallboard to wood studs. Fine thread and high and low thread types are generally used for fastening to metal studs. Some screws are made for use in either wood or metal. A Phillip's or Robertson drive head allows the drywall screw to be countersunk without tearing the surface of the wallboard.

5.7.0 Drive Screws

Drive screws do not require that the hole be tapped. They are installed by hammering the screw into a drilled or punched hole of the proper size. Drive screws are mostly used to fasten parts that will not be exposed to much pressure. A typical use of drive screws is to attach permanent name plates on electric motors and other types of equipment. *Figure 41* shows typical drive screws.

6.0.0 ◆ HAMMER-DRIVEN PINS AND STUDS

Hammer-driven pins or threaded studs (*Figure 42*) can be used to fasten wood or steel to concrete or block without the need to predrill holes. The

WOOD-TO-WOOD   WOOD-TO-METAL   METAL-TO-METAL
                                                103F39.EPS

Figure 39 ◆ Typical deck screws.

ELECTRICAL LEVEL ONE—TRAINEE MODULE 26103

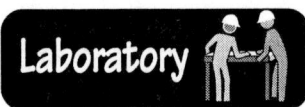
Instructor's Notes:

_____

_____

_____

_____

_____

_____

_____

_____

_____

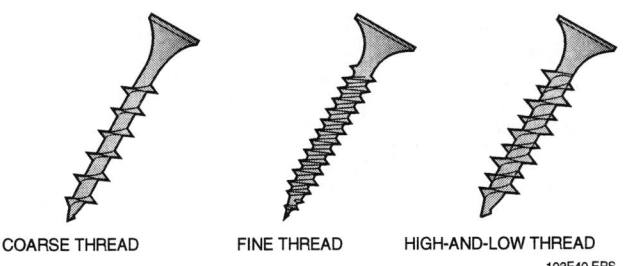

COARSE THREAD   FINE THREAD   HIGH-AND-LOW THREAD

103F40.EPS

Figure 40 ◆ Drywall screws.

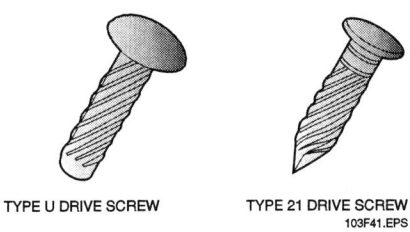

TYPE U DRIVE SCREW   TYPE 21 DRIVE SCREW

103F41.EPS

Figure 41 ◆ Drive screws.

pin or threaded stud is inserted into a hammer-driven tool designed for its use. The pin or stud is inserted in the tool point end out with the washer seated in the recess. The pin or stud is then positioned against the base material where it is to be fastened and the drive rod of the tool tapped lightly until the striker pin contacts the pin or stud. Following this, the tool's drive rod is struck using heavy blows with about a two-pound engineer's hammer. The force of the hammer blows is transmitted through the tool directly to the head of the fastener, causing it to be driven into the concrete or block. For best results, the drive pin or stud should be embedded a minimum of ½" in hard concrete to 1¼" in softer concrete block.

7.0.0 ◆ POWDER-ACTUATED TOOLS AND FASTENERS

Powder-actuated tools (*Figure 43*) can be used to drive a wide variety of specially designed pin and

THREADED STUD

DRIVE PINS

DRIVE ROD

103F42.EPS

Figure 42 ◆ Hammer-driven pins and installation tool.

Classroom

Discuss hammer-driven pins and studs and their applications. Pass around examples for the trainees to examine.

Audiovisual

Show Transparency 43 (Figure 42).

Demonstration

Demonstrate how to install hammer-driven pins.

Discuss powder-actuated tools and their applications.

Show Transparency 44 (Figure 43).

Discuss the safety precautions involved with the use of powder-actuated tools.

INSTALLATION TOOL

3/8"-16 SMOOTH 3/8"-16 KNURLED

THREADED STUDS

SMOOTH KNURLED

DRIVE PINS

POWDER LOADS

103F43.EPS

Figure 43 ◆ Powder-actuated installation tool and fasteners.

threaded stud-type fasteners into masonry and steel. These tools look and fire like a gun and use the force of a detonated gunpowder load (typically 22, 25, or 27 caliber) to drive the fastener into the material. The depth to which the pin or stud is driven is controlled by the density of the base material in which the pin or stud is being installed and by the power level or strength of the cased powder load.

Powder loads and their cases are designed for use with specific types and/or models of powder-actuated tools and are not interchangeable. Typically, powder loads are made in 12 increasing power or load levels used to achieve the proper penetration. The different power levels are identified by a color-code system and load case types. Note that different manufacturers may use different color codes to identify load strength. Power level 1 is the lowest power level while 12 is the highest. Higher number power levels are used

when driving into hard materials or when a deeper penetration is needed. Powder loads are available as single-shot units for use with single-shot tools. They are also made in multi-shot strips or disks for semiautomatic tools.

WARNING!
Powder-actuated fastening tools are to be used only by trained and licensed operators and in accordance with the tool operator's manual. You must carry your license with you whenever you are using a powder-actuated tool.

OSHA Standard 29 CFR 1926.302(e) governs the use of powder-actuated tools and states that only those individuals who have been trained in the op-

3–28

Instructor's Notes:

_____
_____
_____
_____
_____
_____
_____
_____
_____

eration of the particular powder-actuated tool in use be allowed to operate it. Authorized instructors available from the various powder-actuated tool manufacturers generally provide such training and licensing. Trained operators must take precautions to protect both themselves and others in the area when using a powder-actuated driver tool:

- Always use the tool in accordance with the published tool operation instructions. The instructions should be kept with the tool. Never attempt to override the safety features of the tool.
- Never place your hand or other body parts over the front muzzle end of the tool.
- Use only fasteners, powder loads, and tool parts specifically made for use with the tool. Use of other materials can cause improper and unsafe functioning of the tool.
- Operators and bystanders must wear eye and hearing protection along with hard hats. Other personal safety gear, as required, must also be used.
- Always post warning signs that state *Powder-Actuated Tool in Use* within 50 feet of the area where tools are used.
- Prior to using a tool, make sure it is unloaded and perform a proper function test. Check the functioning of the unloaded tool as described in the published tool operation instructions.
- Do not guess before fastening into any base material; always perform a center punch test.
- Always make a test firing into a suitable base material with the lowest power level recommended for the tool being used. If this does not set the fastener, try the next higher power level. Continue this procedure until the proper fastener penetration is obtained.
- Always point the tool away from operators or bystanders.
- Never use the tool in an explosive or flammable area.
- Never leave a loaded tool unattended. Do not load the tool until you are prepared to complete the fastening. Should you decide not to make a fastening after the tool has been loaded, always remove the powder load first, then the fastener. Always unload the tool before cleaning, servicing, or when changing parts, prior to work breaks, and when storing the tool.
- Always hold the tool perpendicular to the work surface and use the spall (chip or fragment) guard or stop spall whenever possible.
- Always follow the required spacing, edge distance, and base material thickness requirements.
- Never fire through an existing hole or into a weld area.

- In the event of a misfire, always hold the tool depressed against the work surface for at least 30 seconds. If the tool still does not fire, follow the published tool instructions. Never carelessly discard or throw unfired powder loads into a trash receptacle.
- Always store the powder loads and unloaded tool under lock and key.

8.0.0 ◆ MECHANICAL ANCHORS

Mechanical anchors are devices used to give fasteners a firm grip in a variety of materials, where the fasteners by themselves would otherwise have a tendency to pull out. Anchors can be classified in many ways by different manufacturers. In this module, anchors have been divided into five broad categories:

- One-step anchors
- Bolt anchors
- Screw anchors
- Self-drilling anchors
- Hollow-wall anchors

8.1.0 One-Step Anchors

One-step anchors are designed so that they can be installed through the mounting holes in the component to be fastened. This is because the anchor and the drilled hole into which it is installed have the same size diameter. They come in various diameters ranging from ¼" to 1¼" with lengths ranging from 1¾" to 12". Wedge, stud, sleeve, one-piece, screw, and nail anchors (*Figure 44*) are common types of one-step anchors.

8.1.1 Wedge Anchors

Wedge anchors are heavy-duty anchors supplied with nuts and washers. The drill bit size used to drill the hole is the same diameter as the anchor. The depth of the hole is not critical as long as the minimum length recommended by the manufacturer is drilled. After the hole is blown clean of dust and other material, the anchor is inserted into the hole and driven with a hammer far enough so that at least six threads are below the top surface of the component. Then, the component is fastened by tightening the anchor nut to expand the anchor and tighten it in the hole.

8.1.2 Stud Bolt Anchors

Stud bolt anchors are heavy-duty threaded anchors. Because this type of anchor is made to

List the common types of mechanical anchors. Refer the trainees to *Appendix A* for an overview.

Discuss one-step anchors and their applications. Pass around examples for the trainees to examine.

Show Transparency 45 (Figure 44).

Powder-Actuated Fasteners

Powder-actuated fasteners can be used directly against the surface to be fastened or they can be used with special extension rods for overhead or remote fastening in tight locations.

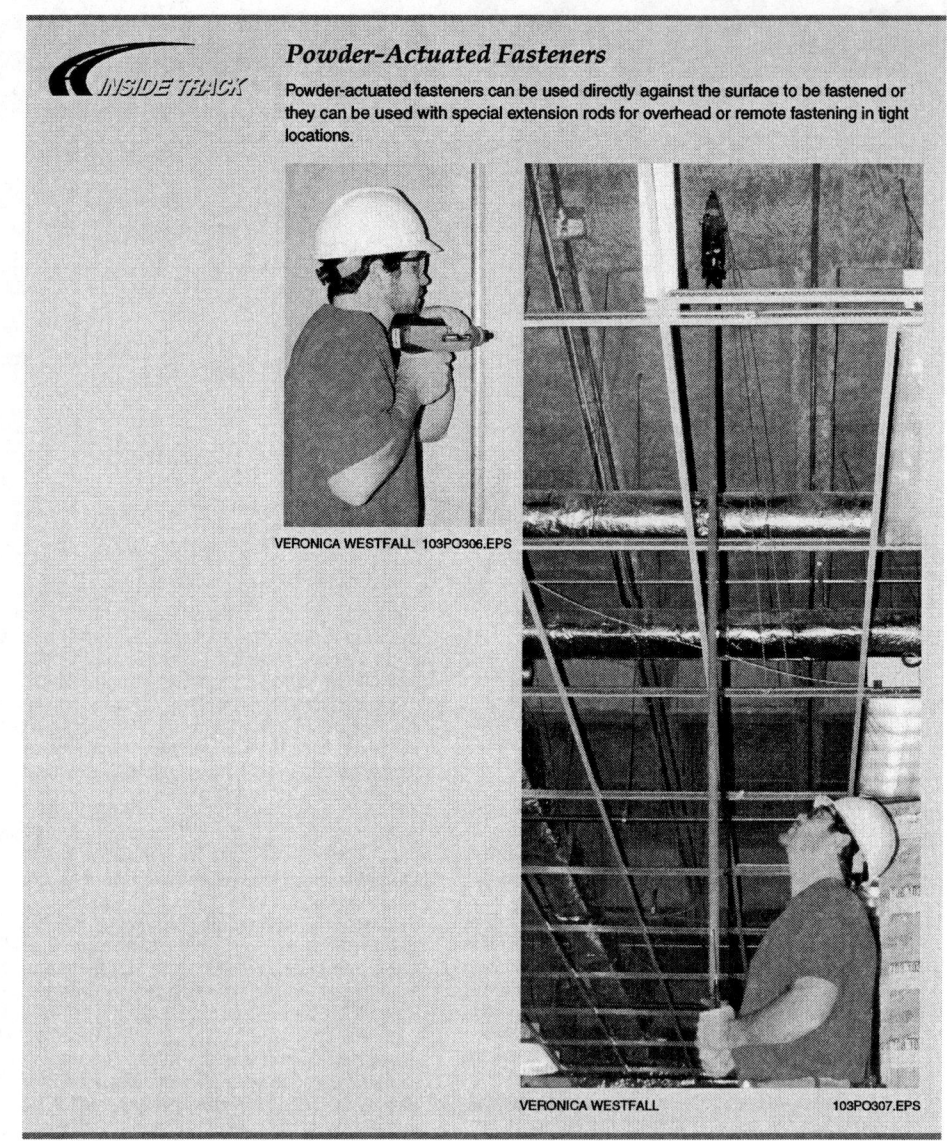

VERONICA WESTFALL   103PO306.EPS

VERONICA WESTFALL   103PO307.EPS

Instructor's Notes:

_____

_____

_____

_____

_____

_____

_____

_____

_____

Figure 44 ◆ One-step anchors.

bottom in its mounting hole, it is a good choice to use when jacking or leveling of the fastened component is needed. The depth of the hole drilled in the masonry must be as specified by the manufacturer in order to achieve proper expansion. After the hole is blown clean of dust and other material, the anchor is inserted in the hole with the expander plug end down. Following this, the anchor is driven into the hole with a hammer (or setting tool) to expand the anchor and tighten it in the hole. The anchor is fully set when it can no longer be driven into the hole. The component is fastened using the correct size and thread bolt for use with the anchor stud.

8.1.3 Sleeve Anchors

Sleeve anchors are multi-purpose anchors. The depth of the anchor hole is not critical as long as the minimum length recommended by the manufacturer is drilled. After the hole is blown clean of dust and other material, the anchor is inserted into the hole and tapped until flush with the component. Then, the anchor nut or screw is tightened to expand the anchor and tighten it in the hole.

8.1.4 One-Piece Anchors

One-piece anchors are multi-purpose anchors. They work on the principle that as the anchor is driven into the hole, the spring force of the expansion mechanism is compressed and flexes to fit the size of the hole. Once set, it tries to regain its original shape. The depth of the hole drilled in the masonry must be at least ½" deeper than the required embedment. The proper depth is crucial. Overdrilling is as bad as underdrilling. After the hole is blown clean of dust and other material, the anchor

is inserted through the component and driven with a hammer into the hole until the head is firmly seated against the component. It is important to make sure that the anchor is driven to the proper embedment depth. Note that manufacturers also make specially designed drivers and manual tools that are used instead of a hammer to drive one-piece anchors. These tools allow the anchors to be installed in confined spaces and help prevent damage to the component from stray hammer blows.

8.1.5 Hammer-Set Anchors

Hammer-set anchors are made for use in concrete and masonry. There are two types: nail and screw. An advantage of the screw-type anchors is that they are removable. Both types have a diameter the same size as the anchoring hole. For both types, the anchor hole must be drilled to the diameter of the anchor and to a depth of at least ¼" deeper than that required for embedment. After the hole is blown clean of dust and other material, the anchor is inserted into the hole through the mounting holes in the component to be fastened; then the screw or nail is driven into the anchor body to expand it. It is important to make sure that the head is seated firmly against the component and is at the proper embedment.

8.2.0 Bolt Anchors

Bolt anchors are designed to be installed flush with the surface of the base material. They are used in conjunction with threaded machine bolts or screws. In some types, they can be used with threaded rod. Drop-in, single and double expansion, and caulk-in anchors (*Figure 45*) are commonly used types of bolt anchors.

Discuss bolt anchors and their applications. Pass around examples for the trainees to examine.

Show Transparency 46 (Figure 45).

STANDARD   HOLLOW-SET   SINGLE   DOUBLE   LEAD (CAULK-IN)

DROP-IN   **EXPANSION**

103F45.EPS

Figure 45 ◆ Bolt anchors.

8.2.1 Drop-In Anchors

Drop-in anchors are typically used as heavy-duty anchors. There are two types of drop-in anchors. The first type, made for use in solid concrete and masonry, has an internally threaded expansion anchor with a preassembled internal expander plug. The anchor hole must be drilled to the specific diameter and depth specified by the manufacturer. After the hole is blown clean of dust and other material, the anchor is inserted into the hole and tapped until it is flush with the surface. Following this, a setting tool supplied with the anchor is driven into the anchor to expand it. The component to be fastened is positioned in place and fastened by threading and tightening the correct size machine bolt or screw into the anchor.

The second type, called a *hollow set drop-in anchor,* is made for use in hollow concrete and masonry base materials. Hollow set drop-in anchors have a slotted, tapered expansion sleeve and a serrated expansion cone. They come in various lengths compatible with the outer wall thickness of most hollow base materials. They can also be used in solid concrete and masonry. The anchor hole must be drilled to the specific diameter specified by the manufacturer. When installed in hollow base materials, the hole is drilled into the cell or void. After the hole is blown clean of dust and other material, the anchor is inserted into the hole and tapped until it is flush with the surface. Following this, the component to be fastened is positioned in place; then the proper size machine bolt or screw is threaded into the anchor and tightened to expand the anchor in the hole.

8.2.2 Single- and Double-Expansion Anchors

Single- and double-expansion anchors are both made for use in concrete and other masonry. The double-expansion anchor is used mainly when fastening into concrete or masonry of questionable strength. For both types, the anchor hole must be drilled to the specific diameter and depth specified by the manufacturer. After the hole is blown clean of dust and other material, the anchor is inserted into the hole, threaded cone end first. It is then tapped until it is flush with the surface. Following this, the component to be fastened is positioned in place; then the proper size machine bolt or screw is threaded into the anchor and tightened to expand the anchor in the hole.

8.2.3 Lead (Caulk-In) Anchors

Lead (caulk-in) anchors are a cast-type anchor used in concrete and masonry. They consist of an internally threaded expander cone with a series of vertical internal ribs and a lead sleeve. The vertical internal ribs prevent the cone from turning in the sleeve as the anchor is tightened. The anchor hole must be drilled to the specific diameter and depth specified by the manufacturer. However, in weak or soft masonry, a slightly deeper hole can be drilled to countersink the anchor below the surface. After the hole is blown clean of dust and other material, the anchor is inserted into the hole, threaded cone end first. Following this, a setting tool supplied with the anchor is driven into the anchor to expand it. The component to be fastened is positioned in place and fastened by threading and tightening the correct size machine bolt or screw into the anchor.

8.3.0 Screw Anchors

Screw anchors are lighter-duty anchors made to be installed flush with the surface of the base material. They are used in conjunction with sheet metal, wood, or lag screws depending on the an-

---

[Classroom icon]

Discuss screw anchors and their applications. Pass around examples for the trainees to examine.

Instructor's Notes:

_____

_____

_____

_____

_____

_____

_____

_____

FIBER   LEAD   PLASTIC

103F46A.EPS

VERONICA WESTFALL   103F46B.EPS

Figure 46 ◆ Screw anchors and screws.

chor type. Fiber, lead, and plastic anchors are common types of screw anchors (*Figure 46*). The lag shield anchor used with lag screws was described earlier in this module.

Fiber, lead, and plastic anchors are typically used in concrete and masonry. Plastic anchors are also commonly used in wallboard and similar base materials. The installation of all types is simple. The anchor hole must be drilled to the diameter specified by the manufacturer. The minimum depth of the hole must equal the anchor length. After the hole is blown clean of dust and other material, the anchor is inserted into the hole and tapped until it is flush with the surface. Following this, the component to be fastened is positioned in place; then the proper type and size screw is driven through the component mounting hole and into the anchor to expand the anchor in the hole.

8.4.0 Self-Drilling Anchors

Some anchors made for use in masonry are self-drilling anchors. *Figure 47* is typical of those in common use. This fastener has a cutting sleeve that is first used as a drill bit and later becomes the expandable fastener itself. A rotary hammer is used to drill the hole in the concrete using the anchor sleeve as the drill bit. After the hole is drilled, the anchor is pulled out and the hole cleaned. This is followed by inserting the anchor's expander plug into the cutting end of the sleeve. The anchor sleeve and expander plug are driven back into the hole with the rotary hammer until they are flush with the surface of the concrete. As the fastener is hammered down, it hits the bottom, where the tapered expander causes the fastener to expand and

lock into the hole. The anchor is then snapped off at the shear point with a quick lateral movement of the hammer. The component to be fastened can then be attached to the anchor using the proper size bolt.

8.5.0 Guidelines for Drilling Anchor Holes in Hardened Concrete or Masonry

When selecting masonry anchors, regardless of the type, always take into consideration and follow the manufacturer's recommendations pertaining to hole diameter and depth, minimum embedment in concrete, maximum thickness of material to be fastened, and the pullout and shear load capacities.

When installing anchors and/or anchor bolts in hardened concrete, make sure the area where the equipment or component is to be fastened is smooth so that it will have solid footing. Uneven footing might cause the equipment to twist, warp, not tighten properly, or vibrate when in operation. Before starting, carefully inspect the rotary hammer or hammer drill and the drill bit(s) to ensure they are in good operating condition. Be sure to use the type of carbide-tipped masonry or percussion drill bits recommended by the drill/hammer or anchor manufacturer because these bits are made to take the higher impact of the masonry materials. Also, it is recommended that the drill or hammer tool depth gauge be set to the depth of the hole needed. The trick to using masonry drill bits is not to force them into the material by pushing down hard on the drill. Use a little pressure

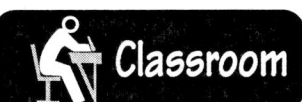
Classroom

Discuss self-drilling anchors and their applications. Pass around examples for the trainees to examine.

Show Transparency 48 (Figure 47).

FASTENERS AND ANCHORS—TRAINEE MODULE 26103   3–33

ROTARY HAMMER

CHUCK

SHEAR POINT

CUTTING SLEEVE

CUTTING TEETH

TAPERED EXPANDER

CONCRETE

AFTER
DRILLING

WITH
FASTENER
IN PLACE

THREADS OF
FASTENER
INSTALLED

103F47.EPS

Figure 47 ◆ Self-drilling anchor.

Review the procedure for installing anchors in concrete.

Show Transparency 49 (Figure 48).

Emphasize the safety precautions necessary when drilling in concrete.

and let the drill do the work. For large holes, start with a smaller bit, then change to a larger bit.

The methods for installing the different types of anchors in hardened concrete or masonry were briefly described in the sections above. Always install the selected anchors according to the manufacturer's directions. Here is an example of a typical procedure used to install many types of expansion anchors in hardened concrete or masonry. Refer to *Figure 48* as you study the procedure.

> **WARNING!**
> Drilling in concrete generates noise, dust, and flying particles. Always wear safety goggles, ear protectors, and gloves. Make sure other workers in the area also wear protective equipment.

Step 1 Drill the anchor bolt hole the same size as the anchor bolt. The hole must be deep enough for six threads of the bolt to be below the surface of the concrete (see *Figure 48, Step 1*). Clean out the hole using a squeeze bulb.

Step 2 Drive the anchor bolt into the hole using a hammer (*Figure 48, Step 2*). Protect the threads of the bolt with a nut that does not allow any threads to be exposed.

Step 3 Put a washer and nut on the bolt, and tighten the nut with a wrench until the anchor is secure in the concrete (*Figure 48, Step 3*).

8.6.0 Hollow-Wall Anchors

Hollow-wall anchors are used in hollow materials such as concrete plank, block, structural steel,

3–34

Instructor's Notes:

_____

_____

_____

_____

_____

_____

_____

_____

_____

STEP 1   STEP 2   STEP 3

103F48.EPS

Figure 48 ◆ Installing an anchor bolt in hardened concrete.

 Safety

Be sure to wear safety goggles whenever you tackle any fastening project, regardless of how small the job may seem. Remember, you can never replace lost eyesight.

wallboard, and plaster. Some types can also be used in solid materials. Toggle bolts, sleeve-type wall anchors, wallboard anchors, and metal drive-in anchors are common anchors used when fastening to hollow materials.

When installing anchors in hollow walls or ceilings, regardless of the type, always follow the manufacturer's recommendations pertaining to use, hole diameter, wall thickness, grip range (thickness of the anchoring material), and the pullout and shear load capacities.

8.6.1 Toggle Bolts

Toggle bolts (*Figure 49*) are used to fasten equipment, hangers, supports, and similar items into hollow surfaces such as walls and ceilings. They consist of a slotted bolt or screw and spring-loaded wings. When inserted through the item to be fastened, then through a predrilled hole in the wall or ceiling, the wings spring apart and provide a firm hold on the inside of the hollow wall or ceiling as the bolt is tightened. Note that the hole drilled in the wall or ceiling should be just large enough for

the compressed wing-head to pass through. Once the toggle bolt is installed, be careful not to completely unscrew the bolt because the wings will fall off, making the fastener useless. Screw-actuated plastic toggle bolts are also made. These are similar to metal toggle bolts, but they come with a pointed screw and do not require as large a hole. Unlike the metal version, the plastic wings remain in place if the screw is removed.

Toggle bolts are used to fasten a part to hollow block, wallboard, plaster, panel, or tile. The following general procedure can be used to install toggle bolts.

 WARNING!
Follow all safety precautions.

Step 1 Select the proper size drill bit or punch and toggle bolt for the job.

 Classroom

Discuss various types of hollow-wall anchors and their applications. Pass around examples for the trainees to examine.

 Safety

Emphasize that safety goggles must be worn when installing fasteners.

 Audiovisual

Show Transparency 50 (Figure 49).

 Demonstration

Demonstrate how to install toggle bolts.

ROUND HEAD   MUSHROOM HEAD   FLAT HEAD   BUTTON HEAD   SLOTTED HEX HEAD

PLASTIC TOGGLE

11"

FIXTURE HANGER TOGGLE BOLT   TIE-WIRE TOGGLE BOLT   REGULAR TOGGLE BOLT

103F49.EPS

Figure 49 ◆ Toggle bolts.

Discuss the "Think About It." Explain that over-tightening can result in damaged material and stripped threads.

Discuss sleeve-type wall anchors.

Show Transparency 51 (Figure 50).

THINK ABOUT IT

Toggle Bolts

What will happen if you fasten a toggle bolt too tightly?

Step 2 Check the toggle bolt for damaged or dirty threads or a malfunctioning wing mechanism.

Step 3 Drill a hole completely through the surface to which the part is to be fastened.

Step 4 Insert the toggle bolt through the opening in the item to be fastened.

Step 5 Screw the toggle wing onto the end of the toggle bolt, ensuring that the flat side of the toggle wing is facing the bolt head.

Step 6 Fold the wings completely back and push them through the drilled hole until the wings spring open.

Step 7 Pull back on the item to be fastened in order to hold the wings firmly against the inside surface to which the item is being attached.

Step 8 Tighten the toggle bolt with a screwdriver until it is snug.

8.6.2 Sleeve-Type Wall Anchors

Sleeve-type wall anchors (*Figure 50*) are suitable for use in concrete, block, plywood, wallboard, hollow tile, and similar materials. The two types made are standard and drive. The standard type is commonly used in walls and ceilings and is installed by drilling a mounting hole to the required diameter. The anchor is inserted into the hole and tapped until the gripper prongs embed in the base material. Following this, the anchor's screw is tightened to draw the anchor tight against the inside of the wall or ceiling. Note that the drive-type anchor is hammered into the material without the need for drilling a mounting hole. After the anchor is installed, the anchor screw is removed, the component being fastened is positioned in place, then the screw is reinstalled through the mounting hole in the component and into the anchor. The screw is tightened into the anchor to secure the component.

3–36

Instructor's Notes:

_____

_____

_____

_____

_____

_____

_____

_____

_____

STANDARD

DRIVE

SLEEVE-TYPE · WALLBOARD · METAL DRIVE-IN

103F50.EPS

Figure 50 ◆ Sleeve-type, wallboard, and metal drive-in anchors.

Sleeve-Type Drive Anchors

💡 **THINK ABOUT IT**

What happens when you remove the screw when a sleeve-type drive anchor is in place?

VERONICA WESTFALL   103PO308.EPS

8.6.3 Wallboard Anchors

Wallboard anchors (*Figure 50*) are self-drilling medium- and light-duty anchors used for fastening in wallboard. The anchor is driven into the wall with a Phillip's head manual or cordless screwdriver until the head of the anchor is flush with the wall or ceiling surface. Following this, the component being fastened is positioned over the anchor, then secured with the proper size sheet metal screw driven into the anchor.

8.6.4 Metal Drive-In Anchors

Metal drive-in anchors (*Figure 50*) are used to fasten light to medium loads to wallboard. They have two pointed legs that stay together when the anchor is hammered into a wall and spread out against the inside of the wall when a No. 6 or 8 sheet metal screw is driven in.

9.0.0 ◆ EPOXY ANCHORING SYSTEMS

Epoxy resin compounds can be used to anchor threaded rods, dowels, and similar fasteners in solid concrete, hollow wall, and brick. For one manufacturer's product, a two-part epoxy is packaged in a two-chamber cartridge that keeps the resin and hardener ingredients separated until use. This cartridge is placed into a special tool similar to a caulking gun. When the gun handle is pumped, the epoxy resin and hardener components are mixed within the gun; then the epoxy is ejected from the gun nozzle.

To use the epoxy to install an anchor in solid concrete (*Figure 51*), a hole of the proper size is drilled in the concrete and cleaned using a nylon (not metal) brush. Following this, a small amount of epoxy is dispensed from the gun to make sure that the resin and hardener have mixed properly. This is indicated

Discuss wallboard and metal drive-in anchors.

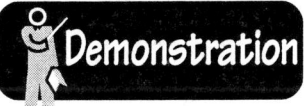

Demonstrate the installation of various anchors.

Discuss the "Think About It." Explain that nothing happens; the anchor remains in place.

Have the trainees practice installing selected anchors and toggle bolts.

Discuss epoxy anchoring systems and their applications.

Show Transparency 52 (Figure 51).

Discuss the "Think About It." Explain that toggle bolts are designed for use in hollow materials, while expandable anchors are used in solid materials.

Discuss the "Think About It." Explain that toggle bolts would be the best choice in this application, but a heavy-duty expansion anchor could also be used.

Nuts and Bolts

Expandable anchors and toggle bolts each do the same job: they fasten objects on hollow walls and from ceilings. However, each is installed differently. Why is this?

Installation Requirements

In a college dormitory, battery-powered emergency lights were anchored to sheetrock hallway ceilings with sheetrock screws, with no additional support. These fixtures weigh 8–10 pounds each and might easily have fallen out of the ceiling, causing severe injury. When the situation was discovered, the contractor had to remove and replace dozens of fixtures.

The Bottom Line: Incorrect anchoring methods can be both costly and dangerous.

Ceiling Installations

In the dormitory problem discussed above, which of the following fasteners could have been used to safely secure the emergency lights?

VERONICA WESTFALL   103PO309.EPS

Instructor's Notes:

_____
_____
_____
_____
_____
_____
_____
_____
_____
_____

Once mixed, epoxy has a limited working time. Therefore, mix exactly what you need and work quickly. After the working time is up, epoxy requires a specific curing time. Always give epoxy its recommended curing time; since epoxy is so strong and sets so quickly, you'll be tempted to stress the bond before it's fully cured.

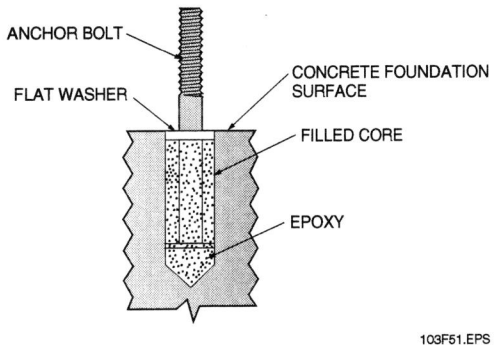

Figure 51 ◆ Fastener anchored in epoxy.

Use the Proper Tool for the Application

To avoid damaging fasteners, use the correct tool for the job. For example, don't use pliers to install bolts, and don't use a screwdriver that is too large or too small.

Putting It All Together

You are installing a surface-mounted panel. What fasteners would you use in each of the following types of wall construction?

- Sheetrock wall with wood studs
- Sheetrock wall with metal studs
- Concrete masonry units (hollow block)
- Poured concrete

Discuss the "Think About It."

by the epoxy being of a uniform color. The gun nozzle is then placed into the hole, and the epoxy is injected into the hole until half the depth of the hole is filled. Following this, the selected fastener is pushed into the hole with a slow twisting motion to make sure that the epoxy fills all voids and crevices, then

is set to the required plumb (or level) position. After the recommended cure time for the epoxy has elapsed, the fastener nut can be tightened to secure the component or fixture in place.

The procedure for installing a fastener in a hollow wall or brick using epoxy is basically the same

FASTENERS AND ANCHORS—TRAINEE MODULE 26103

3–39

FASTENERS AND ANCHORS—INSTRUCTOR'S GUIDE MODULE 26103

3–39

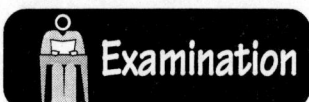
as described above. The difference is that the epoxy is first injected into an anchor screen to fill the screen, then the anchor screen is installed into the drilled hole. Use of the anchor screen is necessary to hold the epoxy intact in the hole until the anchor is inserted into the epoxy.

Summary

Fasteners and anchors are used for a variety of tasks in the electrical craft. In this module, you learned about various types of fasteners and anchors and their uses. Basic installation procedures were also included. Selecting the correct fastener or anchor for a particular job is required to perform high-quality work. In the electrical craft, it is important to be familiar with the correct terms used to describe fasteners and anchors. Using the proper technical terms helps avoid confusion and improper selection. Installation techniques for fasteners and anchors may vary depending on the job. Make sure to check the project specifications and manufacturer's information when installing any fastener or anchor.

New fasteners and anchors are being developed every day. Your local distributor/manufacturer is an excellent source of information.

Review Questions

1. The quality of some fasteners can be determined by the _____ on the head of a bolt or screw.
   a. number of grooves cut
   b. number of sides
   c. length of the lines
   d. grade markings

2. The purpose of a jam nut is to _____.
   a. hold a piece of material stationary
   b. lock a standard nut in place
   c. stop rotation of a machine quickly for safety reasons
   d. compress a lock washer in place

3. Washers are used to _____.
   a. distribute the load over a larger area
   b. attach an item to a hollow surface
   c. anchor materials that expand due to temperature changes
   d. allow the bolts to expand with temperature changes

4. Torque is normally expressed in _____.
   a. pounds per square inch (psi)
   b. gallons per minute (gpm)
   c. foot pounds (ft. lbs.)
   d. cubic feet per minute (cpm)

5. When tightening bolts, nuts, or screws, always use the proper _____.
   a. tightening sequence
   b. bolt pattern
   c. ratchet wrench
   d. lubrication

6. What type of fasteners are commonly used as lifting devices?
   a. Toggle bolts
   b. Anchor bolts
   c. J-bolts
   d. Eye bolts

7. When installing a panel in concrete using a lag screw, you would also use a _____.
   a. drop-in anchor
   b. lag shield
   c. caulk-in anchor
   d. double-expansion anchor

8. When fastening a device or fixture to concrete using concrete/masonry screws, the screws _____.
   a. need not be installed in predrilled holes because they are self-tapping
   b. are installed using specially designed carbide drill bits and installation tools made for use with the screws
   c. are installed by hammering them into a drilled or punched hole of the proper size
   d. are installed in a hole drilled with a standard masonry drill bit

9. When using powder-actuated tools and fasteners, _____.
   a. higher-numbered power level loads are used to drive into soft materials
   b. use only powder loads that are specifically designed for use with the installation tool
   c. use only fasteners that are specifically designed for use with the installation tool
   d. Both b and c.

10. Wedge and sleeve anchors are classified as _____ anchors.
   a. bolt
   b. hollow-wall
   c. one-step
   d. screw

ELECTRICAL LEVEL ONE—TRAINEE MODULE 26103

Instructor's Notes:

_____

_____

_____

_____

_____

_____

_____

_____

_____

ELECTRICAL LEVEL ONE—INSTRUCTOR'S GUIDE MODULE 26103

Gary Edgington, EdgeWel Electric, Inc.

Gary Edgington started his own electrical contracting business at the age of 24, and has developed it into a successful organization that employs dozens of electricians and is able to handle any type of electrical job. Gary credits part of his success to the mentors who helped him early in his career and taught him valuable lessons that have guided him ever since. Here is his success story as he told it:

I got into electrical work because I really wanted to work with my hands. I had an early childhood interest in electrical things that continued through high school.

I got into the trade in 1973 with no formal training. I started on-the-job training with Harry Kinsey, a seasoned electrical contractor who did commercial and residential work. He was really my first role model. He was a greatly admired individual with a very good reputation. I learned from him that if you're going to do a job, be the best at it, and take pride in your work.

Afterwards, I worked for Des Moines, Iowa-based AMF Corporation as an industrial electrician. During this time, I also attended college and studied industrial electronics because it gives you theory and basic electrical knowledge, which allows you to resolve problems better and to troubleshoot on your own. My role model at AMF was Mr. George McCoig. He was a retired Air Force electrician. He was the most highly respected electrician at AMF. He taught me that knowledge is important and that you don't have to know everything, you just have to know how to find the answers to everything. I was only 20 years old at the time, but I guess I had a knack for motor controls and so I found myself in the industrial field. I was assigned to George as his helper. George's job was to teach me electrical systems.

After George, I worked for a man named Bernard Sharples. He was an electrician from Manchester, England, and was second only in reputation to George at AMF. Bernard took me under his wing and he not only recognized my abilities, but helped me expand them. He reminded me of what I knew, not what I didn't know. Bernard was the type of person who brought out the best in other people.

I got my master's (contractor's) license in 1978 after being in the field 5 years. This gave me everything I really needed to be an electrical contractor. I was 24 years old. I started the company I own now with a partner, but after 3 years, I became sole owner and grew a business that started with two people to one with over 35 people. We have a 12,000 sq. ft. facility and do every kind of job. That's really our claim to fame; we do everything from single-family houses to industrial to commercial to telephone data work.

Today, most of my work entails estimating, design, and management. In 1994, I became involved with NCCER's standardized craft training. I already had 9 years worth of teaching experience behind me, so I got involved because of my commitment to training. I'm a person who believes that training costs nothing; in the long run, you get back much more than you invest in it.

Trade Terms Introduced in This Module

American Society for Testing of Materials (ASTM): An organization that publishes specifications and standards relating to fasteners.

Clearance: The amount of space between the threads of bolts and their nuts.

Foot pounds (ft. lbs.): The normal method used for measuring the amount of torque being applied to bolts or nuts.

Inch pounds (in. lbs.): A method of measuring the amount of torque applied to small bolts or nuts that require measurement in smaller increments than foot pounds.

Key: A machined metal part that fits into a keyway and prevents parts such as gears or pulleys from rotating on a shaft.

Keyway: A machined slot in a shaft and on parts such as gears and pulleys that accepts a key.

Nominal size: A means of expressing the size of a bolt or screw. It is the approximate diameter of a bolt or screw.

Society of Automotive Engineers (SAE): An organization that publishes specifications and standards relating to fasteners.

Thread classes: Threads are distinguished by three classifications according to the amount of tolerance the threads provide between the bolt and nut.

Thread identification: Standard symbols used to identify threads.

Thread standards: An established set of standards for machining threads.

Tolerance: The amount of difference allowed from a standard.

Torque: The turning force applied to a fastener.

Unified National Coarse (UNC) thread: A standard type of coarse thread.

Unified National Extra Fine (UNEF) thread: A standard type of extra-fine thread.

Unified National Fine (UNF) thread: A standard type of fine thread.

Instructor's Notes:

_____

_____

_____

_____

_____

_____

_____

_____

_____

Mechanical Anchors and Their Uses

Anchor Type	Typically Used In	Use With Fastener	Typical Working Load Range*
One-Step Anchors			
Wedge	Concrete **Stone	None	Light, medium, and heavy duty
Stud	Concrete **Stone, solid brick and block	None	Light, medium, and heavy duty
Sleeve	Concrete, solid brick and block **Stone, hollow brick and block	None	Light and medium duty
One-piece	Concrete, solid block **Stone, solid and hollow brick, hollow block	None	Light and medium duty
Hammer-set	Concrete, solid block **Stone, solid and hollow brick, hollow block	None	Light duty
Bolt Anchors			
Drop-in	Concrete **Stone, solid brick	Machine screw or bolt	Light, medium, and heavy duty
Hollow-set drop-in	Concrete, solid brick and block **Stone, hollow brick and block	Machine screw or bolt	Light and medium duty
Single-expansion	Concrete, solid brick and block **Stone, hollow brick and block	Machine screw or bolt	Light and medium duty
Double-expansion	Concrete, solid brick and block **Stone, hollow brick and block	Machine screw or bolt	Light and medium duty
Lead (caulk-in)	Concrete, solid brick and block **Stone, hollow brick and block	Machine screw or bolt	Light and medium duty

Anchor Type	Typically Used In	Use With Fastener	Typical Working Load Range*
Screw Anchors			
Lag shield	Concrete **Stone, solid and hollow brick and block	Lag screw	Light and medium duty
Fiber	Concrete, stone, solid brick and block **Hollow brick and block, wallboard	Wood, sheet metal, or lag screw	Light and medium duty
Lead	Concrete, solid brick and block **Hollow brick and block, wallboard	Wood or sheet metal screw	Light duty
Plastic	Concrete, stone, solid brick and block **Hollow brick and block, wallboard	Wood or sheet metal screw	Light duty
Hollow-Wall Anchors			
Toggle bolts	Concrete, plank, hollow block, wallboard, plywood/paneling	None	Light and medium duty
Plastic toggle bolts	Wallboard, plywood/paneling **Hollow block, structural tile	Wood or sheet metal screw	Light duty
Sleeve-type wall	Wallboard, plywood/paneling **Hollow block, structural tile	None	Light duty
Wallboard	Wallboard	Sheet metal screw	Light duty
Metal drive-in	Wallboard	Sheet metal screw	Light duty

*Anchor working loads given in the table are defined below. These are approximate loads only. Actual allowable loads depend on such factors as the anchor style and size, base material strength, spacing and edge distance, and the type of service load applied. Always consult the anchor manufacturer's product literature to determine the correct type of anchor and size to use for a specific application.

• Light duty—Less than 400 lbs.
• Medium duty—400 to 4,000 lbs.
• Heavy duty—Above 4,000 lbs.

**Indicates use may be suitable depending on the application.

Instructor's Notes:

Answers to Review Questions

Answer	Section
1. d	2.1.4
2. b	2.3.1
3. a	2.4.0
4. c	2.5.1
5. a	2.5.2
6. d	4.1.0
7. b	5.2.0
8. b	5.3.0
9. d	7.0.0
10. c	8.1.0

Additional Resources

For advanced study of topics covered in this task module, the following books are suggested:

American Electrician's Handbook, Latest Edition. New York: McGraw-Hill.

Carpentry, Latest Edition. New York: Delmar Publishers.

The Sheet Metal Toolbox Manual, Latest Edition. New York: Prentice Hall.

Instructor's Notes:

_____

_____

_____

_____

_____

_____

_____

_____

NCCER CRAFT TRAINING USER UPDATES

The NCCER makes every effort to keep these textbooks up-to-date and free of technical errors. We appreciate your help in this process. If you have an idea for improving this textbook, or if you find an error, a typographical mistake, or an inaccuracy in the NCCER's Craft Training textbooks, please write us, using this form or a photocopy. Be sure to include the exact module number, page number, a detailed description, and the correction, if applicable. Your input will be brought to the attention of the Technical Review Committee. Thank you for your assistance.

Instructors – If you found that additional materials were necessary in order to teach this module effectively, please let us know so that we may include them in the Equipment/Materials list in the Instructor's Guide.

Write: Curriculum Revision and Development Department
National Center for Construction Education and Research
P.O. Box 141104, Gainesville, FL 32614-1104

Fax: 352-334-0932

E-mail: curriculum@nccer.org

Craft _____   Module Name _____

Copyright Date _____   Module Number _____   Page Number(s) _____

Description _____

_____

_____

_____

(Optional) Correction _____

_____

_____

(Optional) Your Name and Address _____

_____

_____

TEACHING TIPS

Section 2.2.0 *Think About It—Installing Fasteners*

Trainees should be concerned with appearance, decorative finish, strength of fastener and material to be fastened, proper installation, amount of torque required, appropriate personal protective equipment (particularly safety goggles), and using the correct tool(s) for the job.

Section 9.0.0 *Think About It—Putting It All Together*

- *Sheetrock wall with wood studs*—Use wood screws if in studs; if not, use toggle bolts.
- *Sheetrock wall with metal studs*—Use sheet metal screws if in studs; if not, use toggle bolts.
- *Concrete masonry units (hollow block)*—May use hollow-wall anchors, expansion anchors, or powder-actuated fasteners.
- *Poured concrete*—Use expansion anchors or powder-actuated fasteners.

Electrical Theory One
26104-00

MODULE OVERVIEW

This course introduces the electrical trainee to basic electrical theory.

PREREQUISITES

Please refer to the Course Map in the Trainee Module. Prior to training with this module, it is recommended that the trainee shall have successfully completed the following modules:

Core Curriculum; Electrical Level One, Modules 26101 through 26103

LEARNING OBJECTIVES

Upon completion of this module, the trainee will be able to:

1. Recognize what atoms are and how they are constructed.
2. Define voltage and identify the ways in which it can be produced.
3. Explain the difference between conductors and insulators.
4. Define the units of measurement that are used to measure the properties of electricity.
5. Explain how voltage, current, and resistance are related to each other.
6. Using the formula for Ohm's law, calculate an unknown value.
7. Explain the different types of meters used to measure voltage, current, and resistance.
8. Using the power formula, calculate the amount of power used by a circuit.

PERFORMANCE OBJECTIVES

Under supervision of the instructor, the trainee should be able to:

1. Use the formula for Ohm's law to calculate voltage, current, and resistance.
2. Given different resistors, identify the correct resistance value and tolerance using the color code.
3. Draw basic voltmeter and ohmmeter circuits and explain how they operate.
4. Use the power formula to calculate the amount of power used by a circuit.
5. Use a variation of the power formula to calculate the maximum current a resistor can carry based on the resistor's value and power rating.

NCCER STANDARDIZED CRAFT TRAINING PROGRAM

The National Center for Construction Education and Research (NCCER) provides a standardized national program of accredited craft training. Key features of the program include instructor certification, competency-based training, and performance testing. The program provides trainees, instructors, and companies with a standard form of recognition through a National Craft Training Registry. The program is described in full in the *Guidelines for Accreditation*, published by the NCCER. For more information on standardized craft training, contact the NCCER at P.O. Box 141104, Gainesville, FL 32614-1104, 352-334-0911, visit our Web site at www.NCCER.org, or e-mail info@NCCER.org.

HOW TO USE THIS ANNOTATED INSTRUCTOR'S GUIDE

Each page presents two sections of information. The larger section displays each page exactly as it appears in the Trainee Module. The narrow column ties suggested trainee and instructor actions to each page and provides icons to call your attention to material, safety, audiovisual, or testing requirements. The bottom of each page includes space for your notes.

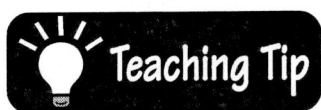

If you see the Teaching Tip icon, that means there is a teaching tip associated with this section. Also refer to the suggested teaching tips at the end of the module.

SAFETY CONSIDERATIONS

Ensure that the trainees are equipped with appropriate personal protective equipment.

PREPARATION

Before teaching this module, you should review the Module Outline, Learning and Performance Objectives, and the Materials and Equipment List. Be sure to allow ample time to prepare your own training or lesson plan and gather all required equipment and materials.

MATERIALS AND EQUIPMENT LIST

Materials:
Transparencies

Markers/chalk

Copy of the latest edition of the *National Electrical Code*®

Various types of resistors

Module Examinations*

Performance Profile Sheets*

Equipment:
Overhead projector and screen

Whiteboard/chalkboard

Appropriate personal protective equipment

*Located in the Test Booklet packaged with this Annotated Instructor's Guide.

ADDITIONAL RESOURCES

This module is intended to present thorough resources for task training. The following reference works are suggested for both instructors and motivated trainees interested in further study. These are optional materials for continued education rather than for task training.

Electronics Fundamentals, Latest Edition. Upper Saddle River, NJ: Prentice Hall.

Principles of Electric Circuits, Latest Edition. Upper Saddle River, NJ: Prentice Hall.

NOTES

The designations "National Electrical Code," "NE Code," and "NEC," where used in this document, refer to the *National Electrical Code*®, which is a registered trademark of the National Fire Protection Association, Quincy, MA. All National Electrical Code (NEC) references in this module refer to the 1999 edition of the NEC.

If you feel that additional math instruction would be helpful, Prentice Hall offers a basic math textbook entitled *Fundamentals of Electrical and Mechanical Mathematics*. It covers the basic math requirements for electrical trainees and may be ordered by contacting Prentice Hall Customer Service at 1-800-922-0579.

TEACHING TIME FOR THIS MODULE

An outline for use in developing your lesson plan is presented below. Note that each Roman numeral in the outline equates to one session of instruction. Each session has a suggested time period of 2½ hours. This includes 10 minutes at the beginning of each session for administrative tasks and one 10-minute break during the session. Approximately 7½ hours are suggested to cover *Electrical Theory One*. You will need to adjust the time required for hands-on activity and testing based on your class size and resources.

Topic	**Planned Time**

Session I. Introduction to Electrical Theory; Conductors and Insulators; Electric Charge and Current

A. Introduction to Electrical Theory   _____

B. Conductors and Insulators   _____

   1. The Atom   _____

      a. The Nucleus   _____

      b. Electrical Charges   _____

   2. Conductors and Insulators   _____

C. Electric Charge and Current   _____

   1. Current Flow   _____

   2. Voltage   _____

Session II. Resistance; Schematic Representation of Circuit Elements; Resistors; Measuring Voltage, Current, and Resistance

A. Resistance   _____

   1. Characteristics of Resistance   _____

   2. Ohm's Law   _____

B. Schematic Representation of Circuit Elements   _____

C. Resistors   _____

   1. Resistor Color Codes   _____

D. Measuring Voltage, Current, and Resistance   _____

   1. Basic Meter Operation   _____

   2. Voltmeter   _____

   3. Ammeter   _____

   4. Ohmmeter   _____

Session III. Electrical Power; Module Examination and Performance Testing

A. Electrical Power   _____

   1. Power Equation   _____

   2. Power Rating of Resistors   _____

B. Module Examination   _____

   1. Trainees must score 70% or higher to receive recognition from the NCCER.

   2. Record the testing results on Craft Training Report Form 200 and submit the results to the Training Program Sponsor.

C. Performance Testing   _____

   1. Trainees must perform each task to the satisfaction of the instructor to receive recognition from the NCCER.

   2. Record the testing results on Craft Training Report Form 200 and submit the results to the Training Program Sponsor.

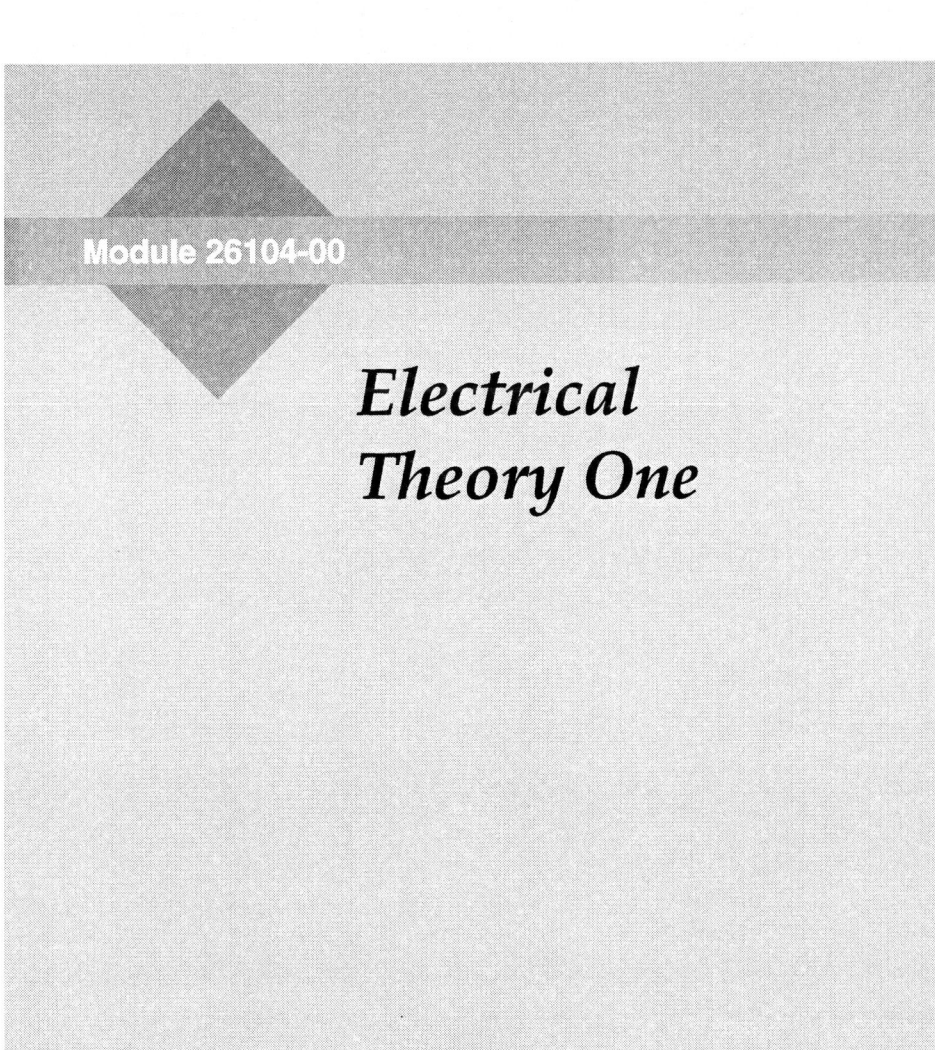

Module 26104-00

Electrical
Theory One

Course Map

This course map shows all of the task modules in the first level of the Electrical curricula. The suggested training order begins at the bottom and proceeds up. Skill levels increase as a trainee advances on the course map. The training order may be adjusted by the local Training Program Sponsor.

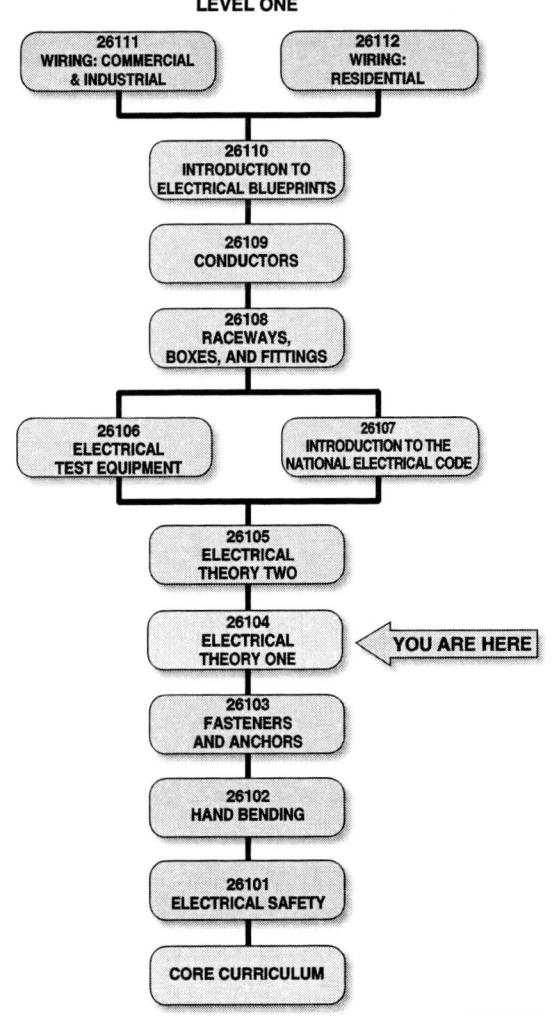

LEVEL ONE

- 26111 WIRING: COMMERCIAL & INDUSTRIAL
- 26112 WIRING: RESIDENTIAL
- 26110 INTRODUCTION TO ELECTRICAL BLUEPRINTS
- 26109 CONDUCTORS
- 26108 RACEWAYS, BOXES, AND FITTINGS
- 26106 ELECTRICAL TEST EQUIPMENT
- 26107 INTRODUCTION TO THE NATIONAL ELECTRICAL CODE
- 26105 ELECTRICAL THEORY TWO
- 26104 ELECTRICAL THEORY ONE ← YOU ARE HERE
- 26103 FASTENERS AND ANCHORS
- 26102 HAND BENDING
- 26101 ELECTRICAL SAFETY
- CORE CURRICULUM

104CMAP.EPS

Assign reading of Module 26104.

4–iii

Materials

Ensure you have everything required to teach the course. Check the Materials and Equipment List at the front of this Instructor's Guide.

Instructor's Notes:

Electrical Theory One

Objectives

Upon completion of this module, the trainee will be able to:

1. Recognize what atoms are and how they are constructed.
2. Define voltage and identify the ways in which it can be produced.
3. Explain the difference between conductors and insulators.
4. Define the units of measurement that are used to measure the properties of electricity.
5. Explain how voltage, current, and resistance are related to each other.
6. Using the formula for Ohm's Law, calculate an unknown value.
7. Explain the different types of meters used to measure voltage, current, and resistance.
8. Using the power formula, calculate the amount of power used by a circuit.

Prerequisites

Successful completion of the following Task Modules is recommended before beginning study of this Task Module: Core Curricula; Electrical Level One, Modules 26101 through 26103.

Required Trainee Materials

1. Trainee Task Module
2. Copy of the latest edition of the *National Electrical Code*
3. Appropriate Personal Protective Equipment

1.0.0 ◆ INTRODUCTION TO ELECTRICAL THEORY

As an electrician, you must work with a force that cannot be seen. However, electricity is there on the job, every day of the year. It is necessary that you understand the forces of electricity so that you will be safe on the job. The first step is a basic understanding of the principles of electricity.

The relationships among current, voltage, resistance, and power in a basic direct current (DC) series circuit are common to all types of electrical circuits. This module provides a general introduction to the electrical concepts used in Ohm's Law. It also presents the opportunity to practice applying these basic concepts to DC series circuits. In this way, you can prepare for further study in electrical and electronics theory and maintenance techniques. By practicing these techniques for all combinations of DC circuits, you will be prepared to work on any DC circuits you might encounter.

Note: The designations "National Electrical Code," "NE Code," and "NEC," where used in this document, refer to the *National Electrical Code®*, which is a registered trademark of the National Fire Protection Association, Quincy, MA. *All National Electrical Code (NEC) references in this module refer to the 1999 edition of the NEC.*

Show Transparency 1, Course Objectives.

Show Transparency 2, Performance Profile Tasks.

Explain that terms shown in bold (blue) are defined in the Glossary at the back of this module.

4–1

Discuss the "Think About It." Emphasize that a professional takes pride in knowing as much as possible about his/her chosen field.

Review basic atomic theory.

Show Transparency 3 (Figure 1).

2.0.0 ◆ CONDUCTORS AND INSULATORS

2.1.0 The Atom

The atom is the smallest part of an element that enters into a chemical change, but it does so in the form of a charged particle. These charged particles are called *ions,* and are of two types—positive and negative. A positive ion may be defined as an atom that has become positively charged. A negative ion may be defined as an atom that has become negatively charged. One of the properties of charged ions is that ions of the same charge tend to repel one another, whereas ions of unlike charge will attract one another. The term *charge* can be taken to mean a quantity of electricity that is either positive or negative.

The structure of an atom is best explained by a detailed analysis of the simplest of all atoms, that of the element hydrogen. The hydrogen atom in *Figure 1* is composed of a nucleus containing one proton and a single orbiting electron. As the electron revolves around the nucleus, it is held in this orbit by two counteracting forces. One of these forces is called *centrifugal force,* which is the force that tends to cause the electron to fly outward as it travels around its circular orbit. The second force acting on the electron is *electrostatic force.* This force tends to pull the electron in toward the nucleus and is provided by the mutual attraction between the positive nucleus and the negative electron. At some given radius, the two forces will balance each other, providing a stable path for the electron.

- A proton (+) repels another proton (+).
- An electron (−) repels another electron (−).
- A proton (+) attracts an electron (−).

Basically, an atom contains three types of sub-atomic particles that are of interest in electricity: electrons, protons, and neutrons.

The protons and neutrons are located in the center, or nucleus, of the atom, and the electrons travel about the nucleus in orbits.

Because protons are relatively heavy, the repulsive force they exert on one another in the nucleus of an atom has little effect.

The attracting and repelling forces on charged materials occur because of the electrostatic lines of force that exist around the charged materials. In a negatively charged object, the lines of force of the excess electrons add to produce an electrostatic field that has lines of force coming into the object from all directions. In a positively charged object, the lines of force of the excess protons add to produce an electrostatic field that has lines of force going out of the object in all directions. The electrostatic fields either aid or oppose each other to attract or repel.

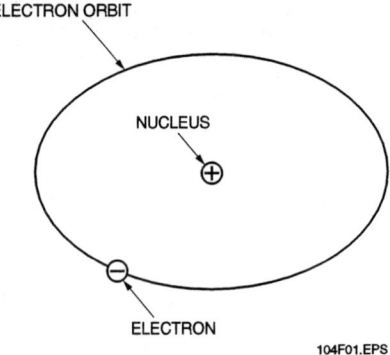

104F01.EPS

Figure 1 ◆ Hydrogen atom.

4–2

ELECTRICAL LEVEL ONE—TRAINEE MODULE 26104

Instructor's Notes:

Electrical Charges

THINK ABOUT IT

Think about the things you come in contact with every day. Where do you see or find examples of electrostatic attraction?

Discuss the "Think About It." Answers will vary, from static cling to shocks when touching metal objects.

Audiovisual

Show Transparency 4 (Figure 2).

Discuss the differences between conductors and insulators.

Show Transparency 5 (Figure 3).

2.1.1 The Nucleus

The nucleus is the central part of the atom. It is made up of heavy particles called protons and neutrons. The proton is a charged particle containing the smallest known unit of positive electricity. The neutron has no electrical charge. The number of protons in the nucleus determines how the atom of one element differs from the atom of another element.

Although a neutron is actually a particle by itself, it is generally thought of as an electron and proton combined and is electrically neutral. Since neutrons are electrically neutral, they are not considered important to the electrical nature of atoms.

2.1.2 Electrical Charges

The negative charge of an electron is equal but opposite to the positive charge of a proton. The charges of an electron and a proton are called *electrostatic charges*. The lines of force associated with each particle produce electrostatic fields. Because of the way these fields act together, charged particles can attract or repel one another. The Law of Electrical Charges states that particles with like charges repel each other and those with unlike charges attract each other. This is shown in *Figure 2*.

2.2.0 Conductors and Insulators

The difference between atoms, with respect to chemical activity and stability, depends on the number and position of the electrons included within the atom. In general, the electrons reside in groups of orbits called *shells*. The shells are arranged in steps that correspond to fixed energy levels.

The number of electrons in the outermost shell determines the valence of an atom. For this reason, the outer shell of an atom is called the valence shell, and the electrons contained in this shell are called *valence electrons* (*Figure 3*). The valence of an atom determines its ability to gain or lose an electron, which in turn determines the chemical and electrical properties of the atom. An atom that is lacking only one or two electrons from its outer shell will easily gain electrons to complete its shell, but a large amount of energy is required to free any of its electrons. An atom having a relatively small number of electrons in its outer shell in comparison to the number of electrons required to fill the shell will easily lose its electrons.

It is the valence electrons that we are most concerned with in electricity. These are the electrons that are easiest to break loose from their parent atom. Normally, a conductor has three or less valence electrons, an insulator has five or more

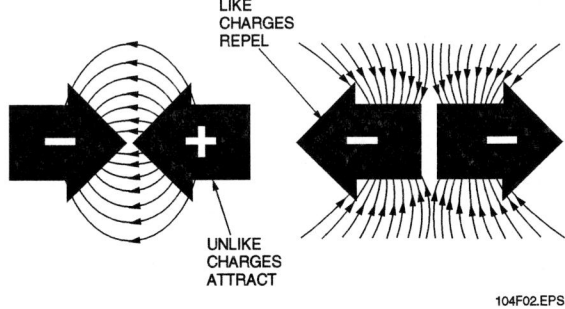

LIKE CHARGES REPEL

UNLIKE CHARGES ATTRACT

104F02.EPS

Figure 2 ✦ Law of electrical charges.

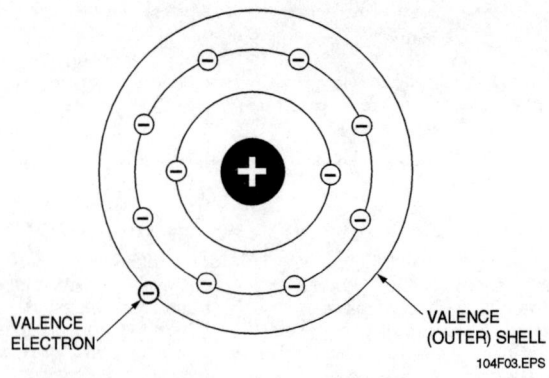

VALENCE
ELECTRON

VALENCE
(OUTER) SHELL

104F03.EPS

Figure 3 ◆ Valence shell and electrons.

THINK ABOUT IT

Conductors

Why do some substances conduct? What happens inside a conductor? What makes a good conductor?

INSIDE TRACK

Insulative Materials Become Conductive When Wet

While pure (distilled) water is an insulator, even trace levels of minerals make it into a conductor.

valence electrons, and semiconductors usually have four valence electrons.

All the elements of which matter is made may be placed into one of three categories: conductors, insulators, and semiconductors.

Conductors, for example, are elements such as copper and silver that will conduct a flow of electricity very readily. Because of their good conducting abilities, they are formed into wire and used whenever it is desired to transfer electrical energy from one point to another.

Insulators, on the other hand, do not conduct electricity to any great degree and are used when it is desirable to prevent the flow of electricity. Compounds such as porcelain and plastic are good insulators.

Materials such as germanium and silicon are not good conductors but cannot be used as insulators either, since their electrical characteristics fall between those of conductors and those of insulators. These in-between materials are classified as *semiconductors.* As you will learn later in your training, semiconductors play a crucial role in electronic circuits.

3.0.0 ◆ ELECTRIC CHARGE AND CURRENT

An electric charge has the ability to do the work of moving another charge by attraction or repulsion. The ability of a charge to do work is called its *potential.* When one charge is different from another, there must be a difference in potential between them. The sum of the difference of potential of all the charges in the electrostatic field is referred to as *electromotive force (emf)* or *voltage.* Voltage is frequently represented by the letter *E.*

Instructor's Notes:

Electric charge is measured in coulombs. An electron has 1.6×10^{-19} coulombs of charge. Therefore, it takes 6.25×10^{18} electrons to make up one coulomb of charge, as shown below.

$$\frac{1}{1.6 \times 10^{-19}} = 6.25 \times 10^{18} \text{ electrons}$$

If two particles, one having charge Q_1 and the other charge Q_2, are a distance (d) apart, then the force between them is given by Coulomb's Law, which states that the force is directly proportional to the product of the two charges and inversely proportional to the square of the distance between them:

$$\text{Force} = \frac{k \times Q_1 \times Q_2}{d^2}$$

If Q_1 and Q_2 are both positive or both negative, then the force is positive; it is repulsive. If Q_1 and Q_2 are of opposite charges, then the force is negative; it is attractive. k equals a constant with a value of 10^9.

3.1.0 Current Flow

The movement of the flow of electrons is called *current*. To produce current, the electrons are moved by a potential difference. Current is represented by the letter *I*. The basic unit in which current is measured is the ampere, also called the *amp*. The symbol for the ampere is *A*. One ampere of current is defined as the movement of one coulomb past any point of a conductor during one second of time. One coulomb is equal to 6.25×10^{18} electrons; therefore, one ampere is equal to 6.25×10^{18} electrons moving past any point of a conductor during one second of time.

The definition of current can be expressed as an equation:

$$I = \frac{Q}{T}$$

Where:

I = current (amperes)
Q = charge (coulombs)
T = time (seconds)

Charge differs from current in that Q is an accumulation of charge, while I measures the intensity of moving charges.

In a conductor, such as copper wire, the free electrons are charges that can be forced to move with relative ease by a potential difference. If a potential difference is connected across two ends of a copper wire, as shown in *Figure 4*, the applied

Discuss Coulomb's law and current flow.

Show Transparency 6 (Figure 4).

The Magic of Electricity

The flow of electrons occurs at close to the speed of light, about 186,000 miles per second. How long does it take the light from the end of a flashlight to reach the floor? If you ran a light circuit from Maine to California and flipped the switch, how long would it take for the light to come on?

Current Flow

Why do you need two wires to use electrical devices? Why can't current simply move to a lamp and be released as light energy?

ELECTRON FLOW

COPPER WIRE CONDUCTOR   FREE ELECTRONS IN MOTION

$-Q_1$ $+Q_2$

$-$ $+$

BATTERY CELL

104F04.EPS

Figure 4 ◆ Potential difference causing electric current.

voltage forces the free electrons to move. This current is a flow of electrons from the point of negative charge ($-$) at one end of the wire, moving through the wire to the positive charge ($+$), at the other end. The direction of the electron flow is

from the negative side of the battery, through the wire, and back to the positive side of the battery. The direction of current flow is therefore from a point of negative potential to a point of positive potential.

3.2.0 Voltage

The force that causes electrons to move is called *voltage, potential difference,* or *electromotive force (emf).* One **volt (V)** is the potential difference between two points for which one coulomb of electricity will do one **joule (J)** of work. A **battery** is one of several means of creating voltage. It chemically creates a large reserve of free electrons at the negative ($-$) terminal. The positive ($+$) terminal has electrons chemically removed and will therefore accept them if an external path is provided from the negative ($-$) terminal. When a battery is no longer able to chemically deposit electrons at the negative ($-$) terminal, it is said to be dead, or in need of recharging. Batteries are normally rated in volts. Large batteries are also rated in ampere-hours, where one ampere-hour is a current of one amp supplied for one hour.

Voltage

Why is voltage called *electrical potential?* Why are we interested in potential charges—why not simply measure the actual flow of current? Can there be very high potential and no current, or a very small current? What happens when lightning strikes in terms of electrical potential?

Instructor's Notes:

_____

_____

_____

_____

_____

_____

_____

_____

_____

Joule's Law

While other scientists of the 19th century were experimenting with batteries, cells, and circuits, James Joule was theorizing about the relationship between heat and energy. He discovered, contrary to popular belief, that work did not just move heat from one place to another; work, in fact, generated heat. Furthermore, he demonstrated that over time, a relationship existed between the temperature of water and electric current. These ideas formed the basis for the concept of energy. In his honor, the modern unit of energy was named the *joule*.

The Visual Language of Electricity

Learning to read circuit diagrams is like learning to read a book—first you learn to read the letters, then you learn to read the words, and before you know it, you are reading without paying attention to the individual letters anymore. Circuits are the same way—you will struggle at first with the individual pieces, but before you know it, you will be reading a circuit without even thinking about it. Studying the table below will help you to understand the fundamental language of electricity.

What's Measured	Unit of Measurement	Symbol	Ohm's Law Symbol
Amount of current	Amp	A	I
Electrical power	Watts	W	P
Force of current	Volt	V	E
Resistance to current	Ohm	Ω	R

4.0.0 ◆ RESISTANCE

4.1.0 Characteristics of Resistance

Resistance is directly related to the ability of a material to conduct electricity. Conductors have very low resistance; insulators have very high resistance.

Resistance can be defined as the opposition to current flow. To add resistance to a circuit, electrical components called resistors are used. A resistor is a device whose resistance to current flow is a known, specified value. Resistance is measured in ohms and is represented by the symbol R in equations. One ohm is defined as that amount of resistance that will limit the current in a conductor to one ampere when the voltage applied to the conductor is one volt. The symbol for an ohm is Ω.

The resistance of a wire is proportional to the length of the wire, inversely proportional to the cross-sectional area of the wire, and dependent upon the kind of material of which the wire is made. The relationship for finding the resistance of a wire is:

$$R = \rho \frac{L}{A}$$

Where:

R = resistance (ohms)

L = length of wire (feet)

A = area of wire (circular mils, CM, or cm^2)

ρ = specific resistance (ohm-CM/ft. or microhm-CM)

A *mil* equals 0.001 inch; a circular mil is the cross-sectional area of a wire one mil in diameter.

The specific resistance is a constant that depends on the material of which the wire is made. *Table 1* shows the properties of various wire conductors.

Table 1 shows that at 75°F, a one-mil diameter, pure annealed copper wire that is one foot long has a resistance of 10.351 ohms; while a one-mil diameter, one-foot-long aluminum wire has a resistance of 16.758 ohms. Temperature is important in determining the resistance of a wire. The hotter a wire, the greater its resistance.

Demonstrate how to calculate resistance.

Have the trainees practice resistance calculations.

4.2.0 Ohm's Law

Ohm's Law defines the relationship between current, voltage, and resistance. There are three ways to express Ohm's Law mathematically.

- The current in a circuit is equal to the voltage applied to the circuit divided by the resistance of the circuit:

$$I = \frac{E}{R}$$

- The resistance of a circuit is equal to the voltage applied to the circuit divided by the current in the circuit:

$$R = \frac{E}{I}$$

Table 1 Conductor Properties

Metal	Specific Resistance (Resistance of 1 CM/ft. in ohms)	
	32°F or 0°C or	75°F or 23.8°C
Silver, pure annealed	8.831	9.674
Copper, pure annealed	9.39	10.351
Copper, annealed	9.59	10.505
Copper, hard-drawn	9.81	10.745
Gold	13.216	14.404
Aluminum	15.219	16.758
Zinc	34.595	37.957
Iron	54.529	62.643

- The applied voltage to a circuit is equal to the product of the current and the resistance of the circuit:

$$E = I \times R = IR$$

Where:

$$I = \text{current (amperes)}$$
$$R = \text{resistance (ohms)}$$
$$E = \text{voltage or emf (volts)}$$

If any two of the quantities E, I, or R are known, the third can be calculated.

The Ohm's Law equations can be memorized and practiced effectively by using an Ohm's Law circle, as shown in *Figure 5*. To find the equation for E, I, or R when two quantities are known, cover the unknown third quantity. The other two quantities in the circle will indicate how the covered quantity may be found.

Example 1:
Find I when E = 120V and R = 30Ω.

$$I = \frac{E}{R}$$
$$I = \frac{120V}{30Ω}$$
$$I = 4A$$

This formula shows that in a DC circuit, current (I) is directly proportional to voltage (E) and inversely proportional to resistance (R).

Example 2:
Find R when E = 240V and I = 20A.

Voltage Matters

Standard household voltage is different the world over, from 100V in Japan to 600V in Bombay, India. Many countries have no standard voltage; for example, France varies from 110V to 360V. If you were to plug a 120V hair dryer into England's 240V, you would burn out the dryer. Use basic electric theory to explain exactly what would happen to destroy the hair dryer.

Using Ohm's Law

Study the resistance values in *Table 1*. Assuming the same length and area of wire, approximately how much will the current decrease from copper annealed wire to aluminum wire to iron wire?

Instructor's Notes:

	LETTER SYMBOL	UNIT OF MEASUREMENT
CURRENT	I	AMPERES (A)
RESISTANCE	R	OHMS (Ω)
VOLTAGE	E	VOLTS (V)

104F05.EPS

Figure 5 ◆ Ohm's Law circle.

$$R = \frac{E}{I}$$

$$R = \frac{240V}{20A}$$

$$R = 12\Omega$$

Example 3:
Find E when I = 15A and R = 8Ω.

$$E = I \times R$$
$$E = 15A \times 8\Omega$$
$$E = 120V$$

5.0.0 ◆ SCHEMATIC REPRESENTATION OF CIRCUIT ELEMENTS

A simple electric circuit is shown in both pictorial and schematic forms in *Figure 6*. The schematic diagram is a shorthand way to draw an electric circuit, and circuits are usually represented in this way. In addition to the connecting wire, three components are shown symbolically: the battery, the switch, and the lamp. Note the positive (+) and negative (−) markings in both the pictorial

and schematic representations of the battery. The schematic components represent the pictorial components in a simplified manner. A schematic diagram is one that shows, by means of graphic symbols, the electrical connections and functions of the different parts of a circuit.

The standard graphic symbols for commonly used electrical and electronic components are shown in *Figure 7*.

6.0.0 ◆ RESISTORS

The function of a resistor is to offer a particular resistance to current flow. For a given current and known resistance, the change in voltage across the component, or voltage drop, can be predicted using Ohm's Law. Voltage drop refers to a specific amount of voltage used, or developed, by that component. An example is a very basic circuit of a 10V battery and a single resistor in a series circuit. The voltage drop across that resistor is 10V because it is the only component in the circuit and all voltage must be dropped across that resistor. Similarly, for a given applied voltage, the current that flows may be predetermined by selection of the

PICTORIAL DIAGRAM   **SCHEMATIC DIAGRAM**

104F06.EPS

Figure 6 ◆ Simple electrical symbols.

Ammeter	—Ⓐ—	Motor (DC)	—Ⓑ—
Battery	—+⊦⊦⊦-—	Resistor (fixed)	—ⱽⱽⱽ—
Capacitor (fixed)	—⊦⊦—	Resistor (variable)	—ⱽⱽⱽ—
Capacitor (variable)	—⊬—	Rheostat	—ⱽⱽⱽ—
Circuit breaker	—⌒—	Switch	—∘⌿∘—
Crystal	—⊏▯⊐—	Semiconductor diode	—+▷⊢-
Fuse	—⌒—	Transformer (general)	⅗ ⅗
Generator (AC)	—Ⓐ/—	Transformer (iron-core)	⅗⏐⅗
Generator (DC)	Ⓖ	Transistor (NPN)	B⊂Ⓔ
Ground	⏚ or ⌿	Transistor (PNP)	B⊂Ⓔ
Inductor (air-core)	⌒⌒⌒	Voltmeter	—Ⓥ—
Inductor (iron-core)	⌒⌒⌒	Wattmeter	—Ⓦ—
Inductor (tapped)	●⌒●⌒●	Wires (connected)	—•—
Lamp	⊓	Wires (unconnected)	—⊦—
Motor (AC)	⊘	Zener diode	—▷⌿—

104F07.EPS

Figure 7 ◆ Standard schematic symbols.

Discuss the "Think About It." Draw a simple schematic on the whiteboard/chalkboard.

Show Transparency 10 (Figure 8A).

Drawing a Schematic

Draw a schematic diagram showing a voltage source, switch, motor, and fuse.

resistor value. The required power dissipation largely dictates the construction and physical size of a resistor.

The two most common types of electronic resistors are *wire-wound* and *carbon composition construction*. A typical wire-wound resistor consists of a length of nickel wire wound on a ceramic tube and covered with porcelain. Low-resistance connecting wires are provided, and the resistance value is usually printed on the side of the component. *Figure 8* illustrates the construction of typical

resistors. Carbon composition resistors are constructed by molding mixtures of powdered carbon and insulating materials into a cylindrical shape. An outer sheath of insulating material affords mechanical and electrical protection, and copper connecting wires are provided at each end. Carbon composition resistors are smaller and less expensive than the wire-wound type. However, the wire-wound type is the more rugged of the two and is able to survive much larger power dissipations than the carbon composition type.

Instructor's Notes:

_____

_____

_____

_____

_____

_____

_____

_____

_____

Using Your Intuition

Learning the meanings of various electrical symbols may seem overwhelming, but if you take a moment to study *Figure 7*, you will see that most of them are intuitive—that is, they are shaped (in a symbolic way) to represent the actual object. For example, the battery shows + and −, just like an actual battery. The motor has two arms that suggest a spinning rotor. The transformer shows two coils. The resistor has a jagged edge to suggest pulling or resistance. Connected wires have a black dot that reminds you of solder. Unconnected wires simply cross. The fuse stretches out in both directions as though to provide extra slack in the line. The circuit breaker shows a line with a break in it. The capacitor shows a gap. The variable resistor has an arrow like a swinging compass needle. As you learn to read schematics, take the time to make mental connections between the symbol and the object it represents.

Teaching Tip

Use *Figure 7* to create schematic flashcards. Divide the class into teams and test them on symbol recognition.

104F08A.EPS

Classroom

Pass around resistors for the trainees to examine.

VERONICA WESTFALL

104F08B.EPS

Figure 8 ◆ Common resistors.

Show Transparency 11 (Figure 9).

Discuss resistor color codes.

Show Transparencies 12 and 13 (Figures 10 and 11).

Most resistors have standard fixed values, so they can be termed *fixed resistors*. Variable resistors, also known as *adjustable resistors*, are used a great deal in electronics. Two common symbols for a variable resistor are shown in *Figure 9*.

A variable resistor consists of a coil of closely wound insulated resistance wire formed into a partial circle. The coil has a low-resistance terminal at each end, and a third terminal is connected to a movable contact with a shaft adjustment facility. The movable contact can be set to any point on a connecting track that extends over one (uninsulated) edge of the coil.

Using the adjustable contact, the resistance from either end terminal to the center terminal may be adjusted from zero to the maximum coil resistance.

Another type of variable resistor is known as a *decade resistance box*. This is a laboratory compo-

Figure 9 ◆ Symbols used for variable resistors.

nent that contains precise values of switched series-connected resistors.

6.1.0 Resistor Color Codes

Because carbon composition resistors are physically small (some are less than 1 cm in length), it is not convenient to print the resistance value on the side. Instead, a color code in the form of colored bands is employed to identify the resistance value and tolerance. The color code is illustrated in *Figure 10*. Starting from one end of the resistor, the first two bands identify the first and second digits of the resistance value, and the third band indicates the number of zeros. An exception to this is when the third band is either silver or gold, which indicates a 0.01 or 0.1 multiplier, respectively. The fourth band is always either silver or gold, and in this position, silver indicates a ± 10% tolerance and gold indicates a ± 5% tolerance. Where no fourth band is present, the resistor tolerance is ± 20%.

We can put this information to practical use by determining the range of values for the carbon resistor in *Figure 11*.

The color code for this resistor is as follows:

- Brown = 1, black = 0, red = 2, gold = a tolerance of ± 5%
- First digit = 1, second digit = 0, number of zeros (2) = 1,000Ω

0	BLACK	7	VIOLET
1	BROWN	8	GREY
2	RED	9	WHITE
3	ORANGE	0.1	GOLD
4	YELLOW	0.01	SILVER
5	GREEN	5%	GOLD – TOLERANCE
6	BLUE	10%	SILVER – TOLERANCE

104F10.EPS

Figure 10 ◆ Resistor color codes.

ELECTRICAL LEVEL ONE—TRAINEE MODULE 26104

Instructor's Notes:

_____

_____

_____

_____

_____

_____

_____

_____

_____

Figure 11 ◆ Sample color codes on a fixed resistor.

Since this resistor has a value of 1,000Ω ± 5%, the resistor can range in value from 950Ω to 1,050Ω.

7.0.0 ◆ MEASURING VOLTAGE, CURRENT, AND RESISTANCE

Working with electricity requires making accurate measurements. This section will discuss the basic meters used to measure voltage, current, and resistance: the voltmeter, ammeter, and ohmmeter.

WARNING!

Only qualified individuals may use these meters. Consult your company's safety policy for applicable rules.

7.1.0 Basic Meter Operation

When troubleshooting or testing equipment, you will need various meters to check for proper circuit voltages, currents, and resistances and to determine if the wiring is defective. Meters are used in repairing, maintaining, and troubleshooting electrical circuits and equipment. The best and most expensive measuring instrument is of no use to you unless you know what you are measuring and what each reading indicates. Remember that the purpose of a meter is to measure quantities existing within a circuit. For this reason, when the meter is connected to a circuit, it must not change the condition of the circuit.

The three basic electrical quantities discussed in this section are current, voltage, and resistance. Actually, it is really current that causes the meter to respond even when voltage or resistance is being measured. In a basic meter, the measurement of current can be calibrated to indicate almost any electrical quantity based on the principle of Ohm's law. The amount of current that flows through a meter is determined by the voltage ap-

plied to the meter and the resistance of the meter, as stated by I = E/R.

For a given meter resistance, different values of applied voltage will cause specific values of current to flow. Although the meter actually measures current, the meter scale can be calibrated in units of voltage. Similarly, for a given applied voltage, different values of resistance will cause specific values of current to flow; therefore, the meter scale can also be calibrated in units of resistance rather than current. The same holds true for power, since power is proportional to current, as stated by P = EI. It is on this principle that the meter was developed and its construction allows for the measurement of various parameters by actually measuring current.

You must understand the purpose and function of each individual piece of test equipment and any limitations associated with it. It is also extremely important that you understand how to safely use each piece of equipment. If you understand the capabilities of the test equipment, you can better use the equipment, better understand the indications on the equipment, and know what substitute or backup meters can be used.

7.2.0 Voltmeter

A simple voltmeter consists of the meter movement in series with the internal resistance of the voltmeter itself. For example, a meter with a 50-microamp (μA) meter movement and a 1,000Ω internal resistance can be used to directly measure voltages up to 0.05V, as shown in *Figure 12*. (The prefix micro means one-millionth.) When the meter is placed across the voltage source, a current determined by the internal resistance of the meter flows through the meter movement. A voltmeter's internal resistance is typically high to minimize meter loading effects on the source.

To measure larger voltages, a multiplier resistor is used. This increased series resistance limits the current that can flow through the meter movement, thus extending the range of the meter.

To avoid damage to the meter movement, the following precautions should be observed when using a voltmeter:

- Always set the full-scale voltage of the meter to be larger than the expected voltage to be measured.
- Always ensure that the internal resistance of the voltmeter is much greater than the resistance of the component to be measured. This means that the current it takes to drive the voltmeter (about 50 microamps) should be a negligible fraction of

Introduce basic meter operation.

Review the operation of a basic voltmeter.

Show Transparency 14 (Figure 12).

$$I_M = \frac{0.025V}{1000\Omega} = 25\mu A$$

104F12.EPS

Figure 12 ◆ Simple voltmeter.

the current flowing through the circuit element being measured.

- If you are unsure of the level of the voltage to be measured, take a reading at the highest range of the voltmeter and progressively (step-by-step) lower the range until the reading is obtained.

In most commercial voltmeters, the internal resistance is expressed by the ohms-per-volt rating of the meter. A typical meter has a rating of 20,000 ohms-per-volt with a 50-microamp movement. This quantity tells what the internal resistance of the meter is on any particular full-scale setting. In general, the meter's internal resistance is the ohms-per-volt rating multiplied by the full-scale voltage. The higher the ohms-per-volt rating, the higher the internal resistance of the meter, and the smaller the effect of the meter on the circuit.

7.3.0 Ammeter

A current meter, usually called an *ammeter,* is used by placing the meter in series with the wire through which the current is flowing. This method of connection is shown in *Figure 13.* Notice how the magnitude of load current will flow through the ammeter. Because of this, an ammeter's internal resistance must be low to minimize the circuit-loading effects as seen by the source. Also, high current magnitudes flowing through an ammeter can damage it. For this reason, ammeter shunts are employed to reduce the ammeter circuit current to a fraction of the current flowing through the load.

To avoid damage to the meter movement, the following precautions should be observed when taking current measurements with an ammeter:

- Always check the polarity of the ammeter. Make certain that the meter is connected to the circuit so that electrons flow into the negative

104F13.EPS

Figure 13 ◆ Ammeter connection.

lead and out of the positive lead. It is easy to tell which is the positive lead because it is normally red. The negative lead is usually black.

- Always set the full-scale deflection of the meter to be larger than the expected current. To be safe, set the full-scale current several times larger than the expected current, and then slowly increase the meter sensitivity to the appropriate scale.
- Always connect the ammeter in series with the circuit element through which the current to be measured is flowing. Never connect the ammeter in parallel. When an ammeter is connected across a constant-potential source of appreciable voltage, the low internal resistance of the meter bypasses the circuit resistance. This results in the application of the source voltage directly to the meter terminals. The resulting excess current will burn out the meter coil.

7.4.0 Ohmmeter

An ohmmeter is used to measure resistance and check continuity. The deflection of the pointer of an ohmmeter is controlled by the amount of battery current passing through the coil. Current flow

ELECTRICAL LEVEL ONE—TRAINEE MODULE 26104

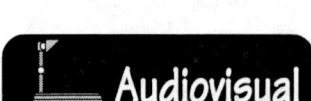

Review the operation of a basic ammeter.

Show Transparency 15 (Figure 13).

Review the operation of a basic ohmmeter.

Instructor's Notes:

ELECTRICAL LEVEL ONE—INSTRUCTOR'S GUIDE MODULE 26104

depends on the applied voltage and the circuit resistance. By applying a constant source voltage to the circuit under test, the resultant current flow depends only on circuit resistance. This magnitude of current will create meter movement. By knowing the relationship between current and resistance, an ohmmeter's scale can be calibrated to indicate circuit resistance based on the magnitude of current for a constant source voltage. Refer to *Figure 14*, a simple ohmmeter circuit.

8.0.0 ◆ ELECTRICAL POWER

Power is defined as the rate of doing work. This is equivalent to the rate at which energy is used or dissipated. Electrons passing through a resistance dissipate energy in the form of heat. In electrical circuits, power is measured in units called watts (W). The power in watts equals the rate of energy conversion. One watt of power equals the work done in one second by one volt of potential difference in moving one coulomb of charge. One coulomb per second is an ampere; therefore, power in watts equals the product of amperes times volts.

The work done in an electrical circuit can be useful work or it can be wasted work. In both cases, the rate at which the work is done is still measured in power. The turning of an electric motor is useful work. On the other hand, the heating

Figure 14 ◆ Simple ohmmeter circuit.

Using an Ohmmeter

An ohmmeter has its own battery to test the resistance or continuity of a circuit. Therefore, the circuit must be deenergized because the ohmmeter is calibrated for its own power source.

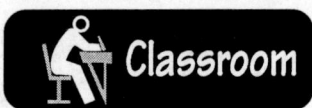

Discuss the "Think About It." The hair dryer (heating element) and light bulb (filament) are resistors.

Resistors

Which of the following items are resistors?

- Hair dryer
- Incandescent light bulb
- Switch
- Receptacle
- Circuit breaker

Discuss the "Think About It." Encourage the widest possible range of answers, from cars to calculators.

Power

We take electrical power for granted, never stopping to think how surprising it is that a flow of submicroscopic electrons can pump thousands of gallons of water or illuminate a skyscraper. Our lives now constantly rely on the ability of the electron to do work. Think about your day up to this moment. How has electrical power shaped your experience?

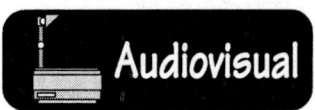

Show Transparency 17 (Figure 15).

of wires or resistors in a circuit is wasted work, since no useful function is performed by the heat.

The unit of electrical work is the joule. This is the amount of work done by one coulomb flowing through a potential difference of one volt. Thus, if five coulombs flow through a potential difference of one volt, five joules of work are done. The time it takes these coulombs to flow through the potential difference has no bearing on the amount of work done.

It is more convenient when working with circuits to think of amperes of current rather than coulombs. As previously discussed, one ampere equals one coulomb passing a point in one second. Using amperes, one joule of work is done in one second when one ampere moves through a potential difference of one volt. This rate of one joule of work in one second is the basic unit of power, and is called a *watt*. Therefore, a watt is the power used when one ampere of current flows

through a potential difference of one volt, as shown in *Figure 15*.

Mechanical power is usually measured in units of horsepower (hp). To convert from horsepower to watts, multiply the number of horsepower by 746. To convert from watts to horsepower, divide the number of watts by 746. Conversions for common units of power are given in *Table 2*.

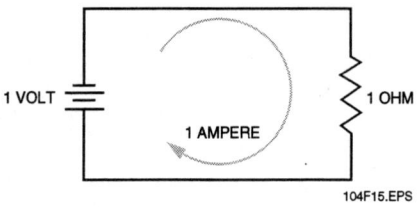

104F15.EPS

Figure 15 ◆ One watt.

Measuring Watts

Electricians are less interested in measuring watts than in measuring amperage and resistance. Who would be most interested in power measurements? What is a common example of a device used to measure watts?

Discuss the "Think About It." Utility companies are concerned with the power measurements made by electric meters.

ELECTRICAL LEVEL ONE—TRAINEE MODULE 26104

Instructor's Notes:

_____

_____

_____

_____

_____

_____

_____

_____

_____

ELECTRICAL LEVEL ONE—INSTRUCTOR'S GUIDE MODULE 26104

Table 2 Conversion Table

1,000 watts (W)	= 1 kilowatt (kW)
1,000,000 watts (W)	= 1 megawatt (MW)
1,000 kilowatts (kW)	= 1 megawatt (MW)
1 watt (W)	= 0.00134 horsepower (hp)
1 horsepower (hp)	= 746 watts (W)

The kilowatt-hour (kWh) is commonly used for large amounts of electrical work or energy. (The prefix *kilo* means one thousand.) The amount is calculated simply as the product of the power in kilowatts multiplied by the time in hours during which the power is used. If a light bulb uses 300W or 0.3kW for 4 hours, the amount of energy is 0.3 × 4, which equals 1.2kWh.

Very large amounts of electrical work or energy are measured in megawatts (MW). (The prefix *mega* means one million.)

Electricity usage is figured in kilowatt-hours of energy. The power line voltage is fairly constant at 120V. Suppose the total load current in the main line equals 20A. Then the power in watts from the 120V line is:

$$P = 120V \times 20A$$

$$P = 2,400W \text{ or } 2.4kW$$

If this power is used for five hours, then the energy of work supplied equals:

$$2.4 \times 5 = 12kWh$$

8.1.0 Power Equation

When one ampere flows through a difference of two volts, two watts must be used. In other words, the number of watts used is equal to the number of amperes of current times the potential difference. This is expressed in equation form as:

$$P = I \times E \text{ or } P = IE$$

Where:

P = power used in watts

I = current in amperes

E = potential difference in volts

The equation is sometimes called *Ohm's Law for power*, because it is similar to *Ohm's Law*. This equation is used to find the power consumed in a circuit or load when the values of current and voltage are known. The second form of the equa-

tion is used to find the voltage when the power and current are known:

$$E = \frac{P}{I}$$

The third form of the equation is used to find the current when the power and voltage are known:

$$I = \frac{P}{E}$$

Using these three equations, the power, voltage, or current in a circuit can be calculated whenever any two of the values are already known.

Example 1:
Calculate the power in a circuit where the source of 100V produces 2A in a 50Ω resistance.

$$P = IE$$

$$P = 2 \times 100$$

$$P = 200W$$

This means the source generates 200W of power while the resistance dissipates 200W in the form of heat.

Example 2:
Calculate the source voltage in a circuit that consumes 1,200W at a current of 5A.

$$E = \frac{P}{I}$$

$$E = \frac{1,200}{5}$$

$$E = 240V$$

Example 3:
Calculate the current in a circuit that consumes 600W with a source voltage of 120V.

$$I = \frac{P}{E}$$

$$I = \frac{600}{120}$$

$$I = 5A$$

Components that use the power dissipated in their resistance are generally rated in terms of power. The power is rated at normal operating voltage, which is usually 120V. For instance, an appliance that draws 5A at 120V would dissipate 600W. The rating for the appliance would then be 600W/120V.

Demonstrate how to use the power equation.

Have the trainees practice using the power equation.

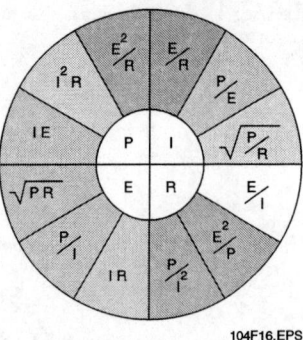

104F16.EPS

Figure 16 ◆ Expanded Ohm's Law circle.

To calculate I or R for components rated in terms of power at a specified voltage, it may be convenient to use the power formula in different forms. There are actually three basic power formulas, but each can be rearranged into three other forms for a total of nine combinations:

$P = IE$	$P = I^2R$	$P = \dfrac{E^2}{R}$
$I = \dfrac{P}{E}$	$R = \dfrac{P}{I^2}$	$R = \dfrac{E^2}{P}$
$E = \dfrac{P}{I}$	$I = \sqrt{\dfrac{P}{R}}$	$E = \sqrt{PR}$

Note that all of these formulas are based on Ohm's Law ($E = IR$) and the power formula ($P = I \times E$). *Figure 16* shows all of the applicable power, voltage, resistance, and current equations.

8.2.0 Power Rating of Resistors

If too much current flows through a resistor, the heat caused by the current will damage or destroy the resistor. This heat is caused by I^2R heating, which is power loss expressed in watts. Therefore, every resistor is given a wattage, or power rating, to show how much I^2R heating it can take before it burns out. This means that a resistor with a power rating of one watt will burn out if it is used in a circuit where the current causes it to dissipate heat at a rate greater than one watt.

If the power rating of a resistor is known, the maximum current it can carry is found by using an equation derived from $P = I^2R$:

$$P = I^2R \text{ becomes } I^2 = P/R,$$
$$\text{which becomes } I = \sqrt{P/R}$$

Using this equation, find the maximum current that can be carried by a 1Ω resistor with a power rating of 4W:

$$I = \sqrt{P/R} = \sqrt{4/1} = 2 \text{ amperes}$$

If such a resistor conducts more than 2 amperes, it will dissipate more than its rated power and burn out.

Power ratings assigned by resistor manufacturers are usually based on the resistors being mounted in an open location where there is free air circulation, and where the temperature is not higher than 104°F (40°C). Therefore, if a resistor is mounted in a small, crowded, enclosed space, or where the temperature is higher than 104°F, there is a good chance it will burn out even before its power rating is exceeded. Also, some resistors are designed to be attached to a chassis or frame that will carry away the heat.

Summary

The relationships among current, voltage, resistance, and power are consistent for all types of DC circuits and can be calculated using Ohm's law and Ohm's law for power. Understanding and being able to apply these concepts is necessary for effective circuit analysis and troubleshooting.

Putting It All Together

Notice the common electrical devices in the building you're in. What is their wattage rating? How much current do they draw? How would you test their voltage or amperage?

Explain that the power formula can be used to derive power, voltage, current, or resistance.

Show Transparency 18 (Figure 16).

Discuss resistor power ratings.

Discuss the "Think About It." Answers will vary depending upon the building devices.

Instructor's Notes:

Review Questions

1. A negative charge that is a type of subatomic particle is a(n) _____.
   a. proton
   b. neutron
   c. electron
   d. atom

2. A proton has a _____ charge.
   a. negative
   b. positive
   c. neutral
   d. Both b and c.

3. Like charges _____ each other.
   a. attract
   b. repel
   c. have no effect on
   d. complement

4. An electron has _____ coulombs of charge.
   a. 1.6×10^{-9}
   b. 1.6×10^{9}
   c. 1.6×10^{19}
   d. 1.6×10^{-19}

5. The quantity that Ohm's Law does not express a relationship for in an electrical circuit is _____.
   a. charge
   b. resistance
   c. voltage
   d. current

6. The color band that represents tolerance on a resistor is the _____.
   a. 4th band
   b. 3rd band
   c. 2nd band
   d. 1st band

7. A resistor with a color code of red/red/orange indicates a value of _____.
   a. 22,000 ohms
   b. 66 ohms
   c. 223 ohms
   d. 220 ohms

8. An ammeter is placed in _____ with the circuit being tested.
   a. parallel
   b. series
   c. compound
   d. None of the above.

9. The basic unit of power is the _____ .
   a. volt
   b. ampere
   c. coulomb
   d. watt

10. The power in a circuit with 120 volts and 5 amps is _____.
    a. 24 watts
    b. 600 watts
    c. 6,000 watts
    d. ¼ watt

Classroom

Have the trainees complete the Review Questions and go over the answers prior to administering the Module Examination.

Examination

Administer the Module Examination and Performance Test. Be sure to record the testing results on Craft Training Report Form 200 and submit the results to the Training Program Sponsor.

Charles Michael (Mike) Holt, Mike Holt Enterprises

Mike Holt's story is one of a high school dropout who became a nationally respected and well-educated master electrician, electrical contractor, author, speaker, and educator. He is recognized as an expert on the NEC and is sought after by major corporations and government agencies as a consultant and lecturer. Along the way, Mike fathered seven children, was twice crowned National Barefoot Waterskiing Champion, and set five national waterskiing records.

How did you choose a career in the electrical field?
I was always fascinated by electricity. I read books about Thomas Edison when I was young and I could just envision myself doing electrical experiments. I tried getting into the high school electricity program, but the class was full. I had to leave high school in the eleventh grade, and after a couple of years I got a job as an electrician's helper.

Tell us about your apprenticeship experience.
I wasn't able to get into the electrician's union because none of my family belonged, so I took a job as a helper in a non-union shop. There was no formal training program, and I really envied the electricians who had gone to apprenticeship school. I didn't let it stop me, though. I bought every book I could afford and studied on my own, and eventually was able to get my journeyman's license.

What positions have you held in the industry?
Since I became a journeyman, I've been a foreman, project manager, estimator, electrical inspector, and instructor. I had my own electrical contracting business for five years, but gave that up to devote full time to developing training materials. Now I spend my time developing new training programs, writing books and articles, consulting, and conducting seminars.

What factor or factors contributed most to your success?
I would have to say my personal commitment and sense of responsibility. I wanted to be an electrician and was proud when I got my journeyman's license through self-study. I did not have a lot of people encouraging me early in my career, so I had to do it on my own. Some of my coworkers even put me down because I studied for my journeyman's license before and after work and during breaks, but I didn't let that stop me.

A short time after I left high school, I came to realize that success depends on education. I got my GED certificate and continued my education, eventually enrolling in the University of Miami's MBA program.

Instructor's Notes:

_____

_____

_____

_____

_____

_____

_____

_____

_____

Trade Terms Introduced in This Module

Ammeter: An instrument for measuring electrical current.

Ampere (A): A unit of electrical current. For example, one volt across one ohm of resistance causes a current flow of one ampere.

Atom: The smallest particle to which an element may be divided and still retain the properties of the element.

Battery: A DC voltage source consisting of two or more cells that convert chemical energy into electrical energy.

Circuit: A complete path for current flow.

Conductor: A material that offers very little resistance to current flow.

Coulomb: An electrical charge equal to 6.25×10^{18} electrons or 6,250,000,000,000,000,000 electrons. A coulomb is the common unit of quantity used for specifying the size of a given charge.

Current: The movement, or flow, of electrons in a circuit. Current (I) is measured in amperes.

Electron: A negatively charged particle that orbits the nucleus of an atom.

Insulator: A material that offers resistance to current flow.

Joule (J): A unit of measurement that represents one newton-meter (Nm), which is a unit of measure for doing work.

Kilo: A prefix used to indicate one thousand; for example, one kilowatt is equal to one thousand watts.

Matter: Any substance that has mass and occupies space.

Mega: A prefix used to indicate one million; for example, one megawatt is equal to one million watts.

Micro: A prefix used to indicate one-millionth; for example, one microwatt is equal to one-millionth of a watt.

Neutrons: Electrically neutral particles (neither positive nor negative) that have the same mass as a proton and are found in the nucleus of an atom.

Nucleus: The center of an atom. It contains the protons and neutrons of the atom.

Ohm (Ω): The basic unit of measurement for resistance.

Ohmmeter: An instrument used for measuring resistance.

Ohm's Law: A statement of the relationships among current, voltage, and resistance in an electrical circuit: current (I) equals voltage (E) divided by resistance (R). Generally expressed as a mathematical formula: $I = E/R$.

Power: The rate of doing work or the rate at which energy is used or dissipated. Electrical power is the rate of doing electrical work. Electrical power is measured in watts.

Protons: The smallest positively charged particles of an atom. Protons are contained in the nucleus of an atom.

Resistance: An electrical property that opposes the flow of current through a circuit. Resistance (R) is measured in ohms.

Resistor: Any device in a circuit that resists the flow of electrons.

Schematic: A type of drawing in which symbols are used to represent the components in a system.

Series circuit: A circuit with only one path for current flow.

Valence shell: The outermost ring of electrons that orbit about the nucleus of an atom.

Volt (V): The unit of measurement for voltage (electromotive force). One volt is equivalent to the force required to produce a current of one ampere through a resistance of one ohm.

Voltage: The driving force that makes current flow in a circuit. Voltage (E) is also referred to as *electromotive force* or *potential.*

Voltage drop: The change in voltage across a component that is caused by the current flowing through it and the amount of resistance opposing it.

Voltmeter: An instrument for measuring voltage. The resistance of the voltmeter is fixed. When the voltmeter is connected to a circuit, the current passing through the meter will be directly proportional to the voltage at the connection points.

Watt (W): The basic unit of measurement for electrical power.

Instructor's Notes:

_____

_____

_____

_____

_____

_____

_____

_____

Answers to Review Questions

Answer	Section
1. c	Terms/2.1.0
2. b	Terms/2.1.0
3. b	2.1.0
4. d	3.0.0
5. a	4.2.0
6. a	6.1.0
7. a	6.1.0
8. b	7.3.0
9. d	8.0.0
10. b	8.1.0

Additional Resources

For advanced study of topics covered in this task module, the following books are suggested:

Electronics Fundamentals, Latest Edition. New York: Prentice Hall.
Principles of Electric Circuits, Latest Edition. New York: Prentice Hall.

Instructor's Notes:

_____

_____

_____

_____

_____

_____

_____

_____

_____

The NCCER makes every effort to keep these textbooks up-to-date and free of technical errors. We appreciate your help in this process. If you have an idea for improving this textbook, or if you find an error, a typographical mistake, or an inaccuracy in the NCCER's Craft Training textbooks, please write us, using this form or a photocopy. Be sure to include the exact module number, page number, a detailed description, and the correction, if applicable. Your input will be brought to the attention of the Technical Review Committee. Thank you for your assistance.

Instructors – If you found that additional materials were necessary in order to teach this module effectively, please let us know so that we may include them in the Equipment/Materials list in the Instructor's Guide.

Write: Curriculum Revision and Development Department
National Center for Construction Education and Research
P.O. Box 141104, Gainesville, FL 32614-1104

Fax: 352-334-0932

E-mail: curriculum@nccer.org

Craft _____   Module Name _____

Copyright Date _____   Module Number _____   Page Number(s) _____

Description _____

_____

_____

_____

(Optional) Correction _____

_____

_____

(Optional) Your Name and Address _____

_____

_____

TEACHING TIPS

Section 2.2.0 *Think About It—Conductors*

Conductors are materials in which current can be established with relative ease. Conductors have a large number of free electrons that are capable of moving from one atom to another in the material. It is these free electrons that make current flow possible.

Section 3.1.0 *Think About It—The Magic of Electricity*

It takes only a fraction of a second for the light from a flashlight to strike the floor. The distance from Maine to California is approximately 2,500 miles. Since electrons travel at about 186,000 miles per second, it would take about 0.013 second (2,500 ÷ 186,000) for the light to come on in California.

Section 3.1.0 *Think About It—Current Flow*

Two wires are needed to provide a closed path for current flow through an electrical device such as a lamp. These wires, one connected to the input of the device and the other to its output, must be connected to a battery or other source of potential difference in order for current to flow in the circuit. Light produced by an incandescent lamp depends upon the current flow through the lamp filament. The current flow causes the filament element to heat and incandesce, thus converting electrical energy to light.

Section 3.2.0 *Think About It—Voltage*

Voltage is called *electrical potential* because it has the capability or potential for doing work. An electrical potential can exist without current flow. For example, an electrical potential of 12V exists between the negative and positive terminals of a fully charged 12V battery when it is disconnected from a circuit. However, when connected to an external circuit, the battery has the potential difference needed to cause current flow through the circuit. Capacitors connected to electric high-voltage power transmission lines provide an example of a device where high electrical potential exists with no or little current as a result of the electrical energy being stored by the capacitors.

The ionization of atoms within turbulent clouds, or between clouds and the earth, causes cloud centers to become oppositely charged, creating an electrical potential between them. Lightning arcs occur when the buildup of this electrical potential exceeds the natural electrical insulator quality of the surrounding air. The result is a high-current electrical discharge between the charged bodies as they act to equalize their respective charges, thereby dissipating the electrical potential between them.

Section 4.2.0 *Think About It—Voltage Matters*

Doubling the voltage applied to the hair dryer will double the amount of current that flows through the fixed resistance of the hair dryer's heating element. This will most likely cause the heating element to burn out.

Section 4.2.0 *Think About It—Ohm's Law*

As shown in *Table 1*, the resistance of copper annealed wire is somewhat lower than that of aluminum wire, and greatly lower than that of iron wire. The current flow through each type of wire is inversely proportional to the wire's resistance. The higher the resistance, the lower the current flow. Therefore, the copper annealed wire will carry the most current, followed by the aluminum wire, and then the iron wire.

Electrical Theory Two
26105-00

MODULE OVERVIEW

This course introduces the electrical trainee to circuit calculations involving the application of Ohm's and Kirchhoff's laws.

PREREQUISITES

Please refer to the Course Map in the Trainee Module. Prior to training with this module, it is recommended that the trainee shall have successfully completed the following modules:

Core Curriculum; Electrical Level One, Modules 26101 through 26104

LEARNING OBJECTIVES

Upon completion of this module, the trainee will be able to:

1. Explain the basic characteristics of a series circuit.
2. Explain the basic characteristics of a parallel circuit.
3. Explain the basic characteristics of a series-parallel circuit.
4. Calculate, using Kirchhoff's voltage law, the voltage drop in series, parallel, and series-parallel circuits.
5. Calculate, using Kirchhoff's current law, the total current in parallel and series-parallel circuits.
6. Find the total amount of resistance in a series circuit.
7. Find the total amount of resistance in a parallel circuit.
8. Find the total amount of resistance in a series-parallel circuit.

PERFORMANCE OBJECTIVES

Under supervision of the instructor, the trainee should be able to:

1. Calculate the total resistance for selected series, parallel, and series-parallel circuits.
2. Use Kirchhoff's current law to calculate the total and unknown currents in parallel and series-parallel circuits.
3. Use Kirchhoff's voltage law to calculate voltage drops in series, parallel, and series-parallel circuits.

NCCER STANDARDIZED CRAFT TRAINING PROGRAM

The National Center for Construction Education and Research (NCCER) provides a standardized national program of accredited craft training. Key features of the program include instructor certification, competency-based training, and performance testing. The program provides trainees, instructors, and companies with a standard form of recognition through a National Craft Training Registry. The program is described in full in the *Guidelines for Accreditation*, published by the NCCER. For more information on standardized craft training, contact the NCCER at P.O. Box 141104, Gainesville, FL 32614-1104, 352-334-0911, visit our Web site at www.NCCER.org, or e-mail info@NCCER.org.

HOW TO USE THIS ANNOTATED INSTRUCTOR'S GUIDE

Each page presents two sections of information. The larger section displays each page exactly as it appears in the Trainee Module. The narrow column ties suggested trainee and instructor actions to each page and provides icons to call your attention to material, safety, audiovisual, or testing requirements. The bottom of each page includes space for your notes.

If you see the Teaching Tip icon, that means there is a teaching tip associated with this section. Also refer to the suggested teaching tips at the end of the module.

SAFETY CONSIDERATIONS

Ensure that the trainees are equipped with appropriate personal protective equipment.

PREPARATION

Before teaching this module, you should review the Module Outline, Learning and Performance Objectives, and the Materials and Equipment List. Be sure to allow ample time to prepare your own training or lesson plan and gather all required equipment and materials.

MATERIALS AND EQUIPMENT LIST

Materials:

Transparencies

Markers/chalk

Copy of the latest edition of the *National Electrical Code®*

Module Examinations*

Performance Profile Sheets*

Equipment:

Overhead projector and screen

Whiteboard/chalkboard

Appropriate personal protective equipment

*Located in the Test Booklet packaged with this Annotated Instructor's Guide.

ADDITIONAL RESOURCES

This module is intended to present thorough resources for task training. The following reference works are suggested for both instructors and motivated trainees interested in further study. These are optional materials for continued education rather than for task training.

Electronics Fundamentals, Latest Edition. Upper Saddle River, NJ: Prentice Hall.

Principles of Electric Circuits, Latest Edition. Upper Saddle River, NJ: Prentice Hall.

NOTES

The designations "National Electrical Code," "NE Code," and "NEC," where used in this document, refer to the *National Electrical Code®*, which is a registered trademark of the National Fire Protection Association, Quincy, MA. All National Electrical Code (NEC) references in this module refer to the 1999 edition of the NEC.

If you feel that additional math instruction would be helpful, Prentice Hall offers a basic math textbook entitled *Fundamentals of Electrical and Mechanical Mathematics*. It covers the basic math requirements for electrical trainees and may be ordered by contacting Prentice Hall Customer Service at 1-800-922-0579.

TEACHING TIME FOR THIS MODULE

An outline for use in developing your lesson plan is presented below. Note that each Roman numeral in the outline equates to one session of instruction. Each session has a suggested time period of 2½ hours. This includes 10 minutes at the beginning of each session for administrative tasks and one 10-minute break during the session. Approximately 7½ hours are suggested to cover *Electrical Theory Two*. You will need to adjust the time required for hands-on activity and testing based on your class size and resources.

Topic	Planned Time

Session I. Resistive Circuits

A. Resistive Circuits _____

   1. Resistances in Series _____

   2. Resistances in Parallel _____

      a. Simplified Formulas _____

   3. Series-Parallel Circuits _____

      a. Reducing Series-Parallel Circuits _____

Session II. Applying Ohm's Law to Resistive Circuits

A. Voltage and Current in Series Circuits _____

B. Voltage and Current in Parallel Circuits _____

C. Voltage and Current in Series-Parallel Circuits _____

Session III. Kirchhoff's Laws; Module Examination and Performance Testing

A. Kirchhoff's Laws _____

   1. Kirchhoff's Current Law _____

   2. Kirchhoff's Voltage Law _____

   3. Loop Equations _____

B. Module Examination _____

   1. Trainees must score 70% or higher to receive recognition from the NCCER.

   2. Record the testing results on Craft Training Report Form 200 and submit the results to the Training Program Sponsor.

C. Performance Testing _____

   1. Trainees must perform each task to the satisfaction of the instructor to receive recognition from the NCCER.

   2. Record the testing results on Craft Training Report Form 200 and submit the results to the Training Program Sponsor.

Module 26105-00

Electrical Theory Two

Course Map

This course map shows all of the task modules in the first level of the Electrical curricula. The suggested training order begins at the bottom and proceeds up. Skill levels increase as a trainee advances on the course map. The training order may be adjusted by the local Training Program Sponsor.

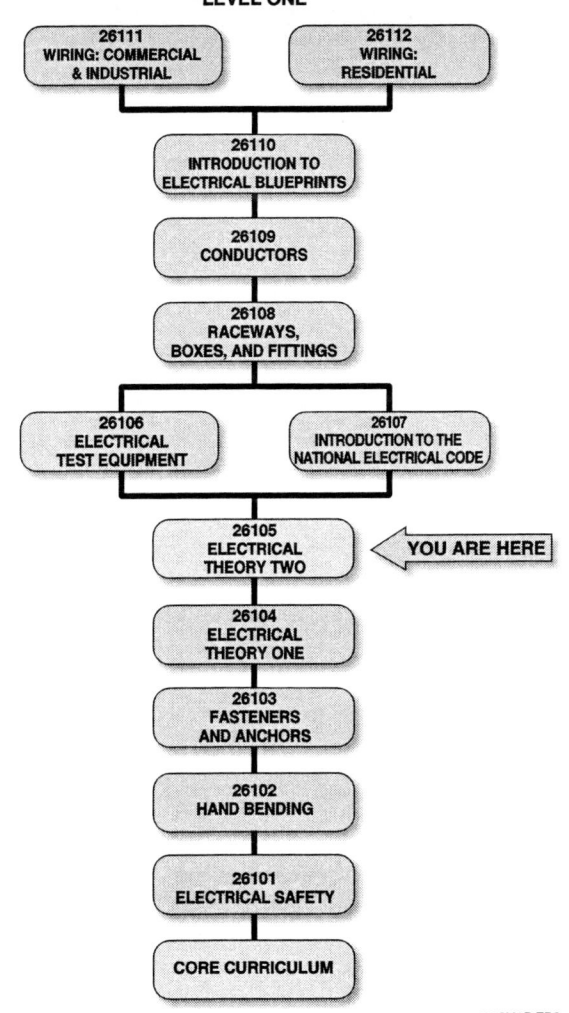

LEVEL ONE

26111
WIRING: COMMERCIAL & INDUSTRIAL

26112
WIRING: RESIDENTIAL

26110
INTRODUCTION TO ELECTRICAL BLUEPRINTS

26109
CONDUCTORS

26108
RACEWAYS, BOXES, AND FITTINGS

26106
ELECTRICAL TEST EQUIPMENT

26107
INTRODUCTION TO THE NATIONAL ELECTRICAL CODE

26105
ELECTRICAL THEORY TWO ◁ YOU ARE HERE

26104
ELECTRICAL THEORY ONE

26103
FASTENERS AND ANCHORS

26102
HAND BENDING

26101
ELECTRICAL SAFETY

CORE CURRICULUM

105CMAP.EPS

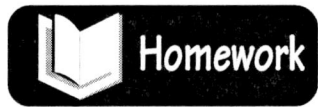

Assign reading of Module 26105.

5—iii

MODULE 26105 CONTENTS

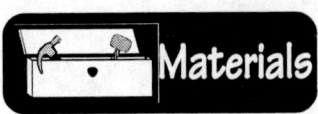

Ensure you have everything required to teach the course. Check the Materials and Equipment List at the front of this Instructor's Guide.

Instructor's Notes:

_____

_____

_____

_____

_____

_____

_____

_____

_____

Electrical Theory Two

Objectives

Upon completion of this module, the trainee will be able to:

1. Explain the basic characteristics of a series circuit.
2. Explain the basic characteristics of a parallel circuit.
3. Explain the basic characteristics of a series-parallel circuit.
4. Calculate, using Kirchhoff's voltage law, the voltage drop in series, parallel, and series-parallel circuits.
5. Calculate, using Kirchhoff's current law, the total current in parallel and series-parallel circuits.
6. Find the total amount of resistance in a series circuit.
7. Find the total amount of resistance in a parallel circuit.
8. Find the total amount of resistance in a series-parallel circuit.

Prerequisites

Successful completion of the following Task Modules is recommended before beginning study of this Task Module: Core Curricula; Electrical Level One, Modules 26101 through 26104.

Required Trainee Materials

1. Trainee Task Module

Note: The designations "National Electrical Code," "NE Code," and "NEC," where used in this document, refer to the *National Electrical Code®*, which is a registered trademark of the National Fire Protection Association, Quincy, MA. *All National Electrical Code (NEC) references in this module refer to the 1999 edition of the NEC.*

2. Copy of the latest edition of the *National Electrical Code*
3. Appropriate Personal Protective Equipment

1.0.0 ◆ INTRODUCTION

Ohm's law was explained in the *Electrical Theory One*. This fundamental concept is now going to be used to analyze more complex series circuits, parallel circuits, and series-parallel circuits. This module will explain how to calculate resistance, current, and voltage in these complex circuits. Ohm's law will be used to develop a new law for voltage and current determination. This law, called *Kirchhoff's law*, will become the new foundation for analyzing circuits.

2.0.0 ◆ RESISTIVE CIRCUITS

2.1.0 Resistances in Series

A series circuit is a circuit in which there is only one path for current flow. Resistance is measured in ohms (Ω). In the series circuit shown in *Figure 1*, the current (I) is the same in all parts of the circuit. This means that the current flowing through R_1 is the same as the current flowing through R_2 and R_3, and it is also the same as the current supplied by the battery.

When resistances are connected in series as in this example, the total resistance in the circuit is

Show Transparency 1, Course Objectives.

Show Transparency 2, Performance Profile Tasks.

Explain that terms shown in bold (blue) are defined in the Glossary at the back of this module.

Discuss basic series circuits.

Show Transparency 3 (Figure 1).

5–1

105F01.EPS

Figure 1 ◆ Series circuit.

equal to the sum of the resistances of all the parts of the circuit:

$$R_T = R_1 + R_2 + R_3$$

Where:

R_T = total resistance
$R_1 + R_2 + R_3$ = resistances in series

Example 1:
The circuit shown in *Figure 2(A)* has 50Ω, 75Ω, and 100Ω resistors in series. Find the total resistance of the circuit.

Add the values of the three resistors in series:

$$R_T = R_1 + R_2 + R_3 = 50 + 75 + 100 = 225Ω$$

Example 2:
The circuit shown in *Figure 2(B)* has three lamps connected in series with the resistances shown. Find the total resistance of the circuit.

Add the values of the three lamp resistances in series:

$$R_T = R_1 + R_2 + R_3 = 20 + 40 + 60 = 120Ω$$

2.2.0 Resistances in Parallel

The total resistance in a parallel resistive circuit is given by the formula:

$$R_T = \frac{1}{\dfrac{1}{R_1} + \dfrac{1}{R_2} + \dfrac{1}{R_3} + \dfrac{1}{R_n}}$$

Where:

R_T = total resistance in parallel
R_1, R_2, R_3, and R_n = branch resistances

Example 1:
Find the total resistance of the 2Ω, 4Ω, and 8Ω resistors in parallel shown in *Figure 3*.

(A) (B)

105F02.EPS

Figure 2 ◆ Total resistance.

Series Circuits

Simple series circuits are seldom encountered in practical wiring. The only simple series circuit you may recognize is older strands of Christmas lights, in which the entire string went dead when one lamp burned out. Think about what the actual wiring of a series circuit would look like in household receptacles. How would the circuit physically be wired? What kind of illumination would you get if you wired your household receptacles in series and plugged half a dozen lamps into those receptacles?

Instructor's Notes:

_____

_____

_____

_____

_____

_____

_____

_____

_____

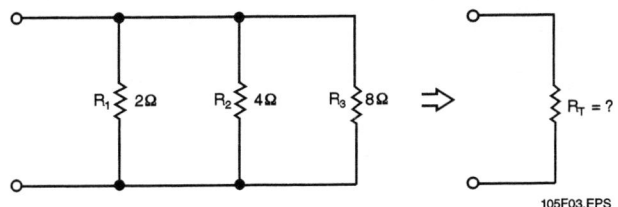

105F03.EPS

Figure 3 ◈ Parallel branch.

Write the formula for the three resistances in parallel:

$$R_T = \frac{1}{\frac{1}{R_1} + \frac{1}{R_2} + \frac{1}{R_3}}$$

Substitute the resistance values:

$$R_T = \frac{1}{\frac{1}{2} + \frac{1}{4} + \frac{1}{8}}$$

$$R_T = \frac{1}{0.5 + 0.25 + 0.125}$$

$$R_T = \frac{1}{0.875}$$

$$R_T = 1.14\Omega$$

Note that when resistances are connected in parallel, the total resistance is always less than the resistance of any single branch.

In this case:

$$R_T = 1.14\Omega < R_1 = 2\Omega, R_2 = 4\Omega, \text{ and } R_3 = 8\Omega$$

Example 2:

Add a fourth parallel resistor of 2Ω to the circuit in *Figure 3*. What is the new total resistance, and what is the net effect of adding another resistance in parallel?

Write the formula for four resistances in parallel:

$$R_T = \frac{1}{\frac{1}{R_1} + \frac{1}{R_2} + \frac{1}{R_3} + \frac{1}{R_4}}$$

Substitute values:

$$R_T = \frac{1}{\frac{1}{2} + \frac{1}{4} + \frac{1}{8} + \frac{1}{2}}$$

$$R_T = \frac{1}{0.5 + 0.25 + 0.125 + 0.5}$$

$$R_T = \frac{1}{1.375}$$

$$R_T = 0.73\Omega$$

The net effect of adding another resistance in parallel is a reduction of the total resistance from 1.14Ω to 0.73Ω.

2.2.1 Simplified Formulas

The total resistance of *equal* resistors in parallel is equal to the resistance of one resistor divided by the number of resistors:

$$R_T = \frac{R}{N}$$

Where:

R_T = total resistance of equal resistors in parallel
R = resistance of one of the equal resistors
N = number of equal resistors

If two resistors with the same resistance are connected in parallel, the equivalent resistance is half of that value, as shown in *Figure 4*.

The two 200Ω resistors in parallel are the equivalent of one 100Ω resistor; the two 100Ω resistors are the equivalent of one 50Ω resistor; and the two 50Ω resistors are the equivalent of one 25Ω resistor.

When any two *unequal* resistors are in parallel, it is often easier to calculate the total resistance by multiplying the two resistances and then dividing the product by the sum of the resistances:

$$R_T = \frac{R_1 \times R_2}{R_1 + R_2}$$

Where:

R_T = total resistance of unequal resistors in parallel
R_1, R_2 = two unequal resistors in parallel

ELECTRICAL THEORY TWO—TRAINEE MODULE 26105

5–3

Teaching Tip

Discuss the "Think About It."

INSIDE TRACK

Parallel Circuits

Most practical circuits are wired in parallel, like the pole lights shown here.

WALTER JOHNSON   105PO501.EPS

THINK ABOUT IT

Parallel Circuits

An interesting fact about circuits is the drop in resistance in a parallel circuit as more resistors are added. But this fact does not mean that you can add an endless number of devices, such as lamps, in a parallel circuit. Why not?

105F04.EPS

Figure 4 ◆ Equal resistances in a parallel circuit.

Instructor's Notes:

_____

_____

_____

_____

_____

_____

_____

_____

_____

Example 1:
Find the total resistance of a 6Ω (R₁) resistor and an 18Ω (R₂) resistor in parallel:

$$R_T = \frac{R_1 \times R_2}{R_1 + R_2} = \frac{6 \times 18}{6 + 18} = \frac{108}{24} = 4.5\Omega$$

Example 2:
Find the total resistance of a 100Ω (R₁) resistor and a 150Ω (R₂) resistor in parallel:

$$R_T = \frac{R_1 \times R_2}{R_1 + R_2} = \frac{100 \times 150}{100 + 150} = \frac{15,000}{250} = 60\Omega$$

2.3.0 Series-Parallel Circuits

To find current, voltage, and resistance in series circuits and parallel circuits is fairly easy. When working with either type, use only the rules that apply to that type. In a series-parallel circuit, some parts of the circuit are series connected and other parts are parallel connected. Thus, in some parts the rules for series circuits apply, and in other parts, the rules for parallel circuits apply. To analyze or solve a problem involving a series-parallel circuit, it is necessary to recognize which parts of the circuit are series connected and which parts are parallel connected. This is obvious if the circuit is simple. Many times, however, the circuit must be redrawn, putting it into a form that is easier to recognize.

In a series circuit, the current is the same at all points. In a parallel circuit, there are one or more points where the current divides and flows in separate branches. In a series-parallel circuit, there are both separate branches and series loads. The easiest way to find out whether a circuit is a series, parallel, or series-parallel circuit is to start at the negative terminal of the power source and trace the path of current through the circuit back to the positive terminal of the power source. If the current does not divide anywhere, it is a series circuit. If the current divides into separate branches, but there are no series loads, it is a parallel circuit. If the current divides into separate branches and there are also series loads, it is a series-parallel circuit. *Figure 5* shows electric lamps connected in series, parallel, and series-parallel circuits.

After determining that a circuit is series-parallel, redraw the circuit so that the branches and the series loads are more easily recognized. This is especially helpful when computing the total resistance of the circuit. *Figure 6* shows resistors connected in a series-parallel circuit and the equivalent circuit redrawn to simplify it.

2.3.1 Reducing Series-Parallel Circuits

Very often, all that is known about a series-parallel circuit is the applied voltage and the values of the individual resistances. To find the voltage drop across any of the loads or the current in any of the branches, the total circuit current must usually be known. But to find the total current, the total resistance of the circuit must be known. To find the total

Discuss series-parallel circuits and explain how they can be redrawn to simplify calculations.

Show Transparencies 7 and 8 (Figures 5 and 6).

SERIES CIRCUIT

PARALLEL CIRCUIT

SERIES-PARALLEL CIRCUIT

CURRENT FLOW

105F05.EPS

Figure 5 ◆ Series, parallel, and series-parallel circuits.

Figure 6 ◆ Redrawing a series-parallel circuit.

Series-Parallel Circuits

Explain *Figure 6*. Which resistors are in series and which are in parallel?

Discuss the "Think About It."

Show Transparency 9 (Figure 7).

Explain how to calculate the total resistance in a series-parallel circuit.

Have the trainees practice calculating the total resistance in a series-parallel circuit.

resistance, reduce the circuit to its simplest form, which is usually one resistance that forms a series circuit with the voltage source. This simple series circuit has the equivalent resistance of the series-parallel circuit it was derived from, and also has the same total current. There are four basic steps in reducing a series-parallel circuit:

- If necessary, redraw the circuit so that all parallel combinations of resistances and series resistances are easily recognized.
- For each parallel combination of resistances, calculate its effective resistance.
- Replace each of the parallel combinations with one resistance whose value is equal to the effec-

tive resistance of that combination. This provides a circuit with all series loads.
- Find the total resistance of this circuit by adding the resistances of all the series loads.

Examine the series-parallel circuit shown in *Figure 7* and reduce it to an equivalent series circuit.

In this circuit, resistors R_2 and R_3 are connected in parallel, but resistor R_1 is in series with both the battery and the parallel combination of R_2 and R_3. The current I_T leaving the negative terminal of the voltage source travels through resistor R_1 before it is divided at the junction of resistors R_1, R_2, and R_3 (Point A) to go through the two branches formed by resistors R_2 and R_3.

Figure 7 ◆ Reducing a series-parallel circuit.

Instructor's Notes:

_____

_____

_____

_____

_____

_____

_____

_____

_____

Given the information in *Figure 7*, calculate the resistance of R_2 and R_3 in parallel and the total resistance of the circuit, R_T.

The total resistance of the circuit is the sum of R_1 and the equivalent resistance of R_2 and R_3 in parallel. To find R_T, first find the resistance of R_2 and R_3 in parallel. Because the two resistances have the same value of 20Ω, the resulting equivalent resistance is 10Ω. Therefore, the total resistance (R_T) is 15Ω ($5\Omega + 10\Omega$).

2.4.0 Applying Ohm's Law

2.4.1 Voltage and Current in Series Circuits

In resistive circuits, unknown circuit parameters can be found by using Ohm's law and the techniques for determining equivalent resistance. Ohm's law may be applied to an entire series circuit or to the individual parts of the circuit. When it is used on a particular part of a circuit, the voltage across that part is equal to the current in that part multiplied by the resistance of that part.

For example, given the information in *Figure 8*, calculate the total resistance (R_T) and the total current (I_T).

To find R_T:

$$R_T = R_1 + R_2 + R_3$$
$$R_T = 20 + 50 + 120$$
$$R_T = 190\Omega$$

To find I_T using Ohm's law:

$$I_T = \frac{E_T}{R_T}$$
$$I_T = \frac{95}{190}$$
$$I_T = 0.5A$$

Find the voltage across each resistor. In a series circuit, the current is the same; that is, $I = 0.5A$ through each resistor:

$$E_1 = IR_1 = 0.5(20) = 10V$$
$$E_2 = IR_2 = 0.5(50) = 25V$$
$$E_3 = IR_3 = 0.5(120) = 60V$$

The voltages E_1, E_2, and E_3 found for *Figure 8* are known as *voltage drops* or *IR drops*. Their effect is to reduce the voltage that is available to be applied across the rest of the components in the circuit. The sum of the voltage drops in any series circuit is always equal to the voltage that is applied to the circuit. The total voltage (E_T) is the same as the applied voltage and can be verified in this example ($E_T = 10 + 25 + 60$ or $95V$).

2.4.2 Voltage and Current in Parallel Circuits

A parallel circuit is a circuit in which two or more components are connected across the same voltage source, as illustrated in *Figure 9*. The resistors R_1, R_2, and R_3 are in parallel with each other and with the battery. Each parallel path is then a

105F08.EPS

Figure 8 ◆ Calculating voltage drops.

Voltage Drops

Calculating voltage drops is not just a schoolroom exercise. It is important to know the voltage drop when sizing circuit components. What would happen if you sized a component without accounting for a substantial voltage drop in the circuit?

Demonstrate how to use Ohm's law to solve for unknown values in series circuits.

Show Transparency 10 (Figure 8).

Demonstrate how to use Ohm's law to solve for unknown values in parallel circuits.

Show Transparency 11 (Figure 9).

Discuss the "Think About It."

105F09.EPS

Figure 9 ◆ Parallel circuit.

Show examples of how to solve for an unknown branch current.

Show Transparency 12 (Figure 10).

branch with its own individual current. When the total current I_T leaves the voltage source E, part I_1 of the current I_T will flow through R_1, part I_2 will flow through R_2, and the remainder I_3 will flow through R_3. The branch currents I_1, I_2, and I_3 can be different. However, if a voltmeter is connected across R_1, R_2, and R_3, the respective voltages E_1, E_2, and E_3 will be equal to the source voltage E.

The total current I_T is equal to the sum of all branch currents.

This formula applies for any number of parallel branches whether the resistances are equal or unequal.

Using Ohm's law, each branch current equals the applied voltage divided by the resistance between the two points where the voltage is applied. Hence, for each branch in *Figure 9* we have the following equations:

$$\text{Branch 1: } I_1 = \frac{E_1}{R_1} = \frac{E}{R_1}$$

$$\text{Branch 2: } I_2 = \frac{E_2}{R_2} = \frac{E}{R_2}$$

$$\text{Branch 3: } I_3 = \frac{E_3}{R_3} = \frac{E}{R_3}$$

With the same applied voltage, any branch that has less resistance allows more current through it than a branch with higher resistance.

Example 1:
The two branches R_1 and R_2, shown in *Figure 10(A)*, across a 110V power line draw a total line current of 20A. Branch R_1 takes 12A. What is the current I_2 in branch R_2?

Transpose to find I_2 and then substitute given values:

$$I_T = I_1 + I_2$$
$$I_2 = I_T - I_1$$
$$I_2 = 20 - 12 = 8A$$

Example 2:
As shown in *Figure 10(B)*, the two branches R_1 and R_2 across a 240V power line draw a total line current of 35A. Branch R_2 takes 20A. What is the current I_1 in branch R_1?

Transpose to find I_1 and then substitute given values:

$$I_T = I_1 + I_2$$
$$I_1 = I_T - I_2$$
$$I_1 = 35 - 20 = 15A$$

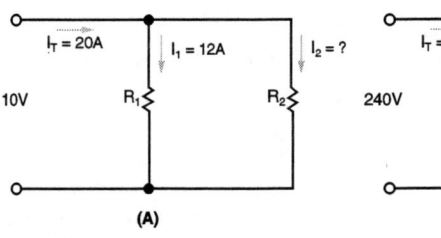

105F10.EPS

Figure 10 ◆ Solving for an unknown current.

ELECTRICAL LEVEL ONE—TRAINEE MODULE 26105

Instructor's Notes:

_____

_____

_____

_____

_____

_____

_____

_____

ELECTRICAL LEVEL ONE—INSTRUCTOR'S GUIDE MODULE 26105

Figure 11 ◆ Series-parallel circuit.

Figure 12 ◆ Simplified series-parallel circuit.

2.4.3 Voltage and Current in Series-Parallel Circuits

Series-parallel circuits combine the elements and characteristics of both the series and parallel configurations. By properly applying the equations and methods previously discussed, the values of individual components of the circuit can be determined. *Figure 11* shows a simple series-parallel circuit with a 1.5V battery.

The current and voltage associated with each component can be determined by first simplifying the circuit to find the total current, and then working across the individual components.

This circuit can be broken into two components: the series resistances R_1 and R_2, and the parallel resistances R_3 and R_4.

R_1 and R_2 can be added together to form the equivalent series resistance R_{1+2}:

$$R_{1+2} = R_1 + R_2$$
$$R_{1+2} = 0.5k\Omega + 0.5k\Omega$$
$$R_{1+2} = 1k\Omega$$

R_3 and R_4 can be totaled using either the general reciprocal formula or, since there are two resistances in parallel, the product over sum method. Both methods are shown below.

$$R_{3+4} = \frac{1}{\frac{1}{R_3} + \frac{1}{R_4}} = \frac{1}{\frac{1}{1k\Omega} + \frac{1}{1k\Omega}}$$

$$= \frac{1}{\frac{2}{1,000\Omega}} = \frac{1}{.002} = 500\Omega$$

$$R_{3+4} = \frac{R_3 \times R_4}{R_3 + R_4} = \frac{1k\Omega \times 1k\Omega}{1k\Omega + 1k\Omega}$$

$$= \frac{1,000,000\Omega}{2,000\Omega} = 500\Omega$$

The equivalent circuit containing the R_{1+2} resistance of 1kΩ and the R_{3+4} resistance of 500Ω is shown in *Figure 12*.

Using the Ohm's law relationship that total current equals voltage divided by circuit resistance, the circuit current can be determined. First, however, total circuit resistance must be found. Since the simplified circuit consists of two resistances in series, they are simply added together to obtain total resistance.

$$R_T = R_{1+2} + R_{3+4}$$
$$R_T = 1k\Omega + 500\Omega$$
$$R_T = 1.5k\Omega$$

Applying this to the current/voltage equation:

$$I_T = \frac{E_T}{R_T}$$

$$I_T = \frac{1.5V}{1.5k\Omega}$$

$$I_T = 1mA \text{ or } 0.001A$$

Now that the total current is known, voltage drops across individual components can be determined:

$$E_{R1} = I_T R_1 = 1mA \times 0.5k\Omega = 0.5V$$
$$E_{R2} = I_T R_2 = 1mA \times 0.5k\Omega = 0.5V$$

Since the total voltage equals the sum of all voltage drops, the voltage drop from A to B can be determined by subtraction:

$$E_T = E_{R1} + E_{R2} + E_{A+B}$$
$$E_T - E_{R1} - E_{R2} = E_{A+B}$$
$$1.5V - 0.5V - 0.5V = E_{A+B} = 0.5V$$

Since R_3 and R_4 are in parallel, some of the total current must pass through each resistor. R_3 and R_4

Demonstrate how to use Ohm's law to solve for unknown values in series-parallel circuits.

Show Transparencies 13 and 14 (Figures 11 and 12).

Have the trainees practice series-parallel circuit calculations.

are equal, so the same current should flow through each branch. Using the relationship:

$$I = \frac{E}{R}$$

$$I_{R3} = \frac{E_{R3}}{R_3} \qquad\qquad I_{R4} = \frac{E_{R4}}{R_4}$$

$$I_{R3} = \frac{0.5V}{1k\Omega} \qquad\qquad I_{R4} = \frac{0.5V}{1k\Omega}$$

$$I_{R3} = 0.5mA \qquad\qquad I_{R4} = 0.5mA$$

$$0.5mA + 0.5mA = 1mA$$

Therefore, the total current for the circuit passes through R_1 and R_2 and is evenly divided between R_3 and R_4.

3.0.0 ◆ KIRCHHOFF'S LAWS

Kirchhoff's laws provide a simple, practical method of solving for unknown parameters in a circuit.

3.1.0 Kirchhoff's Current Law

In its most general form, Kirchhoff's current law, which is also called *Kirchhoff's first law*, states that at any point in a circuit, the total current entering that point must equal the total current leaving that point. For parallel circuits, this implies that the current in a parallel circuit is equal to the sum of the currents in each branch.

When using Kirchhoff's laws to solve circuits, it is necessary to adopt conventions that determine the algebraic signs for current and voltage terms. A convenient system for current is to consider all current flowing into a branch point as positive, and all current directed away from that point as negative.

As an example, in *Figure 13*, the currents can be written as:

$$I_A + I_B - I_C = 0$$

or

$$5A + 3A - 8A = 0$$

Currents I_A and I_B are positive terms because these currents flow into P, but I_C, directed out of P, is negative.

For a circuit application, refer to Point C at the top of the diagram in *Figure 14*. The 6A I_T into Point C divides into the 2A I_3 and 4A I_{4+5}, both directed out. Note that I_{4+5} is the current through R_4 and R_5. The algebraic equation is:

$$I_T - I_3 - I_{4+5} = 0$$

Substituting the values for each current:

$$6A - 2A - 4A = 0$$

For the opposite direction, refer to Point D at the bottom of *Figure 14*. Here, the branch currents into Point D combine to equal the mainline current I_T returning to the voltage source. Now, I_T is directed out from Point D, with I_3 and I_{4+5} directed in. The algebraic equation is:

105F13.EPS

Figure 13 ◆ Kirchhoff's current law.

105F14.EPS

Figure 14 ◆ Application of Kirchhoff's current law.

ELECTRICAL LEVEL ONE—TRAINEE MODULE 26105

Instructor's Notes:

$$I_3 + I_{4+5} - I_T = 0$$
$$2A + 4A - 6A = 0$$

Note that at either Point C or Point D, the sum of the 2A and 4A branch currents must equal the 6A total line current. Therefore, Kirchhoff's current law can also be stated as:

$$I_{IN} = I_{OUT}$$

For *Figure 14*, the equations for current can be written as shown below.

At Point C:

$$6A = 2A + 4A$$

At Point D:

$$2A + 4A = 6A$$

Kirchhoff's current law is really the basis for the practical rule in parallel circuits that the total line current must equal the sum of the branch currents.

3.2.0 Kirchhoff's Voltage Law

Kirchhoff's voltage law states that the algebraic sum of the voltages around any closed path is zero.

Referring to *Figure 15*, the sum of the voltage drops around the circuit must equal the voltage applied to the circuit:

$$E_A = E_1 + E_2 + E_3$$

Where:

E_A = voltage applied to the circuit

E_1, E_2, and E_3 = voltage drops in the circuit

Another way of stating this law is that the algebraic sum of the voltage rises and voltage drops must be equal to zero. A voltage source is consid-

ered a voltage rise; a voltage across a resistor is a voltage drop. (For convenience in labeling, letter subscripts are shown for voltage sources and numerical subscripts are used for voltage drops.) This form of the law can be written by transposing the right members to the left side:

$$\text{Voltage applied} - \text{sum of voltage drops} = 0$$

Substitute letters:

$$E_A - E_1 - E_2 - E_3 = 0$$
$$E_A - (E_1 + E_2 + E_3) = 0$$

3.3.0 Loop Equations

Any closed path is called a *loop*. A loop equation specifies the voltages around the loop. Refer to *Figure 16*.

Consider the inside loop A, C, D, B, A, including the voltage drops E_1, E_3, and E_2, and the source E_T.

Figure 15 ◆ Kirchhoff's voltage law.

Figure 16 ◆ Loop equation.

Demonstrate how to use Kirchhoff's voltage law to solve for unknown circuit values.

 Audiovisual

Show Transparency 17 (Figure 15).

Have the trainees practice using Kirchhoff's voltage law to solve for unknown circuit values.

Discuss loop equations.

Show Transparency 18 (Figure 16).

In a clockwise direction, starting at Point A, the algebraic sum of the voltages is:

$$-E_1 - E_3 - E_2 + E_T = 0$$

or

$$-30V - 120V - 90V + 240V = 0$$

Voltages E_1, E_3, and E_2 have a negative value, because there is a decrease in voltage seen across each of the resistors in a clockwise direction. However, the source E_T is a positive term because an increase in voltage is seen in that same direction.

For the opposite direction, going counterclockwise in the same loop from Point A, E_T is negative while E_1, E_2, and E_3 have positive values. Therefore:

$$-E_T + E_2 + E_3 + E_1 = 0$$

or

$$-240V + 90V + 120V + 30V = 0$$

When the negative term is transposed, the equation becomes:

$$240V = 90V + 120V + 30V$$

In this form, the loop equation shows that Kirchhoff's voltage law is really the basis for the practical rule in series circuits that the sum of the voltage drops must equal the applied voltage.

For example, determine the voltage E_B for the circuit shown in *Figure 17*. The direction of the current flow is shown by the arrow. First mark the polarity of the voltage drops across the resistors and trace the circuit in the direction of the current flow starting at Point A. Then write the voltage equation around the circuit:

$$-E_3 - E_B - E_2 - E_1 + E_A = 0$$

Solve for E_B:

$$E_B = E_A - E_3 - E_2 - E_1$$
$$E_B = 15V - 2V - 6V - 3V$$
$$E_B = 4V$$

Since E_B was found to be positive, the assumed direction of current is in fact the actual direction of current.

In its most general form, Kirchhoff's voltage law (also called *Kirchhoff's second law*) states that the algebraic sum of all the potential differences in a closed loop is equal to zero. A closed loop means any completely closed path consisting of wire, resistors, batteries, or other components. For series circuits, this implies that the sum of the voltage drops around the circuit is equal to the applied voltage. For parallel circuits, this implies that the voltage drops across all branches are equal.

Summary

The relationships among current, voltage, resistance, and power in Ohm's law are the same for both DC series and DC parallel circuits. Understanding and being able to apply these concepts is necessary for effective circuit analysis and troubleshooting. DC series-parallel circuits also have these fundamental relationships. Since DC series-parallel circuits are a combination of simple series and parallel circuits, Kirchhoff's voltage and cur-

105F17.EPS

Figure 17 ◆ Applying Kirchhoff's voltage law.

Putting It All Together

THINK ABOUT IT
Draw four 60W lamps in parallel with a 120V power source. What is the amperage in the circuit? What would happen to the amperage if we doubled the voltage?

Show Transparency 19 (Figure 17).

Teaching Tip

Discuss the "Think About It."

Instructor's Notes:

rent laws will apply. Calculating I, E, R, and P for series-parallel circuits is no more difficult than calculating these values for simple series or parallel circuits. However, for series-parallel circuits, these calculations require more careful circuit analysis in order to use Ohm's law correctly.

Review Questions

1. The formula for calculating the total resistance in a series circuit with three resistors is _____.
   a. $R_T = R_1 + R_2 + R_3$
   b. $R_T = R_1 - R_2 - R_3$
   c. $R_T = R_1 \times R_2 \times R_3$
   d. $R_T = \dfrac{1}{\dfrac{1}{R_1} + \dfrac{1}{R_2} + \dfrac{1}{R_3}}$

2. The formula for calculating the total resistance in a parallel circuit with three resistors is _____.
   a. $R_T = R_1 + R_2 + R_3$
   b. $R_T = R_1 - R_2 - R_3$
   c. $R_T = R_1 \times R_2 \times R_3$
   d. $R_T = \dfrac{1}{\dfrac{1}{R_1} + \dfrac{1}{R_2} + \dfrac{1}{R_3}}$

3. The total resistance in *Figure 18* is _____.
   a. 1,035Ω
   b. 129Ω
   c. 100Ω
   d. 157Ω

4. Find the total resistance in a series circuit with three resistances of 10Ω, 20Ω, and 30Ω.
   a. 15Ω
   b. 1Ω
   c. 20Ω
   d. 60Ω

5. In a parallel circuit, the voltage across each path is equal to the _____.
   a. total circuit resistance times path current
   b. source voltage minus path voltage
   c. path resistance times total current
   d. source voltage

6. The value for total current in *Figure 19* is _____ amps.
   a. 1.25
   b. 2.50
   c. 5
   d. 10

7. A resistor of 32Ω is in parallel with a resistor of 36Ω, and a 54Ω resistor is in series with the pair. When 350V is applied to the combination, the current through the 54Ω resistor is _____ amps.
   a. 2.87
   b. 3.26
   c. 5.86
   d. 4.94

Figure 18 ◆ Series-parallel circuit.

Have the trainees complete the Review Questions and go over the answers prior to administering the Module Examination.

Show Transparencies 20 and 21 (Figures 18 and 19).

Administer the Module Examination and Performance Test. Be sure to record the testing results on Craft Training Report Form 200 and submit the results to the Training Program Sponsor.

$E_T = 100V$ $R_1 = 40\Omega$ $R_2 = 40\Omega$

105F19.EPS

Figure 19 ◆ Parallel circuit.

8. A 242Ω resistor is in parallel with a 180Ω resistor, and a 420Ω resistor is in series with the combination. A current of 22mA flows through the 242Ω resistor. The current through the 180Ω resistor is _____ mA.
   a. 29.6
   b. 40.2
   c. 19.8
   d. 36.4

9. Two 24Ω resistors are in parallel, and a 42Ω resistor is in series with the combination. When 78V is applied to the three resistors, the voltage drop across the 42Ω resistor is _____ volts.
   a. 55.8
   b. 60.5
   c. 65.3
   d. 49.8

10. A 19Ω resistor is in parallel with an 18Ω resistor, and in series with the two is a 14Ω resistor. When 70V is applied to this combination, the current through the 19Ω resistor is _____ amps.
    a. 2.06
    b. 1.03
    c. 1.46
    d. 2.37

Instructor's Notes:

_____

_____

_____

_____

_____

_____

_____

_____

James Mitchem, *TIC—The Industrial Company*

Jim Mitchem serves as a troubleshooter for a large electrical contractor. During his career in the electrical industry, he worked his way up from apprentice to technical services manager.

How did you become an electrician?
Quite by accident. A couple of years after college, I was working as a relief operator in a plant when the lead electrician retired, creating a vacancy. I liked the idea that electricians were expected to use their knowledge and initiative to keep the place running. I applied and was accepted as a trainee.

How did you get your training?
I took an electrical apprenticeship course by correspondence, and I was fortunate enough to work with good people who helped me along. I worked in an environment that exposed me to a variety of equipment and applications, and just about everyone I've ever worked with has taught me something. Now I'm passing my knowledge on to others.

What kinds of work have you done in your career?
I've worked as an apprentice, journeyman, instrument and controls technician, instrument fitter, foreman, general foreman, superintendent, and startup engineer. Each of these positions required that I learn new skills, both technical and managerial. My experience in many disciplines and types of projects has given me a high level of credibility with my employer and our clients.

Now I act as a technical resource and troubleshooter for job sites and in-house functions such as safety, quality assurance, and training. I visit job sites to help solve problems and help out with commissioning and startup.

What factor or factors have contributed the most to your success?
There are several factors. Two very important ones have been a desire to learn and a willingness to do whatever is asked of me. I also keep an eye on the big picture. When I'm on a job, I'm not just pulling wire, I'm building a power plant or whatever the project is. I also think it has helped me to remain with the same employer for 18 years.

Any advice for apprentices just beginning their careers?
Keep learning! And don't depend on others to train you. Take the initiative to buy or borrow books and trade journals. Take licensing tests and do whatever is necessary to keep your licenses current. Finally, make sure you know your own personal and professional values and work with a company that shares those values.

Trade Terms Introduced in This Module

Kirchhoff's current law (KCL): The statement that the total amount of current flowing through a parallel circuit is equal to the sum of the amounts of current flowing through each current path.

Kirchhoff's voltage law (KVL): The statement that the sum of all the voltage drops in a circuit is equal to the source voltage of the circuit.

Parallel circuits: Circuits containing two or more parallel paths through which current can flow.

Series circuits: Circuits with only one path for current flow.

Series-parallel circuits: Circuits that contain both series and parallel current paths.

Instructor's Notes:

_____

_____

_____

_____

_____

_____

_____

_____

_____

Answers to Review Questions

Answer	Section
1. a	2.1.0
2. d	2.2.0
3. c	2.3.1
4. d	2.1.0
5. d	2.4.2
6. c	2.4.2
7. d	2.4.3
8. a	2.4.3
9. b	3.3.0
10. c	3.3.0

Additional Resources

For advanced study of topics covered in this task module, the following books are suggested:

Electronics Fundamentals, Latest Edition. New York: Prentice Hall.
Principles of Electric Circuits, Latest Edition. New York: Prentice Hall.

Instructor's Notes:

_____

_____

_____

_____

_____

_____

_____

_____

NCCER CRAFT TRAINING USER UPDATES

The NCCER makes every effort to keep these textbooks up-to-date and free of technical errors. We appreciate your help in this process. If you have an idea for improving this textbook, or if you find an error, a typographical mistake, or an inaccuracy in the NCCER's Craft Training textbooks, please write us, using this form or a photocopy. Be sure to include the exact module number, page number, a detailed description, and the correction, if applicable. Your input will be brought to the attention of the Technical Review Committee. Thank you for your assistance.

Instructors – If you found that additional materials were necessary in order to teach this module effectively, please let us know so that we may include them in the Equipment/Materials list in the Instructor's Guide.

Write: Curriculum Revision and Development Department
National Center for Construction Education and Research
P.O. Box 141104, Gainesville, FL 32614-1104

Fax: 352-334-0932

E-mail: curriculum@nccer.org

Craft _____ Module Name _____

Copyright Date _____ Module Number _____ Page Number(s) _____

Description _____

_____

_____

_____

(Optional) Correction _____

_____

_____

(Optional) Your Name and Address _____

_____

_____

_____

TEACHING TIPS

Section 2.2.0 *Think About It – Series Circuits*

Receptacles wired in series would have the output terminal of one receptacle wired to the input terminal of the next receptacle, and so on. Assuming the source voltage remains the same, the amount of light produced by each lamp in a series circuit depends on the total number of lamps installed in the circuit. This is because each lamp added to the circuit increases the total circuit resistance. A higher circuit resistance decreases the total current flow through the series circuit and therefore the current through each lamp, causing the amount of light produced by each lamp to decrease.

Section 2.2.1 *Think About It – Parallel Circuits*

Adding additional lamps in parallel increases the total current in a parallel circuit. The total number of devices or lamps that can be connected in parallel is limited by the ampacity of the conductors and the rating of the overcurrent device (fuse or circuit breaker) protecting the circuit. For example, if each lamp has a resistance of 240Ω, you can connect a maximum of 30 lamps in parallel in a 120V, 15A circuit. Thirty 240Ω lamps connected in parallel have a total resistance of 8Ω ($240V \div 30$). The total current in this circuit is equal to 15A ($120V \div 8\Omega$).

Section 2.3.1 *Think About It – Series-Parallel Circuits*

Resistors R_4 and R_5 are in series with each other and in parallel with R_2. This combination (R_4, R_5, and R_2) is in series with R_1 and R_3.

Section 2.4.2 *Think About It – Voltage Drops*

Sizing a component without accounting for a substantial voltage drop across another component in a series circuit can result in specifying a component with too much resistance and/or the capability of handling a higher voltage than necessary. In a parallel circuit, the voltage drop across all branches is equal so there is less chance of error.

Section 3.3.0 *Think About It – Putting It All Together*

A 60W lamp across 120V draws 0.5A of current ($I = 60W \div 120V$). With four 60W lamps connected in parallel across a 120V source, the total circuit current is 2A. If the source voltage is raised from 120V to 240V, the total circuit current would decrease to 1A (0.25A per lamp).

Electrical Test Equipment

26106-00

MODULE OVERVIEW

This course introduces the electrical trainee to the operation and applications of various types of electrical test equipment.

PREREQUISITES

Please refer to the Course Map in the Trainee Module. Prior to training with this module, it is recommended that the trainee shall have successfully completed the following modules:

Core Curriculum; Electrical Level One, Modules 26101 through 26105

LEARNING OBJECTIVES

Upon completion of this module, the trainee will be able to:

1. Explain the operation of and describe the following pieces of test equipment:
   - Ammeter
   - Voltmeter
   - Ohmmeter
   - Volt-ohm-milliammeter (VOM)
   - Wattmeter
   - Megohmmeter
   - Frequency meter
   - Power factor meter
   - Continuity tester
   - Voltage tester
   - Recording instruments
   - Cable-length meters
2. Explain how to read and convert from one scale to another using the above test equipment.
3. Explain the importance of proper meter polarity.
4. Define frequency and explain the use of a frequency meter.
5. Explain the difference between digital and analog meters.

PERFORMANCE OBJECTIVES

Under supervision of the instructor, the trainee should be able to:

1. Demonstrate the procedure for safely using a clamp-on ammeter.
2. Connect a multimeter to a circuit and measure the source voltage.
3. Demonstrate the procedure for safely using a voltage tester.

NCCER STANDARDIZED CRAFT TRAINING PROGRAM

The National Center for Construction Education and Research (NCCER) provides a standardized national program of accredited craft training. Key features of the program include instructor certification, competency-based training, and performance testing. The program provides trainees, instructors, and companies with a standard form of recognition through a National Craft Training Registry. The program is described in full in the *Guidelines for Accreditation*, published by the NCCER. For more information on standardized craft training, contact the NCCER at P.O. Box 141104, Gainesville, FL 32614-1104, 352-334-0911, visit our Web site at www.NCCER.org, or e-mail info@NCCER.org.

HOW TO USE THIS ANNOTATED INSTRUCTOR'S GUIDE

Each page presents two sections of information. The larger section displays each page exactly as it appears in the Trainee Module. The narrow column ties suggested trainee and instructor actions to each page and provides icons to call your attention to material, safety, audiovisual, or testing requirements. The bottom of each page includes space for your notes.

 If you see the Teaching Tip icon, that means there is a teaching tip associated with this section. Also refer to the suggested teaching tips at the end of the module.

SAFETY CONSIDERATIONS

Ensure that the trainees are equipped with appropriate personal protective equipment. Stress the importance of following all safety precautions and procedures when working with all types of electrical test equipment.

PREPARATION

Before teaching this module, you should review the Module Outline, Learning and Performance Objectives, and the Materials and Equipment List. Be sure to allow ample time to prepare your own training or lesson plan and gather all required equipment and materials.

MATERIALS AND EQUIPMENT LIST

Materials:
Transparencies

Markers/chalk

Copy of the latest edition of the *National Electrical Code*®

Calculator

Pencils and scratch paper

Module Examinations*

Performance Profile Sheets*

Equipment:
Overhead projector and screen

Whiteboard/chalkboard

Appropriate personal protective equipment

Clamp-on ammeter

Volt-ohm-milliammeters (both analog and digital)

Wattmeter

Megohmmeter

Cable-length meter

Frequency meter

Power factor meter

Continuity tester

Voltage tester

Recording instruments

*Located in the Test Booklet packaged with this Annotated Instructor's Guide.

ADDITIONAL RESOURCES

This module is intended to present thorough resources for task training. The following reference works are suggested for both instructors and motivated trainees interested in further study. These are optional materials for continued education rather than for task training.

Electronics Fundamentals, Latest Edition. Upper Saddle River, NJ: Prentice Hall.

Principles of Electric Circuits, Latest Edition. Upper Saddle River, NJ: Prentice Hall.

NOTES

The designations "National Electrical Code," "NE Code," and "NEC," where used in this document, refer to the *National Electrical Code®*, which is a registered trademark of the National Fire Protection Association, Quincy, MA. All National Electrical Code (NEC) references in this module refer to the 1999 edition of the NEC.

If you feel that additional math instruction would be helpful, Prentice Hall offers a basic math textbook entitled *Fundamentals of Electrical and Mechanical Mathematics*. It covers the basic math requirements for electrical trainees and may be ordered by contacting Prentice Hall Customer Service at 1-800-922-0579.

TEACHING TIME FOR THIS MODULE

An outline for use in developing your lesson plan is presented below. Note that each Roman numeral in the outline equates to one session of instruction. Each session has a suggested time period of 2½ hours. This includes 10 minutes at the beginning of each session for administrative tasks and one 10-minute break during the session. Approximately 7½ hours are suggested to cover *Electrical Test Equipment*. You will need to adjust the time required for hands-on activity and testing based on your class size and resources.

Topic	Planned Time
Session I. Analog Meters; Ammeter; Laboratory; Voltmeter; Ohmmeter; **Volt-Ohm-Milliammeter**	
A. Analog Meters	_____
1. d'Arsonval Meter Movement	_____
B. Ammeter	_____
C. Laboratory	_____
Under instructor supervision, have the trainees demonstrate the procedure for safely using a clamp-on ammeter.	
D. Voltmeter	_____
E. Ohmmeter	_____
F. Volt-Ohm-Milliammeter	_____
1. Specifications	_____
2. Overload Protection	_____
3. Making Measurements	_____
a. Measuring DC Voltage, 0–250 Millivolts	_____
b. Measuring DC Voltage, 0–1 Volt	_____
c. Measuring DC Voltage, 0–2.5 Through 0–500 Volts	_____
d. Measuring DC Voltage, 0–1,000 Volts	_____
e. Measuring AC Voltage, 0–2.5 Through 0–500 Volts	_____
f. Measuring AC Voltage, 0–1,000 Volts	_____
g. Measuring Output Voltage	_____
h. Measuring Decibels	_____
4. Direct Current Measurements	_____
a. Voltage Drop	_____
b. Measuring Direct Current, 0–50 Microamperes	_____
c. Measuring Direct Current, 0–1 Through 0–500 Milliamperes	_____
d. Measuring Direct Current, 0–10 Amperes	_____
e. Zero Ohm Adjustment	_____
f. Measuring Resistance	_____

Session II. Digital Meters; Laboratory; Wattmeter; Megohommeter (Megger); Frequency Meter; Power Factor Meter

    A. Digital Meters   _____

       1. Features   _____

       2. Operation   _____

          a. Dual Function ON/Clear Button   _____

          b. Measuring Voltage   _____

          c. Measuring Current   _____

          d. Measuring Resistance   _____

          e. Diode/Continuity Test   _____

          f. Transistor Junction Test   _____

          g. Display Test   _____

       3. Maintenance   _____

          a. Battery Replacement   _____

          b. Fuse Replacement   _____

          c. Calibration   _____

    B. Laboratory   _____

    Under instructor supervision, have the trainees connect a multimeter to a circuit and measure the source voltage.

    C. Wattmeter   _____

    D. Megohmmeter (Megger)   _____

       1. Safety Precautions   _____

    E. Frequency Meter   _____

       1. Phase Rotation Tester   _____

       2. Infrared Sensing Device   _____

       3. Cable-Length Meter   _____

       4. Harmonic Test Set   _____

    F. Power Factor Meter   _____

Session III. Continuity Tester; Voltage Tester; Laboratory; Recording Instruments; Safety; Module Examination and Performance Testing

    A. Continuity Tester   _____

       1. Audio Continuity Tester   _____

       2. Visual Continuity Tester   _____

    B. Voltage Tester   _____

       1. Wiggy®   _____

          a. Principle of Operation   _____

    C. Laboratory   _____

    Under instructor supervision, have the trainees demonstrate the procedure for safely using a voltage tester.

    D. Recording Instruments   _____

       1. Strip-Chart Recorders   _____

          a. Typical Strip-Chart Recorder   _____

E. Safety
  1. Use of High-Voltage Protection Equipment   _____
  2. General Testing of Electrical Equipment   _____
    a. Example Policy – Work Permits or Test Permits   _____
    b. Example Policy – Authorized Employees for Testing   _____
    c. Example Policy – Circuit Isolation   _____
    d. Example Policy – Clearances   _____
    e. Electrical Circuit Tests   _____
  3. Meter Loading Effects   _____
F. Module Examination   _____
  1. Trainees must score 70% or higher to receive recognition from the NCCER.
  2. Record the testing results on Craft Training Report Form 200 and submit the results to the Training Program Sponsor.
G. Performance Testing   _____
  1. Trainees must perform each task to the satisfaction of the instructor to receive recognition from the NCCER.
  2. Record the testing results on Craft Training Report Form 200 and submit the results to the Training Program Sponsor.

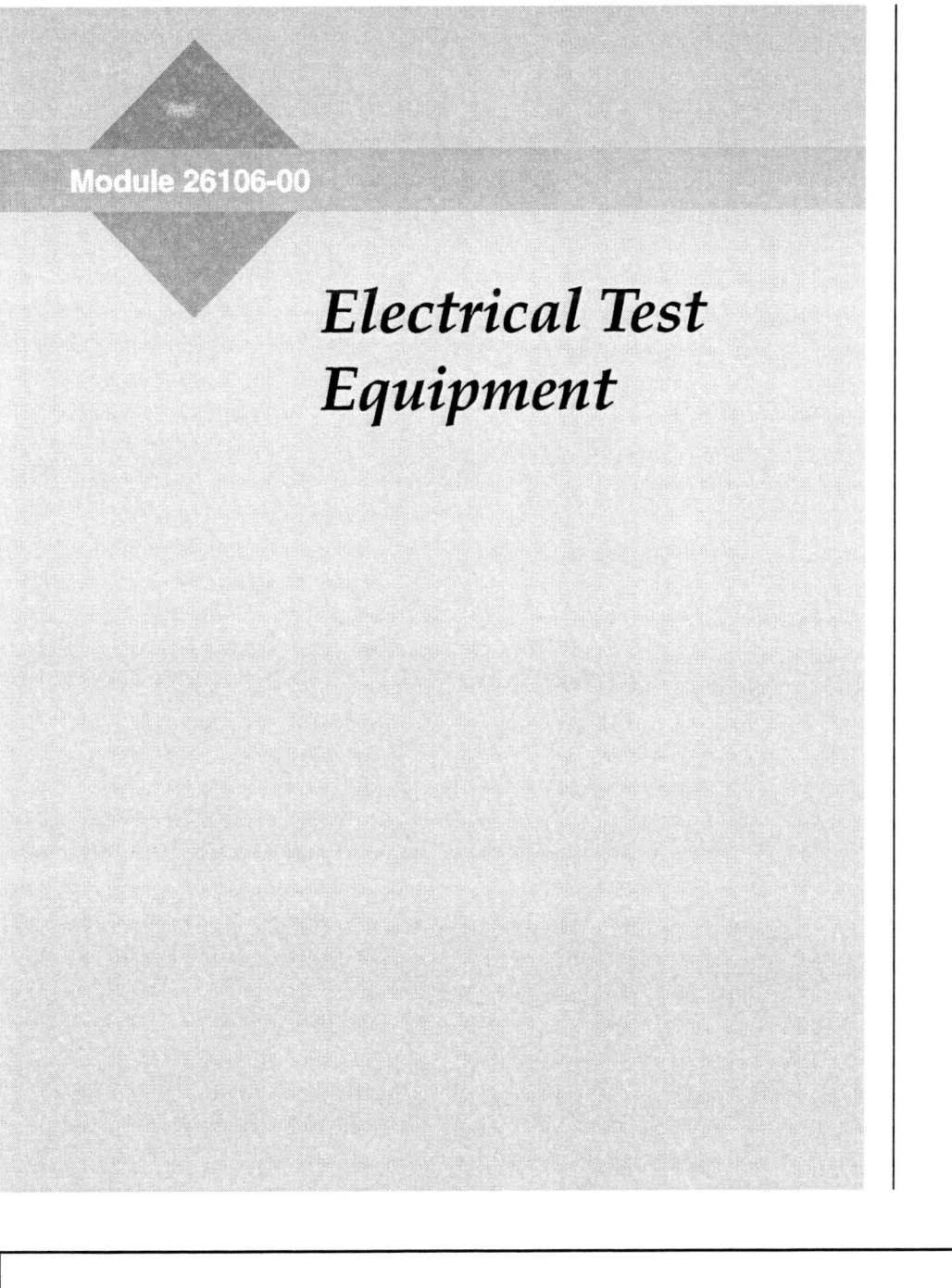

Module 26106-00

Electrical Test Equipment

Instructor's Notes:

Course Map

This course map shows all of the task modules in the first level of the Electrical curricula. The suggested training order begins at the bottom and proceeds up. Skill levels increase as a trainee advances on the course map. The training order may be adjusted by the local Training Program Sponsor.

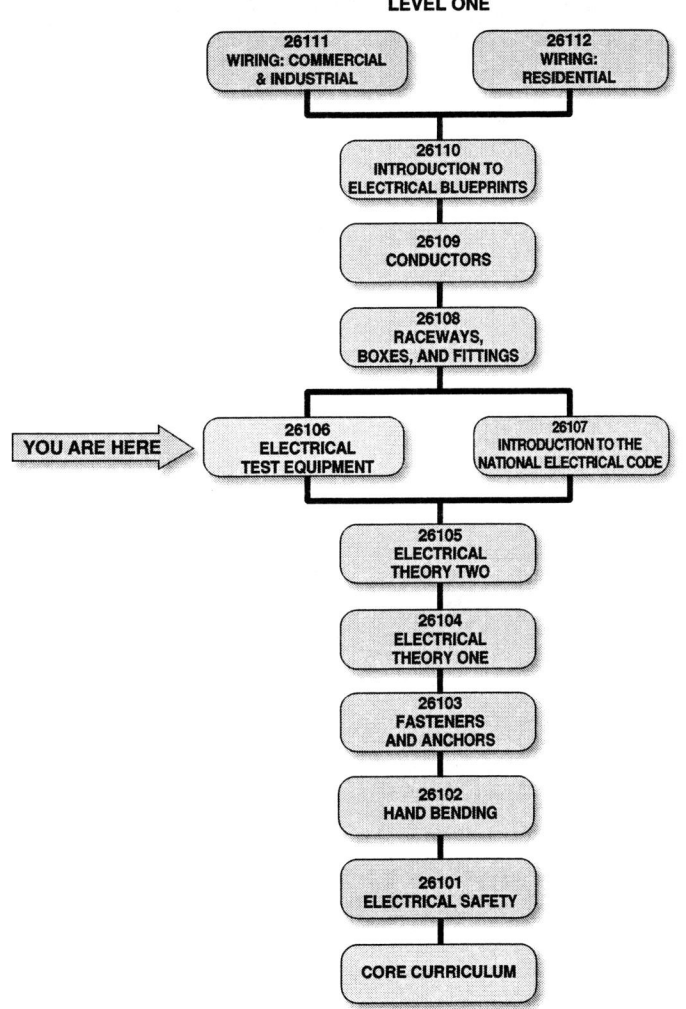

LEVEL ONE

26111
WIRING: COMMERCIAL & INDUSTRIAL

26112
WIRING: RESIDENTIAL

26110
INTRODUCTION TO ELECTRICAL BLUEPRINTS

26109
CONDUCTORS

26108
RACEWAYS, BOXES, AND FITTINGS

YOU ARE HERE →

26106
ELECTRICAL TEST EQUIPMENT

26107
INTRODUCTION TO THE NATIONAL ELECTRICAL CODE

26105
ELECTRICAL THEORY TWO

26104
ELECTRICAL THEORY ONE

26103
FASTENERS AND ANCHORS

26102
HAND BENDING

26101
ELECTRICAL SAFETY

CORE CURRICULUM

106CMAP.EPS

Assign reading of Module 26106.

6–iii

Instructor's Notes:

_____

_____

_____

_____

_____

_____

_____

_____

_____

Figures

Table

Ensure you have everything required to teach the course. Check the Materials and Equipment List at the front of this Instructor's Guide.

Electrical Test Equipment

Objectives

Upon completion of this module, the trainee will be able to:

1. Explain the operation of and describe the following pieces of test equipment:
   - Ammeter
   - Voltmeter
   - Ohmmeter
   - Volt-ohm-milliammeter (VOM)
   - Wattmeter
   - Megohmmeter
   - Frequency meter
   - Power factor meter
   - Continuity tester
   - Voltage tester
   - Recording instruments
   - Cable-length meters
2. Explain how to read and convert from one scale to another using the above test equipment.
3. Explain the importance of proper meter polarity.
4. Define frequency and explain the use of a frequency meter.
5. Explain the difference between digital and analog meters.

Prerequisites

Successful completion of the following Task Modules is recommended before beginning study of this Task Module: Core Curricula; Electrical Level One, Modules 26101 through 26105.

Required Trainee Materials

1. Trainee Task Module
2. Copy of the latest edition of the *National Electrical Code*
3. Appropriate Personal Protective Equipment

1.0.0 ◆ INTRODUCTION

The use of electronic test instruments and meters generally involves these three applications:

- Verifying proper operation of instruments and associated equipment
- Calibrating electronic instruments and associated equipment
- Troubleshooting electrical/electronic circuits and equipment

For these applications, specific test equipment is selected to analyze circuits and to determine specific characteristics of discrete components.

The test equipment an electrician chooses for a specific task depends on the type of measurement and the level of accuracy required. Additional factors that may influence selection include:

- Whether the test equipment is portable
- The amount of information that the test equipment provides

Note: The designations "National Electrical Code," "NE Code," and "NEC," where used in this document, refer to the *National Electrical Code*®, which is a registered trademark of the National Fire Protection Association, Quincy, MA. *All National Electrical Code (NEC) references in this module refer to the 1999 edition of the NEC.*

Show Transparency 1, Course Objectives.

Show Transparency 2, Performance Profile Tasks.

Discuss the applications of electrical test equipment.

- The likelihood that the test equipment may damage the circuit or component being tested (some test equipment can generate enough voltage or current to damage an instrument or electronic circuit)

This module will focus on some of the test equipment that you will be required to use in your job as an electrician. The intent is to familiarize you with the use and operation of such equipment and to provide you with practical experience involving that equipment. Upon completion of this module, you should be able to select the appropriate test equipment and effectively use that equipment to perform an assigned task.

2.0.0 ◆ METERS

The functioning of conventional electrical measuring instruments is based upon electromechanical principles. Their mechanical components usually work on direct current (DC). Mechanical frequency meters are an exception. A meter that measures alternating current (AC) has a built-in rectifier to change the AC to DC and resistors to correct for the various ranges.

Today, many meters are solid-state digital systems; they are superior because they have no moving parts. These meters will work in any position, unlike mechanical meters, which must remain in one position to be read accurately.

2.1.0 d'Arsonval Meter Movement

In 1882, a Frenchman named Arsene d'Arsonval invented the galvanometer. This meter used a stationary permanent magnet and a moving coil (Figure 1) to indicate current flow on a calibrated scale. The early galvanometer was very accurate but could only measure very small currents. Over the following years, many improvements were made that extended the range of the meter and increased its ruggedness. The d'Arsonval meter movement is the most commonly used meter movement today.

A moving-coil meter movement operates on the electromagnetic principle. In its simplest form, the moving-coil meter uses a coil of very fine wire wound on a light aluminum frame. A permanent magnet surrounds the coil. The aluminum frame is mounted on pivots to allow it and the coil to rotate freely between the poles of the permanent magnet. When current flows through the coil, it becomes magnetized, and the polarity of the coil is such that it is repelled by the field of the permanent magnet. This will cause the coil frame to rotate on its pivots, and the distance it rotates is determined by the amount of current that flows through the coil. By attaching a pointer to the coil frame and adding a calibrated scale, the amount of current flowing through the meter can be measured.

The d'Arsonval meter movement uses this same principle of operation (Figure 2). As the cur-

Figure 1 ◆ Moving-coil meter movement.

Instructor's Notes:

_____

_____

_____

_____

_____

_____

_____

_____

rent flow increases, the magnetic field around the coil increases, the amount of coil rotation increases, and the pointer swings farther across the meter scale.

3.0.0 ◆ AMMETER

The ammeter is used to measure current. Most models will measure only small amounts of current. The typical range is in microamperes, µA (0.000001 ampere) or milliampere, mA (0.001 amperes). Very few ammeters can measure more than 10mA. To increase the range to the ampere level, a shunt is used. To measure above 10mA, a shunt with an extremely low resistance is placed in series with the load, and the meter is connected across the shunt to measure the resulting voltage drop proportional to current flow. A shunt has a

very large wattage rating in order to carry a large current.

The meter is connected in parallel with the shunt (*Figure 3*). Shunts located inside the meter case (internal shunts) are generally used to measure values up to 30 amps; shunts located away from the meter case (external shunts) with leads going to the meter are generally used to measure values greater than 30 amps. Above 30 amps of current, the heat generated could damage the meter if an internal shunt were used. The use of a shunt allows the ammeter to derive current in amps by actually measuring the voltage drop across the shunt. Ammeter connections are shown in *Figure 4*.

Never connect an ammeter in parallel with a load. Because of the low resistance in the ammeter, this will cause a short circuit, probable damage to the

Discuss the operation of a basic ammeter.

Show Transparencies 5 and 6 (Figures 3 and 4).

Emphasize the importance of *never* connecting an ammeter in parallel with a load.

Figure 2 ◆ d'Arsonval meter movement.

Measuring Amperage

A common example of measuring amperage is checking for a balanced flow of current on the two legs of a service entry.

Figure 3 ◆ Ammeter shunt.

106F03.EPS

$$I_T = I_{R_1} + I_{R_2}$$

Figure 4 ◆ Ammeter connections.

106F04.EPS

Ammeter Connections

Using basic concepts of series and parallel circuits, explain why an ammeter would be damaged when connected in parallel but not when connected in series. What are some practical examples of connecting an ammeter in series with a load?

meter and/or circuit, and personal injury. When connecting an ammeter in a DC circuit, you must observe proper polarity. In other words, you must connect the negative terminal of the meter to the negative or low potential point in the circuit, and connect the positive terminal of the meter to the positive or high potential point in the circuit (*Figure 5*). Current must flow through the meter from minus (−) to plus (+). If you connect the meter with the polarities reversed, the meter coil will move in the opposite direction, and the pointer might strike the left retaining pin. You will not obtain a current reading, and you might bend the pointer of the meter.

It is not very practical to use an ammeter that has only one range; therefore, a multirange ammeter needs to be discussed. A multirange ammeter is one containing a basic meter movement and several shunts that can be connected across the meter movement. See *Figure 6*.

A range switch is normally used to select the particular shunt for the desired current range. Sometimes, however, separate terminals for each range are mounted on the meter case. Some multirange ammeters have only one set of values on the scale, even though they measure several different current ranges. For example, if the scale is calibrated in values from 0 to 1 milliamp (mA), and the range switch is in the 1mA position, read

LOW-RESISTANCE SHUNT

106F05.EPS

Figure 5 ◆ DC ammeter.

the current directly. However, if the range switch is in the 10mA position, multiply the scale reading by 10 to find the amount of current flowing through the circuit. See *Figure 7*.

Instructor's Notes:

_____

_____

_____

_____

_____

_____

_____

_____

_____

The range of this 1mA meter movement
has been extended to measure 0-10mA,
0-100mA, and 0-1A by using
multiple shunts.

RANGE SWITCH

RANGE
SWITCH

10mA

100mA

1A

A range switch provides the simplest way of
setting the meter to the desired range.

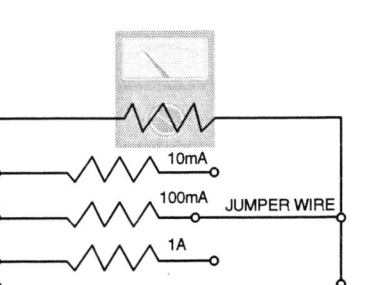

10mA

100mA   JUMPER WIRE

1A

10   100   1
mA   mA   Amp

1 MILLIAMPERE

When separate terminals are used to select the desired range, a jumper must be connected from
the positive terminal to connect the shunt across
the meter movement.

106F06.EPS

Figure 6 ◆ Multirange ammeter.

Some multirange current meters have only one set of
values marked on the scale.

RANGE SWITCH

10mA

1mA   100mA

To find the current flowing in the circuit, multiply
the scale reading by the range switch setting:
current = 0.7 × 10mA = 7mA.

106F07.EPS

Figure 7 ◆ Multirange single-scale ammeter.

Some multirange current meters have a set of values for each range switch position.

RANGE SWITCH

To find the current flowing in the circuit, read the meter scale that corresponds to the position of the range switch: current = 7mA.

106F08.EPS

Figure 8 ◆ Multirange multiscale ammeter.

Other current meters have a separate set of values on the calibrated scale that correspond to the different positions of the range switch. In this case, be sure that you read the set of values that correspond to the position of the range switch (*Figure 8*).

When measuring AC current at levels greater than one ampere, a clamp-on ammeter is often used. This meter also measures DC. The clamp-on ammeter may have a mechanical movement or a digital readout. These meters clamp over the wire and do not break the insulation. This type of ammeter senses current flow by measuring the magnetic field surrounding the conductor. When using this type of ammeter, be sure to measure only one conductor at a time. This type of meter will often measure up to 1,000 amperes at 600 volts. A digital clamp-on ammeter is shown in *Figure 9*.

4.0.0 ◆ VOLTMETER

The basic current meter movement already covered, whether AC or DC, can also be used to measure voltage (electromotive force or emf). The meter coil has a fixed resistance, and, therefore, when current flows through the coil, a voltage drop will be developed across this resistance. According to Ohm's law, the voltage drop will be directly proportional to the amount of current flowing through the coil. Also, the amount of current flowing through the coil is directly proportional to the amount of voltage applied to it. Therefore, by calibrating the meter scale in units of voltage instead of current, the voltage in a circuit can be measured.

Since a basic current meter movement has a low coil resistance and low current-handling capabilities, its use as a voltmeter is very limited. In fact,

the maximum voltage that could be measured with a one milliamp meter movement is one volt (*Figure 10*).

The voltage range of a meter movement can be extended by adding a resistor, called a *multiplier resistor*, in series. The value of this resistor must be such that, when added to the meter coil resistance, the total resistance limits the current to the full-

WALTER JOHNSON   106F09.EPS

Figure 9 ◆ Clamp-on ammeter.

6–6

Instructor's Notes:

_____

_____

_____

_____

_____

_____

_____

_____

_____

Since the voltage across the meter coil resistance is proportional to the current flowing through the coil, the 1mA current meter movement can measure voltage directly by calibrating the meter scale in the units of voltage that produce the current through the coil.

$$E = I_R R_M = 0.001 \times 1000 = 1V$$

Figure 10 ◆ 1mA meter movement.

By connecting a multiplier resistor in series with the meter resistance, the range of a basic meter movement can be extended to measure voltages higher than the $I_M R_M$ voltage drop across the meter coil.

Figure 11 ◆ Adding multiplier resistor.

Voltmeters

Again using basic circuit theory, explain why a voltmeter must be connected in parallel with the circuit. What would you be measuring if the voltmeter were connected in series?

scale current rating of the meter for any applied voltage *(Figure 11)*.

Voltmeters must be in parallel with the circuit component being measured. On the higher ranges, the amount of current flowing through the meter is much lower due to its very high total resistance. However, an inaccurate reading will result if the voltmeter is placed in series rather than in parallel with a circuit component. When connecting a DC voltmeter, always observe the proper polarity. The negative lead of the meter must be connected to the negative or low potential end of the component, and the positive lead to the positive or high potential end of the component. As was the case

Show Transparencies 11 and 12 (Figures 10 and 11).

Discuss the "Think About It."

Discuss multirange voltmeters.

Show Transparencies 13 and 14 (Figures 12 and 13).

Explain the operation of a basic ohmmeter.

Show Transparency 15 (Figure 14).

when using an ammeter, if you connect a voltmeter to the component with opposing polarities, the meter coil will move to the left, and the pointer could be bent. In an AC circuit, the voltage constantly reverses polarity, so there is no need to observe polarity when connecting the voltmeter to a component in an AC circuit.

As with ammeters, it is impractical to have a voltmeter that will only measure one range of voltages; therefore, multirange voltmeters also need to be discussed. To make a voltmeter capable of measuring multiple ranges, an electrician needs several multiplier resistors that are switch-selectable for the different ranges desired. Reading the scale of a voltmeter is as simple as reading the scale of an ammeter. Some multirange voltmeters have only one range of values marked on the scale, and the scale reading must be multiplied by the range switch setting to obtain the correct voltage (*Figure 12*).

Other voltmeters have different ranges on the scale for each setting of the range switch. In this case, be sure that you read the set of values that correspond to the position of the range switch (*Figure 13*).

Some digital voltmeters are autoranging. These types of voltmeters do not have a range switch. The internal construction of the meter itself will select the proper resistance for the current being detected. However, when using a voltmeter that is *not* autoranging, always start with the highest voltage setting and work down until the indication reads somewhere between half-scale and three-quarter scale. This will give a more accurate reading and will prevent damage to the voltmeter.

5.0.0 ◆ OHMMETER

An ohmmeter is a device that measures the resistance of a circuit or component. It can also be used to locate open circuits or shorted circuits. Basically, an ohmmeter consists of a DC current meter movement, a low-voltage DC power source (usually a battery), and current-limiting resistors, all of which are connected in series (*Figure 14*).

This combination of devices allows the meter to calculate resistance by deriving it using Ohm's law. Before measuring the resistance of an un-

To find the voltage across a component, multiply the scale reading by the setting of the range switch: voltage = 0.6 × 10 = 6.0V.

106F12.EPS

Figure 12 ◆ Multirange single-scale voltmeter.

To find the voltage across a component, simply read the meter scale that corresponds to the position of the range switch: voltage = 6V.

106F13.EPS

Figure 13 ◆ Multirange multiscale voltmeter.

Instructor's Notes:

Voltmeter Applications

Common uses of a voltmeter range from testing a car battery to testing for deenergized high-voltage circuits before performing maintenance. Remember that when testing for voltage, you are testing for the presence of electrical potential rather than for the flow of current. What common applications can you think of?

Discuss the "Think About It." Answers will vary, from circuit troubleshooting to motor maintenance.

Figure 14 ◆ Ohmmeter schematic.

known resistor or electrical circuit, the test leads are shorted together. This causes current to flow through the meter movement and the pointer deflects to the right. This means that there is zero resistance present across the input terminals of the ohmmeter, and when zero resistance is present, the pointer will deflect full scale. Therefore, full-scale deflection of an ohmmeter indicates zero resistance. Most ohmmeters have a zero adjustment knob. This is used to correct for the fact that as the

batteries of the meter age, their output voltage decreases. As the voltage drops, the current through the circuit decreases, and the meter will no longer deflect full scale. By correcting for this change before each use, the internal resistance of the meter, along with the resistance of the leads, is lowered and nulled to deflect the pointer full scale.

After the ohmmeter is adjusted to zero, it is ready to be connected in a circuit to measure resistance. The circuit must be verified as deenergized by using a voltmeter prior to taking a reading with an ohmmeter. If the circuit were energized, its voltage could cause a damaging current to flow through the meter. This can cause damage to the meter and/or circuit as well as personal injury.

When making resistance measurements in circuits, each component in the circuit can be tested individually by removing the component from the circuit and connecting the ohmmeter leads across it. Actually, the component does not have to be totally removed from the circuit. Usually, the part can be effectively isolated by disconnecting one of its leads from the circuit. However, this method can still be somewhat time consuming. Some manufacturers provide charts that list the resistance readings that should be obtained from various test points to a reference point in the equipment. There are usually many parts of the circuit between the test point and the reference point, so if you get an abnormal reading, you must begin checking smaller groups of components, or individual components, to isolate the defective one. If resistance charts are not available, be very careful to ensure that other components are not in parallel with the component being tested.

Emphasize the importance of deenergizing the circuit before using an ohmmeter.

Using an Ohmmeter

Ohmmeters are commonly used to check for the mechanical integrity of circuits. For example, suppose that you wanted to determine if a wire in a house circuit had been broken by a sheetrock screw. How would you connect an ohmmeter to test this house circuit and what reading would you get from a broken circuit?

Discuss the "Think About It."

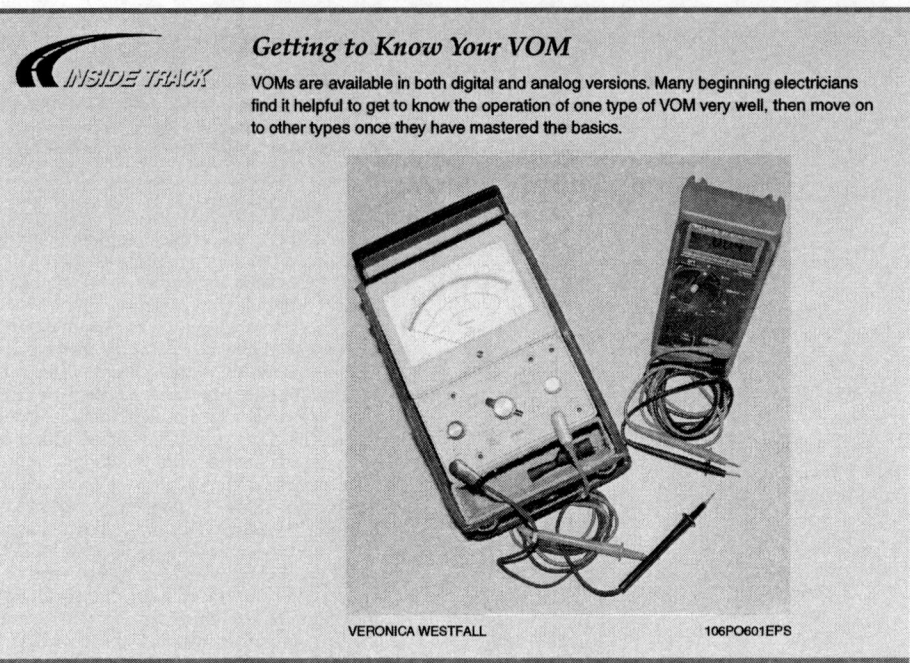

Getting to Know Your VOM

VOMs are available in both digital and analog versions. Many beginning electricians find it helpful to get to know the operation of one type of VOM very well, then move on to other types once they have mastered the basics.

VERONICA WESTFALL   106PO601EPS

Discuss the features of a typical VOM.

Show Transparency 16 (Figure 15).

6.0.0 ◆ VOLT-OHM-MILLIAMMETER

The volt-ohm-milliammeter (VOM), also known as the *multimeter*, is a multipurpose instrument. It is a combination of the three previous meters discussed: the milliammeter, the voltmeter, and the ohmmeter. One common analog multimeter is shown in *Figure 15*. There are many different models of this basic multimeter. To prevent having to discuss each and every meter, one version will be explained here. Any controls or functions on your meter that are not covered here should be reviewed in the applicable owner's manual.

The typical volt-ohm-milliammeter is a rugged, accurate, compact, easy-to-use instrument. The instrument can be used to make accurate measurements of DC and AC voltages, current, resistance, decibels, and output voltage. The output voltage function is used for measuring the AC component of a mixture of AC and DC voltages. This occurs primarily in amplifier circuits.

This meter has the following features: a 0–1 volt DC range, 0–500 volt DC and AC ranges, a TRANSIT position on the range switch, rubber

plug bumpers on the bottom of the case to reduce sliding, and an externally accessible battery and fuse compartment.

6.1.0 Specifications

The specifications of the example multimeter are:

- DC voltage:
  Sensitivity: 20KΩ per volt
  Accuracy: 1¾% of full scale
- AC voltage:
  Sensitivity: 5KΩ per volt
  Accuracy: 3% of full scale
- DC current:
  250mV to 400mV drop
  Accuracy: 1¾% of full scale
- Resistance:
  Accuracy: 1.75° of arc
  Nominal open circuit voltage 1.5V (9V on the 10KΩ ohm range)
- Nominal short circuit current:
  1Ω range: 1.25mA

ELECTRICAL LEVEL ONE—TRAINEE MODULE 26106

Instructor's Notes:

_____

_____

_____

_____

_____

_____

_____

_____

_____

RESISTANCE SCALE

AC VOLTAGE SCALE

DC VOLTAGE/ DC CURRENT SCALE

2.5 VOLTS AC VOLTAGE/ AC CURRENT SCALE

ZERO OHM ADJUSTMENT KNOB

FUNCTION SWITCH

JACKS

RANGE SWITCH

106F15.EPS

Figure 15 ◈ Multimeter.

Check Your Connections

When using a VOM, make sure your leads match the function. The meter can be in the voltage function and even display volts, but if the leads are in the amp jack, the meter will short circuit if connected to a voltage source.

Stress the importance of proper meter connections.

100Ω range: 1.25mA
10KΩ range: 75mA

• Meter frequency response:
Up to 100kHz

6.2.0 Overload Protection

In the example multimeter, a 1A, 250V fuse is provided to protect the circuits on the ohmmeter ranges. It also protects the milliampere ranges from excessive overloads. If the instrument fails to indicate, the fuse may be burned out. The fuse is mounted in a holder in the battery and fuse compartment. A spare fuse is located in a well between the + terminal of the D cell and the side of the

case. Access to the compartment is obtained by loosening the single captivating screw on the compartment cover. To replace a burned-out fuse, remove it from the holder and replace it with a fuse of the exact same type. When removing the fuse from its holder, first remove the battery.

In addition to the fuse, a varistor protects the indicating instrument circuit. The varistor limits the current through the moving coil in case of overload.

The fuse and varistor will prevent serious damage to the meter in most cases of accidental overload. However, no overload protection system is completely foolproof, and misapplication on high-voltage circuits can damage the instrument. Care

Discuss meter overload protection.

Review the storage and use guidelines for VOMs.

Review the safety precautions associated with the use of VOMs.

Demonstrate how to make DC voltage tests in various ranges using an analog VOM. (The instructor may need to revise the given procedures based upon the actual meter available for the demonstration.)

and caution should always be exercised to protect both you and the VOM.

Beside the actual steps used in making a measurement, some other points to consider while using a multimeter are:

• Keep the instrument in a horizontal position when storing and away from the edge of a workbench, shelf, or other area where it may be knocked off and damaged.

• Avoid rapid or extreme temperature changes. For example, do not leave the meter in your truck during hot or cold weather. Rapid and extreme temperature changes will advance the aging of the meter components and adversely affect meter life and accuracy.

• Avoid overloading the measuring circuits of the instrument. Develop a habit of checking the range position before connecting the test leads to a circuit. Even slight overloads can damage the meter. Even though it may not be noticeable in blown fuses or a bent needle, damage has been done. Slight overloads will advance the aging of components, again causing changes in meter life and accuracy.

• Place the range switch in the TRANSIT position when the instrument is not in use or when it is being moved. This reduces the swinging of the pointer when the meter is carried. Every meter does not have a TRANSIT position, but if the meter does, it should be used. Random, uncontrolled swings of the meter movement may damage the movement, bend the needle, or reduce its accuracy.

• If the meter has not been used for a long period of time, rotate the function and range switches in both directions to wipe the switch contacts. Most switch contacts are plated with copper or silver. Over a period of time, these materials will oxidize (tarnish). This will create a high resistance through the switch, causing a large inaccuracy. Rotating through the switch positions will clean the tarnish off and provide good electrical contact.

6.3.0 Making Measurements

 CAUTION

When using the meter as a millivoltmeter, care must be taken to prevent damage to the indicating instrument from excessive voltage. Before using the 250mV range, first use the 1.0V range to determine that the voltage measured is no greater than 250mV (or 0.25VDC).

6.3.1 Measuring DC Voltage, 0–250 Millivolts

Step 1 Set the function switch to +DC.

Step 2 Plug the black test lead into the − (COMMON) jack and the red test lead into the +50μA/250mV jack.

Step 3 Set the range switch to 50μA (dual position with 50V).

Step 4 Connect the black test lead to the negative side of the circuit being measured and the red test lead to the positive side of the circuit.

Step 5 Read the voltage on the black scale marked DC and use the figures marked 0–250. Read directly in millivolts.

6.3.2 Measuring DC Voltage, 0–1 Volt

Step 1 Set the function switch to −DC. Plug the black test lead into the − (COMMON) jack and the red test lead into the +1V jack.

Step 2 Set the range switch to 1V (dual position with 2.5V).

Step 3 Connect the black test lead to the negative side of the circuit being measured and the red test lead to the positive side of the circuit.

Step 4 Read the voltage on the black scale marked DC. Use the figures marked 0–10 and divide the reading by 10.

6.3.3 Measuring DC Voltage, 0–2.5 Through 0–500 Volts

 WARNING!

Be extremely careful when working with higher voltages. Do not touch the instrument test leads while power is on in the circuit being measured.

Step 1 Set the function switch to +DC.

Step 2 Set the range switch to one of the five voltage range positions marked 2.5V, 10V, 50V, 250V, or 500V. When in doubt as to the voltage present, always use the highest voltage range as a protection for the instrument. If the voltage is within a lower range, the switch may be set for the lower range to obtain a more accurate reading.

6–12

ELECTRICAL LEVEL ONE—TRAINEE MODULE 26106

Instructor's Notes:

_____

_____

_____

_____

_____

_____

_____

_____

_____

Step 3 Plug the black test lead into the −(COMMON) jack and the red test lead into the + jack.

Step 4 Connect the black test lead to the negative side of the circuit being measured and the red test lead to the positive side of the circuit.

Step 5 Read the voltage on the black scale marked DC. For the 2.5V range, use the 0–250 figures and divide by 100. For the 10V, 50V, and 250V ranges, read the figures directly. For the 500V range, use the 0–50 figures and multiply by 10.

6.3.4 Measuring DC Voltage, 0–1,000 Volts

 WARNING!
Be extremely careful when working with higher voltages. Do not touch the instrument test leads while power is on in the circuit being measured.

Step 1 Set the function switch to +DC.

Step 2 Set the range switch to 1,000V (dual position with 500V).

Step 3 Plug the black test lead into the −(COMMON) jack and the red test lead into the 1,000V jack.

Step 4 Be sure power is off in the circuit being measured and all capacitors have been discharged. Connect the black test lead to the negative side of the circuit being measured and the red test lead to the positive side of the circuit.

Step 5 Turn on the power in the circuit being measured.

Step 6 Read the voltage using the 0–10 figures on the black scale marked DC. Multiply the reading by 100.

6.3.5 Measuring AC Voltage, 0–2.5 Through 0–500 Volts

 WARNING!
Be extremely careful when working with higher voltages. Do not touch the instrument or test leads while power is on in the circuit being measured.

 CAUTION
When measuring line voltage such as from a 120V, 240V, or 480V source, be sure that the range switch is set to the proper voltage position.

Step 1 Set the function switch to AC.

Step 2 Set the range switch to one of the five voltage range positions marked 2.5V, 10V, 50V, 250V, or 500V. When in doubt as to the actual voltage present, always use the highest voltage range as a protection to the instrument. If the voltage is within a lower range, the switch may be set for the lower range to obtain a more accurate reading.

Step 3 Plug the black test lead into the − (COMMON) jack and the red test lead into the + jack.

Step 4 Connect the test leads across the voltage source (in parallel with the circuit).

Step 5 Turn on the power in the circuit being measured.

Step 6 For the 2.5V range, read the value directly on the AC scale marked 2.5V. For the 10V, 50V, and 250V ranges, read the red scale marked AC and use the black figures immediately above the scale. For the 500V range, read the red scale marked AC and use the 0–50 figures. Multiply the reading by 10.

6.3.6 Measuring AC Voltage, 0–1,000 Volts

WARNING!
Be extremely careful when working with higher voltages. Do not touch the instrument or test leads while power is on in the circuit being measured.

Step 1 Set the function switch to AC.

Step 2 Set the range switch to 1,000V (dual position with 500V).

Step 3 Plug the black test lead into the − (COMMON) jack and the red test lead into the 1,000V jack.

Demonstrate how to make AC voltage tests in various ranges using an analog VOM. (Adjust given procedures as necessary to reflect actual meter in use.)

Point out all applicable safety procedures during the demonstration.

Demonstrate how to measure output voltage using an analog VOM. (Adjust given procedure as necessary to reflect actual meter in use.)

Warn the trainees not to apply the output jack to a DC circuit with a voltage in excess of 400V.

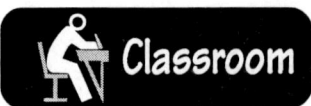

Describe and/or demonstrate how to make decibel measurements.

Step 4 Be sure the power is off in the circuit being measured and that all capacitors have been discharged. Connect the test leads to the circuit.

Step 5 Turn on the power in the circuit being measured.

Step 6 Read the voltage on the red scale marked AC. Use the 0–10 figures and multiply by 100.

6.3.7 Measuring Output Voltage

It is often desired to measure the AC component of an output voltage where both AC and DC voltage levels exist. This occurs primarily in amplifier circuits. The meter has a 0.1µF, 400V capacitor in series with the OUTPUT jack. The capacitor blocks the DC component of the current in the test circuit but allows the AC or desired component to pass on to the indicating instrument circuit. The blocking capacitor may alter the AC response at low frequencies but is usually ignored at audio frequencies.

> **CAUTION**
>
> When using OUTPUT, do not apply it to a circuit whose DC voltage component exceeds the 400V rating of the blocking capacitor.

Step 1 Set the function switch to AC.

Step 2 Plug the black test lead into the – (COMMON) jack and the red test lead into the OUTPUT jack.

Step 3 Set the range switch to one of the range positions marked 2.5V, 10V, 50V, or 250V.

Step 4 Connect the test leads across the circuit being measured, with the black test lead attached to the ground side.

Step 5 Turn on the power in the test circuit. Read the output voltage on the appropriate AC voltage scale. For the 2.5V range, read the value directly on the AC scale marked 2.5V. For the 10V, 50V, or 250V ranges, use the red scale marked AC and read the black figures immediately above the scale.

6.3.8 Measuring Decibels

For some applications, mockup audio frequency voltages are measured in terms of decibels. The decibel scale (dB) at the bottom of the dial is marked from −20 to +10.

Step 1 To measure decibels, read the dB scale in accordance with instructions for measuring AC. For example, when the range switch is set to the 2.5V position, read the dB scale directly.

Step 2 The dB readings on the scale are referenced to a 0dB power level of .001W across 600Ω, or .775VAC across 600Ω.

Step 3 For the 10V range, read the dB scale and add +12dB to the reading. For the 50V range, read the dB scale and add +26dB to the reading. For the 250V range, read the dB scale and add +40dB to the reading.

Step 4 If the 0dB reference level is .006W across 500Ω, subtract +7dB from the reading.

6.4.0 Direct Current Measurements

6.4.1 Voltage Drop

The voltage drop across the meter on all milli-ampere current ranges is approximately 250mV measured at the jacks. An exception is the 0–500mA range with a drop of approximately 400mV. This voltage drop will not affect current measurements. In some transistor circuits, however, it may be necessary to compensate for the added voltage drop when making measurements.

6.4.2 Measuring Direct Current, 0–50 Microamperes

> **CAUTION**
>
> Never connect the test leads directly across voltage when the meter is used as a current-indicating instrument. Always connect the instrument in series with the load across the voltage source.

Step 1 Set the function switch to +DC.

Step 2 Plug the black test lead into the – (COMMON) jack and the red test lead into the +50µA/250mV jack.

Step 3 Set the range switch to 50µA (dual position with 50V).

Step 4 Open the circuit in which the current is being measured. Connect the instrument in

6–14

Instructor's Notes:

_____

_____

_____

_____

_____

_____

_____

_____

_____

Voltage Testing

Voltage testing covers a very wide range of values. For example, if you test the voltage of a residential oil-fired, hot water furnace, the thermostat circuit will be 24V, the circulating pumps will be 120V, and the ignition transformer output will be 10,000V. Notice that the range of voltages on this one appliance can put you outside the range of many multimeters. The photo shown here depicts a voltage reading being taken of a low-voltage control device.

WALTER JOHNSON   106PO602.EPS

series with the circuit. Connect the red test lead to the positive side and the black test lead to the negative side.

Step 5 Read the current on the black DC scale. Use the 0–50 figures to read directly in microamperes.

Note

In all direct current measurements, be certain the power to the circuit being tested has been turned off before disconnecting test leads and restoring circuit **continuity.**

6.4.3 Measuring Direct Current, 0–1 Through 0–500 Milliamperes

Step 1 Set the function switch to +DC.

Step 2 Plug the black test lead into the − (COMMON) jack and the red test lead into the + jack.

Step 3 Set the range switch to one of the four range positions (1mA, 10mA, 100mA, or 500mA).

Step 4 Open the circuit in which the current is being measured. Connect the VOM in series with the circuit. Connect the red test lead to the positive side and the black test lead

Demonstration

Demonstrate how to make direct current measurements using an analog VOM. (Adjust given procedures as necessary to reflect actual meter in use.)

Demonstrate how to adjust an analog VOM to zero prior to making a resistance test.

Demonstrate how to measure resistance using an analog VOM.

Review the applicable safety precautions for making resistance tests.

to the negative side of the part of the circuit you are measuring.

Step 5 Read the current in milliamperes on the black DC scale. For the 1mA range, use the 0–10 figures and divide by 10. For the 10mA range, use the 0–10 figures and multiply by 10. For the 500mA range, use the 0–50 figures and multiply by 10.

6.4.4 Measuring Direct Current, 0–10 Amperes

Step 1 Plug the black test lead into the −10A jack and the red test lead into the +10A jack.

Step 2 Set the range switch to 10A (dual position with 10mA).

Step 3 Open the circuit in which the current is being measured. Connect the instrument in series with the circuit. Connect the red test lead to the positive side and the black test lead to the negative side.

Note
The function switch has no effect on polarity for the 10A range.

Step 4 Read the current on the black DC scale. Use the 0–10 figures to read directly in amperes.

CAUTION
When using the 10A range, never remove a test lead from its panel jack while current is flowing through the circuit. Otherwise, damage may occur to the plug and jack.

6.4.5 Zero Ohm Adjustment

When resistance is measured, the VOM batteries furnish power for the circuit. Since batteries are subject to variation in voltage and internal resistance, the instrument must be adjusted to zero prior to measuring a resistance.

Step 1 Set the range switch to the desired ohms range.

Step 2 Plug the black test lead into the − (COMMON) jack and the red test lead into the + jack.

Step 3 Connect the ends of the test leads to short the VOM resistance circuit.

Step 4 Rotate the ZERO OHM control until the pointer indicates zero ohms. If the pointer cannot be adjusted to zero, one or both of the batteries must be replaced.

Step 5 Disconnect the ends of the test leads and connect them to the component being measured.

6.4.6 Measuring Resistance

CAUTION
Before measuring resistance, be sure power is off to the circuit being tested. Disconnect the component from the circuit before measuring its resistance.

Step 1 Set the range switch to one of the resistance range positions:
- Use R × 1 for resistance readings from 0 to 200 ohms.
- Use R × 100 for resistance readings from 200 to 20,000 ohms.
- Use R × 10,000 for resistance readings above 20,000 ohms.

Step 2 Set the function switch to either the −DC or +DC position. The operation is the same in either position.

Current Testing

Levels of current to be tested vary from milliamperes to thousands of amps. In modern communications and consumer electronic devices, you will be measuring for mA. In household circuits you usually test 15A to 50A circuits or higher when testing the service entrance. A high amperage test would be performed on an industrial service entrance, which might go up to 12,000A.

Instructor's Notes:

_____

_____

_____

_____

_____

_____

_____

_____

_____

Step 3 Adjust the ZERO OHM control for each resistance range.
- Observe the reading on the OHMS scale at the top of the dial. Note that the OHMS scale reads from right to left for increasing values of resistance.
- To determine the actual resistance value, multiply the reading by the factor at the switch position (K on the OHMS scale equals one thousand).

Step 4 If there is a forward and backward resistance such as in diodes, the resistance should be relatively low in one direction (for forward polarity) and higher in the opposite direction.

CAUTION
Check that the OHMS range being used will not damage any of the semiconductors.

Step 5 If the purpose of the resistance measurement is to check a semiconductor in or out of a circuit (forward and reverse bias resistance measurements), check the following prior to making the measurement:
- The polarity of the voltage at the input jacks is identical to the input jack markings. Therefore, be certain that the polarity of the test leads is correct for the application.
- Ensure that the range selected will not damage the semiconductor (use R × 100 or below).
- Refer to the meter specifications and review the limits of the semiconductor according to the manufacturer's ratings.
  – If the semiconductor is a silicon diode or conventional silicon transistor, no precautions are normally required.
  – If the semiconductor material is germanium, check the ratings of the device and refer to its specifications.

Step 6 Rotate the function switch between the two DC positions to reverse polarity. This will determine if there is a difference between the resistance in the two directions.

Step 7 The resistance of such diodes will measure differently from one resistance range to an-

other on the same VOM with the function switch in a given position. For example, a diode that measures 80Ω on the R × 1 range may measure 300Ω on the R × 100 range. The difference in values is a result of the diode characteristic and does not indicate any fault in the VOM.

7.0.0 ◆ DIGITAL METERS

Digital meters have revolutionized the test equipment world. Improved accuracy is very easily attainable, more functions can be incorporated into one meter, and both autoranging and automatic polarity indication can be used. Technically, digital multimeters are classified as electronic multimeters; however, digital multimeters do not use a meter movement. Instead, a digital meter's input circuit converts a current into a digital signal, which is then processed by electronic circuits and displayed numerically on the meter face.

A major limitation with many meters that use meter movements is that the scale reading must be estimated if the meter pointer falls between scale divisions. Digital multimeters eliminate the need to estimate these readings by displaying the reading as a numerical display.

With digital meters, technicians must revise the way the indications are viewed. For example, if a technician were reading the AC voltage on a normal wall outlet with an analog voltmeter, any indication within the range of 120VAC would be considered acceptable. But, when reading with a digital meter, the technician might think something was wrong if the meter showed a reading of 114.53VAC. Bear in mind that the digital meter is very precise in its reading, sometimes more precise than is called for, or is usable. Also, be aware that the indicated parameter may change with the range used. This is primarily due to the change in accuracy and where the meter is rounding off.

There are many types of digital multimeters. Some are bench-type multimeters, while others are designed to be handheld. Most types of digital multimeters have an input impedance of 10 megohms and above. They are very sensitive to small changes in current and are therefore very accurate.

An example of a digital meter is shown in *Figure 16*. The internal operation of this meter is basically the same as other digital meters. The following paragraphs discuss the operation and use of this particular meter. For specific instructions, always refer to the owner's manual supplied with your meter.

Provide an overview of digital meters and their applications.

Show Transparency 17 (Figure 16).

Digital Multimeter Classifications

Newer digital multimeters are rated for safety according to voltage and current limitations and fault interrupting capacity. Never use a multimeter outside of the limits specified by the manufacturer.

VERONICA WESTFALL   106PO603.EPS

Check Your Meter Before Using It

Never assume that a blank meter represents a reading of zero volts. Some digital meters automatically cut off the power to preserve the battery.

Similarly, if you press the HOLD button on some meters when you are reading 0V, it will lock on that reading and will continue to read 0V, regardless of the actual voltage present. Always check the meter for proper operation before using it.

Discuss the "Think About It."

Digital Meters

Most applications do not require the precision of a digital meter. Identify some applications in which precision is important. When wouldn't it be important?

Instructor's Notes:

_____
_____
_____
_____
_____
_____
_____
_____
_____
_____

Figure 16 ◆ Digital meter.

7.1.0 Features

The example meter offers the following features:

- *Autorange/manual range modes*—The meter features autoranging for all measurement ranges. Press the RANGE button to enter manual range mode. A flashing symbol may be used to show that you are in the manual range mode. Press the RANGE button as required to select the desired range. To switch back to auto range, press the ON/CLEAR button once (clear mode) or select another function.
- *Automatic off*—The example meter turns itself off after one hour of non-use. The current draw while the meter is turned off does not affect battery life. If the meter turns itself off while a parameter is being monitored, press the ON/CLEAR button to turn it on again. To protect against electrical damage, the meter also turns itself off if a test lead is inserted into the 10A jack while the meter is in any mode other than A ⎓ or A ∼.
- *Dangerous voltage indication*—The meter shows the symbol for any range over 20V. In the autoranging mode, the meter also beeps when it changes to any range over 20V.

- *Overload display*—The meter displays OL and a rapidly flashing decimal point (position determined by range) when the input is too large to display.
- *Audible acknowledgment*—The meter acknowledges each press of a button or actuation of the selector switch with a beep.

7.2.0 Operation

This section will discuss the use of various controls and explain how measurements should be taken.

7.2.1 Dual Function ON/CLEAR Button

Press the ON/CLEAR button to turn the meter on. Operation begins in the autorange mode, and the range for maximum resolution is selected automatically. Press the ON/CLEAR button again to turn the meter off.

7.2.2 Measuring Voltage

Step 1 Select V ⎓ or V ∼.

Step 2 Connect the test leads as shown in *Figure 17*.

Step 3 Observe the voltage reading on the display. Depending on the range, the meter displays units in mV or V.

To avoid shock hazard or meter damage, do not apply more than 1,500VDC or 1,000VAC to the meter input or between any input jack and earth ground.

7.2.3 Measuring Current

Step 1 Select A ⎓ or A ∼.

Step 2 Insert the meter in series with the circuit with the red lead connected to either:
- The mA jack for input up to 200 milliamps
- The 10A jack for input up to 20 amps

Step 3 Make hookups as shown in *Figures 18* and *19*.

Step 4 Observe the current reading.

 Note

The meter shuts itself off if a test lead is inserted into the 10A jack when the meter is in any function other than A ⎓ or A∼.

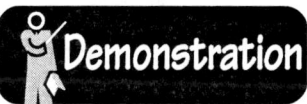

Review the features of a typical digital VOM.

Demonstrate how to make voltage and current measurements using a digital VOM. (Revise procedures as necessary to reflect actual meter in use.)

![Audiovisual]

Show Transparencies 18 through 20 (Figures 17 through 19).

Figure 17 ◆ Measuring voltage.

Figure 18 ◆ Measuring current (mA).

ELECTRICAL LEVEL ONE—TRAINEE MODULE 26106

Instructor's Notes:

ELECTRICAL LEVEL ONE—INSTRUCTOR'S GUIDE MODULE 26106

10A MAXIMUM CONTINUOUS OR 20A FOR 30 SECONDS MAXIMUM

RED (+)

BLACK (−)

RED (+)

BLACK (−)

0-10A

LOAD

LOAD

LOAD

LOAD

106F19.EPS

Figure 19 ◆ Measuring current (amps).

7.2.4 Measuring Resistance

When measuring resistance, any voltage present will cause an incorrect reading. For this reason, the capacitors in a circuit in which resistance measurements are about to be taken should first be discharged.

Step 1 Select Ω (ohms).

Step 2 Connect the test leads as shown in *Figure 20*.

Step 3 Observe the resistance reading.

7.2.5 Diode/Continuity Test

Step 1 Select ▸▸ (diode/continuity).

Step 2 Choose one of the following:
- *Forward bias*—Connect to the diode as shown in *Figure 21(A)*. The meter will display one of the following: the forward voltage drop (V_F) of a good diode (<0.7V), a very low reading for a shorted diode (<0.3V), or OL for an open diode.

- *Reverse bias or open circuit*—Reverse the leads to the diode. The meter displays OL, as shown in *Figure 21(B)*. It does not beep.
- *Continuity*—The meter beeps once if the circuit resistance is less than 150 ohms, as shown in *Figure 21(C)*.

7.2.6 Transistor Junction Test

Test transistors in the same manner as diodes by checking the two diode junctions formed between the base and emitter, and the base and collector of the transistor. *Figure 22* shows the orientation of these effective diode junctions for PNP and NPN transistors. Also check between the collector and emitter to determine if a short is present.

7.2.7 Display Test

To test the LCD display, hold the ON/CLEAR button down when turning on the meter. Verify that the display shows all segments (see *Figure 23*).

IN-CIRCUIT MEASUREMENTS:
DISCHARGE CAPACITORS BEFORE
MEASURING RESISTANCE

BLACK (−)　　RED (+)

106F20.EPS

Figure 20 ◆ Measuring resistance.

Understanding Diodes

A diode is a two-terminal semiconductor device that behaves like a check valve. *Forward bias* means that you are testing that the diode conducts properly when it is closed. *Reverse bias* means that current does not flow through the diode.

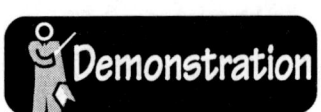

Demonstrate the maintenance procedures required for the proper care of a VOM.

Show Transparency 25 (Figure 24).

7.3.0 Maintenance

The following sections discuss the necessary maintenance for a multimeter.

7.3.1 Battery Replacement

Replace the battery as soon as the meter's decimal point starts blinking during normal use; this indicates that <100 hours of battery life remain. Remove the case back and replace the battery with the same or equivalent 9V alkaline battery.

7.3.2 Fuse Replacement

The meter uses two input protection fuses for the mA and 10A inputs. Remove the case back to gain access to the fuses. Replace with the same type only. The large fuse should be readily available. A

spare for the smaller fuse is included in the case. If necessary, this fuse must be reordered from the factory.

7.3.3 Calibration

Have a qualified technician calibrate the meter once a year.

Step 1 Remove the case back (*Figure 24*). Turn the meter ON and select V ⎓.

Step 2 Apply +1.900VDC +/−0.0001V to the V-n input (negative to COM).

Step 3 Adjust the DC control through the hole in the circuit board for a display of +1.900V.

Step 4 Select V.

ELECTRICAL LEVEL ONE—TRAINEE MODULE 26106

Instructor's Notes:

_____

_____

_____

_____

_____

_____

_____

_____

_____

Figure 21 ◆ Diode/continuity test.

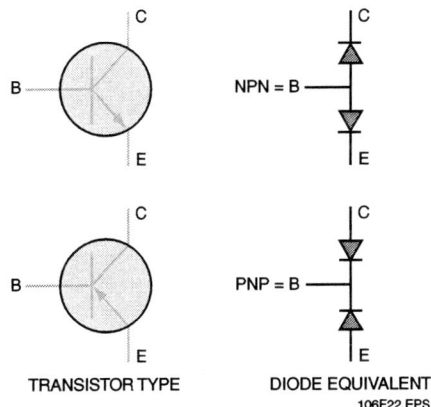

TRANSISTOR TYPE   DIODE EQUIVALENT

106F22.EPS

Figure 22 ◆ Transistor junction test.

Step 5 Apply +1.900VAC +/−0.002VAC@ 60Hz to the V-n input.

Step 6 Adjust the AC control through the hole in the circuit board for a display of +1.900V.

Step 7 Reassemble the meter.

106F23.EPS

Figure 23 ◆ Display test.

8.0.0 ◆ WATTMETER

Rather than performing two measurements and then calculating power, a power-measuring meter called a *wattmeter* can be connected into a circuit to measure power. The power can be read directly from the scale of this meter. Not only does a wattmeter simplify power measurements, but it has two other advantages:

First, voltage and current in an AC circuit are not always in phase; current sometimes either leads or lags the voltage (this is known as the *power factor*). When this happens, multiplying the voltage times the current results in apparent power, not true power. Therefore, in an AC circuit, measuring the voltage and current and then multiplying them can

Discuss the purpose and operation of a basic wattmeter.

BATTERY

SPARE
FUSE
IN HOLDER

FOOT OF
METER

DC CONTROL   AC CONTROL

106F24.TIF

Figure 24 ◆ Meter maintenance.

Show Transparency 26
(Figure 25).

often result in an incorrect value of power dissipation by the circuit. However, the wattmeter takes the power factor into account and always indicates true power.

Second, voltmeters and ammeters consume power. The amount consumed depends on the levels of the voltage and the current in the circuit, and it cannot be accurately predicted. Therefore, very accurate power measurements cannot be made by measuring voltage and current and then calculating power. However, some wattmeters compensate for their own power losses so that only the power dissipated in the circuit is measured. If the wattmeter is not compensated, the power that is dissipated is sometimes marked on the meter, or it can easily be determined so that a very accurate measurement can be made. Typically, the accuracy of a wattmeter is within 1 percent.

The basic wattmeter consists of two stationary coils connected in series and one movable coil (*Figure 25*). The moving coil, wound with many turns of fine wire, has a high resistance.

The stationary coils, wound with a few turns of a larger wire, have a low resistance. The interaction of the magnetic fields around the different coils will cause the movable coil and its pointer to

CURRENT
COIL

VOLTAGE
COIL

SOURCE
(CONNECTED IN
PARALLEL)

LOAD (CONNECTED IN SERIES)

106F25.EPS

Figure 25 ◆ Wattmeter schematic.

ELECTRICAL LEVEL ONE—TRAINEE MODULE 26106

Instructor's Notes:

Wattmeter

Wattmeters are less commonly used than multimeters, since the electrician is usually interested in the existence of electrical potential or the flow of current. The meter used to measure power consumption is the watt-hour meter attached to the service entrance.

VERONICA WESTFALL   106PO604.EPS

rotate in proportion to the voltage across the load and the current through the load. Thus, the meter indicates E times I, or power.

The two circuits in the wattmeter will be damaged if too much current passes through them. This fact is of special importance because the reading on the meter does not tell the user that the coils are being overheated. If an ammeter or voltmeter is overloaded, the pointer will indicate beyond full-scale deflection. In a wattmeter, both the current and potential (voltage) circuits may be carrying such an overload that their insulation is burning, and yet the pointer may only be partway up the scale. A low-power factor circuit will give a low reading on the wattmeter even when the current and potential circuits are loaded to their maximum safe limits.

9.0.0 ◆ MEGOHMMETER (MEGGER)

An ordinary ohmmeter cannot be used for measuring resistances of several million ohms, such as those found in conductor insulation or between motor or transformer windings, and so on. To ad-

equately test these types of very high resistances, it is necessary to use a much higher potential than is furnished by the battery of an ohmmeter. For this purpose, a megger is used. There are three types of meggers: hand, battery, and electric.

The megger is similar to a moving-coil meter except that it has two windings (coils). See *Figure 26*. Coil A is in series with resistor R_2 across the output of the generator.

This coil is wound so that it causes the pointer to move toward the high-resistance end of the scale when the generator is in operation. Winding B is in series with R_1 and R_X (the unknown resistance to be measured). This winding is wound so that it causes the pointer to move toward the low or zero-resistance end of the scale when the generator is in operation.

When an extremely high resistance appears across the input terminals of the megger, the current through coil A causes the pointer to read infinity. Conversely, when a relatively low resistance appears across the input terminals, the current through coil B causes the pointer to deflect toward zero. The pointer stops at a point on the

Describe the basic operation and applications of a megohmmeter.

Show Transparencies 27 and 28 (Figures 26 and 27).

Caring for Your Meter

Like the other meters discussed here, a megger is a sensitive instrument. Treat it with care and keep it in its case when not in use.

VERONICA WESTFALL

106PO605.EPS

Figure 26 ◆ Megger schematic.

106F26.EPS

Instructor's Notes:

106F27.EPS

Figure 27 ◆ Digital readout megger.

scale determined by the current through coil B, which is controlled by R_X.

Digital meggers (*Figure 27*) use the same operational principles. Instead of having a scaled meter movement, these meters give the value of resistance in a digital readout display. The digital readout makes reading the measurement much easier and helps to eliminate errors.

To avoid excessive test voltages, most hand meggers are equipped with friction clutches. When the generator is cranked faster than its rated speed, the clutch slips, and the generator speed and output voltage are maintained at their rated values.

9.1.0 Safety Precautions

> **WARNING!**
> When a megger is used, the generator voltage is present on the test leads. This voltage could be hazardous to you or the equipment you are testing. *NEVER TOUCH THE TEST LEADS WHILE THE TESTER IS BEING USED.* Isolate the item you are testing from the circuit before using the megger. Protect all parts of the test subject from contact by others.

When using a megger, you could be injured or cause damage to the equipment being worked on if you do not observe the following minimum safety precautions:

- Use meggers on high-resistance measurements *only* (such as insulation measurements or to check two separate conductors in a cable).
- *Never* touch the test leads while the handle is being cranked.

- Deenergize and verify the deenergization of the circuit completely before connecting the meter.
- Disconnect the item being checked from other circuitry, if possible, *before* using the meter.
- After the test, ground the tested circuit to discharge any energy that may be left in the circuit.

10.0.0 ◆ FREQUENCY METER

Frequency is the number of cycles completed each second by a given AC voltage, and it is usually expressed in hertz (one hertz = one cycle per second). The frequency meter is used in AC power-producing devices such as generators to ensure that the correct frequency is being produced. Failure to produce the correct frequency will result in excess heat and component damage.

There are two common types of frequency meters. One operates with a set of reeds having natural vibration frequencies that respond in the range being tested. The reed with a natural frequency closest to that of the current being tested will vibrate most strongly when the meter operates. The frequency is read from a calibrated scale.

A moving-disk frequency meter works with two coils, one of which is a magnetizing coil whose current varies inversely with the frequency. A disk with a pointer mounted between the coils turns in the direction determined by the stronger coil. Solid-state frequency meters are also available.

10.1.0 Phase Rotation Tester

A phase rotation tester is an indicator that allows you to see the phase rotation of incoming current.

10.2.0 Infrared Sensing Device

An infrared sensing device is an optical device that measures the infrared heat emitted from an object.

10.3.0 Cable-Length Meter

A cable-length meter measures the length and condition of a cable by sending a signal down the cable and then reading the signal that is reflected back.

10.4.0 Harmonic Test Set

A harmonic is a sine wave whose frequency is a whole number multiple of the original base frequency. For example, a standard 60Hz sine wave may have second and third harmonics at 120Hz and 180Hz, respectively. Harmonics may be caused by

Stress the safety precautions associated with the use of megohmmeters.

Discuss the operation and applications of frequency meters.

Discuss the operation and applications of power factor meters. Demonstrate how to calculate the power factor if a power factor meter is not available.

Show Transparency 29 (Figure 28).

Discuss the operation and applications of continuity testers.

various circuit loads such as fluorescent lights and by certain three-phase transformer connections.

The results of harmonics are heating in wiring and a voltage or current that cannot be detected by most digital meters.

11.0.0 ◆ POWER FACTOR METER

The power factor is the ratio of true (actual) power to apparent power. The power factor of a circuit or piece of equipment may be found by using an ammeter, wattmeter, and voltmeter. To calculate the power factor, divide the wattmeter reading (true power) by the product of the ammeter and voltmeter readings (apparent power or EI). The ideal power factor is 1.

$$\text{Power Factor} = \frac{\text{True Power (Wattmeter Reading)}}{\text{Apparent Power (EI)}}$$

It is not necessary to calculate these readings if a power factor meter is available to read the power factor directly (*Figure 28*). This meter indicates the equivalent of pure resistance or unit power factor, which is a one-to-one ratio.

12.0.0 ◆ CONTINUITY TESTER

An ohmmeter can be used to test continuity, but this means carrying an expensive, often bulky test instrument with you. Pocket-type continuity testers are just as reliable and much more compact and portable.

There are two types of continuity testers: audio and visual. These are specialized devices for identifying conductors in a conduit by checking continuity.

12.1.0 Audio Continuity Tester

This type of tester is used to ring out wires in conduit runs. At one end of the conduit run, select one wire, strip off a little insulation, and connect that wire to the conduit. At the other end of the conduit run, clip one lead of the tester to the conduit. Touch the other lead to one wire at a time until the audible alarm sounds, which indicates continuity (a closed circuit). Then, put matching tags on the wire. Continue this procedure, one wire at a time, to identify the other wires.

12.2.0 Visual Continuity Tester

This type of tester is useful if you are working in an area where background noise might make it hard to hear the audio tester. The procedure is the same. When the proper wire is tested, the light will come on.

13.0.0 ◆ VOLTAGE TESTER

A voltage tester is a simple aid that determines whether there is a potential difference between two points. It *does not* calculate the value of that difference. If the actual value of the potential difference is needed, use a voltmeter.

Figure 28 ◆ Power factor meter.

Instructor's Notes:

_____

_____

_____

_____

_____

_____

_____

_____

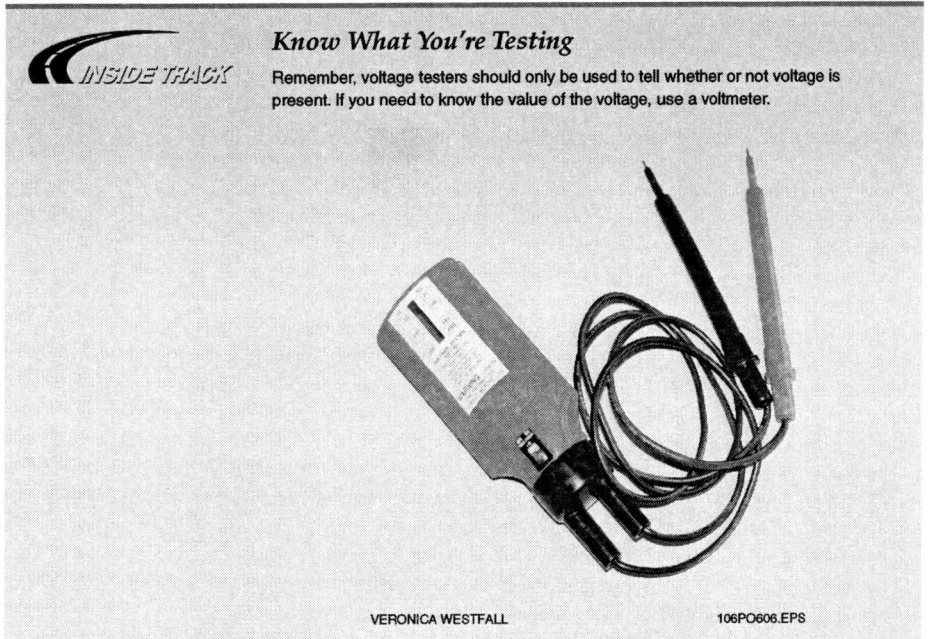

Know What You're Testing

Remember, voltage testers should only be used to tell whether or not voltage is present. If you need to know the value of the voltage, use a voltmeter.

VERONICA WESTFALL   106PO606.EPS

13.1.0 Wiggy®

Pocket-type voltage testers are inexpensive and portable. They can easily be carried in your tool pouch, eliminating the need to carry a delicate voltmeter on the job. Simple neon testers are becoming quite popular. Another type is known as a *Wiggy®* (see *Figure 29*).

13.1.1 Principle of Operation

The operation of a Wiggy® is fairly simple. The basic component is a solenoid. When a current flows through the coil, it will produce a magnetic field, which will pull the plunger down against a spring. The spring will limit how far the plunger can be drawn into the cylinder. As the current increases, the plunger will move farther. The amount of current depends upon the potential difference applied across the coil. A pointer on the plunger indicates the potential difference on the scale.

The scale on the tester has voltage indications for AC on one side of the pointer and DC on the other side. The AC scale indicates 120, 240, 480,

CAUTION

This is an approximation only. If you need to know the actual value of the voltage, use a voltmeter.

CAUTION

To avoid damage to the tester, do not use a Wiggy® above its stated range.

and 600 volts. The DC scale indicates 125, 240, and 600 volts.

To use a voltage tester, place the probes across a possible source of voltage. If there is voltage present, current flows through the coil inside the tester, creating a magnetic field. The magnetized coil pulls the indicator along the scale until it reaches a point corresponding to the approximate

Discuss the operation and applications of voltage testers.

Show Transparency 30 (Figure 29).

Demonstrate how to use a voltage tester to check for the presence of voltage.

Emphasize the safety precautions associated with the use of voltage testers.

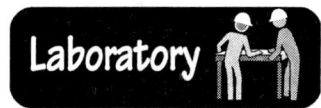

Have the trainees practice using a voltage tester.

106F29.EPS

Figure 29 ◆ Wiggy® voltage tester.

voltage. If there is no voltage, there is no current flow, so the indicator on the scale does not move, and no voltage reading is displayed.

The range of voltage and the type of current (AC and/or DC) that a voltage tester is capable of measuring are usually indicated on the scales that display the reading. The scales for each type of current are marked accordingly.

The methods used to show voltage readings vary from one type of voltage tester to another. For example, there are voltage testers that have lights to indicate the approximate amount of voltage registered. Both lights and scales have relatively broad readout ranges because these voltage testers can only indicate approximate values, not precise voltage measurements.

A voltage tester should always be checked before each use to make sure that it is in good condition and is operating correctly. The external check of the tester should include a careful inspection of the insulation on the leads for cracks or frayed areas. Faulty leads constitute a safety hazard, so they must be replaced. As a check to make sure that the voltage tester is operating correctly, the probes of the tester are connected to a

power source that is known to be energized. The voltage indicated on the tester should match the voltage of the power source. If there is no indication, the voltage tester is not operating correctly, and it must be repaired or replaced. It must also be repaired or replaced if it indicates a voltage different from the known voltage of the source.

It is essential to check a voltage tester before use. A faulty voltage tester can be dangerous to the electrician using it and to other personnel. For example, damaged insulation or a cracked casing could expose the electrician to electrical shock. Also, a faulty voltage tester might indicate that power is off when it is really on. This would create a serious safety hazard for personnel involved in equipment repair. The face plate or the scale on the front of the tester should be checked before a voltage tester is used to be sure that the tester can handle the amount of voltage that the power source may contain. Care should be taken when placing the probes of the tester across the power source. A voltage tester is designed to take a quick reading. If the probes are left in the circuit too long, the tester will burn out. A voltage tester should never be connected for more than a few seconds at a time.

Instructor's Notes:

CAUTION

Do not use a voltage tester in hazardous locations.

Voltage testers are used to make sure that power is available when it is needed and to make sure that power has been cut off when it should have been. In a troubleshooting situation, it might be necessary to verify that power is available in order to be sure that lack of power is not the problem. For example, if there were a problem with a power tool, such as a drill, a voltage tester might be used to make sure that power is available to run the drill. A voltage tester might also be used to verify that there is power available to a three-phase motor that will not start.

For safety reasons, it is always necessary to make sure that the power is turned off before working on any electrical equipment. A voltage tester can be used for such a test.

Keep the following in mind when using a voltage tester:

• Check the tester before each use.
• Handle the tester as if it were a calibrated instrument.
• Use good safety practices when operating the tester.
• Do not use circuits expected to be above the scale on the tester.
• Do not use if damage is indicated to the tester.
• Do not use in classified, hazardous areas or in high-frequency circuits.

14.0.0 ◆ RECORDING INSTRUMENTS

The term *recording instrument* describes many instruments that make a permanent record of measured quantities over a period of time. Recording instruments can be divided into three general groups:

• Instruments that record electrical quantities, including potential difference, current, power, resistance, and frequency.
• Instruments that record nonelectrical quantities by electrical means (such as a temperature recorder that uses a potentiometer system to record thermocouple output).
• Instruments that record nonelectrical quantities by mechanical means (such as a temperature recorder that uses a bimetallic element to move a pen across an advancing strip of paper).

It is often necessary to know the conditions that exist in an electrical circuit over a period of time to determine such things as peak loads, voltage fluctuations, etc. It may be neither practical nor economical to assign a worker to watch an indicating instrument and record its readings. An automatic recording instrument can be connected to take continuous readings, and the record can be collected for review and analysis.

Recording instruments are basically the same as the indicating meters already covered, but they have recording mechanisms attached to them. They are generally made of the same parts, use the same electrical mechanisms, and are connected in the same way. The only basic difference is the permanent record.

14.1.0 Strip-Chart Recorders

Strip-chart recorders are the most widely used recording instruments for electrical measurement. Their name comes from the fact that the record is made on a strip of paper, usually four to six inches wide and perhaps up to 60 feet long. These can be used to record either voltage or current. A recording voltmeter is shown in *Figure 30*.

Strip-chart recorders offer several advantages in electrical measurement. The long charts allow the recording to cover a considerable length of time with little attention, and strip-chart recorders

CURRENT-SENSING ATTACHMENT

106F30.EPS

Figure 30 ◆ Recording voltmeter.

Classroom

Discuss the operation and applications of various recording instruments.

Audiovisual

Show Transparency 31 (Figure 30).

Review the basic safety precautions that apply to the use of test equipment.

can be operated at a relatively high speed to provide very detailed records.

14.1.1 Typical Strip-Chart Recorder

A typical strip-chart recorder includes the following parts:

- The frame, which supports the other parts
- The moving system, which consists of the parts that move as a direct result of variations in the quantity being measured
- The graphic record, which is the line traced on the chart
- The chart carriage, which consists of the clock mechanism, timing gears, timing drum, chart spool, rewinding mechanism, and writing table
- A scale on the paper strip, which makes it possible to read the values of the quantity being measured
- The inking system, which consists of a special pen with an ink reservoir
- A case, which provides convenience in installation and removal, as well as in replacement of the chart paper

The moving parts used in recording instruments are basically the same as those used in the indicating instrument used to measure the same electrical quantity. However, the parts of a recording instrument are larger and require more power because of the added friction of the pen on the chart paper.

The paper chart is graduated in two directions. In one direction, the graduations correspond to the indicating scale of the instrument. In the other, they indicate time (seconds, minutes, hours, or days depending on the clock mechanism). The time graduations are uniformly spaced because the chart moves at a constant speed.

The chart carriage includes the entire mechanism for handling the chart paper. At the top, it holds the roll of paper. Just above the roll, and directly beneath the pen, is the writing table. The table is a metal plate that provides a flat surface under the chart paper at all points on the arc of the pen's motion. This prevents the pen from puncturing or tearing the paper.

The heart of the chart carriage is the clock mechanism, which drives the timing drum. The mechanism may be driven by an electric clock. Some of these mechanisms also contain a motor-wound spring clock or a battery-powered clock for a backup in the event of a loss of power.

The timing drum is connected to the clock through a gear unit, which determines the speed

of the drum. Drive pins around the timing drum fit into holes at the edges of the chart paper.

The inking system usually includes an inkwell containing enough ink for a considerable time.

15.0.0 ◆ SAFETY

In the interest of safety, all test equipment should be inspected and tested before being taken to the job site. A thorough visual inspection, checking for broken meters or knobs, damaged plugs, or frayed cords is important.

Perform an operational check. For example:

- On an ohmmeter, short the probes and ensure that you can zero the meter.
- With an oscilloscope, make sure that you can obtain a trace, and if a test signal is available, connect a test probe and check it.
- A voltmeter can be checked against an AC wall receptacle or a battery.

If a meter has a calibration sticker, check to see if it has been calibrated recently. For precise measurements, a recently calibrated meter is a more reliable instrument.

Every person who works with electronic equipment should be constantly alert to the hazards to which personnel may be exposed, and should also be capable of rendering first aid. The hazards considered in this section are: electric shock, burns, and related hazards.

Safety must be the primary responsibility of all personnel. The installation, maintenance, and operation of electrical equipment enforces a stern safety code. Carelessness on the part of the technician or operator can result in serious injury or death due to electrical shock, falls, burns, flying objects, etc. After an accident has occurred, investigation almost invariably shows that it could have been prevented by the exercise of simple safety precautions and procedures. Each person concerned with electrical equipment is responsible for reading and becoming thoroughly familiar with the safety practices and procedures contained in all safety codes and equipment technical manuals *before* performing work on electrical equipment. It is your personal responsibility to identify and eliminate unsafe conditions and unsafe acts that cause accidents.

You must bear in mind that deenergizing main supply circuits by opening supply switches will not necessarily deenergize all circuits in a given piece of equipment. A source of danger that has often been neglected or ignored, sometimes with tragic results, is the input to electrical equipment

Instructor's Notes:

_____

_____

_____

_____

_____

_____

_____

_____

_____

from other sources, such as backfeeds. Moreover, the rescue of a victim shocked by the power input from a backfeed is often hampered because of the time required to determine the source of power and isolate it. Therefore, turn off *all* power inputs before working on equipment and tag and lock out, then check with an operating tester (e.g., wattmeter) to be sure that the equipment is safe to work on. Take the time to be safe when working on electrical circuits and equipment. Carefully study the schematics and wiring diagrams of the entire system, noting what circuits must be deenergized in addition to the main power supply. Remember, electrical equipment commonly has more than one source of power. Be certain that all power sources are deenergized before servicing the equipment. Do not service any equipment with the power on unless absolutely necessary. Remember that the 115V power supply voltage is not a low, relatively harmless voltage but is the voltage that has caused more deaths than any other medium.

Safety can never be stressed enough. There are times when your life literally depends on it. The following is a listing of common-sense safety precautions that must be observed at all times:

- Use only one hand when turning power switches on or off. Keep the doors to switch and fuse boxes closed except when working inside or replacing fuses. Use a fuse puller to remove cartridge fuses, after first making certain that the circuit is dead.
- Your company will make the determination as to whether or not you are qualified to work on an electrical circuit.
- Do not work with energized equipment by yourself; have another person (safety observer), qualified in first aid for electrical shock, present at all times. The person stationed nearby should also know which circuits and switches control the equipment, and that person should be given instructions to pull the switch immediately if anything unforeseen happens.
- Always be aware of the nearness of high-voltage lines or circuits. Use rubber gloves where applicable and stand on *approved* rubber matting. Not all rubber mats are good insulators.
- Inform those in charge of operations as to the circuit on which work is being performed.
- Keep clothing, hands, and feet dry. When it is necessary to work in wet or damp locations, use a dry platform and place a rubber mat or other nonconductive material on top of the wood. Use insulated tools and insulated flashlights of the molded type when required to work on exposed parts.
- Do *not* work on energized circuits unless absolutely necessary.
- All power supply switches or cutout switches from which power could possibly be fed must be secured in the OPEN (safety) position and locked and tagged.
- Never short out, tamper with, or block open an interlock switch.
- Keep clear of exposed equipment; when it is absolutely necessary to work on it, use only one hand as much as possible.
- Avoid reaching into enclosures except when absolutely necessary. When reaching into an enclosure, use rubber blankets to prevent accidental contact with the enclosure.
- Do not use bare hands to remove hot vacuum tubes from their sockets. Wear protective gloves or use a tube puller.
- Use a shorting stick to discharge all high-voltage capacitors.
- Make certain that the equipment is properly grounded. Ground all test equipment to the equipment under test.
- Turn off the power before connecting alligator clips to any circuit.
- When measuring circuits over 300V, do not hold the insulated test prods with bare hands.

15.1.0 Use of High-Voltage Protection Equipment

Anyone working on or near energized circuitry must use special equipment to provide protection from electrical shock. Protective equipment includes gloves, leather sleeves, rubber blankets, and rubber mats. It should be noted that this electrical protective equipment is in addition to the regular protective equipment normally required for maintenance work. Regular protective equipment typically includes hard hats that are rated for electrical resistance, eye protection, safety shoes, and long sleeves.

Gloves that are approved for protection from electrical shock are made of rubber. A separate leather cover protects the rubber from punctures or other damage. They protect the worker by insulating the hands from electrical shock. Gloves are rated as providing protection from certain amounts of voltage. Whenever an individual is going to be working around exposed conductors, the gloves chosen should be rated for at least as much voltage as the conductors are carrying.

Emphasize the importance of appropriate protective equipment when conducting tests on energized circuits. Stress that this type of work is to be done only by qualified personnel.

Rubber sleeves are used along with gloves to provide additional protection. The combination of sleeves and gloves protects the hands and arms from electrical shock.

Rubber blankets and floor mats have many uses. Blankets are used to cover energized conductors while work is going on around them. They might be used to cover the energized main buses in a breaker panel before working on a deenergized breaker. Rubber floor mats are used to insulate workers from the ground. If a worker is standing on a rubber mat and then contacts an energized conductor, the current cannot flow through the body to the ground, so the worker will not get shocked.

Review example testing policies.

15.2.0 General Testing of Electrical Equipment

As stated earlier, personnel must be familiar with the proper use of available test equipment. Also, personnel must be familiar with and use the local instructions governing the testing of electrical circuits to protect both the person and the equipment.

The next section includes an example of instructions from a local utility and covers work permits, test permits, authorized employees for testing, circuit isolation, and relay, instrument, and meter testing. Each topic is discussed individually.

15.2.1 Example Policy—Work Permits or Test Permits

A written work permit must be obtained by an authorized technician or supervisor from the operator in charge before testing equipment under the operator's control. A test permit is required to make a dielectric proof test.

The following must be on the work or test permit:

- Designation of the equipment to be tested
- Scope and limits of tests to be made
- Method of isolation and protection provided

If it is necessary to alter the nature or to extend the scope of tests for which a work permit or test permit has been granted, return the permit to the operator and request a new permit before proceeding with the revised tests. When the test is completed, return the work permit or test permit to the operator, and report whether the results of the test were satisfactory or unsatisfactory.

Whenever a work permit or test permit remains in effect beyond the working hours of the person to whom it was issued and a different person will be in charge of the tests, consult the operator for the proper turnover procedure. The person assuming the responsibility for continuance of the work must verify that the items on the work or test permit are correct.

If a work permit or test permit is issued for testing that is to continue over a period of several days, the permit must be returned to the operator at the end of each working day and picked up at the beginning of the next working day.

15.2.2 Example Policy—Authorized Employees for Testing

Only employees authorized to perform operations at a given station shall:

- Open, remove, close, or replace doors or covers on electrical compartments.
- Remove or replace potential transformer fuses.
- Open or close switches.
- Open or close fuse cutouts or disconnecting potheads.
- Make tests for the presence of potential.
- Place or remove blocks and protective tags.
- Apply or remove grounds or short circuits.
- Attach or remove leads to high- or intermediate-voltage conductors.
- Make operating tests.
- Test links, fuses, switches associated with generators, buses, feeders, transformers, etc. normally shall be removed, replaced, or operated only under the direct supervision of an authorized employee when the associated equipment is live or is available for normal operation.

In some cases, when it is necessary for the equipment to remain in service, or in any case when the associated equipment has been removed from service, the operator-in-charge may authorize the person to whom the work permit was issued to remove and replace test links and fuses and to operate the heel and toe switches and potential switches in low-voltage circuits.

15.2.3 Example Policy—Circuit Isolation

All equipment to be tested that is rated at 350V phase-to-ground or higher must be isolated from all sources of supply by either two breaks, or a single break and a ground, unless test equipment is used that can be operated safely with one break. Where possible, two breaks in series are preferred over one break.

If no potential is to be applied to the equipment, the ground may be applied on the equipment side of the single break.

Instructor's Notes:

If a test potential is to be applied to the equipment, the ground has to be on the far side of the single break. If a test potential of less than 400V is to be applied to the equipment, the ground may be replaced with an approved protective discharge gap on the equipment to be tested. The protective gaps and test leads are to be connected or disconnected while the equipment is grounded. If disconnect switches are used to provide the breaks, the switches shall be locked open, and the operating motor fuses removed.

On equipment operating at 350V or less, a test potential may be applied with only a single break. An open truck-type or elevating-type circuit breaker withdrawn or lowered to the disconnect position or a swinging-type disconnect switch shall be considered as a single break.

If any of the following are the only type of disconnecting device available, short circuits and grounds must be applied at the nearest available point to the device.

- A disconnect switch with the blade prevented from falling closed by an insulating block placed over the jaw
- A disconnect switch in which the separation of the blade and jaw cannot be verified visually
- An oil circuit breaker or oil switch having no associated disconnecting device
- An oil-filled fusible cutout having no provision for disconnecting the leads

Potential transformers that may become energized from the low-voltage side shall be isolated on that side by opening switches, removing links, removing fuses, or by removing connections temporarily. The low-voltage side of the potential transformer shall also be short circuited and grounded.

Equipment that has been subjected to DC high-voltage tests shall be grounded to discharge the residual voltage before any person is permitted to touch the test cables or any current-carrying part of the equipment tested. If the equipment tested has considerable capacitance, then it should be grounded for at least 30 minutes after the test potential is removed. Equipment of low capacitance should be grounded for a length of time equivalent to the time that the potential was applied.

15.2.4 Example Policy—Clearances

Adequate clearances are to be maintained between energized and exposed conductors and personnel (refer to *NEC Section 110-26*). Where DC voltages are involved, clearances specified shall be used with specified voltages considered as DC line-to-ground values.

If adequate clearances cannot be maintained from exposed live parts of apparatus in the normal course of free movement within the area during test, then access to that area shall be restricted by fences and barricades. Signs clearly indicating the hazard shall be posted in conspicuous locations. This requirement applies to equipment in service as well as to equipment to which test voltages are applied.

Whenever there is any question of the adequacy of clearance between the specific area in which work is to be done and exposed live parts of adjacent equipment, a field inspection shall be made by management representatives of the group involved before starting the job. The result of this inspection should be to outline the protection necessary to complete the work safely, including watchers where needed.

15.2.5 Electrical Circuit Tests

Testing of electrical circuits can be required at new installations or existing installations where wiring has been modified or new wiring has been interfaced with existing circuits. It is broken down into two sections:

- Electrical circuit tests performed on AC relay protection and metering circuits include:
  - Point-to-point wire checks to verify that the wiring has been installed according to the diagram of connections
  - Current transformer polarity, impedance, continuity, and where applicable, tap progression and intercore coupling tests
  - Voltage transformer polarity, ratio, and self-induced high-potential (hi-pot) tests
  - Insulation resistance tests
- Electrical circuit tests performed on control, power, alarm, and annunciator circuits include:
  - Insulation resistance tests
  - Operation tests

An important piece of information is the minimum distance allowed when working near energized electrical circuits because large voltages can arc across an air gap. Personnel must maintain a distance that is greater than that arc distance. This is especially true when using a hot stick to open a disconnect. These distances are listed in *Table 1*.

15.3.0 Meter Loading Effects

It is important to understand how a piece of test equipment will affect the circuit to which it is

Table 1 OSHA Table

Voltage Range (Phase-to-Phase) Kilovolts	Minimum Working and Clear Hot Stick Distance
2.1 to 15	2 ft. 0 in.
15.2 to 35	2 ft. 4 in.
35.1 to 46	2 ft. 6 in.
46.1 to 72.5	3 ft. 0 in.
72.6 to 121	3 ft. 4 in.
138 to 145	3 ft. 6 in.
161 to 169	3 ft. 8 in.
230 to 242	5 ft. 0 in.
345 to 362	*7 ft. 0 in.
500 to 552	*11 ft. 0 in.
700 to 765	*15 ft. 0 in.

*For voltages above 345kV, the minimum working and clear hot stick distances may be reduced provided that such distances are not less than the shortest distance between the energized part and a grounded surface.

Note
Clearances listed in *Table 1* apply to qualified personnel.

attached. Even when properly used, a piece of test equipment can put a load on the circuit under testing.

An ammeter becomes a part of the circuit when it is used. Because of this, the meter can sometimes significantly alter the voltages and currents in the circuit. The following example illustrates how this can happen.

In Circuit A on the left in *Figure 31*, the current is calculated to be 10µA. To measure this current using a meter that has an internal resistance of 2,000Ω, the meter would be placed as shown in Circuit B. The 2,000Ω resistance of the meter adds to the 10,000Ω resistance, reducing the current in the circuit to 8.33µA. The presence of the meter, therefore, alters the current flowing in the circuit, resulting in an incorrect reading.

A voltmeter can also alter circuit parameters. The following example illustrates how this can happen.

In Circuit A in *Figure 32*, the voltage across the 40,000Ω resistor is to be measured using a voltmeter with a 1,000Ω internal resistance. We can determine the effect that the meter will have on the circuit when it is connected as shown in Circuit B.

Connecting the voltmeter in parallel with the 40,000Ω resistor, as shown in Circuit B, significantly changes the total resistance in the circuit. The parallel combination of 40,000Ω and 200,000Ω is:

$$R = \frac{40,000 \times 200,000}{40,000 + 200,000} = 33,333\Omega$$

Add the 10KΩ resistance in series for a total circuit resistance of 43,333Ω.

Therefore, with the voltmeter in the circuit, the current is:

$$E \div R = I \text{ or } 10 \div 43,333 = 2.31 \times 10^{-4}\,A$$

This current causes a voltage drop across the combination of the meter and the 40,000Ω resistor of:

$$E = IR \text{ or } (2.31 \times 10^{-4}) \times 33,333 = 7.7V$$

The voltmeter will read 7.7V, whereas the voltage drop across the 40,000Ω resistor without the voltmeter connected is 8V.

106F31.EPS

Figure 31 ◆ Meter loading—Example 1.

ELECTRICAL LEVEL ONE—TRAINEE MODULE 26106

Instructor's Notes:

_____

_____

_____

_____

_____

_____

_____

_____

_____

CIRCUIT A   CIRCUIT B

106F32.EPS

Figure 32 ◆ Meter loading—Example 2.

Putting It All Together

What kind of measuring device would you select or need for the following tasks, and how would you apply the device?

- A short circuit in house wiring
- The output from an AC transformer
- A fuse in house wiring
- A relay

In most cases, the current flowing through the voltmeter movement is negligible compared to the current flowing through the element whose voltage is being measured. When this is the case, the voltmeter has a negligible effect on the circuit.

In most commercial voltmeters, the internal resistance is expressed as the ohms-per-volt rating of the meter. This quantity tells what the internal resistance of the meter is on any particular full-scale setting. In general, the meter's internal resistance is the ohms-per-volt rating times the full-scale voltage. The higher the ohms-per-volt rating, the higher the internal resistance of the meter, and the smaller the effect of the meter on the circuit.

Care should be taken when selecting either an ammeter or a voltmeter in which the meter does not load down the circuit under test. Also, realize that the same circuit under test with a different meter will indicate differently due to the separate loading effects of each meter.

Summary

The use of test equipment is an important part of your job as an electrician. Selecting the proper instrument to be used in a specific application will help you to fully perform your task. You will use the information provided by these instruments to help evaluate the work being accomplished.

Discuss the "Think About It."

Have the trainees complete the Review Questions and go over the answers prior to administering the Module Examination.

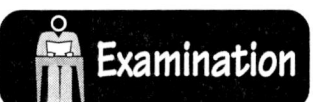

Administer the Module Examination and Performance Test. Be sure to record the testing results on Craft Training Report Form 200 and submit the results to the Training Program Sponsor.

Review Questions

1. An ammeter is used to measure _____.
   a. current
   b. voltage
   c. resistance
   d. insulation value

2. Ammeters use a _____ to measure values higher than 10mA.
   a. multiplier resistor
   b. rectifier bridge
   c. transformer coil
   d. shunt

3. Voltmeters use a _____ to measure values higher than one volt.
   a. multiplier resistor
   b. rectifier bridge
   c. transformer coil
   d. shunt

4. Electromotive force (emf) is measured using a(n) _____.
   a. ammeter
   b. wattmeter
   c. voltmeter
   d. ohmmeter

5. A voltmeter is connected _____ with the circuit being tested.
   a. in parallel
   b. in series
   c. in a sequential configuration
   d. in a looped configuration

6. Which of the following is *not* true regarding an ohmmeter?
   a. It is powered by a DC source, usually a battery.
   b. It contains current-limiting resistors.
   c. It must be used in an energized circuit.
   d. It should be adjusted to zero before each use.

7. Which of the following values can be measured using a VOM?
   a. Decibels
   b. Voltage
   c. Resistance
   d. All of the above.

8. When in doubt as to the voltage present, always use the _____ range to protect the VOM.
   a. lowest
   b. highest
   c. middle
   d. It does not matter with today's precision instruments.

9. Wattmeters are used to measure _____.
   a. power
   b. voltage
   c. resistance
   d. impedance

10. To check the resistance between motor or transformer windings, use a _____.
    a. standard ohmmeter
    b. wattmeter
    c. megger
    d. power factor meter

11. A Wiggy® is used to _____.
    a. measure wattage
    b. measure impedance
    c. test for potential difference
    d. provide an exact voltage reading

12. Recording instruments can be used to measure _____.
    a. power
    b. current
    c. frequency
    d. All of the above.

13. The minimum working distance for a 15kV circuit is _____.
    a. 2 ft. 0 in.
    b. 5 ft. 0 in.
    c. 7 ft. 0 in.
    d. 15 ft. 0 in.

14. Connecting a voltmeter may affect the total _____ in a circuit.
    a. voltage
    b. current
    c. resistance
    d. electromotive force

15. Which of the following is *not* true regarding meter loading?
    a. It may cause different meters to produce different readings.
    b. The higher the ohms-per-volt rating, the smaller the effect of the meter on the circuit.
    c. It may impact both voltage and resistance.
    d. It can be prevented by removing power to the circuit being tested.

Instructor's Notes:

_____
_____
_____
_____
_____
_____
_____
_____

Joe Sullivan, Bay Harbour Electric

Nothing is more important to the electrician than safety. Joe Sullivan recognized this fact and chose to point his career in that direction. Today, he is the safety compliance officer for his company and an OSHA Outreach Program trainer.

What do you think it takes to be a success in the electrical trade?

Training and education. In the electrical trade, there are many industry standards and codes that must be adhered to. Knowledge of these codes and standards is essential. Math skills, organizational skills, and the ability to plan ahead are all required for the electrician. The way you develop these skills is to make a commitment to learning.

Why is good equipment care important in electrical work?

Accidents can be prevented. Keeping your tools and equipment in good condition and following safe work practices in the use of tools and equipment will dramatically reduce the chances of work-related injuries. A person who takes pride in his tools will also take pride in his work.

What do you like most about your job?

It is gratifying to work for a company like Bay Harbour Electric. The company is committed to being a leader of distinction in our industry and our community, and they care about the success of their employees. Knowing that I have the full support of the chairman and president in developing and implementing safety policies and procedures assures me that my company is concerned about the safety, health, and welfare of its employees. When I travel from job site to job site to conduct safety audits and safety training seminars, it is very satisfying to see the safety policies and procedures I developed being put into practice.

Why did you make a career change from field electrician to safety officer?

When I started with Bay Harbour Electric, I had to take the 10-hour safety course, learn CPR and first aid, and attend frequent safety courses and workshops. I also worked on a project that involved international travel and required me to become familiar with the safety codes and practices of other countries. All of this sparked my interest in safety. When the safety compliance officer job opened at Bay Harbour, I applied for it. My employer thought my interest in safety, combined with my electrical experience, was the right mix for the job.

What would you say to someone entering the trade today?

If you are still in high school, and your school has a school-to-work program, take full advantage of it. Doing hands-on work in the trade will help you learn what the trade is really about and what will be expected of you. Also, choose a company that has an accredited apprenticeship program. This will ensure that you have an opportunity to work in the field while you are still in school.

Trade Terms Introduced in This Module

Coil: A number of turns of wire, especially in spiral form, used for electromagnetic effects or for providing electrical resistance.

Continuity: An uninterrupted electrical path for current flow.

d'Arsonval meter movement: A meter movement that uses a permanent magnet and moving coil arrangement to move a pointer across a scale.

Decibel: A unit used to express a relative difference in power between electric signals.

Frequency: The number of cycles completed each second by a given AC voltage; usually expressed in hertz; one hertz = one cycle per second.

Instructor's Notes:

_____

_____

_____

_____

_____

_____

_____

_____

_____

Answers to Review Questions

Answer	Section
1. a	3.0.0
2. d	3.0.0
3. a	4.0.0
4. c	4.0.0
5. a	4.0.0
6. c	5.0.0
7. d	6.0.0
8. b	6.3.3
9. a	8.0.0
10. c	9.0.0
11. c	13.1.1
12. d	14.0.0
13. a	15.2.5
14. c	15.3.0
15. d	15.3.0

Additional Resources

For advanced study of topics covered in this task module, the following books are suggested:

Electronics Fundamentals, Latest Edition. New York: Prentice Hall.

Principles of Electric Circuits, Latest Edition. New York: Prentice Hall.

Instructor's Notes:

_____

_____

_____

_____

_____

_____

_____

_____

_____

NCCER CRAFT TRAINING USER UPDATES

The NCCER makes every effort to keep these textbooks up-to-date and free of technical errors. We appreciate your help in this process. If you have an idea for improving this textbook, or if you find an error, a typographical mistake, or an inaccuracy in the NCCER's Craft Training textbooks, please write us, using this form or a photocopy. Be sure to include the exact module number, page number, a detailed description, and the correction, if applicable. Your input will be brought to the attention of the Technical Review Committee. Thank you for your assistance.

Instructors – If you found that additional materials were necessary in order to teach this module effectively, please let us know so that we may include them in the Equipment/Materials list in the Instructor's Guide.

Write: Curriculum Revision and Development Department
National Center for Construction Education and Research
P.O. Box 141104, Gainesville, FL 32614-1104

Fax: 352-334-0932

E-mail: curriculum@nccer.org

Craft _____ Module Name _____

Copyright Date _____ Module Number _____ Page Number(s) _____

Description _____

_____

_____

_____

(Optional) Correction _____

_____

_____

(Optional) Your Name and Address _____

_____

_____

TEACHING TIPS

Section 3.0.0   *Think About It—Ammeter Connections*

Connecting an ammeter in parallel with a load (resistance) in an energized circuit applies the voltage developed across that load directly across the parallel-connected meter movement and shunt resistor internal to the ammeter. This voltage, which is higher than that for which the ammeter movement is designed, can cause excessive current to flow through the meter movement, damaging the movement.

One example of an ammeter connected in series with a load is the *AMPS* meter installed in the instrument panel of some automobiles. Another example is the ammeter installed in a battery charger used to monitor the rate of current flow as a battery is being charged.

Section 4.0.0   *Think About It—Voltmeters*

A voltmeter must be connected in parallel with the circuit under test so that the series-connected multiplier resistor and meter coil resistance internal to the voltmeter are connected across the voltage being measured. In this manner, the current flow through the multiplier and meter coil resistance produces a meter indication that is directly proportional to the amplitude of the measured voltage. If the voltmeter were connected in series in the circuit like an ammeter, the meter would indicate a voltage equal to the voltage drop developed across the meter resistance by the circuit current flowing through it. Also, when connected in series, it would limit the amount of current flow through the circuit as a result of the large resistance of the multiplier resistor built into the voltmeter.

Section 5.0.0   *Think About It—Using an Ohmmeter*

The instructor should demonstrate this circuit measurement. Connect one end of the deenergized wire to ground using a jumper wire. Measure between the other end of the wire and ground using an ohmmeter. If the wire is broken, the ohmmeter will indicate infinite resistance. If the wire is good, the ohmmeter will indicate a short (0Ω).

Section 7.0.0   *Think About It—Digital Meters*

Precision measurements are required when adjusting and calibrating equipment such as power supplies, or when determining the percentage of voltage or current imbalance that exists in a three-phase system. Precision is not important if making measurements just to determine the absence or presence of voltage or current, as is commonly done when troubleshooting.

Section 15.3.0   *Think About It—Putting It All Together*

- Ohmmeter
- Voltmeter
- Ohmmeter or voltmeter
- Ohmmeter, voltmeter, or ammeter

The instructor should discuss and/or demonstrate ways to use different meters to test the same device under different circumstances.

Introduction to the
National Electrical Code®

26107-00

MODULE OVERVIEW

This course introduces the electrical trainee to the requirements of the *National Electrical Code*®.

PREREQUISITES

Please refer to the Course Map in the Trainee Module. Prior to training with this module, it is recommended that the trainee shall have successfully completed the following modules:

Core Curriculum; Electrical Level One, Modules 26101 through 26106

LEARNING OBJECTIVES

Upon completion of this module, the trainee will be able to:

1. Explain the purpose and history of the *National Electrical Code* (NEC).
2. Describe the layout of the NEC.
3. Explain how to navigate the NEC.
4. Describe the purpose of the National Electrical Manufacturers' Association (NEMA) and the National Fire Protection Association (NFPA).
5. Explain the role of testing laboratories.

PERFORMANCE OBJECTIVES

Under supervision of the instructor, the trainee should be able to:

1. Use *NEC Article 90* to determine the scope of the NEC. State what is covered by the NEC and what is not.
2. Find the definition of the term *feeder* in the NEC.
3. Look up the NEC specifications that you would need to follow if you were installing an outlet near a swimming pool.
4. Find the minimum wire bending space required if two No. 1/0 AWG conductors were to be installed in a junction box or cabinet.

NCCER STANDARDIZED CRAFT TRAINING PROGRAM

The National Center for Construction Education and Research (NCCER) provides a standardized national program of accredited craft training. Key features of the program include instructor certification, competency-based training, and performance testing. The program provides trainees, instructors, and companies with a standard form of recognition through a National Craft Training Registry. The program is described in full in the *Guidelines for Accreditation*, published by the NCCER. For more information on standardized craft training, contact the NCCER at P.O. Box 141104, Gainesville, FL 32614-1104, 352-334-0911, visit our Web site at www.NCCER.org, or e-mail info@NCCER.org.

HOW TO USE THIS ANNOTATED INSTRUCTOR'S GUIDE

Each page presents two sections of information. The larger section displays each page exactly as it appears in the Trainee Module. The narrow column ties suggested trainee and instructor actions to each page and provides icons to call your attention to material, safety, audiovisual, or testing requirements. The bottom of each page includes space for your notes.

If you see the Teaching Tip icon, that means there is a teaching tip associated with this section. Also refer to the suggested teaching tips at the end of the module.

SAFETY CONSIDERATIONS

Ensure that the trainees are equipped with appropriate personal protective equipment. Emphasize that the primary purpose of the *National Electrical Code* is to safeguard persons and property from the hazards arising from the use of electricity.

PREPARATION

Before teaching this module, you should review the Module Outline, Learning and Performance Objectives, and the Materials and Equipment List. Be sure to allow ample time to prepare your own training or lesson plan and gather all required equipment and materials.

MATERIALS AND EQUIPMENT LIST

Materials:
Transparencies

Markers/chalk

Copy of the latest edition of the *National Electrical Code*®

Module Examinations*

Performance Profile Sheets*

Equipment:
Overhead projector and screen

Whiteboard/chalkboard

Appropriate personal protective equipment

*Located in the Test Booklet packaged with this Annotated Instructor's Guide.

ADDITIONAL RESOURCES

This module is intended to present thorough resources for task training. The following reference is suggested for both instructors and motivated trainees interested in further study. This is optional material for continued education rather than for task training.

National Electrical Code Handbook, Latest Edition. Quincy, MA: National Fire Protection Association.

NOTES

The designations "National Electrical Code," "NE Code," and "NEC," where used in this document, refer to the *National Electrical Code*®, which is a registered trademark of the National Fire Protection Association, Quincy, MA. All National Electrical Code (NEC) references in this module refer to the 1999 edition of the NEC.

If you feel that additional math instruction would be helpful, Prentice Hall offers a basic math textbook entitled *Fundamentals of Electrical and Mechanical Mathematics*. It covers the basic math requirements for electrical trainees and may be ordered by contacting Prentice Hall Customer Service at 1-800-922-0579.

TEACHING TIME FOR THIS MODULE

An outline for use in developing your lesson plan is presented below. Note that each Roman numeral in the outline equates to one session of instruction. Each session has a suggested time period of 2½ hours. This includes 10 minutes at the beginning of each session for administrative tasks and one 10-minute break during the session. Approximately 2½ hours are suggested to cover *Introduction to the National Electrical Code*. You will need to adjust the time required for hands-on activity and testing based on your class size and resources.

Topic	Planned Time

Session I. Purpose and History of the NEC; Layout of the NEC; Navigating the NEC; Other Organizations; Module Examination and Performance Testing

A. Purpose and History of the NEC   _____

     1. History   _____

         a. Who is Involved?   _____

B. Layout of the NEC   _____

     1. Types of Rules   _____

     2. NEC Introduction   _____

     3. Body of the NEC   _____

         a. NEC Definitions   _____

     4. Reference Portion of the NEC   _____

         a. Organization of the Chapters   _____

     5. Text in the NEC   _____

C. Navigating the NEC   _____

     1. Equipment for General Use   _____

     2. Special Occupancies   _____

     3. Special Equipment   _____

     4. Special Conditions   _____

     5. Examples of Navigating the NEC   _____

         a. Installing Type SE Cable   _____

         b. Installing Track Lighting   _____

D. Other Organizations   _____

     1. Role of Testing Laboratories   _____

     2. Nationally Recognized Testing Laboratories   _____

     3. National Electrical Manufacturers' Association   _____

E. Module Examination   _____

     1. Trainees must score 70% or higher to receive recognition from the NCCER.

     2. Record the testing results on Craft Training Report Form 200 and submit the results to the Training Program Sponsor.

F. Performance Testing   _____

     1. Trainees must perform each task to the satisfaction of the instructor to receive recognition from the NCCER.

     2. Record the testing results on Craft Training Report Form 200 and submit the results to the Training Program Sponsor.

Module 26107-00

Introduction to the National Electrical Code®

_____

_____

_____

_____

_____

_____

_____

_____

_____

Course Map

This course map shows all of the task modules in the first level of the Electrical curricula. The suggested training order begins at the bottom and proceeds up. Skill levels increase as a trainee advances on the course map. The training order may be adjusted by the local Training Program Sponsor.

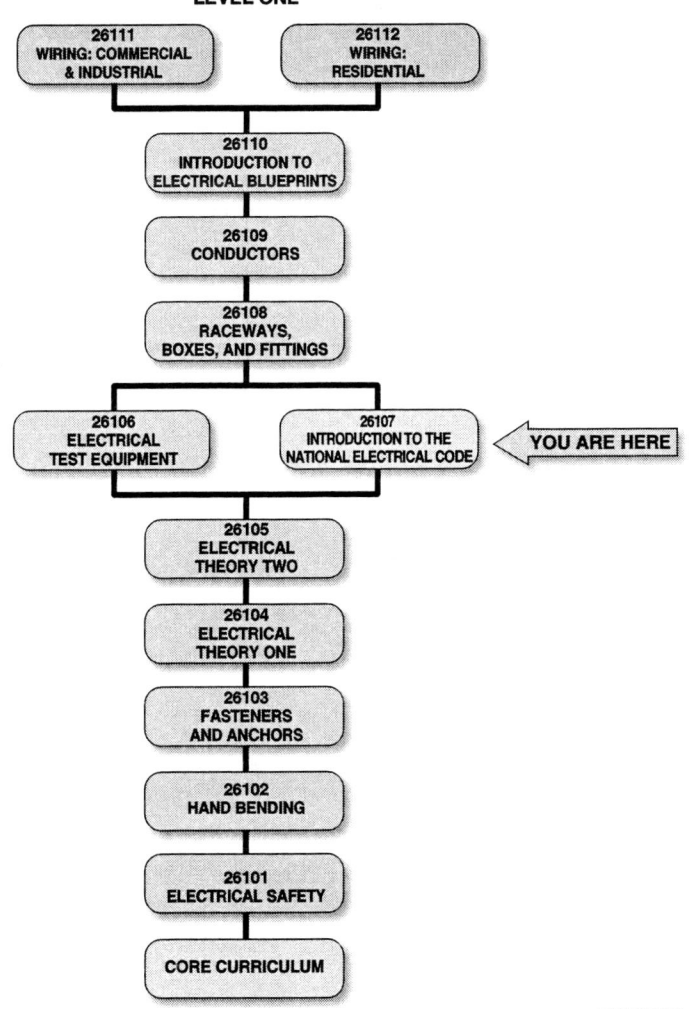

LEVEL ONE

26111
WIRING: COMMERCIAL & INDUSTRIAL

26112
WIRING: RESIDENTIAL

26110
INTRODUCTION TO ELECTRICAL BLUEPRINTS

26109
CONDUCTORS

26108
RACEWAYS, BOXES, AND FITTINGS

26106
ELECTRICAL TEST EQUIPMENT

26107
INTRODUCTION TO THE NATIONAL ELECTRICAL CODE

⇐ YOU ARE HERE

26105
ELECTRICAL THEORY TWO

26104
ELECTRICAL THEORY ONE

26103
FASTENERS AND ANCHORS

26102
HAND BENDING

26101
ELECTRICAL SAFETY

CORE CURRICULUM

107CMAP.EPS

Assign reading of Module 26107.

7–iii

MODULE 26107 CONTENTS

Ensure you have everything required to teach the course. Check the Materials and Equipment List at the front of this Instructor's Guide.

Figures

ELECTRICAL LEVEL ONE—TRAINEE MODULE 26107

Instructor's Notes:

_____

_____

_____

_____

_____

_____

_____

_____

_____

ELECTRICAL LEVEL ONE—INSTRUCTOR'S GUIDE MODULE 26107

Introduction to the National Electrical Code®

Objectives

Upon completion of this module, the trainee will be able to:

1. Explain the purpose and history of the *National Electrical Code* (NEC).
2. Describe the layout of the NEC.
3. Explain how to navigate the NEC.
4. Describe the purpose of the National Electrical Manufacturers' Association (NEMA) and the National Fire Protection Association (NFPA).
5. Explain the role of testing laboratories.

Prerequisites

Successful completion of the following Task Modules is recommended before beginning study of this Task Module: Core Curricula; Electrical Level One, Modules 26101 through 26106.

Required Trainee Materials

1. Trainee Task Module
2. Copy of the latest edition of the *National Electrical Code*

1.0.0 ◆ INTRODUCTION

The *National Electrical Code* (NEC) is published by the National Fire Protection Association (NFPA). The NEC is one of the most important tools for the electrician. When used together with the electrical code for your local area, the NEC provides the minimum requirements for the installation of electrical systems. Unless otherwise specified, always use the latest edition of the NEC as your on-the-job reference. It specifies the minimum provisions necessary for protecting people and property from electrical hazards. In some areas, however, local laws may specify different editions of the NEC, so be sure to use the edition specified by your employer. Also, bear in mind that the NEC only specifies *minimum* requirements, so local or job requirements may be more stringent.

2.0.0 ◆ PURPOSE AND HISTORY OF THE NEC

The primary purpose of the NEC is the practical safeguarding of persons and property from hazards arising from the use of electricity [*NEC Section 90-1(a)*]. A thorough knowledge of the NEC is one of the first requirements for becoming a trained electrician. The NEC is probably the most widely used and generally accepted code in the world. It has been translated into several languages. It is used as an electrical installation, safety, and reference guide in the United States. Many other parts of the world use it as well. Compliance with NEC standards increases the safety of electrical installations—the reason the NEC is so widely used.

Note: The designations "National Electrical Code," "NE Code," and "NEC," where used in this document, refer to the *National Electrical Code®*, which is a registered trademark of the National Fire Protection Association, Quincy, MA. *All National Electrical Code (NEC) references in this module refer to the 1999 edition of the NEC.*

Show Transparency 1, Course Objectives.

Show Transparency 2, Performance Profile Tasks.

Describe the purpose of the NEC.

Explain that items shown in bold (blue) are defined in the Glossary at the back of this module.

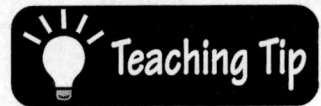

What's wrong with these pictures?

WALTER JOHNSON   107PO701.EPS

WALTER JOHNSON   107PO702.EPS

TIM ELY   107PO703.EPS

Stress the importance of becoming familiar with NEC requirements.

Although the NEC itself states "This Code is not intended as a design specification nor an instruction manual for untrained persons," it does provide a sound basis for the study of electrical installation procedures—under the proper guidance. The NEC has become the standard reference of the electrical construction industry. Anyone involved in electrical work should obtain an up-to-date copy and refer to it frequently.

Whether you are installing a new electrical system or altering an existing one, all electrical work must comply with the current NEC and all local ordinances. Like most laws, the NEC is easier to work with once you understand the language

Instructor's Notes:

_____

_____

_____

_____

_____

_____

_____

_____

_____

The NEC

Why do you think it's necessary to have a standard set of procedures for electrical installations? Find out who does the electrical inspection in your area. Who determines what will be inspected, when it will be inspected, and who will do the inspection?

Discuss the "Think About It."

Discuss the history and structure of the NEC Committee.

Note

This module is *not* a substitute for the NEC. You need to acquire a copy of the most recent edition and keep it handy at all times. The more you know about the NEC, the better an electrician you will become.

and know where to look for the information you need.

2.1.0 History

In 1881, the National Association of Fire Engineers met in Richmond, Virginia. From this meeting came the idea to draft the first *National Electrical Code*. The first nationally recommended electrical code was published by the National Board of Fire Underwriters (now the American Insurance Association) in 1895.

In 1896, the National Electric Light Association (NELA) was working to make the requirements of the fire insurance organizations and electrical utilities fit together. NELA succeeded in promoting a conference that would result in producing a standard national code. The NELA code would serve the interests of the insurance industry, operating concerns, manufacturing, and industry.

The conference produced a set of requirements that was unanimously accepted. In 1897, the first edition of the NEC was published, and the NEC became the first cooperatively produced national code. The organization that produced the NEC was known as the *National Conference on Standard Electrical Rules*. This group became a permanent organization, and its job was to develop the NEC.

In 1911, the NFPA took over administration and control of the NEC. However, the National Bureau of Fire Underwriters (NBFU) continued to publish the NEC until 1962. During the period from 1911 until now, the NEC has experienced several major changes, as well as regular three-year updates. In 1923, the NEC was rearranged and rewritten, and

in 1937, it was editorially revised. In 1959, the NEC was revised to include a numbering system under which each section of each article was identified with an article/section number.

2.1.1 Who Is Involved?

The creation of a universally accepted set of rules is an involved and complicated process. Rules made by a committee have the advantage that they usually do not leave out the interest of any of the groups represented at the committee. However, since the rules must represent the interests and requirements of an assortment of groups, they are often quite complicated and wordy.

In 1949, the NFPA reorganized the NEC committee into its present structure. The present structure consists of a Correlating Committee and twenty Code Making Panels (CMPs).

The Correlating Committee consists of ten principal voting members and six alternates. The principal function of the Correlating Committee is to ensure that:

• No conflict of requirements exists
• Correlation has been achieved
• NFPA regulations governing committee projects have been followed
• A practical schedule of revision and publication is established and maintained

Each of the twenty CMPs have members who are experts on particular subjects and have been assigned certain articles to supervise and revise as required. Members of the CMPs represent such special interest groups as trade associations, electrical contractors, electrical designers and engineers, electrical inspectors, electrical manufacturers and suppliers, electrical testing laboratories, and insurance organizations.

Each panel is structured so that not more than one-third of its members are from a single interest group. The members of the NEC Committee create or revise requirements for the NEC through probing, debating, analyzing, weighing, and reviewing new input. Anyone, including you, can submit proposals to amend the NEC. Sample

Discuss the "Think About It."

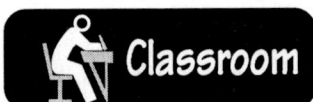

Discuss the layout of the NEC.

Show Transparency 3 (Figure 1).

Code Changes

THINK ABOUT IT

This photograph shows the first code book and the current code book. Code changes occur every three years. Who can suggest changes to the *National Electrical Code?* What might be reasons for submitting changes?

VERONICA WESTFALL   107PO704.EPS

forms for this purpose may be obtained from the Secretary of the Standards Council at NFPA Headquarters. In addition to written proposals, the NFPA also holds meetings to discuss code changes and proposals.

The NFPA membership is drawn from the fields listed above. In addition to publishing the NEC, the duties of the NFPA include the following:

- Developing, publishing, and distributing standards that are intended to minimize the possibility and effects of fire and explosion
- Conducting fire safety education programs for the general public
- Providing information on fire protection, prevention, and suppression
- Compiling annual statistics on causes and occupancies of fires, large-loss fires (over one million dollars), fire deaths, and firefighter casualties
- Providing field service by specialists on electricity, flammable liquids and gases, and marine fire problems

- Conducting research projects that apply statistical methods and operations research to develop computer models and data management systems

3.0.0 ◆ THE LAYOUT OF THE NEC

The NEC begins with a brief history. *Figure 1* shows how the NEC is organized.

3.1.0 Types of Rules

There are two basic types of rules in the NEC: mandatory rules and advisory rules. Here is how to recognize the two types of rules and how they relate to all types of electrical systems:

- *Mandatory rules*—All mandatory rules have the word *shall* in them. The word *shall* means *must*. If a rule is mandatory, you must comply with it.
- *Advisory rules*—All advisory rules have the word *should* in them. The word *should* in this case means *recommended but not necessarily re-*

ELECTRICAL LEVEL ONE—TRAINEE MODULE 26107

Instructor's Notes:

_____

_____

_____

_____

_____

_____

_____

_____

_____

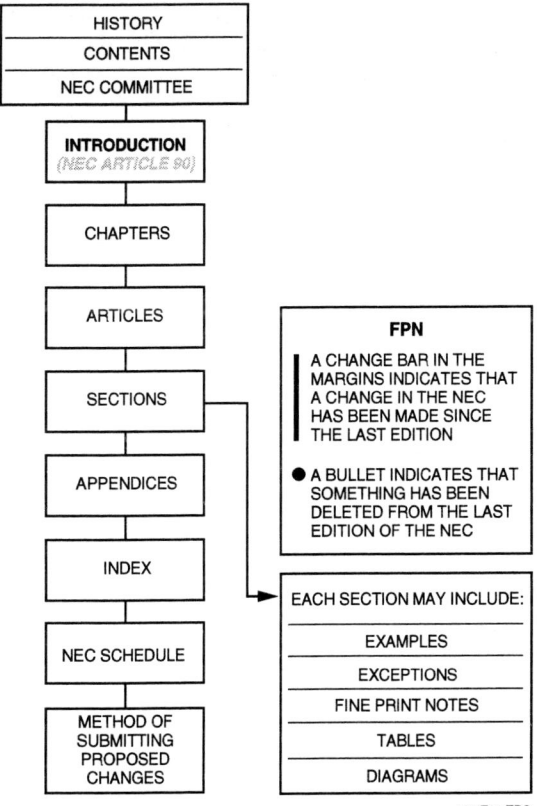

```
HISTORY
CONTENTS
NEC COMMITTEE

INTRODUCTION
(NEC ARTICLE 90)

CHAPTERS

ARTICLES

SECTIONS

APPENDICES

INDEX

NEC SCHEDULE

METHOD OF
SUBMITTING
PROPOSED
CHANGES
```

FPN

| A CHANGE BAR IN THE MARGINS INDICATES THAT A CHANGE IN THE NEC HAS BEEN MADE SINCE THE LAST EDITION

● A BULLET INDICATES THAT SOMETHING HAS BEEN DELETED FROM THE LAST EDITION OF THE NEC

EACH SECTION MAY INCLUDE:

EXAMPLES

EXCEPTIONS

FINE PRINT NOTES

TABLES

DIAGRAMS

107F01.EPS

Figure 1 ◆ The layout of the NEC.

 NEC Layout

Remember, chapters contain a group of articles. An article is a specific subject (i.e., *NEC Article 300, Wiring Methods*). When an article contains a large quantity of information, it's divided into sections, and some sections or rules have exceptions.

quired. If a rule is advisory, compliance is discretionary. If you want to comply with it, do so, but it is not required by the NEC.

Be alert to local amendments to the NEC. Local ordinances may amend the language of the NEC, changing it from *should* to *shall*. This means that you must do in that county or city what may only be recommended in some other area. The office that issues building permits will either sell you a copy of the code that is enforced in that county or tell you where the code is sold.

3.2.0 NEC Introduction

The main body of the text begins with an *Introduction,* also entitled *NEC Article 90.* This introduction gives you an overview of the NEC. Items included in this section are:

- Purpose of the NEC
- Scope of the code book
- Code arrangement
- Code enforcement
- Mandatory rules and explanatory material
- Formal interpretation
- Examination of equipment for safety
- Wiring planning
- Metric units of measurement

3.3.0 The Body of the NEC

The remainder of the book is organized into nine chapters. See *Figure 1* for an overview of the layout. *NEC Chapter 1* contains a complete list of definitions used in the NEC. These definitions are referred to as *NEC Article 100. NEC Article 110* gives the general requirements for electrical installations. It is important for you to be familiar with this general information and the definitions.

3.3.1 NEC Definitions

There are many definitions included in *NEC Article 100.* You should become familiar with the definitions. Here are two that you should become especially familiar with:

- *Labeled*—"Equipment or materials to which has been attached a label or other identifying mark of an organization that is acceptable to the authority having jurisdiction and concerned with product evaluation, that maintains periodic inspection of production of labeled equipment or materials, and by whose labeling the manufacturer indicates compliance with appropriate standards or performance in a specified manner."
- *Listed*—"Equipment or materials included in a list published by an organization acceptable to the authority having jurisdiction and concerned with product evaluation, that maintains periodic inspection of production of listed equipment or materials, and whose listing states either that the equipment or material meets appropriate designated standards or has been tested and found suitable for use in a specified manner."

Besides installation rules, you will also have to be concerned with the type and quality of materi-

The NEC as a Reference Tool

The NEC, although carefully organized, is a highly technical document; as such, it uses many technical words and phrases. It's crucial for you to understand their meaning—if you don't, you'll have trouble understanding the rules themselves. When using the NEC as a reference tool, you may have to refer to a number of articles and sections to find your answer(s). NEC definitions are covered in *NEC Article 100.* Many issues concerning the NEC may be resolved by simply reviewing the definitions.

Underwriters' Laboratories, Inc.

Underwriters' Laboratories, Inc. is an internationally recognized authority on product safety testing and safety certification and standards development. It was established in 1893. The Chicago World's Fair was opened and thanks to Edison's introduction of the electric light bulb, the World's Fair lit up the world. But all was not perfect—wires soon sputtered and crackled, and, ironically, the Palace of Electricity caught fire. The fair's insurance company called in a troubleshooting engineer, who, after careful inspection, found faulty and brittle insulation, worn out and deteriorated wiring, bare wires, and overloaded circuits.

He called for standards in the electrical industry, and then set up a testing laboratory above a Chicago firehouse to do just that. Hence, Underwriters' Laboratories, Inc. (UL), an independent testing organization, was born.

Instructor's Notes:

_____

_____

_____

_____

_____

_____

_____

_____

_____

als that are used in electrical wiring systems. Nationally recognized testing laboratories are product safety certification laboratories. Underwriters' Laboratories, also called *UL,* is one such laboratory. These laboratories establish and operate product safety certification programs to make sure that items produced under the service are safeguarded against reasonable foreseeable risks. Some of these organizations maintain a worldwide network of field representatives who make unannounced visits to manufacturing facilities to counter-check products bearing their seal of approval. The UL label is shown in *Figure 2.*

3.4.0 The Reference Portion of the NEC

NEC Chapters 2 through 9 each contain numerous articles and sections. Each chapter focuses on a general category of electrical application, such as *NEC Chapter 2, Wiring and Protection.* Each article emphasizes a more specific part of that category, such as *NEC Article 210, Branch Circuits, Part A, General Provisions.* Each section gives examples of a specific application of the NEC, such as *NEC Section 210-4, Multiwire Branch Circuits.*

3.4.1 Organization of the Chapters

The chapters of the NEC are organized into three major categories:

* *NEC Chapters 1, 2, 3, 4, and 9*—The first four chapters present the rules for the design and in-

107F02.EPS

Figure 2 ◆ Underwriters' Laboratories label.

stallation of electrical wiring systems. *NEC Chapter 9* contains tables which specify the properties of conductors and rules for the use of conduit to enclose the conductors. The examples in this chapter demonstrate the use of the rules for design given in the first four chapters.
* *NEC Chapters 5, 6, and 7*—These chapters are concerned with special occupancies, equipment, and conditions. Rules in these chapters may modify or amend those in the first four chapters.
* *NEC Chapter 8*—This chapter covers communications systems, such as the telephone and telegraph, as well as radio and television receiving equipment.

3.5.0 Text in the NEC

When you open the NEC, you will notice several different types of text or printing used. Here is an explanation of each type of text:

* *Bold black letters*—Headings for each NEC application.
* *Exceptions*—Exceptions explain the circumstances under which a specific part of the NEC does not apply. Exceptions are written in *italics* under that part of the NEC to which they pertain.
* *Fine print notes*—Fine print notes (FPNs) explain something in an application, suggest other sections to read about the application, or provide tips about the application. These are defined in the text by the term *(FPN)* shown in parentheses before a paragraph in smaller print.
* *Diagrams*—These may be included with explanations to give you a picture of what your application may look like.
* *Tables*—Tables are often included when there is more than one possible application of the NEC. You would use a table to look up the specifications of your application.

4.0.0 ◆ NAVIGATING THE NEC

To locate information for a particular procedure being performed, use the following steps:

Other Testing Laboratories

In addition to UL, there are several other recognized testing laboratories. How many can you name?

Discuss the chapter organization of the NEC.

Review the different types of text in the NEC.

Demonstrate the procedure for locating information in the NEC.

Discuss the "Think About It."

Have the trainees practice finding information in the NEC.

Discuss the NEC chapters dealing with servicing electrical systems.

Discuss the "Think About It."

Discuss the "Think About It."

Step 1 Familiarize yourself with *NEC Articles 90, 100, and 110* to gain an understanding of the material covered in the NEC and the definitions used in it.

Step 2 Turn to the *Table of Contents* at the beginning of the NEC.

Step 3 Locate the chapter that focuses on the desired category.

Step 4 Find the article pertaining to your specific application.

Step 5 Turn to the page indicated. Each application will begin with a bold heading.

Note

An index is provided at the end of the NEC. The index lists specific topics and provides a reference to the location of the material within the NEC. The index is helpful when you are looking for a specific topic rather than a general category.

Once you are familiar with *NEC Articles 90, 100, and 110,* you can move on to the rest of the NEC. There are several key sections used often in servicing electrical systems.

• *Wiring design and protection—NEC Chapter 2* discusses wiring design and protection, the information electrical technicians need most often. It covers the use and identification of grounded conductors, branch circuits, feeders, calculations, services, overcurrent protection, and grounding. This is essential information for all types of electrical systems. This chapter is also a *how-to* chapter. It explains how to provide proper spacing for conductor supports, how to provide temporary wiring, and how to size the proper grounding conductor or electrode. If you run into a problem related to the design or installation of a conventional electrical system, you can probably find a solution for it in this chapter.

• *Wiring methods and materials—NEC Chapter 3* lists the rules on wiring methods and materials. The materials and procedures to use on a particular system depend on the type of building construction, the type of occupancy, the location of the wiring in the building, the type of atmosphere in the building or in the area surrounding the building, mechanical factors, and the relative costs of different wiring methods. The provisions of this chapter apply to all wiring installations except remote control switching (*NEC Article 725*), low-energy power circuits (*NEC Article 725*), signal systems (*NEC Article 725*), and communications systems and conductors (*NEC Article 800*) when these items form an integral part of equipment, such as in motors and motor controllers.

There are four basic wiring methods used in most modern electrical systems. Nearly all wiring methods are a variation of one or more of these methods.

• Sheathed cables of two or more conductors, such as NM cable and BX armored cable (*NEC Articles 330 through 339*)
• Raceway wiring systems, such as rigid and EMT conduit (*NEC Articles 342 through 358*)
• Busways (*NEC Article 364*)
• Cable trays (*NEC Article 318*)

Junction Boxes

Find the rule in the NEC that explains whether a junction box without devices can be supported solely by two or more EMT conduits. Explain the technical terminology in everyday language.

NEC Article 90

After you've familiarized yourself with *NEC Article 90,* explain its intent. What part of electrical installation does it not cover?

Instructor's Notes:

_____

_____

_____

_____

_____

_____

_____

_____

_____

_____

NEC Article 310 gives a complete description of all types of electrical conductors. Electrical conductors come in a wide range of sizes and forms. Be sure to check the working drawings and specifications to see what sizes and types of conductors are required for a specific job. If the conductor type and size are not specified, choose the most appropriate type and size meeting standard NEC requirements.

NEC Articles 318 through 384 give rules for raceways, boxes, cabinets, and raceway fittings. Outlet boxes vary in size and shape, depending on their use, the size of the raceway, the number of conductors entering the box, the type of building construction, and the atmospheric conditions of the area. This chapter should answer most questions on the selection and use of these items.

The NEC does not describe in detail all types and sizes of outlet boxes. However, the manufacturers of outlet boxes provide excellent catalogs showing their products. Collect these catalogs, since these items are essential to your work.

NEC Article 380 covers switches, push-buttons, pilot lamps, receptacles, and convenience outlets you will use to control electrical circuits or to connect portable equipment to electric circuits. Again, get the manufacturers' catalogs on these items. They will provide you with detailed descriptions of each of the wiring devices.

NEC Article 384 covers switchboards and panelboards, including their location, installation methods, clearances, grounding, and overcurrent protection.

4.1.0 Equipment for General Use

NEC Chapter 4 begins with the use and installation of flexible cords and cables, including the trade name, type letter, wire size, number of conductors, conductor insulation, outer covering, and use of each. This chapter also covers fixture wires, again giving the trade name, type letter, and other important details.

NEC Article 410 on lighting fixtures is especially important. It gives installation procedures for fixtures in specific locations. For example, it covers fixtures near combustible material and fixtures in closets. However, the NEC does not describe how many fixtures will be needed in a given area to provide a certain amount of illumination.

NEC Article 430 covers electric motors, including mounting the motor and making electrical connections to it. Motor controls and overload protection are also covered.

NEC Articles 440 through 460 cover air conditioning and heating equipment, transformers, and capacitors.

NEC Article 480 provides requirements related to battery-operated electrical systems. Storage batteries are seldom thought of as part of a conventional electrical system, but they often provide standby emergency lighting service. They may also supply power to security systems that are separate from the main AC electrical system.

4.2.0 Special Occupancies

NEC Chapter 5 covers special occupancy areas. These are areas where the sparks generated by electrical equipment may cause an explosion or fire. The hazard may be due to the atmosphere of the area or the presence of a volatile material in the area. Commercial garages, aircraft hangers, and service stations are typical special occupancy locations.

NEC Articles 500 and 501 cover the different types of special occupancy atmospheres where an explosion is possible. The atmospheric groups

Disconnects
How would you proceed to find the NEC rule for the maximum number of disconnects permitted for a service?

Sealing Requirements
Study *NEC Section 501-5* and summarize the key points. Discuss your interpretation with the rest of the class.

INTRODUCTION TO THE NATIONAL ELECTRICAL CODE—TRAINEE MODULE 26107   7–9

Discuss the three classes of special occupancy location and provide examples of each type.

were established to make it easy to test and approve equipment for various types of uses.

NEC Articles 501-4, 502-4, and 503-3 cover the installation of explosion-proof wiring. An explosion-proof system is designed to prevent the ignition of a surrounding explosive atmosphere when arcing occurs within the electrical system.

There are three main classes of special occupancy location:

- *Class I*—Areas containing flammable gases or vapors in the air. Class I areas include paint spray booths, dyeing plants where hazardous liquids are used, and gas generator rooms (*NEC Article 501*).
- *Class II*—Areas where combustible dust is present, such as grain-handling and storage plants, dust and stock collector areas, and sugar-pulverizing plants (*NEC Article 502*). These are areas where, under normal operating conditions, there may be enough combustible dust in the air to produce explosive or ignitable mixtures.
- *Class III*—Areas that are hazardous because of the presence of easily ignitable fibers or other particles in the air, although not in large enough quantities to produce ignitable mixtures (*NEC Article 503*). Class III locations include cotton mills, rayon mills, and clothing manufacturing plants.

NEC Articles 511 and 514 regulate garages and similar locations where volatile or flammable liquids are used. While these areas are not always considered critically hazardous locations, there may be enough danger to require special precautions in the electrical installation. In these areas, the NEC requires that volatile gases be confined to an area not more than four feet above the floor. So in most cases, conventional raceway systems are permitted above this level. If the area is judged to be critically hazardous, explosion-proof wiring (including seal-offs) may be required.

NEC Article 520 regulates theaters and similar occupancies where fire and panic can cause hazards to life and property. Drive-in theaters do not present the same hazards as enclosed auditoriums, but the projection rooms and adjacent areas must be properly ventilated and wired for the protection of operating personnel and others using the area.

NEC Chapter 5 also covers service stations, bulk storage plants, health care facilities, and mobile homes and parks.

4.3.0 Special Equipment

Residential electrical workers will seldom need to refer to the articles in *NEC Chapter 6*, but this chapter is of great concern to commercial and industrial electrical workers.

NEC Article 600 covers electric signs and outline lighting. *NEC Article 610* applies to cranes and hoists. *NEC Article 620* covers the majority of the electrical work involved in the installation and operation of elevators, dumbwaiters, escalators, and moving walks. The manufacturer is responsible for most of this work. The electrician usually just furnishes a feeder terminating in a disconnect means in the bottom of the elevator shaft. The electrician may also be responsible for a lighting circuit to a junction box midway in the elevator shaft for connecting the elevator cage lighting cable and exhaust fans. The articles in this chapter list most of the requirements for these installations.

NEC Article 630 regulates electric welding equipment. It is normally treated as a piece of industrial power equipment requiring a special power outlet, but there are special conditions that apply to the circuits supplying welding equipment. These are outlined in detail in this chapter.

NEC Article 640 covers wiring for sound recording and similar equipment. This type of equipment normally requires low-voltage wiring. Special outlet boxes or cabinets are usually provided with the equipment, but some items may be mounted in or on standard outlet boxes. Some sound recording systems require direct current. It is supplied from rectifying equipment, batteries, or motor generators. Low-voltage alternating current comes from relatively small transformers connected on the primary side to a 120V circuit within the building.

Other items covered in *NEC Chapter 6* include X-ray equipment (*NEC Article 660*), induction and dielectric heat-generating equipment (*NEC Article 665*), and machine tools (*NEC Article 670*).

If you ever have work that involves *NEC Chapter 6*, study the chapter before work begins. That can save a lot of installation time. Another way to cut down on labor hours and prevent installation error is to acquire a set of rough-in drawings of the equipment being installed. It is easy to install the wrong outlet box or to install the right box in the wrong place. Having a set of rough-in drawings can prevent these simple but costly errors.

4.4.0 Special Conditions

In most commercial buildings, the NEC and local ordinances require a means of lighting public rooms, halls, stairways, and entrances. There must be enough light to allow the occupants to

Discuss *NEC Chapter 6*.

Instructor's Notes:

_____

_____

_____

_____

_____

_____

_____

_____

_____

exit from the building if the general building lighting is interrupted. Exit doors must be clearly indicated by illuminated exit signs.

NEC Chapter 7 covers the installation of emergency systems. These circuits should be arranged so that they can automatically transfer to an alternate source of current, usually storage batteries or gasoline-driven generators. As an alternative in some types of occupancies, you can connect them to the supply side of the main service, so disconnecting the main service switch would not disconnect the emergency circuits. This chapter also covers a variety of other equipment, systems, and conditions that are not easily categorized elsewhere in the NEC.

NEC Chapter 8 is a special category for wiring associated with electronic communications systems including telephone and telegraph, radio and TV, fire and burglar alarms, and community antenna systems.

4.5.0 Examples of Navigating the NEC

4.5.1 Installing Type SE Cable

Suppose you are installing Type SE (service-entrance) cable on the side of a home. You know that this cable must be secured, but you are not sure of the spacing between cable clamps. To find out this information, use the following procedure:

Step 1 Look in the NEC *Table of Contents* and follow down the list until you find an appropriate category. (Or you can use the index at the end of the book.)

Step 2 *NEC Article 230* will probably catch your eye first, so turn to the page where it begins.

Step 3 Scan down through the section numbers until you come to *NEC Section 230-51, Mounting Supports.* Upon reading this section, you will find in paragraph *(a) Service-Entrance Cables* that "Service-entrance cable shall be supported by straps or other approved means within 12 inches (305mm) of every service head, gooseneck, or connection to a raceway or enclosure, and at intervals not exceeding 30 inches (762mm)."

After reading this section, you will know that a cable strap is required within 12 inches of the service head and within 12 inches of the meter base. Furthermore, the cable must be secured in between these two termination points at intervals not exceeding 30 inches.

4.5.2 Installing Track Lighting

Assume that you are installing track lighting in a residential occupancy. The owners want the track located behind the curtain of their sliding glass patio doors. To determine if this is an NEC violation, follow these steps:

Step 1 Look in the NEC *Table of Contents* and find the chapter that contains information about the general application you are working on. *NEC Chapter 4, Equipment for General Use,* covers track lighting.

Step 2 Now look for the article that fits the specific category you are working on. In this case, *NEC Article 410* covers lighting fixtures, lampholders, lamps, and receptacles.

Step 3 Next locate the section within *NEC Article 410* that deals with the specific application. For this example, refer to *Part R, Lighting Track.*

Step 4 Turn to the page listed.

Step 5 Read *NEC Section 410-100, Definitions,* to become familiar with track lighting. Continue down the page with *NEC Section 410-101* and read the information contained therein. Note that paragraph *(c) Locations Not Permitted* under *NEC Section 410-101* states the following: "Lighting track shall not be installed (1) where likely to be subjected to physical damage; (2) in wet or damp locations; (3) where subject to corrosive vapors; (4) in storage battery rooms; (5) in hazardous (classified) locations; (6) where concealed; (7) where extended through walls or partitions; (8) less than 5 feet (1.52m) above the finished floor except where protected from physical damage or track operating at less than 30 volts RMS open-circuit voltage."

Step 6 Read *NEC Section 410-101(c)* carefully. Do you see any conditions that would violate any NEC requirements if the track lighting is installed in the area specified? In checking these items, you will probably note condition (6), "where concealed." Since the track lighting is to be installed behind a curtain, this sounds like an NEC violation. We need to check further.

Step 7 You need the NEC definition of concealed. Therefore, turn to *NEC Article 100, Definitions* and find the main term *concealed.* It reads: "Concealed: Rendered inaccessible by the structure or finish of the building . . ."

Classroom

Review *NEC Chapters 7 and 8.*

Demonstrate how to find the NEC requirements for example installations.

Step 8 Although the track lighting may be out of sight if the curtain is drawn, it will still be readily accessible for maintenance. Consequently, the track lighting is really not concealed according to the NEC definition.

When using the NEC to determine electrical installation requirements, keep in mind that you will nearly always have to refer to more than one section. Sometimes the NEC itself refers the reader to other articles and sections. In some cases, the user will have to be familiar enough with the NEC to know what other sections pertain to the installation at hand. It can be a confusing situation, but time and experience in using the NEC will make it much easier. A pictorial road map of some NEC topics is shown in *Figures 3* and *4*.

5.0.0 ◆ OTHER ORGANIZATIONS

5.1.0 The Role of Testing Laboratories

As mentioned earlier, testing laboratories are an integral part of the development of the NEC. The NFPA and other organizations provide testing laboratories to conduct research into electrical equipment and its safety. These laboratories perform extensive testing of new products to make sure they are built to NEC standards for electrical and fire safety. These organizations receive statistics and reports from agencies all over the United States concerning electrical shocks and fires and their causes. Upon seeing trends developing concerning the association of certain equipment and dangerous situations or circumstances, this equip-

ment will be specifically targeted for research. All the reports from these laboratories are used in the generation of changes or revisions to the NEC.

5.2.0 Nationally Recognized Testing Laboratories

Nationally Recognized Testing Laboratories (NRTL) are product safety certification laboratories. They establish and operate product safety certification programs to make sure that items produced under the service are safeguarded against reasonably foreseeable risks. NRTLs maintain a worldwide network of field representatives who make unannounced visits to factories to check products bearing their safety marks.

5.3.0 National Electrical Manufacturers' Association

The National Electrical Manufacturers' Association (NEMA) was founded in 1926. It is made up of companies that manufacture equipment used for generation, transmission, distribution, control, and utilization of electric power. The objectives of NEMA are to maintain and improve the quality and reliability of products; to ensure safety standards in the manufacture and use of products; and to develop product standards covering such matters as naming, ratings, performance, testing, and dimensions. NEMA participates in developing the NEC and the *National Electrical Safety Code* and advocates their acceptance by state and local authorities.

Different Interpretations

For as many trainees in class with you today, there will be as many different interpretations of the NEC. However, a difference of opinion is not always a problem— discussing the NEC with your peers will allow you to expand your own understanding of it. In your previous discussion about *NEC Section 501-5*, what differences of interpretation did you have to resolve?

Conformance and Electrical Equipment

What other resources are available for finding information about the use of electrical equipment and materials?

Instructor's Notes:

_____

_____

_____

_____

_____

_____

_____

_____

_____

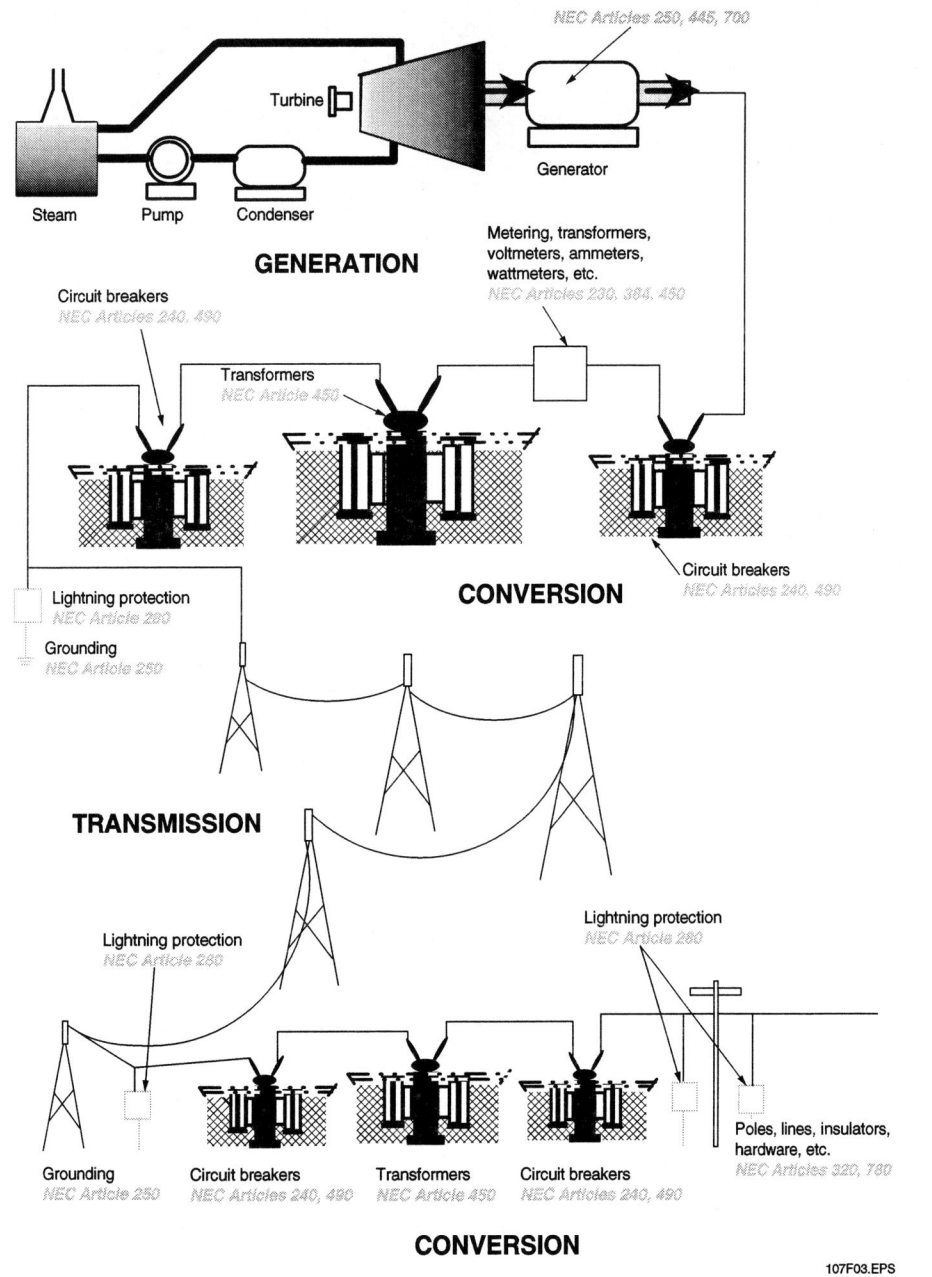

GENERATION

Turbine

Steam   Pump   Condenser

Generator

NEC Articles 250, 445, 700

Metering, transformers,
voltmeters, ammeters,
wattmeters, etc.
NEC Articles 230, 384, 450

Circuit breakers
NEC Articles 240, 490

Transformers
NEC Article 450

CONVERSION

Circuit breakers
NEC Articles 240, 490

Lightning protection
NEC Article 280

Grounding
NEC Article 250

TRANSMISSION

Lightning protection
NEC Article 280

Lightning protection
NEC Article 280

Poles, lines, insulators,
hardware, etc.
NEC Articles 320, 780

Grounding
NEC Article 250

Circuit breakers
NEC Articles 240, 490

Transformers
NEC Article 450

Circuit breakers
NEC Articles 240, 490

CONVERSION

107F03.EPS

Figure 3 ◆ NEC references for power generation and transmission.

INDUSTRIAL AND COMMERCIAL POWER

Metering, transformers, fuses
NEC Articles 230, 240, 384, 450

Capacitors
NEC Article 460

Switchboards
NEC Articles 360, 384

Panelboards
NEC Article 384

Motor controls
NEC Article 430

Motors
NEC Article 430

Busways
NEC Article 364

Disconnect switches
NEC Articles 360, 430

Circuit breakers
NEC Article 240

Transformers
NEC Article 450

DOMESTIC AND GENERAL POWER

Mobile homes and mobile home parks
NEC Article 550

Poles, lines, insulators, hardware
NEC Articles 320, 780

Services
NEC Article 230

Lighting
NEC Article 410

NEC Article 810

Switches
NEC Article 380

Branch circuits
NEC Article 220

TV

Grounding
NEC Article 250

Panelboards
NEC Article 384
Overcurrent protection
NEC Article 240

Wiring methods
NEC Article 300

Isolating switches, fuses
NEC Articles 240, 380

Lightning protection
NEC Article 260

Circuit breakers
NEC Article 240

Transformers
NEC Article 450

DISTRIBUTION PRIMARY/SECONDARY

107F04.EPS

Figure 4 ◆ NEC references for industrial, commercial, and residential power.

Instructor's Notes:

THINK ABOUT IT

Putting It All Together

Look around you at the electrical components and products used and the quality of the work. Do you see any components or products that have not been listed or labeled? If so, how might these devices put you in harm's way? Do you see any code violations?

Summary

The NEC specifies the minimum provisions necessary for protecting people and property from hazards arising from the use of electricity and electrical equipment. As an electrician, you must be aware of how to use and apply the NEC on the job site. Using the NEC will help you to safely install and maintain the electrical equipment and systems you come into contact with.

Review Questions

1. The NEC provides the _____ requirements for the installation of electrical systems.
   a. minimum
   b. most stringent
   c. design specification
   d. complete

2. Which of the following groups is usually *not* represented on the Code Making Panels?
   a. Trade associations
   b. Electrical inspectors
   c. Insurance organizations
   d. Government lobbyists

3. What is the difference between a mandatory rule and an advisory rule?
   a. You must comply with mandatory rules.
   b. Mandatory rules include the word *shall*.
   c. It is recommended that you comply with advisory rules.
   d. All of the above.

4. What is covered in *NEC Article 110?*
   a. Branch circuits
   b. Definitions
   c. General requirements for electrical installations
   d. Wiring design and protection

5. Which NEC chapters cover the design and installation of electrical systems?
   a. 1, 2, and 7
   b. 1, 2, 3, 4, and 9
   c. 6, 7, and 9
   d. 5, 6, 7, and 9

6. Which of the following covers busways?
   a. *NEC Articles 330 through 339*
   b. *NEC Articles 342 through 358*
   c. *NEC Article 364*
   d. *NEC Article 318*

7. Which article of the NEC provides installation procedures for lighting fixtures in specific locations?
   a. *NEC Article 410*
   b. *NEC Article 501*
   c. *NEC Article 364*
   d. *NEC Article 460*

8. Which article of the NEC covers theaters?
   a. *NEC Article 339*
   b. *NEC Article 110*
   c. *NEC Article 430*
   d. *NEC Article 520*

9. *NEC Article 600* covers _____.
   a. track lighting
   b. electric signs and outline lighting
   c. X-ray equipment
   d. emergency lighting systems

10. Which chapter of the NEC covers devices such as radios, televisions, and telephones?
    a. *NEC Chapter 8*
    b. *NEC Chapter 7*
    c. *NEC Chapter 6*
    d. *NEC Chapter 5*

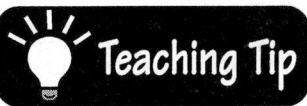 **Teaching Tip**

Discuss the "Think About It."

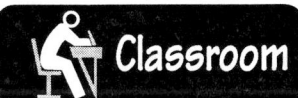 **Classroom**

Have the trainees complete the Review Questions and go over the answers prior to administering the Module Examination.

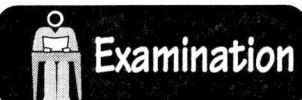 **Examination**

Administer the Module Examination and Performance Test. Be sure to record the testing results on Craft Training Report Form 200 and submit the results to the Training Program Sponsor.

PROFILE IN SUCCESS

James G. Stallcup, Greyboy & Associates

Today, the name James Stallcup is recognized throughout the electrical industry as an author, publisher, teacher, lecturer, and expert on electrical codes.

James Stallcup didn't start out his career as a writer and teacher. Like most others who have achieved success in the industry, he began in a four-year electrical apprenticeship program, which he entered because his father-in-law was an electrical contractor. In addition to his four-year apprenticeship program, James completed 756 hours of industrial electrical training.

Later on, he applied for and obtained a job as an electrical inspector for the city of Fort Worth, Texas. Because of his ability and his extensive training, James was elevated to the position of chief electrical inspector within one year.

James began teaching at the local union apprentice school to supplement his city income, and later began his own continuing education school, where he trained many electricians, corporate maintenance people, and municipal inspectors.

The handouts for his course became the embryo of his first book. Since then, James has written and published more than a dozen books. He also serves on numerous industry committees and has received many industry awards.

This remarkable success story could not have been told were it not for three key factors: proper training, the desire to continue learning as a lifetime endeavor, and James's determination to stretch his personal horizons.

Instructor's Notes:

_____

_____

_____

_____

_____

_____

_____

_____

_____

Trade Terms Introduced in This Module

Articles: The articles are the main topics of the NEC, beginning with *NEC Article 90, Introduction,* and ending with *NEC Article 820, Community Antenna Television and Radio Distribution Systems.*

Chapters: Nine chapters form the broad structure of the NEC.

Examples: Examples in *NEC Chapter 9* provide methods for performing load calculations for various types of buildings, feeders, and branch circuits.

Exceptions: Exceptions follow the applicable sections of the NEC and allow alternative methods to be used under specific conditions.

Fine Print Note (FPN): Explanatory material that follows specific NEC sections.

National Electrical Manufacturers' Association (NEMA): The association that maintains and improves the quality and reliability of electrical products.

National Fire Protection Association (NFPA): The publishers of the NEC. The NFPA develops standards to minimize the possibility and effects of fire.

Nationally Recognized Testing Laboratories (NRTL): Product safety certification laboratories that are responsible for testing and certifying electrical equipment.

Parts: Certain articles in the NEC are subdivided into parts. Parts have letter designations (e.g., Part A).

Sections: Parts and articles are subdivided into sections. Sections have numeric designations that follow the article number and are preceded by a hyphen (e.g., 501-4).

Tables: Tables are located within chapters to provide more detailed information explaining NEC content.

Answers to Review Questions

Answer	Section
1. a	1.0.0
2. d	2.1.1
3. d	3.1.0
4. c	3.3.0
5. b	3.4.1
6. c	4.0.0
7. a	4.1.0
8. d	4.2.0
9. b	4.3.0
10. a	3.4.1/4.4.0

Instructor's Notes:

_____

_____

_____

_____

_____

_____

_____

_____

_____

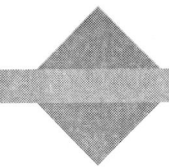

Additional Resources

For advanced study of topics covered in this task module, the following book is suggested:

National Electrical Code Handbook, Latest Edition. Quincy, MA: National Fire Protection Association.

NCCER CRAFT TRAINING USER UPDATES

The NCCER makes every effort to keep these textbooks up-to-date and free of technical errors. We appreciate your help in this process. If you have an idea for improving this textbook, or if you find an error, a typographical mistake, or an inaccuracy in the NCCER's Craft Training textbooks, please write us, using this form or a photocopy. Be sure to include the exact module number, page number, a detailed description, and the correction, if applicable. Your input will be brought to the attention of the Technical Review Committee. Thank you for your assistance.

Instructors – If you found that additional materials were necessary in order to teach this module effectively, please let us know so that we may include them in the Equipment/Materials list in the Instructor's Guide.

Write: Curriculum Revision and Development Department
National Center for Construction Education and Research
P.O. Box 141104, Gainesville, FL 32614-1104

Fax: 352-334-0932

E-mail: curriculum@nccer.org

Craft _____ Module Name _____

Copyright Date _____ Module Number _____ Page Number(s) _____

Description _____

_____

_____

_____

(Optional) Correction _____

_____

_____

(Optional) Your Name and Address _____

_____

_____

TEACHING TIPS

Section 2.0.0 *What's Wrong With These Pictures?*

Top left—Fluorescent fixture is not supported (wire is simply wrapped around the overhead beam).

Bottom Left—Improper box entry (no fittings).

Right—This unsupported PVC conduit melted in the summer heat.

Section 2.0.0 *Think About It—The NEC*

Describe the reasons for standardizing installation procedures, including convenience and the obvious interest to public safety. Have the trainees talk to practicing electricians about their use of the NEC, including what is to be inspected, when it is to be inspected, and who does the inspecting. (Inspection procedures will vary by location.)

Section 2.1.1 *Think About It—Code Changes*

Anyone can suggest a change to the *National Electrical Code®* by submitting a proposal. Suggestions for change are typically a result of safety considerations, but may also be influenced by convenience, cost, or the introduction of new materials or devices.

Section 3.3.0 *Think About It—Other Testing Laboratories*

There are many testing laboratories other than UL. For example, the Canadian Standards Association (CSA) performs product evaluations and tests similar to those performed by UL. Through international agreements, both UL and CSA are authorized to test and certify products for use in both the U.S. and Canadian markets. Other testing laboratories include ITS (Intertek Testing Services), which encompasses various labels including ETL, Warnock Hersey, and Semco; Metlabs; TÜV; Factory Mutual; and others. Encourage the trainees to examine various electrical devices for these and other marks of compliance.

Section 4.0.0 *Think About It—Junction Boxes*

NEC Section 370-23(e) covers raceway supported enclosures. Have the trainees interpret the main points of this section, including the exception.

Section 4.0.0 *Think About It—NEC Article 90*

NEC Article 90 covers the purpose and scope of the NEC and provides a brief overview of its arrangement, enforcement, types of rules, formal interpretation procedures, and other points.

The NEC does not cover certain types of installations, as listed in *NEC Section 90-2(b)*. For example, the NEC does not cover installations in ships, automobiles, or aircraft. It also does not cover installations under the control of electrical utilities, including metering equipment.

Section 4.0.0 *Think About It—Disconnects*

The trainees should follow the steps outlined in Section 4.0.0 of the Trainee Module for navigating the NEC. *NEC Section 230-71* covers the maximum number of disconnects.

Section 4.2.0 *Think About It—Sealing Requirements*

NEC Section 501-5 covers sealing conduit and cable systems and drainage of control equipment, motors, generators, etc. in Class I hazardous locations. Class I locations are those in which flammable gases or vapors are or may be present in the air in quantities sufficient to produce explosive or ignitable mixtures. The trainees should read this section carefully and restate it in their own words.

Section 5.3.0 *Think About It—Conformance and Electrical Equipment*

Other resources for finding information about the use of electrical equipment and materials can include master electricians, manufacturers' catalogs, Web sites, electrical distributors, application engineers, and manufacturers' sales representatives.

Section 5.3.0 *Think About It—Putting It All Together*

Answers will depend upon the building(s) in question. Unlisted devices and/or improper installation can result in short circuits or ground faults, as well as other problems, including faulty operation, environmental dangers, and financial losses, such as rework and possible liability.

Raceways, Boxes, and Fittings
26108-00

MODULE OVERVIEW

This course introduces the electrical trainee to the various types of raceways, boxes, and fittings, including their installation procedures and NEC requirements.

PREREQUISITES

Please refer to the Course Map in the Trainee Module. Prior to training with this module, it is recommended that the trainee shall have successfully completed the following modules:

Core Curriculum; Electrical Level One, Modules 26101 through 26107

LEARNING OBJECTIVES

Upon completion of this module, the trainee will be able to:

1. Describe various types of cable trays and raceways.
2. Identify and select various types and sizes of raceways.
3. Identify and select various types and sizes of cable trays.
4. Identify and select various types of raceway fittings.
5. Identify various methods used to install raceways.
6. Demonstrate knowledge of NEC raceway requirements.
7. Describe procedures for installing raceways and boxes on masonry surfaces.
8. Describe procedures for installing raceways and boxes on concrete surfaces.
9. Describe procedures for installing raceways and boxes in a metal stud environment.
10. Describe procedures for installing raceways and boxes in a wood frame environment.
11. Describe procedures for installing raceways and boxes on drywall surfaces.
12. Recognize safety precautions that must be followed when working with boxes and raceways.

PERFORMANCE OBJECTIVES

Under supervision of the instructor, the trainee should be able to:

1. Identify fittings, boxes, and types of raceways. Reference the appropriate NEC section(s) for these items.
2. Make a conduit-to-box connection.

NCCER STANDARDIZED CRAFT TRAINING PROGRAM

The National Center for Construction Education and Research (NCCER) provides a standardized national program of accredited craft training. Key features of the program include instructor certification, competency-based training, and performance testing. The program provides trainees, instructors, and companies with a standard form of recognition through a National Craft Training Registry. The program is described in full in the *Guidelines for Accreditation,* published by the NCCER. For more information on standardized craft training, contact the NCCER at P.O. Box 141104, Gainesville, FL 32614-1104, 352-334-0911, visit our Web site at www.NCCER.org, or e-mail info@NCCER.org.

HOW TO USE THIS ANNOTATED INSTRUCTOR'S GUIDE

Each page presents two sections of information. The larger section displays each page exactly as it appears in the Trainee Module. The narrow column ties suggested trainee and instructor actions to each page and provides icons to call your attention to material, safety, audiovisual, or testing requirements. The bottom of each page includes space for your notes.

If you see the Teaching Tip icon, that means there is a teaching tip associated with this section. Also refer to the suggested teaching tips at the end of the module.

SAFETY CONSIDERATIONS

Ensure that the trainees are equipped with appropriate personal protective equipment. Emphasize the use of the proper safety procedures when working with the tools and materials used in the installation of raceways, boxes, and fittings.

PREPARATION

Before teaching this module, you should review the Module Outline, Learning and Performance Objectives, and the Materials and Equipment List. Be sure to allow ample time to prepare your own training or lesson plan and gather all required equipment and materials.

MATERIALS AND EQUIPMENT LIST

Materials:
Transparencies
Markers/chalk
Copy of the latest edition of the *National Electrical Code*®
Module Examinations*
Performance Profile Sheets*

Equipment:
Overhead projector and screen
Whiteboard/chalkboard
Appropriate personal protective equipment
Section of EMT
EMT compression fittings
EMT setscrew fittings
Section of rigid metal conduit
Section of uncoated GRC
Section of plastic-coated GRC
Section of aluminum conduit
Section of rigid black conduit
Section of IMC
Sections of EB and DB PVC conduit
Section of LFNC and connectors
Section of flexible metal conduit
Various conduit couplings
Type C, Type L, Type T, and Type X conduit bodies

Threaded weatherproof hubs
Insulating bushings
Offsets
Metal boxes
Nonmetallic boxes
Bushings and locknuts
Sealing fittings and packing material
Various straps
Standoff support
Framing channel
Beam clamp
Access to a job site where trainees can observe a variety of wireway components, including:
    Connectors
    End plates
    Closing plates
    Tee fittings
    Crosses
    Elbows
    Nipples
    Slip fittings
Access to a job site where trainees can observe a variety of cable tray support systems, including:
    Direct rod
    Trapeze mounting
    Center hung support
    Wall mounting
    Pipe rack mounting

*Located in the Test Booklet packaged with this Annotated Instructor's Guide.

ADDITIONAL RESOURCES

This module is intended to present thorough resources for task training. The following reference works are suggested for both instructors and motivated trainees interested in further study. These are optional materials for continued education rather than for task training.

Benfield Conduit Bending Manual, Latest Edition. New York, NY: McGraw-Hill Publishing Company.

National Electrical Code Handbook, Latest Edition. Quincy, MA: National Fire Protection Association.

NOTES

The designations "National Electrical Code," "NE Code," and "NEC," where used in this document, refer to the *National Electrical Code*®, which is a registered trademark of the National Fire Protection Association, Quincy, MA. All National Electrical Code (NEC) references in this module refer to the 1999 edition of the NEC.

If you feel that additional math instruction would be helpful, Prentice Hall offers a basic math textbook entitled *Fundamentals of Electrical and Mechanical Mathematics.* It covers the basic math requirements for electrical trainees and may be ordered by contacting Prentice Hall Customer Service at 1-800-922-0579.

TEACHING TIME FOR THIS MODULE

An outline for use in developing your lesson plan is presented below. Note that each Roman numeral in the outline equates to one session of instruction. Each session has a suggested time period of 2½ hours. This includes 10 minutes at the beginning of each session for administrative tasks and one 10-minute break during the session. Approximately 12½ hours are suggested to cover *Raceways, Boxes, and Fittings.* You will need to adjust the time required for hands-on activity and testing based on your class size and resources.

Topic	Planned Time
Session I. Raceways; Conduit	
A. Raceways	_____
B. Conduit	
1. Conduit as a Ground Path	_____
2. Types of Conduit	_____
a. Electrical Metallic Tubing	_____
b. Rigid Metal Conduit	_____
c. Galvanized Rigid Steel Conduit	_____
d. Plastic-Coated GRC	_____
e. Aluminum Conduit	_____
f. Black Enamel Steel Conduit	_____
g. Intermediate Metal Conduit	_____
h. Rigid Nonmetallic Conduit	_____
i. Liquid-Tight Flexible Nonmetallic Conduit	_____
j. Flexible Metal Conduit	_____
Session II. Metal Conduit Fittings; Boxes; Bushings and Locknuts; Sealing Fittings	
A. Metal Conduit Fittings	_____
1. Couplings	_____
2. Conduit Bodies	_____
a. Type C Conduit Bodies	_____
b. Type L Conduit Bodies	_____
c. Type T Conduit Bodies	_____

        d. Type X Conduit Bodies   _____

        e. Threaded Weatherproof Hubs   _____

    3. Insulating Bushings   _____

        a. Nongrounding Insulating Bushings   _____

        b. Grounding Insulating Bushings   _____

    4. Offset Nipples   _____

B. Boxes   _____

    1. Metal Boxes   _____

        a. Pryouts   _____

        b. Knockouts   _____

    2. Nonmetallic Boxes   _____

C. Bushings and Locknuts   _____

D. Sealing Fittings   _____

Session III. Raceway Supports; Wireways; Field Trip

A. Raceway Supports   _____

    1. Straps   _____

    2. Standoff Supports   _____

    3. Electrical Framing Channels   _____

    4. Beam Clamps   _____

B. Wireways   _____

    1. Auxiliary Gutters   _____

    2. Types of Wireways   _____

        a. Wireway Fittings   _____

        b. Connectors   _____

        c. End Plates   _____

        d. Tees   _____

        e. Crosses   _____

        f. Elbows   _____

        g. Telescopic Fittings   _____

    3. Wireway Supports   _____

        a. Suspended Hangers   _____

        b. Gusset Brackets   _____

        c. Standard Hangers   _____

        d. Wireway Hangers   _____

    4. Other Types of Raceways   _____

        a. Surface Metal and Nonmetallic Raceways   _____

        b. Plugmold Multi-Outlet Systems   _____

        c. Pole Systems   _____

        d. Underfloor Systems   _____

        e. Cellular Metal Floor Raceways   _____

        f. Cellular Concrete Floor Raceways   _____

C. Field Trip   _____

    If possible, take the trainees to a commercial/industrial job site where they may observe various types of wireway components and raceway systems.

Session IV. Cable Trays; Field Trip; Storing Raceways; Handling Raceways; Ducting

A. Cable Trays   _____

   1. Cable Tray Fittings   _____

   2. Cable Tray Supports   _____

     a. Direct Rod Suspension   _____

     b. Trapeze Mounting and Center Hung Support   _____

     c. Wall Mounting   _____

     d. Pipe Rack Mounting   _____

B. Field Trip   _____

If possible, take the trainees to a commercial/industrial job site where they may observe various types of cable tray systems.

C. Storing Raceways   _____

D. Handling Raceways   _____

E. Ducting   _____

Session V. Underground Systems; Making a Conduit-to-Box Connection; Laboratory; Various Construction Procedures; Module Examination and Performance Testing

A. Underground Systems   _____

   1. Duct Line   _____

   2. Duct Materials   _____

   3. Plastic Conduit   _____

   4. Monolithic Concrete Duct   _____

   5. Cable-in-Duct   _____

B. Making a Conduit-to-Box Connection   _____

C. Laboratory   _____

Under instructor supervision, have the trainees practice making a conduit-to-box connection.

D. Various Construction Procedures   _____

   1. Masonry and Concrete Flush-Mount Construction   _____

   2. Metal Stud Environment   _____

   3. Wood Frame Environment   _____

   4. Steel Environment   _____

E. Module Examination   _____

   1. Trainees must score 70% or higher to receive recognition from the NCCER.

   2. Record the testing results on Craft Training Report Form 200 and submit the results to the Training Program Sponsor.

F. Performance Testing   _____

   1. Trainees must perform each task to the satisfaction of the instructor to receive recognition from the NCCER.

   2. Record the testing results on Craft Training Report Form 200 and submit the results to the Training Program Sponsor.

Module 26108-00

Raceways, Boxes, and Fittings

Course Map

This course map shows all of the task modules in the first level of the Electrical curricula. The suggested training order begins at the bottom and proceeds up. Skill levels increase as a trainee advances on the course map. The training order may be adjusted by the local Training Program Sponsor.

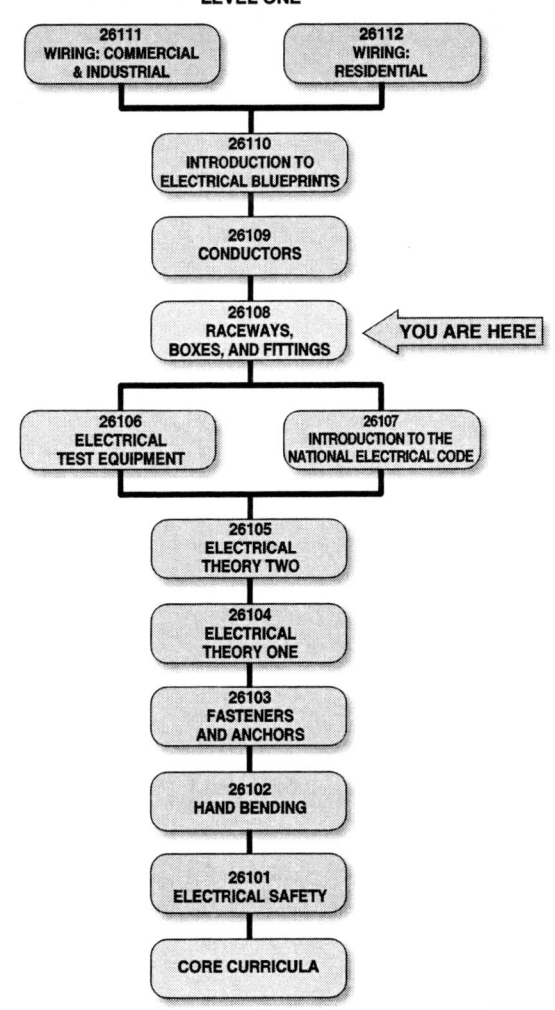

LEVEL ONE

26111
WIRING: COMMERCIAL
& INDUSTRIAL

26112
WIRING:
RESIDENTIAL

26110
INTRODUCTION TO
ELECTRICAL BLUEPRINTS

26109
CONDUCTORS

26108
RACEWAYS,
BOXES, AND FITTINGS ◁ YOU ARE HERE

26106
ELECTRICAL
TEST EQUIPMENT

26107
INTRODUCTION TO THE
NATIONAL ELECTRICAL CODE

26105
ELECTRICAL
THEORY TWO

26104
ELECTRICAL
THEORY ONE

26103
FASTENERS
AND ANCHORS

26102
HAND BENDING

26101
ELECTRICAL SAFETY

CORE CURRICULA

108CMAP.EPS

 Homework

Assign reading of Module 26108.

Instructor's Notes:

_____

_____

_____

_____

_____

_____

_____

_____

_____

Figures

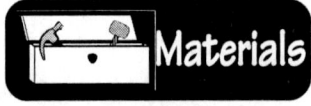

Ensure you have everything required to teach the course. Check the Materials and Equipment List at the front of this Instructor's Guide.

Instructor's Notes:

_____

_____

_____

_____

_____

_____

_____

_____

_____

Raceways, Boxes, and Fittings

Objectives

Upon completion of this module, the trainee will be able to:

1. Describe various types of cable trays and raceways.
2. Identify and select various types and sizes of raceways.
3. Identify and select various types and sizes of cable trays.
4. Identify and select various types of raceway fittings.
5. Identify various methods used to install raceways.
6. Demonstrate knowledge of NEC raceway requirements.
7. Describe procedures for installing raceways and boxes on masonry surfaces.
8. Describe procedures for installing raceways and boxes on concrete surfaces.
9. Describe procedures for installing raceways and boxes in a metal stud environment.
10. Describe procedures for installing raceways and boxes in a wood frame environment.
11. Describe procedures for installing raceways and boxes on drywall surfaces.
12. Recognize safety precautions that must be followed when working with boxes and raceways.

Prerequisites

Successful completion of the following Task Modules is recommended before beginning study of this Task Module: Core Curricula; Electrical Level One, Modules 26101 through 26107.

Required Trainee Materials

1. Trainee Task Module
2. Copy of the latest edition of the *National Electrical Code*
3. Appropriate Personal Protective Equipment
4. *OSHA Electrical Safety Guidelines* (pocket guide)

1.0.0 ◆ INTRODUCTION

Electrical raceways present challenges and new requirements involving proper installation techniques, general understanding of raceway systems, and applications of the NEC to raceway systems. Acquiring quality installation skills for raceway systems requires practice, knowledge, and training.

A presentation of the various types of raceway systems and fittings, basic raceway installation skills, and NEC requirements applicable to raceway systems are included in this module. This module also covers raceway supports and environmental considerations for raceway systems, as well as general raceway information.

Note: The designations "National Electrical Code," "NE Code," and "NEC," where used in this document, refer to the National Electrical Code®, which is a registered trademark of the National Fire Protection Association, Quincy, MA. All National Electrical Code (NEC) references in this module refer to the 1999 edition of the NEC.

Show Transparency 1, Course Objectives.

Show Transparency 2, Performance Profile Tasks.

Introduce raceway systems.

Explain that items shown in bold (blue) are defined in the Glossary at the back of this module.

8–1

Review the NEC requirements for raceway systems.

Define *raceway*.

Define *conduit*.

Explain how equipment is grounded.

Along with the study of this module, the following NEC Articles should be referenced:

- *NEC Article 250—Grounding*
- *NEC Article 318—Cable Trays*
- *NEC Article 345—Intermediate Metal Conduit*
- *NEC Article 346—Rigid Metal Conduit*
- *NEC Article 347—Rigid Nonmetallic Conduit*
- *NEC Article 348—Electrical Metallic Tubing*
- *NEC Article 350—Flexible Metal Conduit*
- *NEC Article 351—Liquid-Tight Flexible Metal Conduit and Flexible Nonmetallic Conduit*
- *NEC Article 362—Wireways*

> **Note**
> Mandatory rules in the NEC are characterized by the use of the word *shall*. Explanatory material is in the form of Fine Print Notes (FPNs). When referencing specific sections of the NEC, always check to see if any exceptions apply.

2.0.0 ◆ RACEWAYS

Raceway is a general term referring to a wide range of circular and rectangular enclosed channels used to house electrical wiring. Raceways can be metallic or nonmetallic and come in different shapes. Depending on the particular purpose for which they are intended, raceways include enclosures such as underfloor raceways, flexible metal conduit, wireways, surface metal raceways, and surface nonmetallic raceways.

3.0.0 ◆ CONDUIT

Conduit is a raceway with a circular cross section, similar to pipe, that contains wires or cables. Conduit is used to provide protection for conductors and route them from one place to another. In addition, conduit makes it easier to replace or add wires to existing structures. Metal conduit also provides a permanent electrical path to ground. This equipment should be listed per the NEC.

3.1.0 Conduit as a Ground Path

For safety reasons, most equipment that receives electrical power and has a metallic frame is grounded. In order to ground the equipment, an electrical connection must be made to connect the metal frame of the electrically powered equipment to the grounding point at the service-

entrance equipment. This is usually done in one or both of the following ways:

- The frame of the equipment is connected to a wire (equipment grounding conductor), which is directly connected to the ground point at the grounding terminal.
- The frame of the equipment is connected (bonded) to a metal conduit or other type of raceway system, which provides an uninterrupted and low-impedance circuit to the ground point at the service-entrance equipment. The metal raceway or conduit acts as the equipment grounding conductor.

> **Note**
> According to *NEC Section 250-96*, metal raceways, cable trays, cable armor, cable sheath, enclosures, frames, fittings, and other metal noncurrent-carrying parts that are to serve as grounding conductors with or without the use of supplementary equipment grounding conductors shall be effectively bonded where necessary to ensure electrical continuity and the capacity to safely conduct any fault current likely to be imposed on them. The purpose of the equipment grounding conductor is to provide a low resistance path to ground for all equipment that receives power. This is done so that if an ungrounded conductor comes in contact with the frame of a piece of equipment, the circuit overcurrent device immediately acts to open the circuit. It also reduces the voltage to ground that would be present on the faulted equipment if a person came in contact with the equipment frame.

3.2.0 Types of Conduit

There are many types of conduit used in the construction industry. The size of conduit to be used is determined by engineering specifications, local codes, and the NEC. Refer to *NEC Chapter 9, Tables 1 through 8 and Appendix C* for conduit fill with various conductors. There are several common types of conduit to examine.

3.2.1 Electrical Metallic Tubing

Electrical metallic tubing (EMT) is the lightest duty tubing available for enclosing and protecting electrical wiring. EMT is widely used for residential, commercial, and industrial wiring systems. It is

Instructor's Notes:

_____

_____

_____

_____

_____

_____

_____

_____

_____

lightweight, easily bent and/or cut to shape, and is the least costly type of metallic conduit. Because the wall thickness of EMT is less than that of rigid conduit, it is often referred to as *thinwall conduit*. A comparison of inside and outside diameters of EMT to rigid metal conduit and intermediate metal conduit (IMC) is shown in *Figure 1*.

NEC Section 348-4 permits the installation of EMT for either exposed or concealed work where it will not be subject to severe physical damage during installation or after construction. Installation of EMT is permitted in wet locations such as outdoors or indoors in dairies, laundries, and canneries using waterproof fittings.

> **Note**
> Refer to *NEC Section 348-5* for restrictions that apply to the use of EMT.

EMT shall not be used (1) where, during installation or afterward, it will be subject to severe physical damage; (2) where protected from corrosion solely by enamel; (3) in cinder concrete or cinder fill where subject to permanent moisture unless protected on all sides by a layer of non-cinder concrete at least two inches thick or unless

the tubing is at least 18 inches under the fill; (4) in any hazardous (classified) locations except as permitted by *NEC Sections 502-4, 503-3, and 504-20;* or (5) for the support of fixtures or other equipment.

In a wet area, EMT and other conduit must be installed to prevent water from entering the conduit system. In locations where walls are subject to regular wash-down (see *NEC Section 300-6*), the entire conduit system must be installed to provide a ¼-inch air space between it and the wall or supporting surface. The entire conduit system is considered to include conduit, boxes, and fittings. To ensure resistance to corrosion caused by wet environments, EMT is galvanized. The term *galvanized* is used to describe the procedure in which the interior and exterior of the conduit is coated with a corrosion-resistant zinc compound.

EMT, being a good conductor of electricity, may be used as an equipment grounding conductor. In order to qualify as an equipment grounding conductor (see *NEC Section 250-68*), the conduit system must be tightly connected at each joint and provide a continuous grounding path from each electrical load to the service equipment. The connectors used in an EMT system ensure electrical and mechanical continuity throughout the system (see *NEC Sections 250-96, 300-10, and 348-10*).

EMT fittings are manufactured in two basic types. One type of fitting is the compression coupling. (See *Figure 2*.)

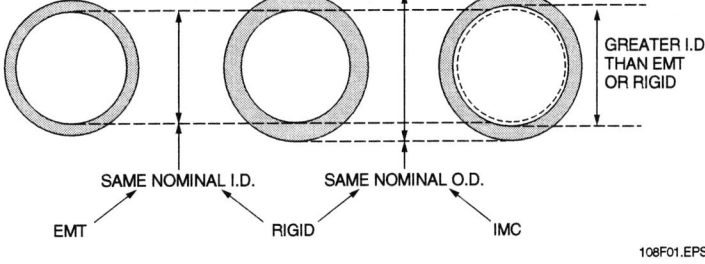

SAME NOMINAL I.D.   SAME NOMINAL O.D.

GREATER I.D. THAN EMT OR RIGID

EMT   RIGID   IMC

108F01.EPS

Figure 1 ◆ EMT conduit comparison.

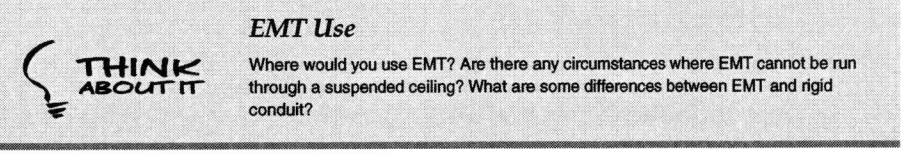

EMT Use

THINK ABOUT IT

Where would you use EMT? Are there any circumstances where EMT cannot be run through a suspended ceiling? What are some differences between EMT and rigid conduit?

RACEWAYS, BOXES, AND FITTINGS—TRAINEE MODULE 26108

8–3

COUPLING   CONNECTOR

108F02.EPS

Figure 2 ◆ Compression fittings.

Note
Support requirements for EMT are also covered in *NEC Section 348-13*. The types of supports will be discussed later in this training module.

Because EMT is too thin for threads, special fittings must be used. For wet or damp locations, compression fittings such as those shown in *Figure 2* are used. These fittings contain a compression ring made of metal that forms a watertight seal.

When EMT compression couplings are used, they must be securely tightened, and when installed in masonry concrete, they must be of the concrete-tight type. If installed in a wet location, they must be the raintight type. Refer to *NEC Article 348.*

EMT fittings for dry locations can be either the setscrew type or the indenting type. To use the setscrew type, the ends of the EMT are inserted into the sleeve and the setscrews are tightened to make the connection. Various types of setscrew fittings are shown in *Figure 3.*

EMT sizes of 2½ inches and larger have the same outside diameter as corresponding sizes of galvanized rigid steel conduit (GRC). GRC threadless connectors may be used to connect EMT.

Note
EMT connectors, although they are the same size as GRC threadless connectors, may not be used to connect GRC.

Both setscrew and compression couplings are available in die-cast or steel construction. Steel couplings are stronger than die-cast couplings and have superior quality.

Support requirements for EMT are presented in *NEC Section 348-13*. As with most other metal con-

VERONICA WESTFALL   108F03.EPS

Figure 3 ◆ Setscrew fittings.

duit, EMT must be supported at least every 10 feet and within 3 feet of each outlet box, junction box, cabinet, fitting, or terminating end of the conduit. An exception to *NEC Section 348-13* allows the fastening of unbroken lengths of EMT to be increased to a distance of 5 feet (1.52m) where structural members do not readily permit fastening within 3 feet (914mm).

Electrical nonmetallic tubing (ENT) is also available. It provides an economical alternative to EMT, but it can only be used behind a 15-minute barrier inside non-fire-rated walls in certain applications.

3.2.2 Rigid Metal Conduit

Rigid metal conduit is conduit that is constructed of metal having sufficient thickness to permit the cutting of pipe threads at each end. Rigid metal conduit provides the best physical protection for conductors of any of the various types of conduit. Rigid metal conduit is supplied in 10-foot lengths including a threaded coupling on one end.

Rigid metal conduit may be fabricated from steel or aluminum. Rigid metal steel conduit may be galvanized, or enamel-coated inside and out. Because of its threaded fittings, rigid metal conduit provides an excellent equipment grounding conductor as defined in *NEC Section 250-118(2)*. A

Instructor's Notes:

_____

_____

_____

_____

_____

_____

_____

_____

_____

EMT Installation

When installing EMT, hook your index finger up through the box to check that the conduit is seated in the connector. If you feel a lip between the conduit and the connector, the conduit is not properly seated.

WALTER JOHNSON   108PO801.EPS

Note

Specific information on rigid metal conduit may be found in *NEC Article 346.*

piece of rigid metal conduit is shown in *Figure 4.* The support requirements for rigid metal conduit are presented in *NEC Section 346-12.*

3.2.3 Galvanized Rigid Steel Conduit

GRC is mostly used in industrial applications. GRC is heavier than EMT and IMC. It is more

WALTER JOHNSON   108F04.EPS

Figure 4 ◆ Rigid metal conduit.

difficult to cut and bend, requires threading of each end, and has a higher purchase price than

Discuss GRC and its applications. Pass around an example for the trainees to examine.

Rigid Metal Conduit Installations

Use rigid metal conduit in hazardous environments such as feed mills or in areas such as outdoor environments where there is a chance of physical abuse or extreme moisture. The NEC does allow EMT to be buried in the ground or in concrete, but galvanized rigid metal conduit is more commonly used.

Discuss plastic-coated GRC and its applications. Pass around an example for the trainees to examine.

EMT and IMC. As a result, the cost of installing GRC is generally higher than the cost of installing EMT and IMC.

3.2.4 Plastic-Coated GRC

Plastic-coated GRC has a thin coating of PVC over the GRC. This combination is useful when an environment calls for the ruggedness of GRC along with the corrosion resistance of PVC. Typical installations where plastic-coated GRC may be required are:

- Chemical plants
- Food plants
- Refineries
- Fertilizer plants
- Paper mills
- Wastewater treatment plants

Plastic-coated GRC requires special threading and bending techniques.

3.2.5 Aluminum Conduit

Aluminum conduit has several characteristics that distinguish it from steel conduit. Because it has better resistance to wet environments and some chemical environments, aluminum conduit generally requires less maintenance in installations such as sewage treatment plants.

Direct burial of aluminum conduit results in a self-stopping chemical reaction on the conduit surface, which forms a coating on the conduit. This coating acts to prevent further corrosion, increasing the life of the installation.

Note
Caution must be exercised to avoid burial of aluminum conduit in soil or concrete that contains calcium chloride. Calcium chloride may interfere with the corrosion resistance of aluminum conduit. Calcium chloride and similar materials are often added to concrete to speed concrete setting. It is important to determine if chlorides are to be used in the concrete prior to installing aluminum conduit. If chlorides are to be used, aluminum conduit must be avoided. Check with local authorities regarding this type of usage.

Since aluminum conduit is lighter than steel conduit, there are some installation advantages to using aluminum. For example, a 10-foot section of 3-inch aluminum conduit weighs about 23 pounds, compared to the 68-pound weight of its steel counterpart.

3.2.6 Black Enamel Steel Conduit

Rigid black enamel steel conduit (often called *black conduit*) is steel conduit that is coated with a black enamel. In the past, this type of conduit was used exclusively for indoor wiring. Black enamel steel conduit is no longer manufactured for sale in the United States. It is mentioned only because it may still be found in existing installations.

Discuss aluminum conduit and its applications. Pass around an example for the trainees to examine.

Discuss black enamel steel conduit and its applications. Pass around an example for the trainees to examine.

Use of Aluminum Conduit

Aluminum conduit is used for special purposes such as high-cycle lines (400 cycles or above); around cooling towers, food service areas, and other applications in which corrosion is a factor; or where magnetic induction is a concern, such as near magnetic resonance imaging (MRI) equipment in hospitals.

Instructor's Notes:

_____

_____

_____

_____

_____

_____

_____

_____

_____

3.2.7 Intermediate Metal Conduit

Intermediate metal conduit (IMC) has a wall thickness that is less than rigid metal conduit but greater than that of EMT. The weight of IMC is approximately ⅔ that of rigid metal conduit. Because of its lower purchase price, lighter weight, and thinner walls, IMC installations are generally less expensive than comparable rigid metal conduit installations. However, IMC installations still have high strength ratings.

Note

Additional information on intermediate metal conduit may be found in *NEC Article 345.*

The outside diameter of a given size of IMC is the same as that of the comparable size of rigid metal conduit. Therefore, rigid metal conduit fittings may be used with IMC. Since the threads on IMC and rigid metal conduit are the same size, no special threading tools are needed to thread IMC. Some electricians feel that threading IMC is more difficult than threading rigid metal conduit because IMC is somewhat harder.

The internal diameter of a given size of IMC is somewhat larger than the internal diameter of the same size of rigid metal conduit because of the difference in wall thickness. Bending IMC is considered easier than bending rigid metal conduit because of the reduced wall thickness. However, bending is sometimes complicated by kinking, which may be caused by the increased hardness of IMC.

The NEC requires that intermediate metal conduit be identified along its length at 5-foot intervals with the letters *IMC. NEC Sections 110-21 and 345-16(c)* describe this marking requirement.

Intermediate metal conduit, like rigid metal conduit, is permitted to act as an equipment grounding conductor, as defined in *NEC Section 250-118(3).*

The use of IMC may be restricted in some jurisdictions. It is important to investigate the requirements of each jurisdiction before selecting any materials.

3.2.8 Rigid Nonmetallic Conduit

The most common type of rigid nonmetallic conduit is manufactured from polyvinyl chloride

Figure 5 ◈ PVC conduit.

(PVC). Because PVC conduit is noncorrosive, chemically inert, and non-aging, it is often used for installation in wet or corrosive environments. Corrosion problems found with steel and aluminum rigid metal conduit do not occur with PVC. However, PVC conduit may deteriorate under some conditions, such as extreme sunlight.

All PVC conduit is marked according to standards established by the National Electrical Manufacturers' Association (NEMA) or Underwriters' Laboratories (UL). A section of PVC conduit is shown in *Figure 5.*

Since PVC conduit is lighter than steel or aluminum rigid conduit, IMC, or EMT, it is considered easier to handle. PVC conduit can usually be installed much faster than other types of conduit because the joints are made up with cement and require no threading.

PVC conduit contains no metal. This characteristic reduces the voltage drop of conductors carrying alternating current in PVC compared to identical conductors in steel conduit.

Because PVC is nonconducting, it cannot be used as an equipment grounding conductor. An equipment grounding conductor sized in accordance with *NEC Table 250-122* must be pulled in each PVC conductor run (except for underground service-entrance conductors).

PVC is available in lengths up to 20 feet. However, some jurisdictions require it to be cut to 10-foot lengths prior to installation. PVC expands and contracts significantly when exposed to changing temperatures. This expansion and contraction characteristic is worse in long runs. To avoid damage to PVC conduit caused by temperature changes, expansion couplings such as the one shown in *Figure 6* are used. The inside of the coupling is sealed with one or more O-rings. This type of coupling may allow up to six inches of movement. Check the requirements of the local jurisdiction prior to installing PVC.

Discuss IMC and its applications. Pass around an example for the trainees to examine.

Discuss PVC conduit and its applications. Pass around examples of Types EB and DB for the trainees to examine.

Show Transparency 5 (Figure 5).

Discuss PVC expansion couplings.

Show Transparency 6 (Figure 6).

108F06.EPS

Figure 6 ◆ PVC expansion coupling.

PVC conduit is manufactured in two types:

- *Type EB*—Thin wall for underground use only when encased in concrete. Also referred to as *Type I*.
- *Type DB*—Thick wall for underground use without encasement in concrete. Also referred to as *Type II*.

Type DB is available in two wall thicknesses, Schedule 40 and Schedule 80.

- Schedule 40 is heavy wall for direct burial in the earth and above-ground installations.
- Schedule 80 is extra heavy wall for direct burial in the earth, above-ground installations for general applications, and installations where the conduit is subject to physical damage.

PVC conduit is affected by higher-than-usual ambient temperatures. Support requirements for PVC are found in *NEC Table 347-8*. As with other conduit, it must be supported within three feet of each termination, but the maximum spacing between supports depends upon the size of the conduit. Some of the regulations for the maximum spacing of supports are:

- ½- to 1-inch conduit: every 3 feet
- 1¼- to 2-inch conduit: every 5 feet
- 2½- to 3-inch conduit: every 6 feet
- 3½- to 5-inch conduit: every 7 feet
- 6-inch conduit: every 8 feet

3.2.9 Liquid-Tight Flexible Nonmetallic Conduit

Liquid-tight flexible nonmetallic conduit (LFNC) was developed as a raceway for industrial equipment where flexibility was required and protec-tion of conductors from liquids was also neces-sary. This is covered by *NEC Article 351, Part B*. Usage of LFNC has been expanded from indus-trial applications to outside and direct burial us-age where listed.

Several varieties of LFNC have been introduced. The first product (LFNC-A) was commonly referred to as *hose*. It consisted of an inner and outer layer of neoprene with a nylon reinforcing web between the layers. A second-generation product (LFNC-B) con-sisted of a smooth wall, flexible PVC with a rigid PVC integral reinforcement rod. The third product (LFNC-C) was a nylon corrugated shape without any integral reinforcements. These three permitted LFNC raceway designs must be flame resistant with fittings approved for installation of electrical con-ductors. Nonmetallic connectors are listed for use and some liquid-tight metallic flexible conduit con-nectors are dual-listed for both metallic and non-metallic liquid-tight flexible conduit.

LFNC is sunlight-resistant and suitable for use at conduit temperatures of 80°C dry and 60°C wet. It is available in ⅜-inch through 4-inch sizes. *NEC Section 351-23(b)* states that LFNC cannot be used where subject to physical damage or in lengths longer than six feet, except where properly secured, where flexibility is required, or as permitted by *NEC Section 351-23(a)(5)*. Also, it cannot be used to contain conductors in excess of 600 volts nominal except as permitted by *NEC Section 600-31*.

Liquid-tight flexible metal conduit is a raceway of circular cross section having an outer liquid-tight, nonmetallic, sunlight-resistant jacket over an inner flexible metal core with associated cou-plings and connectors.

Flex connectors are used to connect flexible conduit to boxes or equipment. They are available in straight, 45°, and 90° configurations (*Figure 7*).

3.2.10 Flexible Metal Conduit

Flexible metal conduit, sometimes called *flex*, may be used for many kinds of wiring systems. Flexi-ble metal conduit is made from a single strip of steel or aluminum, wound and interlocked. It is typically available in sizes from ⅜ inch to 4 inches

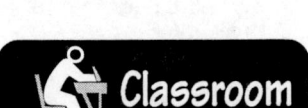

Discuss LFNC and its applications. Pass around an example for the trainees to examine.

Show Transparency 7 (Figure 7).

Discuss the "Think About It."

Applications of PVC

What installations would be suitable for the use of PVC? For what situations would PVC be a poor choice?

ELECTRICAL LEVEL ONE—TRAINEE MODULE 26108

Instructor's Notes:

ELECTRICAL LEVEL ONE—INSTRUCTOR'S GUIDE MODULE 26108

STRAIGHT CONNECTOR

45° CONNECTOR

90° CONNECTOR

108F07.EPS

Figure 7 ◆ Flex connectors.

in diameter. An illustration of flexible metal conduit is shown in *Figure 8.*

Flexible metal conduit is often used to connect equipment or machines that vibrate or move slightly during operation. Also, final connection to equipment having an electrical connection point that is marginally accessible is often accomplished with flexible metal conduit.

Flexible metal conduit is easily bent, but the minimum bending radius is the same as for other types of conduit. It should not be bent more than the equivalent of four quarter bends (360° total)

between pull points (e.g., conduit bodies and boxes). It can be connected to boxes with a flexible conduit connector and to rigid conduit or EMT by using a combination coupling.

Two types of combination couplings are shown in *Figure 9.*

Flexible metal conduit is generally available in two types: nonliquid-tight and liquid-tight. *NEC Articles 350 and 351* cover the uses of flexible metal conduit.

Liquid-tight flexible metal conduit has an outer covering of liquid-tight, sunlight-resistant flexible

Classroom

Discuss flexible metal conduit and its applications. Pass around an example for the trainees to examine.

Audiovisual

Show Transparency 8 (Figure 9).

VERONICA WESTFALL   108F08.EPS

Figure 8 ◆ Flexible metal conduit.

FLEXIBLE TO EMT

FLEXIBLE TO RIGID
108F09.EPS

Figure 9 ◆ Combination couplings.

material that acts as a moisture seal. It is intended for use in wet locations. It is used primarily for equipment and motor connections when movement of the equipment is likely to occur. The number of bends, size, and support requirements for liquid-tight conduit are the same as for all flexible conduit. Fittings used with liquid-tight conduit must also be of the liquid-tight type.

Support requirements for flexible metal conduit are found in *NEC Sections 350-18 and 351-8*. Straps or other means of securing the flexible metal conduit must be spaced every 4½ feet and within 12 inches of each end. (This spacing is closer together than for rigid conduit.) However, at terminals where flexibility is necessary, lengths of up to 36 inches without support are permitted. Failure to provide proper support for flexible conduit can make pulling conductors difficult.

4.0.0 ◆ METAL CONDUIT FITTINGS

A large variety of conduit fittings are available to do electrical work. Manufacturers design and construct fittings to permit a multitude of applications. The type of conduit fitting used in a particular application depends upon the size and type of conduit, the type of fitting needed for the application, the location of the fitting, and the installation method. The requirements and proper applications of boxes and fittings (conduit bodies)

COUPLING   CONDUIT

108F10.EPS

Figure 10 ◆ Conduit and coupling.

are found in *NEC Section 300-15*. Some of the more common types of fittings are examined in the following sections.

4.1.0 Couplings

Couplings are sleeve-like fittings that are typically threaded inside to join two male threaded pieces of rigid conduit or IMC. A piece of conduit with a coupling is shown in *Figure 10*.

Other types of couplings may be used depending upon the location and type of conduit. Several types are shown in *Figure 11*.

4.2.0 Conduit Bodies

Conduit bodies are a separate portion of a conduit or tubing system that provide access through a removable cover(s) to the interior of the system at a

8–10

Classroom

Discuss couplings and their applications. Pass around examples for the trainees to examine.

Audiovisual

Show Transparencies 9 and 10 (Figures 10 and 11).

Instructor's Notes:

THREE-PIECE COUPLING

HINGED COUPLING

CONCRETE-TIGHT SETSCREW

EMT TO RIGID

108F11.EPS

Figure 11 ⬥ Metal conduit couplings.

junction of two or more sections of the system or at a terminal point of the system. They are usually cast and are significantly higher in cost than the stamped steel boxes permitted with EMT. However, there are situations in which conduit bodies are preferable, such as in outdoor locations, for appearance's sake in an exposed location, or to change types or sizes of raceways. Also, conduit bodies do not have to be supported, as do stamped steel boxes. They are also used when elbows or bends would not be appropriate.

NEC Section 370-16(c) states that conduit bodies cannot contain splices, taps, or devices unless they are durably and legibly marked by the manufacturer with their cubic inch capacity. The maximum number of conductors permitted in a conduit body is found using *NEC Table 370-16(b).* (See *Table 1.*)

4.2.1 Type C Conduit Bodies

Type C conduit bodies may be used to provide a pull point in a long conduit run or a conduit run that has bends totaling more than 360°. A Type C conduit body is shown in *Figure 12*.

4.2.2 Type L Conduit Bodies

When referring to conduit bodies, the letter *L* represents an *elbow*. A Type L conduit body is used as a pulling point for conduit that requires a 90° change in direction. The cover is removed, then the wire is pulled out, coiled on the ground or

Table 1   Volume Required per Conductor

Size of Conductor	Free Space Within Box for Each Conductor
No. 18	1.5 cubic inches
No. 16	1.75 cubic inches
No. 14	2.0 cubic inches
No. 12	2.25 cubic inches
No. 10	2.5 cubic inches
No. 8	3.0 cubic inches
No. 6	5.0 cubic inches

VERONICA WESTFALL   108F12.EPS

Figure 12 ⬥ Type C conduit body.

floor, reinserted into the other conduit body's opening, and pulled. The cover and its associated gasket are then replaced. Type L conduit bodies are available with the cover on the back (Type LB), on the sides (Type LL or LR), or on both sides (Type LRL). Several Type L conduit bodies are shown in *Figure 13*.

Explain how to find the maximum number of conductors permitted in a conduit body. Pass around various conduit bodies for the trainees to examine.

Discuss Type C conduit bodies and their applications.

Discuss Type L conduit bodies and their applications.

Explain how to identify Type L conduit bodies.

VERONICA WESTFALL   108F13A.EPS   VERONICA WESTFALL   108F13B.EPS

Figure 13 ◆ Type L conduit bodies and how to identify them.

> **Note**
> The cover and gasket must be ordered separately. Do not assume that these parts come with conduit bodies when they are ordered.

To identify Type L conduit bodies, use the following method:

Step 1   Hold the body like a pistol.

Step 2   Locate the opening on the body:
- If the opening is to the left, it is a Type LL.
- If the opening is to the right, it is a Type LR.
- If the opening is on top (back), it is a Type LB.
- If there are openings on both the left and the right, it is a Type LRL.

4.2.3   Type T Conduit Bodies

Type T conduit bodies are used to provide a junction point for three intersecting conduits and are used extensively in conduit systems. A Type T conduit body is shown in *Figure 14*.

4.2.4   Type X Conduit Bodies

Type X conduit bodies are used to provide a junction point for four intersecting conduits. The removable cover provides access to the interior of the X so that wire pulling and splicing may be performed. A Type X conduit body is shown in *Figure 15*.

VERONICA WESTFALL   108F14.EPS

Figure 14 ◆ Type T conduit body.

VERONICA WESTFALL   108F15.EPS

Figure 15 ◆ Type X conduit body.

4.2.5   Threaded Weatherproof Hub

Threaded weatherproof hubs are used for conduit entering a box in a wet location. *Figure 16* shows a typical threaded weatherproof hub.

Discuss Type T conduit bodies and their applications.

Discuss Type X conduit bodies and their applications.

ELECTRICAL LEVEL ONE—TRAINEE MODULE 26108

Instructor's Notes:

ELECTRICAL LEVEL ONE—INSTRUCTOR'S GUIDE MODULE 26108

Installation of Conduit Bodies

INSIDE TRACK It will be much easier to identify conduit bodies once you begin to see them in use. Here we show liquid-tight nonmetallic conduit entering a Type T conduit body (A) and a Type LB conduit body in an outdoor commercial application (B).

WALTER JOHNSON   (A)   108PO803.EPS   VERONICA WESTFALL   (B)   108PO804.EPS

VERONICA WESTFALL   108F16.EPS

Figure 16 ◆ Threaded weatherproof hubs.

4.3.0 Insulating Bushings

An insulating bushing is either made of nonmetal or has an insulated throat. Insulating bushings are installed on the threaded end of conduit that enters a sheet metal enclosure.

4.3.1 Nongrounding Insulating Bushings

The purpose of a nongrounding insulating bushing is to protect the conductors from being damaged by the sharp edges of the threaded conduit end. *NEC Sections 345-15 and 347-12* state that where a conduit enters a box, fitting, or other enclosure, a bushing must be provided to protect the wire from

abrasion unless the design of the box, fitting, or enclosure is such as to afford equivalent protection. *NEC Section 373-6(c)* references *Section 300-4(f)* that states where ungrounded conductors of No. 4 or larger enter a raceway in a cabinet or box enclosure, the conductors shall be protected by a substantial fitting providing a smoothly rounded insulating surface, unless the conductors are separated from the raceway fitting by substantial insulating material securely fastened in place. An exception is where threaded hubs or bosses that are an integral part of a cabinet, box enclosure, or raceway provide a smoothly rounded or flared entry for conductors. Insulating bushings are shown in *Figure 17*.

4.3.2 Grounding Insulating Bushings

Grounded insulating bushings, usually called *grounding bushings*, are used to protect conductors and also have provisions for connection of an equipment grounding conductor. The ground wire, once connected to the grounding bushing, may be connected to the enclosure to which the conduit is connected. Grounding insulating bushings are shown in *Figure 18*.

4.4.0 Offset Nipples

Offset nipples are used to connect two pieces of electrical equipment in close proximity where a slight offset is required. They come in sizes ranging from ½" to 2" in diameter. See *Figure 19*.

Classroom

Discuss weatherproof hubs and their applications. Pass around examples for the trainees to examine.

Discuss nongrounding and grounding insulating bushings and their applications. Pass around examples for the trainees to examine.

Discuss offset nipples and their applications. Pass around examples for the trainees to examine.

VERONICA WESTFALL   108F17.EPS

Figure 17 ◆ Insulating bushings.

VERONICA WESTFALL   108F19.EPS

Figure 19 ◆ Offset nipples.

VERONICA WESTFALL   108F18.EPS

Figure 18 ◆ Grounding insulating bushings.

5.0.0 ◆ BOXES

A box is installed at each outlet, switch, or junction point for all wiring installations and branch circuits. Boxes are made from either metallic or nonmetallic material. *Figure 20* shows various boxes used in raceway systems.

5.1.0 Metal Boxes

Metal boxes are made from sheet steel. The surface is galvanized to resist corrosion and provide a continuous ground. Refer to *NEC Section 370-40* for information on thickness and grounding pro-

108F20.EPS

Figure 20 ◆ Various boxes used in raceway systems.

ELECTRICAL LEVEL ONE—TRAINEE MODULE 26108

Instructor's Notes:

visions. Metal boxes are made with removable circular sections called *pryouts* or *knockouts*. These circular sections are removed to make openings for conduit or cable connections.

5.1.1 Pryouts

In a pryout, a section is cut completely through the metal but only part of the way around, leaving solid metal tabs at two points. A slot is cut in the center of the pryout. To remove the pryout, a screwdriver is inserted into the slot and twisted to break the solid tabs (*Figure 21*).

5.1.2 Knockouts

Knockouts are pre-punched circular sections that do not include a pryout slot. The knockout is easily removed when sharply hit by a hammer and punch, as shown in *Figure 22*.

Often conduit must enter boxes, cabinets, or panels that do not have pre-cut knockouts. In these cases, a knockout punch can be used to make a hole for the conduit connection. A knockout punch kit is shown in *Figure 23*.

5.2.0 Nonmetallic Boxes

Nonmetallic boxes are made of PVC or Bakelite (a fiber-reinforced plastic). Nonmetallic boxes are often used in corrosive environments. *NEC Section 370-3* covers the use of nonmetallic boxes and the types of conduit, fittings, and grounding requirements for specific applications.

6.0.0 ◆ BUSHINGS AND LOCKNUTS

Conduit is joined to boxes by connectors, adapters, threaded hubs, or locknuts.

Bushings protect the wires from the sharp edges of the conduit. Bushings are usually made of plastic or metal. Some metal bushings have a grounding screw to permit a bonding wire to be installed. Some different types of plastic and metal bushings are shown in *Figure 24*.

Locknuts are used on the inside and outside walls of the box to which the conduit is connected.

108F21.EPS

Figure 21 ◆ Pryout removal.

KNOCKOUT

108F22.EPS

Figure 22 ◆ Knockout removal.

VERONICA WESTFALL   108F23.EPS

Figure 23 ◆ Knockout punch kit.

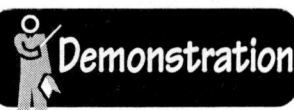

Demonstrate how to remove knockouts and pryouts.

Show Transparencies 12 and 13 (Figures 21 and 22).

Discuss nonmetallic boxes and their applications. Pass around examples for the trainees to examine.

Discuss bushings and locknuts and their applications. Pass around examples for the trainees to examine.

Show Transparencies 14 and 15 (Figures 24 and 25).

Demonstrate how to use a knockout punch.

Using a Punch

To cut a hole with a knockout punch, first measure for the center and drill a pilot hole large enough to insert the drive screw. Turn the drive nut with a wrench until the punch cuts through the box wall.

Removing Knockouts

For concentric or eccentric knockouts, first drive down one section of the smallest ring and cut it in half, then drive down the next section of the ring and cut it in half. Finally, twist off the attached portion of the knockout.

TIM ELY   108PO805.EPS

PLASTIC
INSULATING
BUSHING

METALLIC
BUSHING

INSULATED
METALLIC
BUSHING
108F24.EPS

Figure 24 ◆ Bushings.

ELECTRICAL LEVEL ONE—TRAINEE MODULE 26108

Instructor's Notes:

_____

_____

_____

_____

_____

_____

_____

_____

_____

SEALING LOCKNUT   STANDARD LOCKNUT   STANDARD LOCKNUT   GROUNDING LOCKNUT

108F25.TIF

Figure 25 ◆ Locknuts.

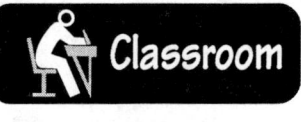

Explain the purpose of sealing fittings. Pass around examples for the trainees to examine.

Show Transparency 16 (Figure 26).

A grounding locknut may be needed if a bonding wire is to be installed. Special sealing locknuts are also used in wet locations. Several types of locknuts are shown in *Figure 25*.

7.0.0 ◆ SEALING FITTINGS

Hazardous locations in manufacturing plants and other industrial facilities involve a wide variety of flammable gases and vapors and ignitable dusts. These hazardous substances have widely differ-

ent flash points, ignition temperatures, and flammable limits requiring fittings that can be sealed. Sealing fittings are installed in conduit runs to minimize the passage of gases, vapors, or flames through the conduit and reduce the accumulation of moisture. They are required by *NEC Article 500* in hazardous locations where explosions may occur. They are also required where conduit passes from a hazardous location of one classification to another or to an unclassified location. Several types of sealing fittings are shown in *Figure 26*.

VERTICAL OR HORIZONTAL

VERTICAL ONLY   ELBOW-SEAL

ROTATING SPOUT   SEALING HUB

WEEP-HOLE PLUG

SEALING COMPOUND   DAM

CORE

DRAIN FITTING CROSS SECTION

Figure 26 ◆ Sealing fittings.

108F26.EPS

RACEWAYS, BOXES, AND FITTINGS—TRAINEE MODULE 26108

8–17

Discuss various types of raceway supports, including straps and standoff supports. Pass around examples for the trainees to examine.

Show Transparency 17 (Figure 27).

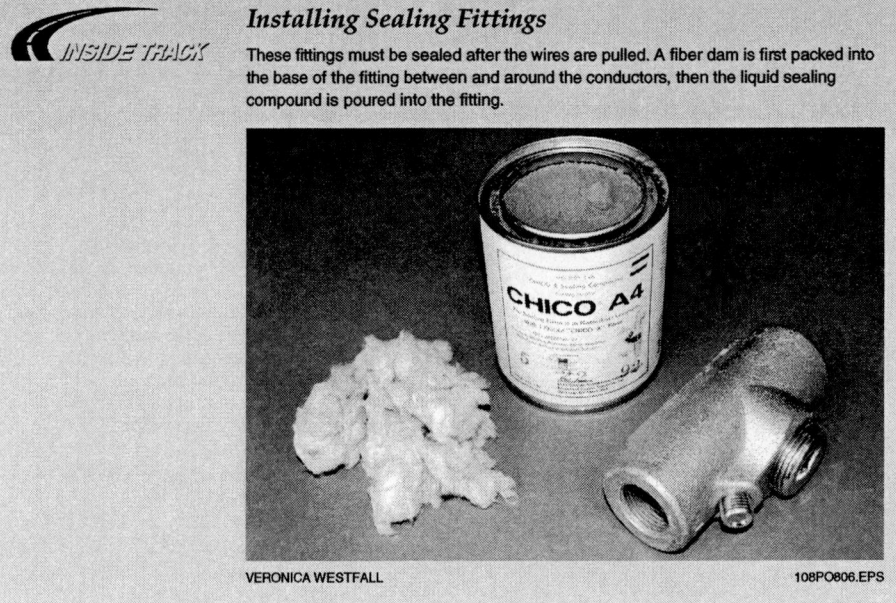

INSIDE TRACK *Installing Sealing Fittings*

These fittings must be sealed after the wires are pulled. A fiber dam is first packed into the base of the fitting between and around the conductors, then the liquid sealing compound is poured into the fitting.

VERONICA WESTFALL   108PO806.EPS

8.0.0 ◆ RACEWAY SUPPORTS

Raceway supports are available in many types and configurations. This section discusses the most common conduit supports found in electrical installations. *NEC Section 300-11* discusses the requirements for branch circuit wiring that is supported from above. Electrical equipment and raceways must have their own supporting methods and may not be supported by the supporting hardware of a fire-rated roof/ceiling assembly.

8.1.0 Straps

Straps are used to support conduit to a surface (see *Figure 27*). The spacing of these supports must conform to the minimum support spacing requirements for each type of conduit. One- and two-hole straps are used for all types of conduit: EMT, GRC, IMC, PVC, and flex. The straps can be flexible or rigid. Two-part straps are used to secure conduit to electrical framing channels (struts). Parallel and right angle beam clamps are also used to support conduit from structural members.

Clamp back straps can also be used with a backplate to maintain the ¼-inch spacing from the surface required for installations in wet locations.

8.2.0 Standoff Supports

The standoff support, often referred to as a *Minerallac* (the name of a manufacturer of this type of support), is used to support conduit away from the supporting structure. In the case of the one-hole and two-hole straps, the conduit must be kicked up wherever a fitting occurs. If standoff supports are used, the conduit is held away from the supporting surface, and no offsets (kicks) are required in the conduit at the fittings. Standoff supports may be used to support all types of conduit including GRC, IMC, EMT, PVC, and flex, as well as tubing installations. A standoff support is shown in *Figure 28*.

Instructor's Notes:

_____

_____

_____

_____

_____

_____

_____

_____

_____

_____

ONE-HOLE STRAP

RIGID STRAP

TWO-HOLE STRAP

CLAMP STRAP

108F27.EPS

Figure 27 ◆ Straps.

8.3.0 Electrical Framing Channels

Electrical framing channels or other similar framing materials are used together with Unistrut-type conduit clamps to support conduit (see *Figure 29*). They may be attached to a ceiling, wall, or other surface or be supported from a trapeze hanger.

8.4.0 Beam Clamps

Beam clamps are used with suspended hangers. The raceway is attached to or laid in the hanger. The hanger is suspended by a threaded rod. One end of the threaded rod is attached to the hanger and the other end is attached to a beam clamp. The beam clamp is then attached to a beam. A beam clamp with wireway support assembly is shown in *Figure 30*.

Discuss framing channels and beam clamps and their applications.

VERONICA WESTFALL   108F28.EPS

Figure 28 ◆ Standoff support.

TRAPEZE HANGER

ELECTRICAL FRAMING CHANNEL

ELECTRICAL FRAMING CHANNEL

WALL MOUNT

108F29.EPS

Figure 29 ◆ Electrical framing channels.

Show Transparencies 18 and 19 (Figures 29 and 30).

Bundling Conductors

When conductors are bundled together in a wireway their magnetic fields tend to cancel, thus minimizing inductive heating in the conductors.

Figure 30 ◆ Beam clamp.

Discuss various types of wireways and their applications.

Discuss the purpose of auxiliary gutters.

Stress NEC requirements.

9.0.0 ◆ WIREWAYS

Wireways are sheet metal troughs provided with hinged or screw-on removable covers. Like other types of raceways, wireways are used for housing electric wires and cables. Wireways are available in various lengths, including 1, 2, 3, 4, 5, and 10 feet. The availability of various lengths allows runs of any exact number of feet to be made without cutting the wireway ducts. Wireways are dealt with specifically in *NEC Article 362.*

As listed in *NEC Section 362-5*, the maximum number of current-carrying conductors at any cross section of a wireway is 30. However, this number does not include conductors for signaling circuits or conductors between a motor and its starter. Also, it is important to know that the sum of the cross-sectional areas of all conductors contained at any cross section must not exceed 20% of the interior cross-sectional area of the wireway. Refer to *NEC Section 362-5* for the exceptions to this rule.

It is also noted in *NEC Section 362-7* that conductors, together with splices and taps, must not fill the wireway to more than 75% of its cross-sectional area. Each wireway is also rated for a maximum permitted conductor size for any single conductor. No conductor larger than the rated size is to be installed in any wireway.

9.1.0 Auxiliary Gutters

Strictly speaking, an auxiliary gutter is a wireway that is intended to add to wiring space at switchboards, meters, and other distribution locations. Auxiliary gutters are dealt with specifically in *NEC Article 374*. Even though the component parts of wireways and auxiliary gutters are identical, you should be familiar with the differences in their use. Auxiliary gutters are used as parts of complete assemblies of apparatus such as switchboards, distribution centers, and control equipment. However, an auxiliary gutter may only contain conductors or busbars, even though it looks like a surface metal raceway that may contain devices and equipment. Unlike auxiliary gutters, wireways represent a type of wiring because they are used to carry conductors between points located considerable distances apart.

The allowable ampacities for insulated conductors in wireways are given in *NEC Tables 310-16 through 310-19.* It should be noted that these tables are used for raceways in general. These NEC tables and the notes are often used to determine if the correct materials are on hand for an installation. They are also used to determine if it is possible to add conductors in an existing wireway. For example, if a wireway is to contain not more than three insulated conductors rated at 0 to 2,000 volts at 60°C to 90°C, the table to use is *NEC Table 310-16*. Where the number of current-carrying conductors in a raceway or cable exceeds three, the allowable ampacities shall be reduced as shown in *NEC Table 310-15(b)(2)(a)*.

In many situations, it is necessary to make extensions from the wireways to wall receptacles

Instructor's Notes:

_____

_____

_____

_____

_____

_____

_____

_____

_____

Figure 31 ◆ Wireway system layout.

Labels in figure:
JUNCTION BOX
4 x 4 WIREWAY 5 FT. LONG
X PULL BOX
CONDUIT END FITTING
90° PULL BOX ELBOW
BRANCH CIRCUIT WIRING
POWER RECEPTACLE
WIREWAY COUPLING
HEATER CONTROL
90° ELBOW
MOTOR DISCONNECT
BOX CONNECTOR COUPLING
POWER PANEL
SWITCHBOARD
SLIP FITTING

108F31.EPS

Discuss the various components in a typical wireway system.

Show Transparency 20 (Figure 31).

Discuss duct-type wireways.

Show Transparencies 21 and 22 (Figures 32 and 33).

and control devices. In these cases, *NEC Section 362-11* specifies that these extensions be made using any wiring method presented in *NEC Article 300* that includes a means for equipment grounding. Finally, as required in *NEC Section 362-12*, wireways must be marked in such a way that their manufacturer's name or trademark will be visible.

As you can see in *Figure 31*, a wide range of fittings is required for connecting wireways to one another and to fixtures such as switchboards, power panels, and conduit.

9.2.0 Types of Wireways

Rectangular duct-type wireways come as either hinged-cover or screw-cover troughs. Typical lengths are 1, 2, 3, 4, 5, and 10 feet. Shorter lengths are also available. Raintight troughs are permitted to be used in environments where moisture is not permitted within the raceway. However, the raintight trough should not be confused with the raintight lay-in wireway, which has a hinged cover. *Figure 32* shows a raintight trough with a removable side cover.

Wireway troughs are exposed when first installed. Whenever possible, they are mounted on the ceilings or walls, although they may sometimes be suspended from the ceiling. Note that in *Figure 33*, the trough has knockouts similar to those found on junction boxes. After the wireway system has been installed, branch circuits are brought from the distribution panels using conduit. The conduit is joined to the wireway at the most convenient knockout possible.

Discuss wireway fittings.

Show Transparency 23 (Figure 34).

Figure 32 ◆ Raintight trough.

108F32.EPS

Wireway components such as trough crosses, 90° internal elbows, and tee connectors serve the same function as fittings on other types of raceways. The fittings are attached to the duct using slip-on connectors. All attachments are made with nuts and bolts or screws. When assembling wireways, always place the head of the bolt on the inside and the nut on the outside so that the conductors will not be resting against a sharp edge. It is usually best to assemble sections of the wireway system on the floor, and then raise the sections into position. An exploded view of a section of wireway is shown in *Figure 34*. Both the wireway fittings and the duct come with screw-on, hinged, or snap-on covers to permit conductors to be laid in or pulled through.

The NEC specifies that wireways may be used only for exposed work. Therefore, they cannot be used in underfloor installations. If they are used for outdoor work, they must be of an approved raintight construction. It is important to note that wireways must not be installed where they are subject to severe physical damage, corrosive vapors, or hazardous locations.

Wireway troughs must be installed so that they are supported at distances not exceeding 5 feet. When specially approved supports are used, the distance between supports must not exceed 10 feet.

9.2.1 Wireway Fittings

Many different types of fittings are available for wireways, especially for use in exposed, dry locations. The following sections explain fittings commonly used in the electrical craft.

9.2.2 Connectors

Connectors are used to join wireway sections and fittings. Connectors are slipped inside the end of a wireway section and are held in place by small

---

《 INSIDE TRACK

Trough or Wireway?

The raintight lay-in wireway has a hinged cover, as shown here. A raintight trough simply has a removable cover.

VERONICA WESTFALL   108PO807.EPS

Instructor's Notes:

_____

_____

_____

_____

_____

_____

_____

_____

_____

TROUGH

"U" CONNECTOR

END CAP

TROUGH COVER

108F33.EPS

Figure 33 ◆ Trough.

45° ELBOW

90° INTERNAL
ELBOW SWEEP

"U" CONNECTOR
(5 PLACES)

ONE-FOOT TROUGH AND
COVER, WITH KNOCKOUTS
(2 PLACES)

ONE-FOOT TROUGH AND
COVER, LESS KNOCKOUTS

TEE CONNECTOR
WITH COVER SWEEP

108F34.EPS

Figure 34 ◆ Wireway sections.

bolts and nuts. Alignment slots allow the connector to be moved until it is flush with the inside surface of the wireway. After the connector is in position, it can be bolted to the wireway. This helps to ensure a strong rigid connection. Connectors have a friction hinge that helps hold the wireway cover open when needed. A connector is shown in *Figure 35*.

**Discuss wireway
connectors.**

**Show Transparency 24
(Figure 35).**

Figure 35 ◆ Connector.

9.2.3 End Plates

End plates, or closing plates, are used to seal the ends of wireways. They are inserted into the end of the wireway and fastened by screws and bolts. End plates contain knockouts so that conduit or cable may be extended from the wireway. An end plate is shown in *Figure 36*.

9.2.4 Tees

Tee fittings are used when a tee connection is needed in a wireway system. A tee connection is used where circuit conductors may branch in different directions. The tee fitting's covers and sides can be removed for access to splices and taps. Tee fittings are attached to other wireway sections using standard connectors. A tee is shown in *Figure 37*.

9.2.5 Crosses

Crosses have four openings and are attached to other wireway sections with standard connectors. The cover is held in place by screws and can be easily removed for laying in wires or for making connections. A cross is shown in *Figure 38*.

Figure 36 ◆ End plate.

9.2.6 Elbows

Elbows are used to make a bend in the wireway. They are available in angles of 22½°, 45°, or 90°, and are either internal or external. They are attached to wireway sections with standard connectors. Covers and sides can be removed for wire installation. The inside corners of elbows are rounded to prevent damage to conductor insulation. An inside elbow is shown in *Figure 39*.

Figure 37 ◆ Tee.

Figure 38 ◆ Cross.

Discuss end plates, tees, crosses, and elbows.

Show Transparencies 25 through 28 (Figures 36 through 39).

Instructor's Notes:

_____

_____

_____

_____

_____

_____

_____

_____

Figure 39 ◆ 90° inside elbow.

HEX NUT

HANGER

108F40.EPS

Figure 40 ◆ Suspended hanger.

9.2.7 Telescopic Fittings

Telescopic or slip fittings may be used between lengths of wireway. Slip fittings are attached to standard lengths by setscrews and usually adjust from ½ inch to 11½ inches. Slip fittings have a removable cover for installing wires and are similar in appearance to a nipple.

9.3.0 Wireway Supports

Wireways should be securely supported where run horizontally at each end and at intervals of no more than 5 feet or for individual lengths greater than 5 feet at each end or joint, unless listed for other support intervals. In no case shall the support distance be greater than 10 feet, in accordance with *NEC Section 362-8*. If possible, wireways can be mounted directly to a surface. Otherwise, wireways are supported by hangers or brackets.

9.3.1 Suspended Hangers

In many cases, the wireway is supported from a ceiling, beam, or other structural member. In such installations, a suspended hanger (*Figure 40*) may be used to support the wireway.

The wireway is attached to or laid in the hanger. The hanger is suspended by a threaded rod. One end of the rod is attached to the hanger with hex nuts. The other end of the rod is attached to a beam clamp or anchor.

9.3.2 Gusset Brackets

Another type of support used to mount wireways is a gusset bracket. This is an L-type bracket that is mounted to a wall. The wireway rests on the

bracket and is attached by screws or bolts. A gusset bracket is shown in *Figure 41*.

9.3.3 Standard Hangers

Standard hangers are made in two pieces. The two pieces are combined in different ways for different installation requirements. The wireway is attached to the hanger by bolts and nuts. A standard hanger is shown in *Figure 42*.

9.3.4 Wireway Hangers

When a larger wireway must be suspended, a wireway hanger may be used. A wireway hanger is made by suspending a piece of strut from a ceiling, beam, or other structural member. The strut is suspended by threaded rods attached to beam clamps or other ceiling anchors, as shown in *Figure 43*.

9.4.0 Other Types of Raceways

In this section, other types of raceways will be discussed. Depending on the particular purpose for which they are intended, raceways include enclosures such as surface metal and nonmetallic raceways, and underfloor raceways.

9.4.1 Surface Metal and Nonmetallic Raceways

Surface metal raceways consist of a wide variety of special raceways designed primarily to carry power and communications wiring to locations on the surface of ceilings or walls of building interiors.

Classroom

Discuss slip fittings.

Discuss wireway supports, including hangers and brackets.

Audiovisual

Show Transparencies 29 through 32 (Figures 40 through 43).

Discuss surface metal and nonmetallic raceways and their applications.

Stress NEC requirements.

RACEWAYS, BOXES, AND FITTINGS—TRAINEE MODULE 26108   8–25

WIREWAY ATTACHES HERE

BRACKET ATTACHES TO WALL HERE

108F41.EPS

Figure 41 ◆ Gusset bracket.

Show Transparencies 33 and 34 (Figures 44 and 45).

108F42.EPS

Figure 42 ◆ Standard hanger.

Installation specifications of both surface metal raceways and surface nonmetallic raceways are listed in detail in *NEC Article 352*. All these raceways must be installed in dry, interior locations. The number of conductors, their amperage, and the allowable cross-sectional area of the conductors, as well as regulations for combination raceways, are specified in *NEC Tables 310-16 through 310-19 and NEC Article 352*.

One use of surface metal raceways is to protect conductors that run to non-accessible outlets.

Surface metal and nonmetallic raceways have been divided into subgroups based on the specific purpose for which they are intended. There are three small surface raceways that are primarily

used for extending power circuits from one point to another. In addition, there are six larger surface raceways that have a much wider range of applications. Typical cross sections of the first three smaller raceways are shown in *Figure 44*.

Additional surface metal raceway designs are referred to as *pancake raceways*, because their flat cross sections resemble pancakes. Their primary use is to extend power, lighting, telephone, or signal wire to locations away from the walls of a room without embedding them under the floor. A pancake raceway is shown in *Figure 45*.

There are also surface metal raceways available that house two or three different conductor raceways. These are referred to as *twinduct* or *tripleduct*. These raceways permit different circuits, such as power and signal, to be placed within the same raceway.

The number and types of conductors permitted to be installed and the capacity of a particular surface raceway must be calculated and matched with NEC requirements, as discussed previously. *NEC Tables 310-16 through 310-19* are used for surface raceways in the same manner in which they are used for wireways. For surface raceway installations with more than three conductors in each raceway, particular reference must be made to *NEC Table 310-15(b)(2)(a)*.

9.4.2 Plugmold Multi-Outlet Systems

Manufacturers offer a wide variety of plugmold multi-outlet surface raceways. Their function is to hold receptacles and other devices within the

Instructor's Notes:

_____

_____

_____

_____

_____

_____

_____

_____

_____

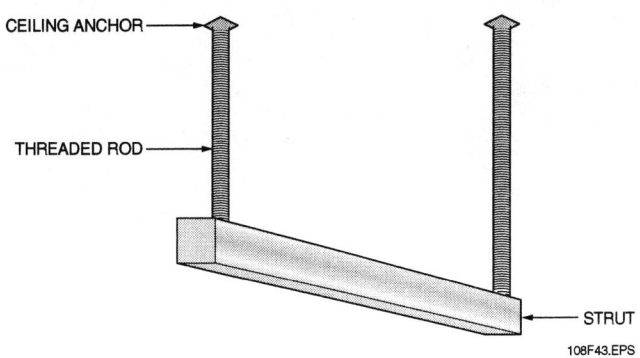

Figure 43 ◈ Wireway hanger.

108F43.EPS

108F44.EPS

Figure 44 ◈ Smaller surface raceways.

Figure 45 ◈ Pancake raceway.

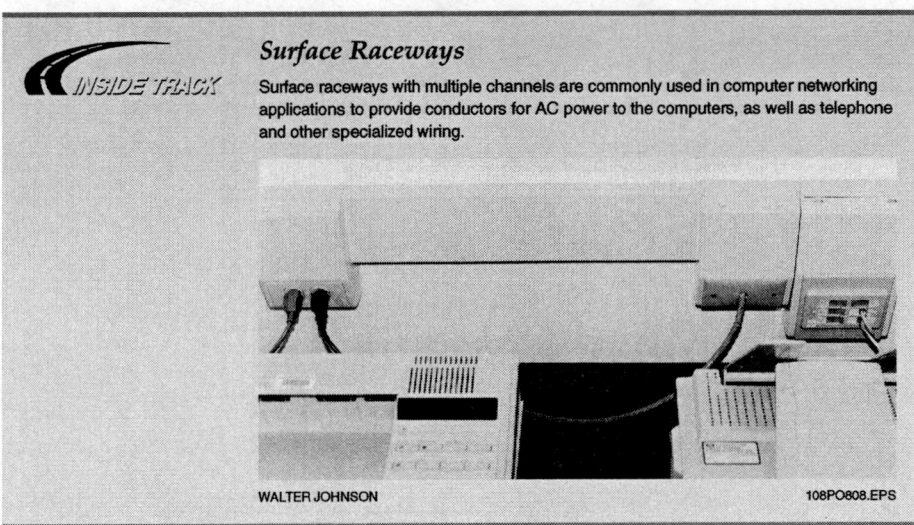

Surface Raceways

Surface raceways with multiple channels are commonly used in computer networking applications to provide conductors for AC power to the computers, as well as telephone and other specialized wiring.

WALTER JOHNSON   108PO808.EPS

Show Transparency 35 (Figure 46).

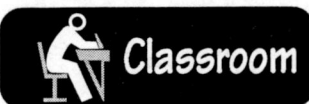

Discuss pole, underfloor, and cellular metal raceways and their applications.

Show Transparency 36 (Figure 48).

raceway. When the surface raceways are used in this manner, the assembly is referred to as a *multi-outlet assembly*.

Plugmold systems are either wired in the field or come pre-wired from the factory. *Figure 46* shows typical plugmold cross sections, wiring configurations, and some of the available fittings for the system.

9.4.3 Pole Systems

There are many situations in which power and other electric circuits have to be carried from overhead wiring systems to devices that are not located near existing wall outlets or control circuits. This type of wiring is typically used in open office spaces where cubicles are provided by temporary dividers. Poles are used to accomplish this. Some common manufacturers' names for these poles include *Tele-Power poles*, *Quick-E poles*, and *Walkerpoles*. The poles usually come in lengths suitable for 10-, 12-, or 15-foot ceilings. *Figure 47* shows a typical pole base.

9.4.4 Underfloor Systems

Underfloor raceway systems were developed to provide a practical means of bringing conductors for lighting, power, and signaling to cabinets and consoles. Underfloor raceways are available in 10-foot lengths and widths of 4 and 8 inches. The sec-

tions are made with inserts spaced every 24 inches. The inserts can be removed for outlet installation. These are explained in *NEC Article 354*.

> **Note**
> Inserts must be installed so that they are flush with the finished grade of the floor.

Junction boxes are used to join sections of underfloor raceways. Conduit is also used with underfloor raceways by using a raceway-to-conduit connector (conduit adapter). A typical underfloor raceway duct with fittings is shown in *Figure 48*.

This wiring method makes it possible to place a desk or table in any location where it will always be over, or very near to, a duct line. The wiring method for lighting and power between cabinets and the raceway junction boxes may be conduit, underfloor raceway, wall elbows, and cabinet connectors. *NEC Article 354* covers the installation of underfloor raceways.

9.4.5 Cellular Metal Floor Raceways

A cellular metal floor raceway is a type of floor construction designed for use in steel-frame

Instructor's Notes:

_____

_____

_____

_____

_____

_____

_____

_____

_____

DESCRIPTION	
3/8" — 2 1/8"	**2200B Base: .040" steel; 3200C Cover: .040" steel; Packed (10) 5ft. lengths of ea. per carton.
— 2 1/8"	** .040" steel; 10ft. lengths. Packed (10) 10 ft lengths per carton.
— 2 1/8"	.040" steel; 10ft. lengths. Packed (10) 10 ft lengths per carton.

ITEM	FITTING SPECIFICATIONS	
WIRE AND DEVICE CLIP (PLATED)	3/4"	Holds conductors in place in cover or base. Also used to hold Plugmold receptacles in cover. 2200WC used as a device clip and as a wire clip
COUPLING (PLATED)	3"	For joining lengths of 2200.

Figure 46 ◆ Plugmold system.

GB Series:
3-wire, 1 circuit; has insulated grounding conductor.

GBA Series:
4-wire, 2 circuit, outlets wired alternately, has insulated grounding conductor.

DGB Series:
3-wire, 1 circuit, duplex outlets, has insulated grounding conductor.

DGBA Series:
4-wire, duplex outlets, 2-circuit grounding. Each Duplex wired alternately. Used where multiple circuits are required, has insulated grounding conductor.

ITEM	FITTING SPECIFICATIONS	
BLANK END FITTING	1"	For closing open end of 2200.
TRIM END FITTING	1 3/16" 2 3/16"	Serves dual purpose of closing open end of 2200 and providing a trim end. 32108T installed in 2200 at door casing.

108F46.EPS

buildings. In these buildings, the members supporting the floor between the beams consist of sheet steel rolled into shapes. These shapes are combined to form cells, or closed passageways, which extend across the building. The cells are of various shapes and sizes, depending upon the structural strength required. The cells of this type of floor construction form the raceways, as shown in *Figure 49*.

Connections to the cells are made using headers that extend across the cells. A header connects only to those cells to be used as raceways for conductors. A junction box or access fitting is necessary at each joint where a header connects to a cell. Two or three separate headers, connecting to different sets of cells, may be used for different systems. For example, light and power, signaling systems, and public telephones would each have a separate header. A special elbow fitting is used to extend the headers up to the distribution equipment on a wall or column. *NEC Article 356* covers the installation of cellular metal floor raceways.

9.4.6 *Cellular Concrete Floor Raceways*

The term *precast cellular concrete floor* refers to a type of floor used in steel-frame, concrete-frame, and wall-bearing construction. In this type of system, the floor members are precast with hollow voids, which form smooth, round cells. The cells form raceways, which can be adapted, using fittings, for use as underfloor raceways. A precast cellular concrete floor is fire-resistant and requires no further fireproofing. The precast reinforced concrete floor members form the structural floor and are supported by beams or bearing walls. Connections to the cells are made with headers that are secured to the precast concrete floor. *NEC Article 358* covers the installation of cellular concrete floor raceways.

10.0.0 ◆ CABLE TRAYS

Cable trays function as a support for conductors and tubing (see *NEC Article 318*). A cable tray has the advantage of easy access to conductors, and thus lends itself to installations where the addition

Audiovisual

Show Transparency 37 (Figure 49).

Classroom

Discuss cellular concrete floor raceways and their applications.

Teaching Tip

If possible, take the trainees to a commercial/industrial job site where they can observe various types of wireway components and raceway systems.

VERONICA WESTFALL   108F47.EPS

Figure 47 ◆ Power pole.

or removal of conductors is a common practice. Cable trays are fabricated from aluminum, steel, and fiberglass. Cable trays are available in two basic forms: ladder and trough. Ladder tray, as the name implies, consists of two parallel channels connected by rungs. Trough consists of two parallel channels (side rails) having a corrugated, ventilated bottom, or a corrugated, solid bottom. (There is also a special center rail cable tray available for use in light-duty applications such as telephone and sound wiring. We will discuss this type of cable tray in more detail in Level 2.)

Cable trays are commonly available in 12- and 24-foot lengths. They are usually available in widths of 6, 9, 12, 18, 24, 30, and 36 inches, and load depths of 4, 6, and 8 inches.

Cable trays may be used in most electrical installations. Cable trays may be used in air handling ceiling space, but only to support the wiring methods permitted in such spaces by *NEC Section 300-22(c)*. Also, cable trays may be used in Class 1, Division 2 locations according to *NEC Section 501-4*. Cable trays may also be used above a suspended ceiling that is not used as an air handling space. Some manufacturers offer an aluminum cable tray that is coated with PVC for installation in caustic environments. A typical cable tray system with fittings is shown in *Figure 50*.

Wire and cable installation in cable trays is defined by the NEC. Read *NEC Article 318* to become familiar with the requirements and restrictions made by the NEC for safe installation of wire and cable in a cable tray.

Metallic cable trays that support electrical conductors must be grounded as required by *NEC Article 250*. Where steel and aluminum cable tray systems are used as an equipment grounding conductor, all of the provisions of *NEC Section 318-7(b)* must be complied with.

WARNING!
Do not stand on, climb in, or walk on a cable tray.

108F48.EPS

Figure 48 ◆ Underfloor raceway duct.

ELECTRICAL LEVEL ONE—TRAINEE MODULE 26108

Classroom

Discuss cable trays and their applications.

Audiovisual

Show Transparency 38 (Figure 50).

Instructor's Notes:

ELECTRICAL LEVEL ONE—INSTRUCTOR'S GUIDE MODULE 26108

CONCRETE

3 1/8"

5"

CHANNEL SUPPORT

108F49.EPS

Figure 49 ◆ Cross section of a cellular floor.

Typical
Cable Trays and
Cable Tray
Components

Nomenclature

1. **Ladder Cable Tray**
2. **Ventilated Trough Cable Tray**
3. **Splice Plates**
4. **90° Horizontal Bend, Ladder Type**
5. **45° Horizontal Bend, Ladder Type**
6. **Horizontal Tee, Ladder Type**
7. **Horizontal Cross, Ladder Type**
8. **90° Vertical Outside Bend, Ladder Type**
9. **45° Vertical Outside Bend, Ladder Type**
10. **30° Vertical Outside Bend, Ladder Type**
11. **VBS-2 Vertical Bend Segment**
12. **Vertical Tee, Ventilated Trough Type**
13. **Left Hand Reducer, Ladder Type**
14. **Box Connector**
15. **Barrier Strip**
16. **Solid Flanged Tray Cover**
17. **Cable Channel, Non-Ventilated**
18. **Cable Channel, 90° Vertical**

108F50.EPS

Figure 50 ◆ Cable tray system.

Cable Trays and Wireways

What is the difference between a wireway and a cable tray? What kinds of conductors would you expect to find in a cable tray as compared to a wireway?

Cable Tray Systems

Cable tray systems must be continuous and grounded. One of the advantages of using a cable tray system is that it makes it easy to expand or modify the wiring system following installation. Unlike conduit systems, wires can be added or changed by simply laying them into (or lifting them out of) the tray.

Discuss cable tray fittings and their applications.

Discuss various types of cable tray supports and their applications.

Show Transparencies 39 through 41 (Figures 51 through 53).

10.1.0 Cable Tray Fittings

Cable tray fittings are part of the cable tray system and provide a means of changing the direction or dimension of the different trays. Some of the uses of horizontal and vertical tees, horizontal and vertical bends, horizontal crosses, reducers, barrier strips, covers, and box connectors are shown in *Figure 50*.

10.2.0 Cable Tray Supports

Cable trays are usually supported in one of five ways: direct rod suspension, trapeze mounting, center hung, wall mounting, and pipe rack mounting.

10.2.1 Direct Rod Suspension

The direct rod suspension method of supporting cable tray uses threaded rods and hanger clamps. One end of the threaded rod is connected to the ceiling or other overhead structure. The other end is connected to hanger clamps that are attached to the cable tray side rails. A direct rod suspension assembly is shown in *Figure 51*.

10.2.2 Trapeze Mounting and Center Hung Support

Trapeze mounting of cable tray is similar to direct rod suspension mounting. The difference is in the method of attaching the cable tray to the threaded rods. A structural member, usually a steel channel or strut, is connected to the vertical supports to provide an appearance similar to a swing or trapeze. The cable tray is mounted to the struc-

THREADED ROD

108F51.EPS

Figure 51 ◆ Direct rod suspension.

tural member. Often, the underside of the channel or strut is used to support conduit. A trapeze mounting assembly is shown in *Figure 52*.

A method that is similar to trapeze mounting is a center hung tray support (*Figure 52*). In this case, only one rod is used and it is centered between the cable tray side rails.

10.2.3 Wall Mounting

Wall mounting is accomplished by supporting the cable tray with structural members attached to the wall. This method of support is often used in tunnels and other underground or sheltered installations where large numbers of conductors interconnect equipment that is separated by long distances. A wall mounting assembly is shown in *Figure 53*.

8–32

ELECTRICAL LEVEL ONE—TRAINEE MODULE 26108

Instructor's Notes:

THREADED ROD

TRAPEZE

THREADED ROD

CENTER HUNG

108F52.EPS

Figure 52 ◆ Trapeze mounting and center hung support.

108F53.EPS

Figure 53 ◆ Wall mounting.

10.2.4 Pipe Rack Mounting

Pipe racks are structural frames used to support piping that interconnects equipment in outdoor industrial facilities. Usually, some space on the rack is reserved for conduit and cable tray. Pipe rack mounting of cable tray is often used when power distribution and electrical wiring is routed over a large area.

11.0.0 ◆ STORING RACEWAYS

Proper and safe methods of storing conduit, wireways, raceways, and cable trays may sound like a simple task, but improper storage techniques can result in wasted time and damage to the raceways, as well as personal injury. There are correct ways to store raceways that will help avoid costly damage, save time in identifying stored raceways, and reduce the chance of personal injury.

Pipe racks are commonly used for storing conduit. The racks provide support to prevent bending, sagging, distorting, scratching, or marring of conduit surfaces. Most racks have compartments where different types and sizes of conduit can be separated for ease of identification and selection. The storage compartments in racks are usually elevated to help avoid damage that might occur at floor level. Conduit that is stored at floor level is easily damaged by people and other materials or equipment in the area.

The ends of stored conduit should be sealed to help prevent contamination and damage. Conduit ends can be capped, taped, or plugged.

Always inspect raceway before storing it to make sure that it is clean and not damaged. It is discouraging to get raceway for a job and find that it is dirty or damaged. Also, make sure that the raceway is stored securely so that when someone comes to get it for a job, it will not fall in any way that could cause injury.

To prevent contamination and corrosion of stored raceway, it should be covered with a tarpaulin or other suitable covering. It should also be separated from non-compatible materials such as hazardous chemicals.

Wireways, surface metal raceways, and cable trays should always be stored off the ground on boards in an area where people will not step on it and equipment will not run over it. Stepping or running over raceway bends the metal and makes it unusable.

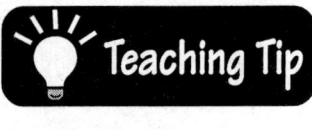

Teaching Tip

If possible, take the trainees to a commercial/industrial job site where they can observe various types of cable tray support systems.

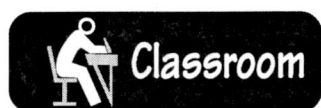

Classroom

Discuss the proper method of raceway storage.

Discuss the proper method of handling raceways.

Discuss ducting and list the various types of underground duct systems and their applications.

Discuss duct lines.

12.0.0 ◆ HANDLING RACEWAYS

Raceway is made to strict specifications. It can be easily damaged by careless handling. From the time raceway is delivered to a job site until the installation is complete, use proper and safe handling techniques. These are a few basic guidelines for handling raceway that will help avoid damaging or contaminating it:

- Never drag raceway off a delivery truck or off other lengths of raceway.
- Never drag raceway on the ground or floor. Dragging raceway can cause damage to the ends.
- Keep the thread protection caps on when handling or transporting conduit raceway.
- Keep raceway away from any material that might contaminate it during handling.
- Flag the ends of long lengths of raceway when transporting it to the job site.
- Never drop or throw raceway when handling it.
- Never hit raceway against other objects when transporting it.
- Always use two people when carrying long pieces of raceway. Make sure that you both stay on the same side and that the load is balanced. Each person should be about ¼ of the length of the raceway from the end. Lift and put down the raceway at the same time.

13.0.0 ◆ DUCTING

In the common vocabulary of the electrical trade, a duct is a single enclosed raceway, or runway, through which conductors or cables can be led. Basically, ducting is a system of ducts. However, underground duct systems include manholes, transformer vaults, and risers.

There are several reasons for running power lines underground rather than overhead. In some situations, an overhead high-voltage line would be dangerous, or the space may not be adequate. For aesthetic reasons, architectural plans may require buried lines throughout a subdivision or a planned community. Tunnels may already exist, or be planned, for carrying steam or water lines. In any of these situations, underground installations are appropriate. Underground cables may be buried directly in the ground or run through conduit.

In underground construction, a duct system provides a safe passageway for power lines, communication cables, or both. In buildings, underfloor raceways and cellular floor raceways are built to provide ducting so that electricity will be available throughout a large area. As an electri-

cian, you need to know the approved methods of constructing underground ducting. You also need to know how to avoid potential electrical hazards in both original construction and maintenance. It is essential to understand the requirements and limitations imposed on running wires through underfloor and cellular floor raceways and ducts.

14.0.0 ◆ UNDERGROUND SYSTEMS

There are five different ways to install cable underground:

- Duct line
- Conductors located in tunnels
- Conductors buried directly in the earth
- PVC
- GRC

The method used will depend on the materials that are available, the number of conductors to be pulled, and the type of wiring to be done (service drop, branch circuit, commercial, etc.).

14.1.0 Duct Line

A duct line consists of at least one subway placed in a trench and covered with earth. Conduit, in some cases, can be classified as a *subway*. Subways may come in a single duct line or multiple duct lines (2, 3, 4, and 6 subways per section). The depth in which the duct will be placed is determined using *NEC Table 300-5*. The conduit subways are encased in concrete or other materials. This provides good mechanical strength and allows for power losses from the cable to be dissipated into the earth.

In underground cable installations, a duct is a buried conduit through which a cable passes. Manholes are set at intervals in an underground duct run. Manholes provide access through throats (sometimes called *chimneys*). At ground level, or street surface level, a manhole cover closes off the manhole area tightly. An individual cable length running underground normally terminates at a manhole, where it is spliced to another length of cable. A duct line may consist of a single conduit or several, each carrying a cable length from one manhole to the next.

Manholes provide room for installing lengths of cable in conduit lines. They are also used for maintenance work and for performing tests. Workers enter a manhole from above. In a two-way manhole, cables enter and leave in only two directions. There are also three-way and four-way manholes. Often manholes are located at the in-

Instructor's Notes:

_____

_____

_____

_____

_____

_____

_____

_____

_____

tersection of two streets so that they can be used for cables leaving in four directions. Manholes are usually constructed of brick or concrete. Their design must provide room for drainage and for workers to move around inside them. A similar opening known as a *handhole* is sometimes provided for splicing on lateral two-way duct lines.

Transformer vaults house power transformers, voltage regulators, network protectors, meters, and circuit breakers. A cable may end at a transformer vault. Other cables end at a customer's substation or terminate as risers that connect with overhead lines.

14.2.0 Duct Materials

Underground duct lines can be made of fiber, vitrified tile, iron conduit, plastic, or poured concrete. The inside diameter of the ducting for a specific job is determined by the size of the cable that will be drawn into the duct. Sizes from two to six inches (inside diameter) are available for most types of ducting.

 WARNING!

Be careful when working with unfamiliar duct materials. In older installations, asbestos/cement duct may have been used. You must be certified to remove or disturb asbestos.

14.3.0 Plastic Conduit

Plastic conduit may be made of PVC (polyvinyl chloride), PE (polyethylene), or styrene. Since this type of conduit is available in lengths up to 20 feet, fewer couplings are needed than with other types of ducting. Plastic conduit is popular because it is easy to install, requires less labor than other types of conduit, and is low in cost.

14.4.0 Monolithic Concrete Duct

Monolithic concrete duct is poured at the job site. Multiple duct lines can be formed using rubber tubing cores on spacers. The cores may be removed after the concrete has set. A die containing steel tubes, known as a *boat*, can also be used to form ducts. It is pulled slowly through the trench on a track as concrete is poured from the top. Poured concrete ducting made by either method is relatively expensive, but offers the advantage of creating a very clean duct interior with no residue that can decay. The rubber core method is especially useful for curving or turning part of a duct system.

14.5.0 Cable-in-Duct

One of the most popular duct types is the cable-in-duct. This type of duct comes from the manufacturer with cables already installed. The duct comes in a reel and can be laid in the trench with ease. The installed cables can be withdrawn in the future, if necessary. This type of duct, because of the form in which it comes, reduces the need for fittings and couplings. It is most frequently used for street lighting systems.

15.0.0 ◆ MAKING A CONDUIT-TO-BOX CONNECTION

A proper conduit-to-box connection is shown in *Figure 54*.

In order to make a good connection, use the following procedure:

Step 1 Thread the external locknut onto the conduit. Run the locknut to the bottom of the threads.

Step 2 Insert the conduit into the box opening.

Step 3 If an inside locknut or grounding locknut is required, screw it onto the conduit inside the box opening.

Step 4 Screw the bushing onto the threads projecting into the box opening. Make sure the bushing is tightened as much as possible.

Step 5 Tighten the external locknut to secure the conduit to the box.

It is important that the bushings and locknuts fit tightly against the box. For this reason, the conduit must enter straight into the box (*Figure 55*). This may require that a box offset or kick be made in the conduit.

16.0.0 ◆ VARIOUS CONSTRUCTION PROCEDURES

16.1.0 Masonry and Concrete Flush-Mount Construction

In a reinforced concrete construction environment, the conduit and boxes must be embedded in the concrete to achieve a flush surface. Ordinary boxes may be used, but special concrete boxes are preferred and are available in depths up

Discuss the various types of material used in underground duct lines.

Discuss PVC, monolithic concrete, and cable-in-duct systems and list the advantages and disadvantages of each type.

Demonstrate how to make a conduit-to-box connection.

Show Transparencies 42 and 43 (Figures 54 and 55).

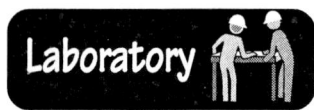

Have the trainees practice making conduit-to-box connections.

Discuss the various procedures for installing boxes in concrete/masonry construction.

EXTERNAL LOCKNUT

BUSHING

GROUNDING LOCKNUT (IF REQUIRED)

108F54.EPS

Figure 54 ◆ Conduit-to-box connection.

OFFSET BEND

108F55.EPS

Figure 55 ◆ Correct entrance angle.

to six inches. These boxes have special ears by which they are nailed to the wooden forms for the concrete. When installing them, stuff the boxes tightly with paper to prevent concrete from seeping in. *Figure 56* shows an installed box.

Flush construction can also be done on existing concrete walls, but this requires chiseling a channel and box opening, anchoring the box and conduit, and then resealing the wall.

To achieve flush construction with masonry walls, the most acceptable method is for the electrician to work closely with the mason laying the blocks. When the construction blocks reach the convenience outlet elevation, boxes are made up as shown in *Figure 57*. The figure shows a raised tile ring or box device cover.

Figure 58 shows the use of a 4-S extension ring installed to bring the box to the masonry surface. *Figure 59* shows a masonry box that needs no extension or deep plaster ring to bring it to the surface.

Note

The electrician must work with the mason to ensure the box is properly grouted and sealed.

Note

Conduit should be installed in the rear knockout of this masonry switch box.

Sections of conduit are then coupled in short (four- or five-foot) lengths. This is done because it is impractical for the mason to maneuver blocks over ten-foot sections of conduit.

ELECTRICAL LEVEL ONE—TRAINEE MODULE 26108

Audiovisual

Show Transparencies 44 through 47 (Figures 56 through 59).

Instructor's Notes:

ELECTRICAL LEVEL ONE—INSTRUCTOR'S GUIDE MODULE 26108

Figure 56 ◆ Concrete flush-mount installation.

WOODEN FORMS

CONCRETE

EARS FOR SUPPORTING DEVICES

SPECIAL EARS FOR NAILING BOX TO FORMS

CONDUIT

108F56.EPS

Figure 57 ◆ Concrete outlet box.

108F57.EPS

16.2.0 Metal Stud Environment

Metal stud walls are a popular method of construction for the interior walls of commercial buildings. Metal stud framing consists of relatively thin metal channel studs, usually constructed of galvanized steel and with an overall dimension the same as standard 2 × 4 wooden

108F58.EPS

Figure 58 ◆ 4-S extension ring used to bring the box to the masonry surface.

108F59.EPS

Figure 59 ◆ Three-gang switch box.

studs. Wiring in this type of construction is relatively easy when compared to masonry.

EMT conduit is the most common type of raceway specified for metal stud wiring. Metal studs usually have some number of pre-punched holes that can be used to route the conduit. If a pre-punched hole is not located where it needs to be, holes can be easily punched in the metal stud with a hole cutter or knockout punch.

CAUTION
Cutting or punching metal studs can create sharp edges. Avoid contact that can result in cuts.

Explain the procedure for routing conduit and installing boxes in metal stud walls.

Stress the importance of wearing gloves when working with metal.

108F60.EPS

Figure 60 ◆ Caddy-fastening devices.

Emphasize the importance of proper safety procedures and protective equipment during conduit and box installation.

Show Transparency 48 (Figure 60).

Discuss the procedures for running conduit and installing boxes in wood frame construction.

Show Transparency 49 (Figure 61).

Boxes can be secured to the metal stud using self-tapping screws or one of the many types of box supports available. EMT conduit is supported by the metal studs using conduit straps or other approved methods. It is important that the conduit be properly supported to facilitate pulling the conductors through the tubing. Boxes are mounted on the metal studs so that the box will be flush with the finished walls. You must know what the finished wall thickness is going to be to properly secure the boxes to the metal studs. For example, if the finished wall will be ⅝-inch drywall, then the box must be fastened so that it protrudes ⅝ of an inch from the metal stud.

WARNING!
When using a screw gun or cordless drill to mount boxes to studs, keep the hand holding the box away from the gun/drill to avoid injury.

Figure 60 shows several examples of clips known as *caddy-fastening devices* that are used in metal stud environments.

16.3.0 Wood Frame Environment

At one time, the use of rigid conduit in partitions and ceilings was a laborious and time-consuming operation. Thinwall conduit makes an easier and far quicker job, largely because of the types of fittings that are specially adapted to it.

Figure 61 shows two methods of running thinwall conduit in these locations: boring timbers and notching them. When boring, holes must be drilled large enough for the tubing to be inserted between the studs. The tubing is cut rather short, calling for multiple couplings. EMT can be bowed quite a bit while threading through holes in studs. Boring is the preferred method.

WARNING!
Always wear safety goggles when boring wood.

NEC Section 300-4 addresses the requirements to prevent physical damage to conductors and cabling in wood members. By keeping the edge of the drilled hole 1¼ inches from the closest edge of the stud, nails are not likely to penetrate the stud far enough to damage the cables. The building codes provide maximum requirements for bored or notched holes in studs.

The exception in the NEC permits IMC, RMC, RNMC, and EMT to be installed through bored holes or laid in notches less than 1¼ inches from the nearest edge without a steel plate or bushing.

Because of its weakening effect upon the structure, notching should be resorted to only where absolutely necessary. Notches should be as narrow as possible and in no case deeper than ⅙ the stock of a bearing timber. A bearing timber sup-

ELECTRICAL LEVEL ONE—TRAINEE MODULE 26108

Instructor's Notes:

ELECTRICAL LEVEL ONE—INSTRUCTOR'S GUIDE MODULE 26108

Horizontal runs of EMT may be supported by openings in framing members at intervals not greater than 10 feet when securely fastened within a distance of 3 feet at each of its termination points.

NEC Section 348-13

EMT may be run through wood joists where the edges of the bored holes are less than 1¹/₄" from the nearest edge of the stud, or where the studs are notched without the need for a steel plate.

NEC Section 300-4(a) and (b)

EMT must be securely fastened in place every 10 feet and within 3 feet of each outlet box device, cabinet conduit body, or other termination.

NEC Section 348-13

Unbroken lengths of EMT can be fastened at a distance of up to 5 feet from a termination point when structural members do not readily permit fastening within 3 feet.

NEC Section 348-13, Exception 1

Where fastening of EMT is impractical in finished buildings or prefinished walls, unbroken lengths of EMT may be fished.

NEC Section 348-13, Exception 2

108F61.EPS

Figure 61 ◆ Installing wire or conduit in a wood-frame building.

ports floor joists or other weight. An additional requirement is for the notch to be covered with a steel reinforcement bracket. This bracket aids in retaining the original strength of the timber.

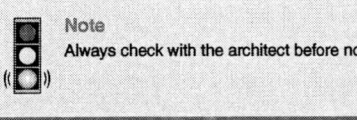

Note

Always check with the architect before notching.

16.4.0 Steel Environment

Electrical installations in buildings where steel beams are the structural framework are most often found in industrial buildings and warehouses. This type of construction is typically of the pre-engineered building where beams and other supports are pre-cut and pre-drilled so that erection of the building is fast and simple.

The interior of the building in most cases will be unfinished, and the wiring will be supported by the metal beams and purlings. Beams and

Stress NEC requirements.

Discuss the procedures for running conduit and installing boxes in steel beam construction.

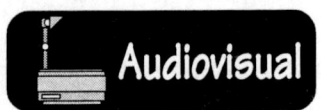
THINK ABOUT IT

Putting It All Together

Think about the effort that goes into the design of a large industrial installation. If you were to design a large complex, such as the one shown here, where would you start and why?

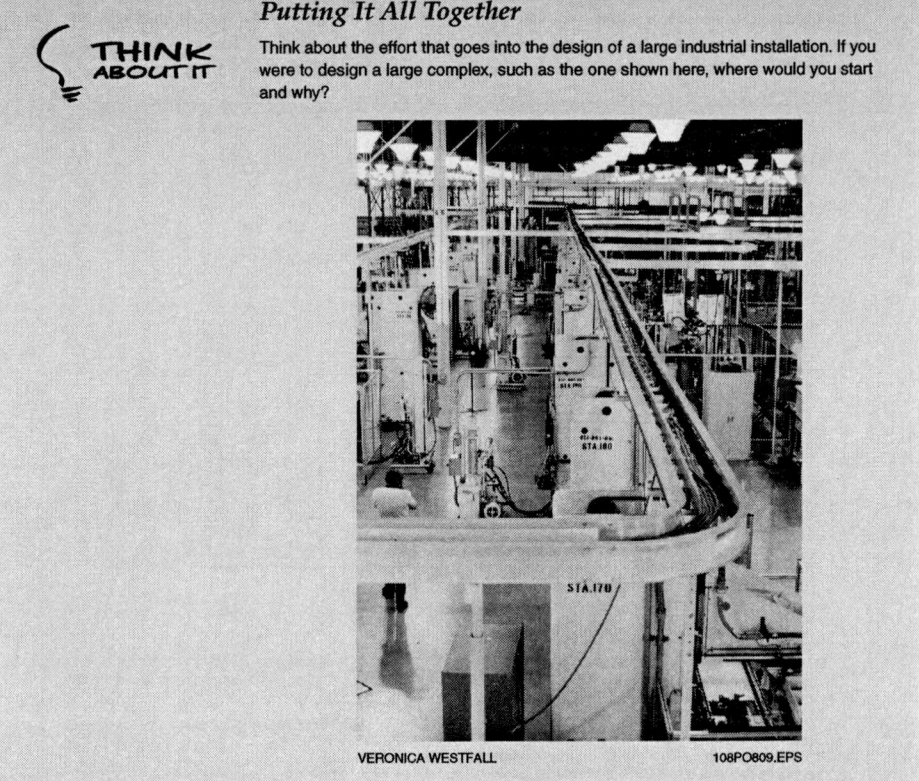

VERONICA WESTFALL   108PO809.EPS

purlings should not be drilled through; consequently, the conduit is supported from the metal beams by anchoring devices designed especially for that purpose. The supports attach to the beams or supports and have clamps to secure the conduit to the structure. All conduit runs should be plumb since they are exposed. Bends should be correct and have a neat and orderly appearance.

Since steel construction usually takes place in buildings where load handling and moving large and heavy items is common, rigid metal conduit is often required. If a large number of conduits are run along the same path, strut-type systems are used. These systems are sometimes referred to as *Unistrut* systems (Unistrut is a manufacturer of these systems). Another manufacturer of strut systems is *B-Line* systems. Both are very similar. These systems use a channel-type member that can support conduits from the ceiling by using threaded rod supports for the channel, as shown in *Figure 62*. Strut channel can also be secured to masonry walls to support vertical runs of conduit, wireways, and various types of boxes.

Summary

This module discussed the various types of raceways, boxes, and fittings, including their uses and procedures for installation. The primary purpose of raceways is to house electric wire used for power distribution, communication, or electronic signal transmission. Raceways provide protection

Instructor's Notes:

_____
_____
_____
_____
_____
_____
_____
_____
_____

Figure 62 ◆ Steel strut system.

108F62.EPS

to the wiring and even a means of identifying one type of wire from another when run adjacent to each other. This process requires proper planning to allow for current needs, future expansion, and a neat and orderly appearance.

Review Questions

1. _____ is the lightest duty and most widely used non-flexible metal conduit available for enclosing and protecting electrical wiring.
   a. Electrical metallic tubing
   b. Rigid metal conduit
   c. Aluminum conduit
   d. Plastic-coated GRC

2. Because the wall thickness of _____ is much less than that of rigid conduit, it is often referred to as *thinwall conduit*.
   a. intermediate metal conduit
   b. electrical metallic tubing
   c. rigid non-metallic conduit
   d. galvanized rigid steel conduit

3. The type of conduit that provides the best physical protection for the wire inside is _____.
   a. flexible metal conduit
   b. rigid metal conduit
   c. electrical metallic tubing
   d. intermediate metal conduit

4. Bending IMC is often easier than bending the same size and quantity of rigid metal conduit because IMC has a slightly _____.
   a. larger outside diameter
   b. larger internal diameter
   c. smaller outside diameter
   d. smaller internal diameter

5. Flexible metal conduit _____.
   a. is made from a single strip of steel or aluminum
   b. can be used in wet locations if lead-covered conductors are installed
   c. can be used in underground locations
   d. is available in sizes up to six inches in diameter

6. Liquid-tight flexible metal conduit can be used in _____ locations.
   a. dry
   b. wet
   c. underground
   d. outdoor

7. A Type LB conduit body has the cover on _____.
   a. the left
   b. the right
   c. the back
   d. both sides

8. When conduit is joined to metal boxes, _____ protect the wires from the sharp edges of the conduit.
   a. washers
   b. locknuts
   c. bushings
   d. couplings

9. _____ are rigid rectangular raceways used for housing electric wires and cables.
   a. Troughs
   b. Gutters
   c. Pull boxes
   d. Conduits

10. Wireways must not be installed where they are subject to _____.
    a. physical damage
    b. corrosion and sunlight
    c. hazardous locations
    d. physical damage, corrosive vapors, and hazardous locations

11. Pancake raceways are a type of _____.
    a. rigid tubing
    b. flexible metal conduit
    c. surface metal raceway
    d. surface nonmetallic raceway

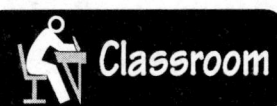

Classroom

Have the trainees complete the Review Questions and go over the answers prior to administering the Module Examination.

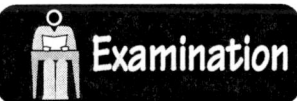

Examination

Administer the Module Examination and Performance Test. Be sure to record the testing results on Craft Training Report Form 200 and submit the results to the Training Program Sponsor.

12. The statement that best describes the use of surface metal raceway systems is a raceway system _____.
    a. designed for concealed wiring within walls and partitions
    b. suited for use in wet or damp locations
    c. mainly used for making the transition from a masonry surface to a wood surface
    d. designed for exposed installations on the surface of walls and ceilings

13. Surface metal raceways are designed primarily to protect conductors that run to devices _____ in a room.
    a. located near the walls and outlets
    b. located on the surface of the walls and ceilings
    c. not located near the walls and outlets
    d. not located near the walls and ceiling

14. When storing raceway, it is best to _____.
    a. lay it on the floor so that there is no chance that it will fall and injure someone
    b. store all raceway together regardless of the type of raceway material
    c. stand it on end so that it does not collect moisture
    d. cover it with a tarpaulin or other suitable covering to prevent contamination and corrosion

15. When transporting metal conduit, it is best to _____.
    a. remove your gloves so that you do not contaminate the conduit
    b. remove the thread protection caps so that they do not get lost
    c. place the metal conduit inside a larger raceway to keep it from bending
    d. keep the thread protection caps on to prevent damage to the threads

Instructor's Notes:

_____

_____

_____

_____

_____

_____

_____

_____

Matt Leitsinger, Beacon Electric Co.

Matt Leitsinger's apprentice training got off on the wrong foot, but he persevered and went on to win the National Craft Olympics in the fourth year of his apprenticeship. Following his apprenticeship, he achieved success as a journeyman, lead electrician, foreman, and instructor.

How did you decide to become an electrician?
It was a natural choice for me. My grandfather, father, brother, uncle, and cousin are all electricians.

What was your apprentice training like?
I started my training while I was still in high school. I would attend classes for half a day and work half a day. After I graduated, I would work a full day and go to class two nights a week for three hours each night. I was fortunate enough to have an excellent instructor and to learn from the NCCER Standardized Craft Training Program. One thing I remember especially is working eight or ten hours in the cold and then going to a nice, warm classroom.

I started off on the wrong foot because I tried to test out of the first year. I didn't pass, and it wound up costing me quite a bit of money. As it turned out, I learned a lot in my first year, so it worked out okay in the end.

The work experience during my apprenticeship was excellent training because it was demanding and complicated work on hospital and research facilities.

In the fourth year of my apprentice training, I qualified for the National Craft Olympics in Hawaii, and I came in first. My employer generously paid for both my wife and me to attend.

What positions have you held since you completed your apprentice training?
I have worked as a journeyman, a lead electrician, and a foreman. In my leadership positions, I have had the opportunity to work on all kinds of jobs, from small residential projects to multimillion-dollar commercial projects. This experience has helped me to become a well-rounded electrician. Currently, I work in our special projects division. I deal directly with clients and perform all kinds of work from simple maintenance to high-voltage terminations.

I have also had the opportunity to participate in training new electricians, both as an advisor to the local vocational school and an instructor at the vocational school and at my own company.

What would you say was the greatest single factor that contributed to your success?
I am highly motivated to continue learning. Also, I make it a point to show up for work every day and do my job. It sounds simple, but it's important that your employer be able to count on you. It really can have a damaging effect on a project if members of the team just don't show up for work. I guess you would call that having a sense of responsibility.

My company has been very supportive and my superiors have helped me at every step along my career path. A career in the electrical industry is not easy, but with the right attitude and training, the rewards will greatly outweigh the effort.

Trade Terms Introduced
in This Module

Accessible: Able to be reached, as for service or repair.

Approved: Meeting the requirements of an appropriate regulatory agency.

Bonding wire: A wire used to make a continuous grounding path between equipment and ground.

Cable trays: Rigid structures used to support electrical conductors.

Conductors: Wires or cables used to carry electrical current.

Conduit: A round raceway, similar to pipe, that houses conductors.

Exposed location: Not permanently closed in by the structure or finish of a building; able to be installed or removed without damage to the structure.

Kick: A bend in a piece of conduit, usually less than 45°, made to change the direction of the conduit.

Raceways: Enclosed channels designed expressly for holding wires, cables, or busbars, with additional functions as permitted in the NEC.

Splice: Connection of two or more conductors.

Tap: Intermediate point on a main circuit where another wire is connected to supply electrical current to another circuit.

Trough: A long, narrow box used to house electrical connections that could be exposed to the environment.

Underwriters' Laboratories (UL): An agency that evaluates and approves electrical components and equipment.

Wireways: Steel troughs designed to carry electrical wire and cable.

Instructor's Notes:

_____

_____

_____

_____

_____

_____

_____

_____

_____

Answers to Review Questions

Answer	Section
1. a	3.2.1
2. b	3.2.1
3. b	3.2.2
4. b	3.2.7
5. a	3.2.10
6. b	3.2.9
7. c	4.2.2
8. c	6.0.0
9. a	Terms/9.0.0
10. d	9.2.0
11. c	9.4.1
12. d	9.4.1
13. b	9.4.1
14. d	11.0.0
15. d	12.0.0

Additional Resources

For advanced study of topics covered in this task module, the following books are suggested:

Benfield Conduit Bending Manual, Latest Edition. New York: McGraw-Hill Publishing Company.

National Electrical Code Handbook, Latest Edition. Quincy, MA: National Fire Protection Association.

Instructor's Notes:

_____

_____

_____

_____

_____

_____

_____

_____

_____

NCCER CRAFT TRAINING USER UPDATES

The NCCER makes every effort to keep these textbooks up-to-date and free of technical errors. We appreciate your help in this process. If you have an idea for improving this textbook, or if you find an error, a typographical mistake, or an inaccuracy in the NCCER's Craft Training textbooks, please write us, using this form or a photocopy. Be sure to include the exact module number, page number, a detailed description, and the correction, if applicable. Your input will be brought to the attention of the Technical Review Committee. Thank you for your assistance.

Instructors – If you found that additional materials were necessary in order to teach this module effectively, please let us know so that we may include them in the Equipment/Materials list in the Instructor's Guide.

Write: Curriculum Revision and Development Department
National Center for Construction Education and Research
P.O. Box 141104, Gainesville, FL 32614-1104

Fax: 352-334-0932

E-mail: curriculum@nccer.org

Craft _____ Module Name _____

Copyright Date _____ Module Number _____ Page Number(s) _____

Description _____

_____

_____

_____

(Optional) Correction _____

_____

_____

(Optional) Your Name and Address _____

_____

_____

TEACHING TIPS

Section 3.2.1   *Think About It—EMT Use*

The requirements governing the use of EMT are covered in *NEC Article 348.* EMT is widely used in both exposed and concealed locations with certain restrictions, as defined in *NEC Section 348-5.* EMT should not be used in installations where it may be subject to physical damage. For such installations, rigid metal conduit provides greater protection against mechanical injury to conductors.

Section 3.2.8   *Think About It—Applications of PVC*

The requirements governing the use of PVC conduit are covered in *NEC Article 347.* PVC conduit is corrosion-proof and may be placed underground as well as above grade, indoors and out. It is commonly used outdoors, especially for service-entrance conductor raceways and wiring to swimming pools and spas. Its use is restricted in certain hazardous locations and where it could be physically damaged or exposed to high temperatures exceeding 122°F. It cannot be used to support fixtures.

Section 10.0.0   *Think About It—Cable Trays and Wireways*

Wireways are fully enclosed troughs, while cable trays are open to allow easy access to the cables. Wireways generally have a smaller capacity for holding conductors. Both wireways and cable trays are used to support large single conductors. Cable trays are also widely used to carry multiconductor cables as well as different types of telecommunications cables.

Section 16.4.0   *Think About It—Putting It All Together*

The design of a large industrial electrical installation starts with a thorough review of the construction drawings (blueprints) for the building and the specifications for the job. Once the job requirements are fully understood, the building and surrounding area should be surveyed to determine information important to both the design and installation. Such information typically includes: the physical location, size, and type of construction for all equipment rooms and other building structures; the location of system grounding points and prime power inputs; the locations where equipment is to be installed; and the best paths for installing the various conduit runs, wireways, and/or cable trays.

Conductors
26109-00

MODULE OVERVIEW

This course introduces the electrical trainee to the various types of conductors, explains how conductors are rated by the NEC, and discusses the different methods used for pulling conductors through conduit runs.

PREREQUISITES

Please refer to the Course Map in the Trainee Module. Prior to training with this module, it is recommended that the trainee shall have successfully completed the following modules:

Core Curriculum; Electrical Level One, Modules 26101 through 26108

LEARNING OBJECTIVES

Upon completion of this module, the trainee will be able to:

1. Explain the various sizes and gauges of wire in accordance with American Wire Gauge standards.
2. Identify insulation and jacket types according to conditions and applications.
3. Describe voltage ratings of conductors and cables.
4. Read and identify markings on conductors and cables.
5. Use the tables in the NEC to determine the ampacity of a conductor.
6. State the purpose of stranded wire.
7. State the purpose of compressed conductors.
8. Describe the different materials from which conductors are made.
9. Describe the different types of conductor insulation.
10. Describe the color coding of insulation.
11. Describe instrumentation control wiring.
12. Describe the equipment required for pulling wire through conduit.
13. Describe the procedure for pulling wire through conduit.
14. Install conductors in conduit.
15. Pull conductors in a conduit system.

PERFORMANCE OBJECTIVES

Under supervision of the instructor, the trainee should be able to:

1. Perform a manual single cable pull:
   - Select the proper pulling rope for the pull.
   - Attach the pulling rope to the cable.
   - Attach the pulling rope to the puller.
   - Pull the cable through the conduit.
2. Perform the feed:
   - Make a pulling head attachment.
   - Apply lubricant to the cable as it is being pulled.
   - Seal the end of the cable after the pull is complete.

NCCER STANDARDIZED CRAFT TRAINING PROGRAM

The National Center for Construction Education and Research (NCCER) provides a standardized national program of accredited craft training. Key features of the program include instructor certification, competency-based training, and performance testing. The program provides trainees, instructors, and companies with a standard form of recognition through a National Craft Training Registry. The program is described in full in the *Guidelines for Accreditation*, published by the NCCER. For more information on standardized craft training, contact the NCCER at P.O. Box 141104, Gainesville, FL 32614-1104, 352-334-0911, visit our Web site at www.NCCER.org, or e-mail info@NCCER.org.

HOW TO USE THIS ANNOTATED INSTRUCTOR'S GUIDE

Each page presents two sections of information. The larger section displays each page exactly as it appears in the Trainee Module. The narrow column ties suggested trainee and instructor actions to each page and provides icons to call your attention to material, safety, audiovisual, or testing requirements. The bottom of each page includes space for your notes.

 If you see the Teaching Tip icon, that means there is a teaching tip associated with this section. Also refer to the suggested teaching tips at the end of the module.

SAFETY CONSIDERATIONS

Ensure that the trainees are equipped with appropriate personal protective equipment. Emphasize the use of the proper safety precautions and procedures when working with the materials and tools used to install conductors.

PREPARATION

Before teaching this module, you should review the Module Outline, Learning and Performance Objectives, and the Materials and Equipment List. Be sure to allow ample time to prepare your own training or lesson plan and gather all required equipment and materials.

MATERIALS AND EQUIPMENT LIST

Materials:
Transparencies

Markers/chalk

Copy of the latest edition of the *National Electrical Code®*

Variety of solid wire conductors

Samples of stranded conductors

Samples of cable, including:
    Type NM
    Type NMC
    Type UF
    Type NMS
    Type MV
    High-voltage
    Type FC
Instrumentation control wiring
    Type FCC
    Type TC
    Type USE

Module Examinations*

Performance Profile Sheets*

Equipment:
Overhead projector and screen

Whiteboard/chalkboard

Appropriate personal protective equipment

Fish tapes

Power fishing system

Basket grip

Wire grips

Manual wire puller

Power puller

Pull lines

Reel cart

Electrician's hand tools

Access to a conduit run

*Located in the Test Booklet packaged with this Annotated Instructor's Guide.

ADDITIONAL RESOURCES

This module is intended to present thorough resources for task training. The following reference is suggested for both instructors and motivated trainees interested in further study. This is optional material for continued education rather than for task training.

National Electrical Code Handbook, Latest Edition. Quincy, MA: National Fire Protection Association.

NOTES

The designations "National Electrical Code," "NE Code," and "NEC," where used in this document, refer to the *National Electrical Code®*, which is a registered trademark of the National Fire Protection Association, Quincy, MA. All National Electrical Code (NEC) references in this module refer to the 1999 edition of the NEC.

If you feel that additional math instruction would be helpful, Prentice Hall offers a basic math textbook entitled *Fundamentals of Electrical and Mechanical Mathematics*. It covers the basic math requirements for electrical trainees and may be ordered by contacting Prentice Hall Customer Service at 1-800-922-0579.

TEACHING TIME FOR THIS MODULE

An outline for use in developing your lesson plan is presented below. Note that each Roman numeral in the outline equates to one session of instruction. Each session has a suggested time period of 2½ hours. This includes 10 minutes at the beginning of each session for administrative tasks and one 10-minute break during the session. Approximately 15 hours are suggested to cover *Conductors*. You will need to adjust the time required for hands-on activity and testing based on your class size and resources.

Topic	Planned Time
Session I. Ampacity; Underground Installations; Wire Size; Conductor Material	
A. Ampacity	_____
1. NEC Ampacity Tables	_____
B. Underground Installations	_____
1. Direct Burial	_____
C. Wire Size	_____
1. AWG System	_____
2. Stranding	_____
3. Compressed Conductors	_____
4. Circular Mils	_____
D. Conductor Material	_____
1. Conductivity	_____
2. Cost	_____
3. Availability	_____
4. Workability	_____
Session II. Conductor Insulation	
A. Thermoplastic	_____
B. Letter Coding	_____
C. Color Coding	_____
D. Wire Ratings	_____
Session III. Fixture Wires; Cables; Instrumentation Control Wiring	
A. Fixture Wires	_____
1. Heating Cables	_____
B. Cables	_____
1. Cable Markings	_____
2. Nonmetallic-Sheathed Cable	_____
3. Type UF Cable	_____
4. Type NMS Cable	_____
5. Type MV Cable	_____
6. High-Voltage Shielded Cable	_____
7. Channel Wire Assemblies	_____

      8. Flat Conductor Cable   _____

      9. Type TC Cable   _____

     10. Type USE Cable   _____

C. Instrumentation Control Wiring   _____

    1. Shields   _____

    2. Grounding   _____

    3. Jackets   _____

Session IV. Fish Tape; Wire Grips; Pull Lines; Safety Precautions; Pulling Equipment

A. Fish Tape   _____

    1. Power Conduit Fishing Systems   _____

    2. Connecting Wire to a Stringline   _____

B. Wire Grips   _____

C. Pull Lines   _____

D. Safety Precautions   _____

E. Pulling Equipment   _____

Session V. Feeding Conductors into Conduit; Conductor Lubrication; Conductor Termination

A. Feeding Conductors into Conduit   _____

B. Conductor Lubrication   _____

C. Conductor Termination   _____

Session VI. Laboratory; Module Examination and Performance Testing

A. Laboratory   _____

Under instructor supervision, have the trainees practice performing a manual cable pull, including selecting and attaching the pulling rope, making a pulling head attachment, applying lubricant to the cable as it is being pulled, and sealing the end of the cable after the pull is complete.

B. Module Examination   _____

    1. Trainees must score 70% or higher to receive recognition from the NCCER.

    2. Record the testing results on Craft Training Report Form 200 and submit the results to the Training Program Sponsor.

C. Performance Testing   _____

    1. Trainees must perform each task to the satisfaction of the instructor to receive recognition from the NCCER.

    2. Record the testing results on Craft Training Report Form 200 and submit the results to the Training Program Sponsor.

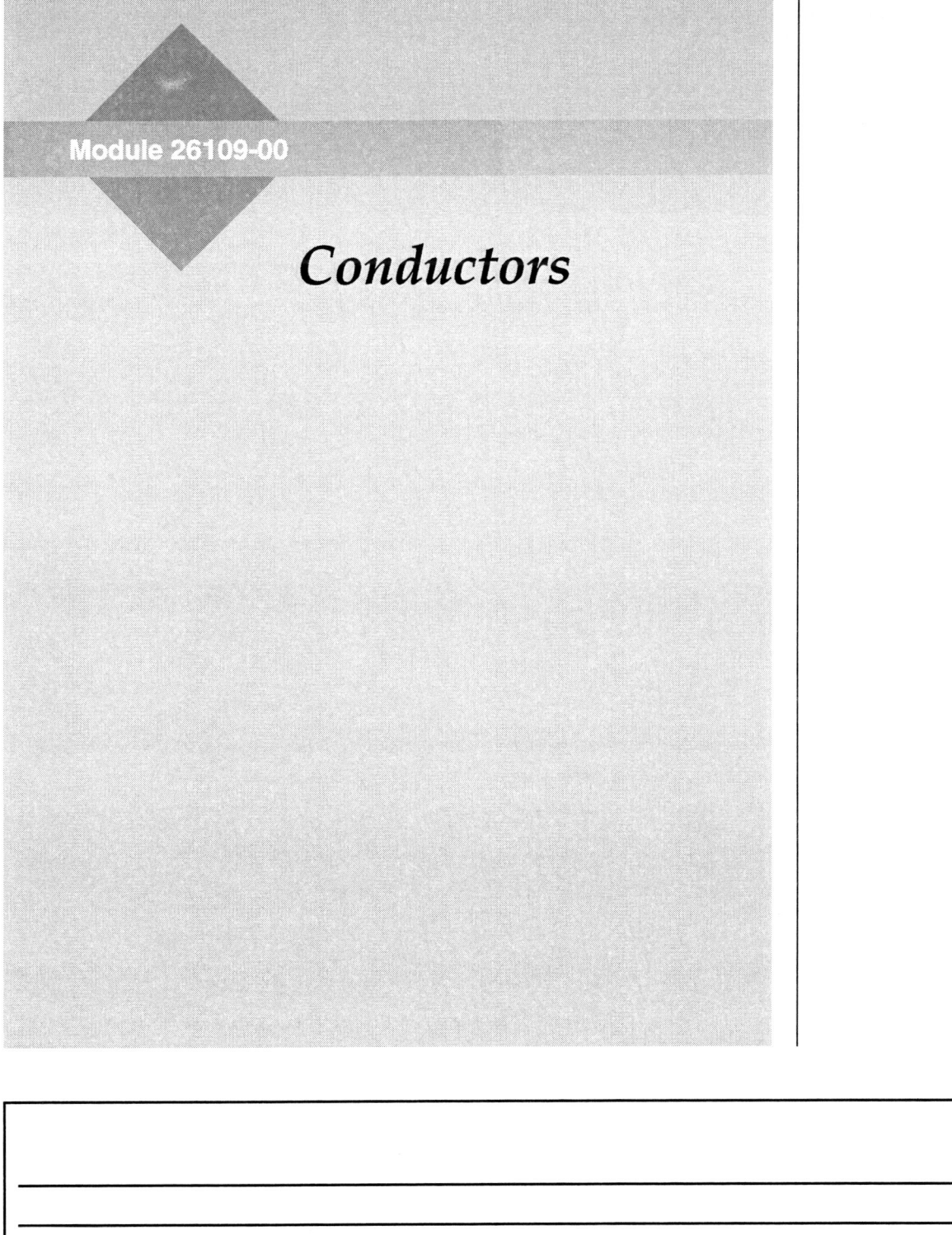

Module 26109-00

Conductors

Course Map

This course map shows all of the task modules in the first level of the Electrical curricula. The suggested training order begins at the bottom and proceeds up. Skill levels increase as a trainee advances on the course map. The training order may be adjusted by the local Training Program Sponsor.

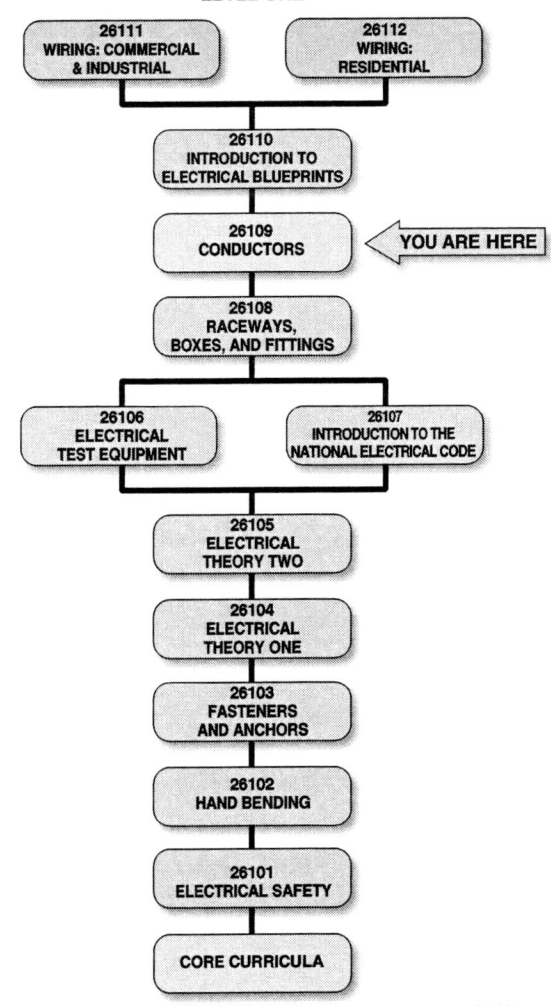

LEVEL ONE

26111
WIRING: COMMERCIAL & INDUSTRIAL

26112
WIRING: RESIDENTIAL

26110
INTRODUCTION TO ELECTRICAL BLUEPRINTS

26109
CONDUCTORS ◁ YOU ARE HERE

26108
RACEWAYS, BOXES, AND FITTINGS

26106
ELECTRICAL TEST EQUIPMENT

26107
INTRODUCTION TO THE NATIONAL ELECTRICAL CODE

26105
ELECTRICAL THEORY TWO

26104
ELECTRICAL THEORY ONE

26103
FASTENERS AND ANCHORS

26102
HAND BENDING

26101
ELECTRICAL SAFETY

CORE CURRICULA

109CMAP.EPS

Assign reading of Module 26109.

9–iii

Instructor's Notes:

_____

_____

_____

_____

_____

_____

_____

_____

_____

Figures

Ensure you have everything required to teach the course. Check the Materials and Equipment List at the front of this Instructor's Guide.

Instructor's Notes:

_____

_____

_____

_____

_____

_____

_____

_____

_____

Conductors

Objectives

Upon completion of this module, the trainee will be able to:

1. Explain the various sizes and gauges of wire in accordance with American Wire Gauge standards.
2. Identify insulation and jacket types according to conditions and applications.
3. Describe voltage ratings of conductors and cables.
4. Read and identify markings on conductors and cables.
5. Use the tables in the NEC to determine the ampacity of a conductor.
6. State the purpose of stranded wire.
7. State the purpose of compressed conductors.
8. Describe the different materials from which conductors are made.
9. Describe the different types of conductor insulation.
10. Describe the color coding of insulation.
11. Describe instrumentation control wiring.
12. Describe the equipment required for pulling wire through conduit.
13. Describe the procedure for pulling wire through conduit.
14. Install conductors in conduit.
15. Pull conductors in a conduit system.

Prerequisites

Successful completion of the following Task Modules is recommended before beginning study of this Task Module: Core Curricula; Electrical Level One, Modules 26101 through 26108.

Required Trainee Materials

1. Trainee Task Module
2. Copy of the latest edition of the *National Electrical Code*
3. Appropriate Personal Protective Equipment

1.0.0 ◆ INTRODUCTION

As an electrician, you will be required to select the proper wire and/or cable for a job. You will also be required to pull this wire or cable through conduit runs in order to terminate it. This module will examine the different types of conductors and conductor insulation. It will also examine how these conductors are rated and classified by the NEC and the different methods used for pulling these conductors through conduit runs.

Show Transparency 1, Course Objectives.

Show Transparency 2, Performance Profile Tasks.

Discuss the importance of understanding how to properly identify, select, and pull conductors.

Note: The designations "National Electrical Code," "NE Code," and "NEC," where used in this document, refer to the *National Electrical Code*®, which is a registered trademark of the National Fire Protection Association, Quincy, MA. All National Electrical Code (NEC) references in this module refer to the 1999 edition of the NEC.

9–1

2.0.0 ◆ CONDUCTORS AND INSULATION

The term *conductor* is used in two ways. It is used to describe the current-carrying portion of a wire or cable, and it is used to describe the wire or cable composed of a current-carrying portion and an outer covering (insulation). In this module, the term *conductor*, if not specified otherwise, will be used to describe the wire assembly, which includes the insulation and the current-carrying portion of the wire. Conductors are uniquely identified by size and insulation material. Size refers to the physical size of the current-carrying portion of the conductor.

NEC Table 310-13 presents application and construction data on the wide range of 600-volt insulated, individual conductors recognized by the NEC, with the appropriate letter designation used to identify each type of insulated conductor. Important data that should be noted in this table are as follows:

- The designation for one thousand circular mils is *kcmil,* which has been substituted for the long-time designation *MCM* in this table and throughout the NEC.
- Type MI (mineral insulated) cable may have either a copper or an alloy steel sheath.
- Type RHW-2 is a conductor insulation made of moisture- and heat-resistant rubber with a 90°C (194°F) rating for use in dry and wet locations.
- Type XHHW-2 is a moisture- and heat-resistant crosslinked synthetic polymer with a 90°C (194°F) rating for use in dry and wet locations.
- The suffix *LS* designates a conductor insulation as low-smoke producing and flame retardant. For example, Type THHN/LS is a THHN conductor with a limited smoke-producing characteristic.
- Type THHW is a moisture- and heat-resistant insulation rated at 75°C (167°F) for wet locations and 90°C (194°F) for dry locations. This is similar to THWN and THHN without the outer nylon covering but with thicker insulation.
- All insulations using asbestos have been deleted from *NEC Table 310-13* because they are no longer made.

2.1.0 Ampacity

Ampacity is the current in amperes a conductor can carry continuously under the conditions of use without exceeding its temperature rating. The ampacity of conductors for given conditions of use are listed in *NEC Tables 310-16 through 310-19.*

2.1.1 NEC Ampacity Tables

NEC Table 310-16 covers conductors rated up to 2,000 volts where not more than three conductors are installed in a raceway or cable or are directly buried in the earth, based on an ambient temperature of 30°C (86°F).

NEC Table 310-17 covers both copper conductors and aluminum or copper-clad aluminum conductors up to 2,000 volts where conductors are used as single conductors in free air, based on an ambient temperature of 30°C (86°F).

NEC Tables 310-18 and 19 apply to conductors rated at 150°C to 250°C (302°F to 482°F), used either in raceway or cable or as single conductors in free air, based on an ambient temperature of 40°C (104°F).

Example:

Determine the ampacity of a No. 12 Cu (copper) THW conductor.

Solution:

25 amps (from *NEC Table 310-16*).

2.2.0 Underground

Any conductor used in a wet location (see definition under *Wet Location* in *NEC Article 100*) must be designated as suitable for wet locations. Any conduit run underground is assumed to be subject to water infiltration and is, therefore, in a wet location, requiring the use of only the listed conductor types.

2.2.1 Direct Burial

Direct burial conductors should be trench-laid without crossovers; slightly snaked to allow for

---

Overheating

What happens when insulation is overheated? What affects the ampacity of an insulated conductor?

Instructor's Notes:

_____

_____

_____

_____

_____

_____

_____

_____

_____

possible earth settlement, movement, or heaving due to frost action; and have cushions and covers of sand or screened fill to protect conductors against sharp objects in trenches or backfill.

2.3.0 Wire Size

Wire sizes are expressed in gauge numbers. The standard system of wire sizes in the United States is the American Wire Gauge (AWG) system.

2.3.1 AWG System

The AWG system uses numbers to identify the different sizes of wire and cable (*Figure 1*). The larger the number, the smaller the cross-sectional area of the wire. The larger the cross-sectional area of the current-carrying portion of a conductor, the higher the amount of current the wire can conduct. The AWG numbers range from 50 to 1; then 0, 00, 000, and 0000 (one aught [1/0], two aught [2/0], three aught [3/0], and four aught [4/0]). Any wire larger than 0000 is identified by its area in circular mils. Wire sizes smaller than No. 18 AWG are usually solid, but may be stranded in some cases. Wire sizes of No. 6 AWG or larger are stranded.

For wire sizes larger than No. 16 AWG, the wire size is marked on the insulation (*Figure 2*).

NEC Chapter 9, Table 8 has descriptive information on wire sizes. Again, note that all wires smaller than No. 6 are available as solid or stranded. Wire sizes of No. 6 or larger are shown only as stranded. Solid wire larger than No. 6 is manufactured; however, the NEC only permits the use of solid wire in a raceway for sizes smaller than No. 8 (*NEC Section 310-3*).

2.3.2 Stranding

According to *NEC Chapter 9, Table 8,* wire sizes No. 18 to No. 2 have seven strands; wire sizes No. 1 to No. 4/0 have 19 strands; and wire sizes between 250 kcmil and 500 kcmil have 37 strands. The purpose of stranding is to increase the flexibility of the wire. Terminating solid wire sizes larger than No. 8 in pull boxes, disconnect switches, and panels would not only be very difficult, but might also result in damage to equipment and wire insulation.

Pulling solid wire in conduit around bends could pose a major problem and cause damage to equipment for wire sizes larger than No. 8. The

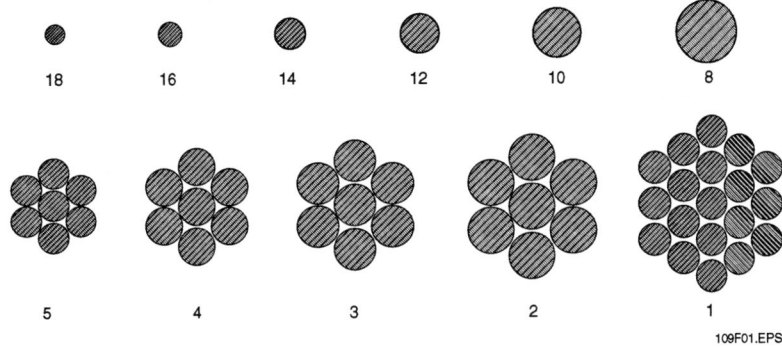

18   16   14   12   10   8

5   4   3   2   1

109F01.EPS

Figure 1 ◆ Comparison of wire sizes (enlarged) from No. 18 to No. 1 AWG.

THINK ABOUT IT

Wire Size

Why is wire size a critical factor in a wiring system? Other than load, what other factors may dictate wire size? What can happen when a wire isn't properly sized for the load?

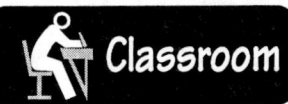

Discuss the AWG wire sizing system.

Show Transparencies 3 and 4 (Figures 1 and 2).

Discuss wire stranding.

Discuss the "Think About It."

WIRE SIZE MARKING

109F02.EPS

Figure 2 ◆ Wire size marking.

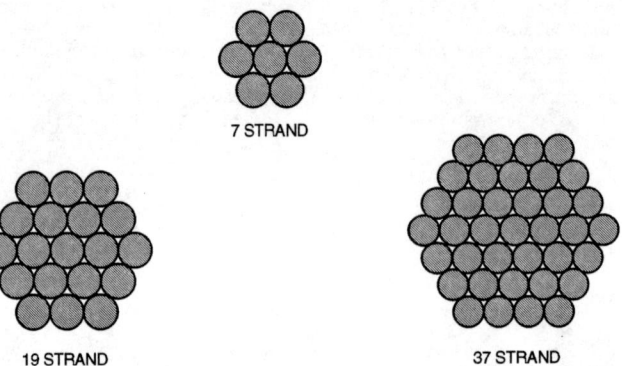

7 STRAND

19 STRAND

37 STRAND

109F03.EPS

Figure 3 ◆ Strand configurations.

Show Transparency 5 (Figure 3).

Discuss the advantages of using compressed conductors.

Show Transparency 6 (Figure 4).

Define *circular mil.*

reason for choosing 7, 19, and 37 strands for stranded conductors is that it is necessary to provide a flexible, almost round conductor. In order for a conductor to be flexible, individual strands must not be too large. *Figure 3* shows how these conductors are configured.

2.3.3 Compressed Conductors

A relatively new entry in the scheme of conductors is compressed aluminum conductors (*Figure 4*). Compressed aluminum conductors are those that have been compressed to reduce the air space between the strands.

The purpose of a compressed conductor is to reduce the overall diameter of the cable so that it may be installed in a conduit that is smaller than that required for standard conductors of the same wire size. Compressed conductors are especially useful when increasing the ampacity of an existing service.

2.3.4 Circular Mils

A circular mil is a circle that has a diameter of 1 mil. A mil is 0.001 inch. When a wire size is 250 kcmil, the cross-sectional area of the current-

9–4

Instructor's Notes:

_____

_____

_____

_____

_____

_____

_____

_____

_____

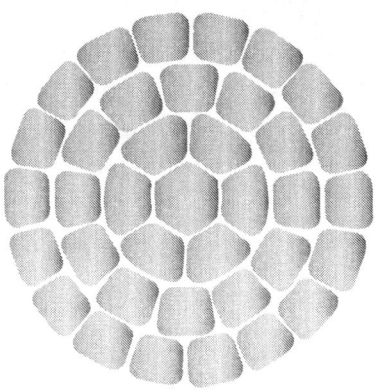

109F04.EPS

Figure 4 ◆ Compressed conductor.

carrying portion of the wire is the same as 250,000 circles having a diameter of 0.001 inch. This may seem to be a rather clumsy way of sizing wire at first; however, the alternative would be to size the wire as a function of its cross-sectional area expressed in square inches.

According to *NEC Chapter 9, Table 8,* the cross-sectional area of a 250 kcmil conductor is 0.260 square inch.

If a conductor is to be sized by cross-sectional area, it is much easier to express the wire size in circular mils (or thousands of circular mils) than in square inches.

2.4.0 Conductor Material

The most common conductor material is copper. Copper is used because of its excellent conductivity (low resistance), ease of use, and value. The value of a material as an ingredient for wire is determined by several factors, including conductivity, cost, availability, and workability.

2.4.1 Conductivity

Conductivity is a word that describes the ease (or difficulty) of travel presented to an electric current by a conductor. If a conductor has a low resistance, it has a high conductivity. Platinum is one of the best conductors since it has very low resistance and high conductivity. Copper has high conductivity and a much lower price than platinum. Aluminum, another material with good conductivity, is also a good choice for conductor material. The conductivity of aluminum is approximately ⅔ that of copper.

2.4.2 Cost

Cost is always an issue that contributes to the selection of a material to be used for a given application. Often, a material that has low cost may be selected as a conductor material even though it has physical properties that are inferior to the more expensive material. Such is the case in the selection of copper over platinum. Here, the cost of platinum is very high, and very little thinking is required to determine that copper is a better choice. The choice between copper and aluminum is often more difficult to make.

2.4.3 Availability

The availability of some material is often a concern when selecting components for a job. As applied to wire, the mining industry often controls the availability of raw materials, which could produce shortages of some material. The availability of a substance such as copper or aluminum affects the price of the finished product (copper or aluminum wire).

2.4.4 Workability

It is a good idea to select a material that requires less expense for tools and is easier to work with. Aluminum conductors are lighter than copper conductors of the same size. They are also much

Classroom

Discuss the factors involved in the selection of conductor material. Cover:
- **Conductivity**
- **Cost**
- **Availability**
- **Workability**

Terminating Aluminum Wire

INSIDE TRACK

Care must be taken to use listed connectors when terminating aluminum wire. All aluminum connections also require the use of anti-oxidizing compound. Some connectors are precoated with compound; others require the addition of it. Be sure to check the connectors before beginning the installation.

Stress the importance of using listed connectors with aluminum wire.

Classroom

Provide an overview of the history of conductor insulation.

List the most commonly used types of thermoplastic insulation.

Discuss letter coding of insulation. Refer the trainees to the NEC Ampacity Tables.

Conductor Insulation

THINK ABOUT IT What are the functions of conductor insulation? Under what conditions does the NEC allow uninsulated conductors?

more flexible than copper conductors and, in general, are easier to work with. However, terminating aluminum conductors often requires special tools and treatment of termination surfaces with an anti-oxidation material. Splicing and terminating aluminum conductors often requires a higher degree of training on the part of the electrician than do similar efforts with copper wire. This is partly due to the fact that aluminum expands and contracts with heat more than copper.

2.5.0 Conductor Insulation

The first attempt to insulate wire was made in the early 1800s during the development of the telegraph. This insulation was designed to provide physical protection rather than electrical protection. Electrical insulation was not an important issue because the telegraph operated at low-voltage DC. This early form of insulation was a substance composed of tarred hemp or cotton fiber and shellac and was used primarily for weatherproofing long distance distribution lines to mines, industrial sites, and railroads.

Some early electrical distribution systems utilized the knob-and-tube technique of installing wire. The wire was often bare and was pulled between and wrapped around ceramic knobs that were affixed to the building structure. When it was necessary to pull wire through structural members, it was pulled through ceramic tubes. The structural member (usually wood) was drilled, the tube was pressed into the hole, and the wire was pulled through the hole in the tube. As dangerous as this may appear, older homes still exist that have knob-and-tube wiring that was installed in the early 1900s and is still operational. Knob-and-tube wiring was revised to use insulated conductors and was in use up to 1957 in some areas.

The grounded or neutral conductor in overhead services may be bare. Furthermore, the concentric grounded conductor in Type SE cable may be bare when used as a service-entrance cable. However, all current-carrying conductors (including the grounded conductor) must be insulated

when used on the inside of buildings, or after the first overcurrent protection device.

2.5.1 Thermoplastic

Thermoplastic is a popular and effective insulation material used on conductors for the present-day market. The following thermoplastics are widely used as insulation materials:

- *Polyvinyl chloride (PVC)*—The base material used for the manufacture of TW and THW insulation.
- *Polyethylene (PE)*—An excellent weatherproofing material used primarily for insulation of control and communications wiring. It is not used for high-voltage conductors (those exceeding 5,000 volts).
- *Cross-linked polyethylene (XLP)*—An improved PE with superior heat- and moisture-resistant qualities. Used for THHN, THWN, and XHHW wiring as well as most high-voltage cables.
- *Nylon*—Primarily used as jacketing material. THHN building wire has an outer coating of nylon.
- *Teflon®*—A high-temperature insulation. Widely used for telephone wiring in a plenum (where other insulated conductors require conduit routing).

2.5.2 Letter Coding

Conductor insulation as applied to building wire is coded by letters. The letters generally, but not always, indicate the type of insulation or its environmental rating. The types of conductor insulation described in this module will be those indicated at the top of *NEC Table 310-16*. The various insulation designations are:

Letter	Description
A	Asbestos
B	Braid
E	Ethylene or Entrance
F	Fluorinated or Feeder
H	Heat-Rated or Flame-Retardant
I	Impregnated
N	Nylon
P	Propylene

Instructor's Notes:

_____

_____

_____

_____

_____

_____

_____

_____

_____

_____

R	Rubber	TBS	Thermoplastic Braided Silicon
S	Silicon or Synthetic	SA	Silicon Asbestos
T	Thermoplastic	AVB	Asbestos Varnished and Braided
U	Underground	SIS	Synthetic Heat-Resistant
V	Varnished Cambric	FEP	Fluorinated Ethylene Propylene
W	Weather-Rated	RHH	Flame-Retardant Heat-Rated Rubber

R Rubber
S Silicon or Synthetic
T Thermoplastic
U Underground
V Varnished Cambric
W Weather-Rated
X Cross-Linked Polyethylene
Z Modified Ethylene Tetrafluorethylene
TW Weather-Rated Thermoplastic (60°C/140°F)
UF Underground Feeder
FEPW Weather-Rated Fluorinated Ethylene Propylene
RH Heat-Rated Rubber (75°C/167°F)
RHW Weather-Rated, Heat-Rated Rubber (75°C/167°F)
THW Weather-Rated, Heat-Rated Thermoplastic (75°C/167°F)
THWN Weather-Rated, Heat-Rated Thermoplastic with Nylon Cover
XHHW Heat-Rated, Flame-Retardant, Weather-Rated, Cross-Linked Polyethylene
USE Underground
ZW Weather-Rated Modified Ethylene Tetrafluorethylene
TA Thermoplastic and Asbestos

TBS Thermoplastic Braided Silicon
SA Silicon Asbestos
AVB Asbestos Varnished and Braided
SIS Synthetic Heat-Resistant
FEP Fluorinated Ethylene Propylene
RHH Flame-Retardant Heat-Rated Rubber

2.5.3 Color Coding

A color code is used to help identify wires by the color of the insulation. Understanding the color code makes it easier to install and properly connect the wires. When two or more wires are used together, as in a cable, the color code is as follows:

- *Two-conductor cable*—One white wire, one black wire, and a grounding wire (usually bare)
- *Three-conductor cable*—One white, one black, one red, and a grounding wire
- *Four-conductor cable*—Same as three-conductor cable plus fourth wire (blue)
- *Five-conductor cable*—Same as four-conductor cable plus fifth wire (yellow)

The grounding conductor may be either green or green with a yellow stripe. Color codes are shown in *Figure 5*.

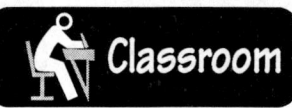

Discuss the color coding of insulation. Provide examples of various conductors for the trainees to examine.

Show Transparency 7 (Figure 5).

Discuss the "Think About It."

Insulation Types

THINK ABOUT IT

Use *NEC Table 310-13* to identify two types of insulation that are suitable for use in wet locations.

Figure 5 ◆ Insulator color codes.

Discuss temperature ratings and correction factors.

Discuss the NEC requirements for fixture wires.

Show Transparency 8 (Figure 6).

Discuss the "Think About It."

Stress the importance of verifying all conductors using a voltmeter.

The NEC does not require color coding of ungrounded conductors in many cases. The ungrounded conductors may be any color with the exception of white, grey, or green; however, it is a good practice to color code conductors as described here. In fact, many construction specifications require color coding. Furthermore, on a four-wire, delta-connected secondary where the midpoint of one phase is grounded to supply lighting and similar loads, the phase conductor having the higher voltage to ground must be identified by an outer finish that is orange in color, by tagging, or by other effective means. Such identification must be placed at each point where a connection is made if the grounded conductor is also present. In most cases, orange tape is used at all termination points when such a condition exists.

2.5.4 Wire Ratings

Conductor selection is based largely on the temperature rating of the wire. This requirement is extremely important and is the basis of safe operation for insulated conductors. As shown in *NEC Table 310-13,* conductors have various ratings, (60°C, 75°C, 90°C, etc.). Since *NEC Tables 310-16 through 310-19* are based on an assumed ambient temperature of 30°C (86°F), conductor ampacities are based on the ambient temperature plus the heat (I^2R) produced by the conductor while carrying current. Therefore, the type of insulation used on the conductor is the first consideration in determining the maximum permitted conductor ampacity.

For example, a No. 3/0 THW copper conductor for use in a raceway has an ampacity of 200 according to *NEC Table 310-16.* In a 30°C ambient temperature, the conductor is subjected to this temperature when it carries no current. Since a THW-insulated conductor is rated at 75°C, this leaves 45°C (75 − 30) for increased temperature due to current flow. If the ambient temperature exceeds 30°C, the conductor maximum load-current rating must be reduced proportionally (see *Correction Factors* at the bottom of *NEC Table 310-16*) so that the total temperature (ambient plus conductor temperature rise due to current flow) will not exceed the temperature rating of the conductor insulation (60°C, 75°C, etc.). For the same reason, conductor maximum load current ratings must be reduced below the ampacity values where more than three conductors are contained in a raceway or cable.

Using the ampacity tables—An important step in circuit design is the selection of the type of conductor to be used (TW, THW, THWN, RHH, THHN, XHHW, etc.). The various types of conductors are covered in *NEC Article 310,* and the ampacities of conductors are given in *NEC Tables 310-16 through 310-19* for the varying conditions of use (e.g., in a raceway, in open air, at normal or higher-than-normal ambient temperatures). Conductors must be used in accordance with the data in these tables and notes.

2.6.0 Fixture Wires

Fixture wire is used for the interior wiring of fixtures and for wiring fixtures to a power source. Guidelines concerning fixture wire are given in *NEC Article 402.* The list of approved types of fixture wire is given in *NEC Table 402-3. Figure 6* shows one example of fixture wire. The wires are composed of insulated conductors with or without an outer jacket. The conductors range in size from No. 18 to No. 10 AWG.

Color Coding Ungrounded Conductors

Although the NEC does not require the use of color-coded ungrounded current-carrying conductors, why might it be a good idea to use them anyway?

Color Coding

Color designations are good indicators, but never trust your life to them. Always protect yourself by testing circuits with a voltmeter.

Instructor's Notes:

_____

_____

_____

_____

_____

_____

_____

_____

_____

_____

CONDUCTOR   INSULATION   TRANSPARENT OUTER JACKET

109F06.EPS

Figure 6 ◆ Fixture wire.

The decision of which fixture wire to use depends primarily upon the operating temperature that is expected within the fixture. Therefore, it is the character of the insulation that will determine the wire selected. For instance, fixture wires insulated with perfluoroalkoxy (PAF) or extruded polytetrafluorethylene (PTF) would be selected if the operating temperature of the fixture is expected to reach a maximum of 482°F. This is the highest operating temperature allowed for any fixture wire.

As indicated by *NEC Section 402-3,* fixture wires are suitable for service at 600 volts unless otherwise specified in *NEC Table 402-3.* The allowed ampacities of fixture wire are given in *NEC Table 402-5.*

Although the primary use for fixture wire is the internal wiring of fixtures, several of the wires listed in *NEC Table 402-3* may be used for wiring remote-control, signaling, or power-limited circuits in accordance with *NEC Section 725-15.* Fixture wires may never be used as substitutes for branch circuit conductors.

2.6.1 Heating Cables

Heating cables are another important group of special wires that are commonly used in the building industry. The purpose of these cables is to produce heat when energized. There are three general categories of heating cables:

- Interior space-heating cables
- De-icing and snow-melting cables
- Pipeline and vessel heating cables

These heating cables consist of a long insulated wire, cabled heating wire, or resistance wire, which produces heat in the wire when it is connected to a power source. The ends of the resistance wire are connected to non-heating lead wires, which extend to a thermostat or junction box.

Guidelines pertaining to heating cable used for interior space heating are given in *NEC Article 424.* The NEC considers the term *heating equipment*

to include any factory-manufactured heating devices such as unit heaters, boilers, and local heating systems, as well as heating cables. Heating cables are specifically covered in *NEC Article 424, Part E.*

Heating cables meant for installation in ceilings are available from the factory in unit lengths from 75 to 1,800 feet, capacities ranging from 200 to 5,000 watts, and voltage ratings of 120, 208, and 240 volts. The cables are usually rated at 2¾ watts per linear foot. The nonheating lead wires are at least seven feet in length and are color coded to indicate the voltage rating. The insulation on the heating wire is designed to be resistant to high temperatures, water absorption, aging, and chemical action.

Heating cables meant for installation in floors have capacities that vary with the cable dimensions and heating wire resistance. They are typically rated at voltages from 120 to 600 volts. There are two popular floor heating cables. In the first type, the resistance wire is covered with a sheath of polyvinyl chloride (PVC). In the second, the resistance wire is encased in a mineral insulation and then covered with a copper sheath. This construction is identical to that of Type MI (mineral-insulated, metal-sheathed cable). In contrast to ceiling heating cables, which are available in specific unit lengths, these heating cables are sold in random lengths, which must be cut and properly terminated at the job site according to the job's heating requirements.

NEC Article 424, Part E specifies guidelines for installing floor and ceiling heating cables; it should be read carefully. Generally, ceiling heating cables are stapled to the ceiling with at least 1½ inches between adjacent runs. According to the NEC, the cables must not come within eight inches of recessed lighting fixtures or within six inches of other metallic materials in the ceiling. Floor heating cables are usually embedded in concrete, with at least a one-inch space between adjacent runs of cable. Spacing must also be maintained between the cable and other metallic

Discuss the three classes of heating cables.

Have the trainees locate the NEC requirements for each type of heating cable.

Cable Selection

There are two factors to be considered when determining the type of cable to be used for a specific application: the type of conductor insulation and the cable jacket. Both must be appropriate for the application.

objects in the floor, unless a grounded metal-clad cable is used.

Guidelines pertaining to de-icing and snow-melting cables are given in *NEC Article 426.* These cables are similar in construction to interior floor heating cables in that they consist of resistance wires insulated with a sheath of PVC or with mineral insulation and a copper sheath. Most de-icing and snow-melting cables are installed by embedding them in concrete or asphalt. As with floor heating cables, these cables are rated at various voltages and must be cut to the desired length.

Guidelines pertaining to pipeline and vessel heating cables are given in *NEC Article 427.* These cables are generally available in two types of construction. The cable can either be insulated resistance wire with nonheating lead wires attached to each end, or it can be in the form of a flat heating tape. The tape consists of a single piece of heating wire doubled over so that the halves of the wire run parallel to each other. The wire is encased in a plastic enclosure, thus producing an assembly that resembles a tape. The nonheating lead wires are attached to the open end of the tape.

The obvious purpose of pipeline and vessel heating cables is to prevent the fluid in the pipeline or vessel from falling below a specified minimum temperature. Flat heating tapes are often used in buildings under construction or not completely enclosed. They can be used on water pipes to keep the water from freezing in cold weather.

2.7.0 Cables

Cables are two or more insulated wires and a grounding wire covered by an outer jacket or sheath. Cable is usually classified by the type of covering it has, either nonmetallic (i.e., plastic) or metallic, also called *armored cable.*

Cable may also be classified according to where it can be used (see *NEC Table 400-4*). Because water is such a good conductor of electricity, moisture on conductors can cause power loss or short circuits. For this reason, cables are classified for either dry, damp, or wet locations. Cables can also be classified regarding exposure to sunlight and rough use.

2.7.1 Cable Markings

All cables are marked to show important properties and uses. Cable markings show the wire size, number of conductors, cable type, and voltage rating. In addition, a marking may be included to signify approved service or applications. This information is printed on nonmetallic cable (*Figure 7*). On metallic cable, marking information is usually included on a tag.

2.7.2 Nonmetallic-Sheathed Cable

Nonmetallic-sheathed cable (Type NM and Type NMC) is widely used for branch circuits and feeders in residential and commercial systems. See *Figure 8* on page 9–12. Both types are commonly called *Romex,* even though the cable manufacturer only calls Type NM cable Romex. Guidelines for the use of nonmetallic-sheathed cable are given in *NEC Article 336.* This cable consists of two or three insulated conductors and one bare conductor enclosed in a nonmetallic sheath. The conductors may be wrapped individually with paper, and the spaces between the conductors may be filled with jute or paper to protect the conductors and help the cable keep its shape. The sheath covering both Type NM cable and Type NMC cable is flame-retardant and moisture-resistant. The sheath covering Type NMC cable has the additional characteristics of being fungus- and corrosion-resistant.

NEC Article 336 lists the allowed and prohibited uses for Type NM cable and Type NMC cable. Both are allowed to be installed in either exposed or concealed work. The primary difference in their applied uses is that Type NM cable is suitable for dry locations only, whereas Type NMC is permitted for dry, moist, damp, or corrosive locations. Since you will probably work with these two cables often, read *NEC Article 336* carefully.

Guidelines for the use of Type SE (service-entrance) cable are given in *NEC Article 338.* The NEC contains no specifications for the construc-

Classroom

Discuss cable classifications and markings.

Audiovisual

Show Transparency 9 (Figure 7A).

Discuss Types NM and NMC cable and pass around examples for the trainees to examine.

Show Transparency 10 (Figure 8A).

9–10 and module number

9–10

ELECTRICAL LEVEL ONE—TRAINEE MODULE 26109

Instructor's Notes:

VERONICA WESTFALL   109F07B.EPS

Figure 7 ◆ Nonmetallic cable markings.

tion of this cable; it is left to UL to determine what types of cable should be approved for this purpose. Currently, service-entrance cable is labeled in sizes No. 12 AWG and larger for copper, and No. 10 AWG and larger for aluminum or copper-clad aluminum, with Types RH, RHW, RHH, or XHHW conductors. If the type designation for the conductor is marked on the outside surface of the cable, the temperature rating of the cable corresponds to the rating of the individual conductor. When this marking does not appear, the temperature rating of the cable is 75°C (167°F). Type SE cable is for above-ground installation.

When used as a service-entrance cable, Type SE must be installed as specified in *NEC Article 230.* Service-entrance cable may also be used as feeder and branch circuit cable. Guidelines for the use of service-entrance cable are given in *NEC Section 338-3.*

2.7.3 Type UF Cable

Guidelines for the use of Type UF (underground feeder and branch circuit) cable are given in *NEC Article 339.* Type UF cable is very similar in appearance, construction, and use to Type NMC cable. The main difference between these two cables is that Type UF cable is suitable for direct burial, whereas Type NMC cable is not.

2.7.4 Type NMS Cable

Refer to *NEC Sections 336-4(c), 336-5, and 336-30* for the applications of Type NMS cable. Type NMS cable is a form of nonmetallic-sheathed cable that contains a factory assembly of power, communications, and signaling conductors enclosed within a moisture-resistant, flame-retardant sheath.

2.7.5 Type MV Cable

Type MV (medium-voltage) cable is covered in *NEC Article 326.* It consists of one or more insulated conductors encased in an outer jacket. This cable is suitable for use with voltages ranging from 2,001 to 35,000 volts. It may be installed in wet and dry locations and may be buried directly in the earth.

2.7.6 High-Voltage Shielded Cable

Shielding of high-voltage cables protects the conductor assembly against surface discharge or burning due to corona discharge in ionized air, which can be destructive to the insulation and jacketing.

Electrostatic shielding of cables makes use of both nonmetallic and metallic materials (*Figures 9 and 10* on pages 9–13 and 9–14).

CONDUCTORS—TRAINEE MODULE 26109   9–11

PLASTIC SHEATH
PAPER OVERWRAP
BARE GROUNDING WIRE

14-2G TYPE NM

TYPE NM CABLE

SOLID PLASTIC
BARE GROUNDING WIRE

14-2G TYPE NMC

TYPE NMC CABLE

109F08A .EPS

WALTER JOHNSON   109F08B.EPS

Figure 8 ◆ Nonmetallic-sheathed cable.

2.7.7 Channel Wire Assemblies

Channel wire assemblies (Type FC) comprise an entire wiring system, which includes the cable, cable supports, splicers, circuit taps, fixture hangers, and fittings (*Figure 11* on page 9–15). Guidelines for the use of this system are given in *NEC Article 363.* Type FC cable is a flat cable assembly with three or four parallel No. 10 special stranded copper conductors. The assembly is installed in an approved U-channel surface metal raceway with one side open. Tap devices can be inserted anywhere along the run. Connections from the tap devices to the flat cable assembly are made by pin-type contacts when the tap devices are fastened in place. The pin-type contacts penetrate the insulation of the cable assembly and contact the multi-stranded conductors in a matched phase sequence. These taps can then be wired to lighting fixtures or power outlets (*Figure 12* on page 9–16).

As indicated in *NEC Section 363-3,* this wiring system is suitable for branch circuits that only supply small appliances and lights. This system is suitable for exposed wiring only and may not be concealed within the building structure. It is ideal for quick branch circuit wiring at field installations.

2.7.8 Flat Conductor Cable

Type FCC (flat conductor) cable comprises an entire branch wiring system similar in many re-

ELECTRICAL LEVEL ONE—TRAINEE MODULE 26109

Discuss Type FC wiring systems and their components. Pass around example components for the trainees to examine.

Show Transparencies 13 and 14 (Figures 11 and 12).

Instructor's Notes:

_____
_____
_____
_____
_____
_____
_____
_____
_____

Figure 9 ◆ Metallic shielding.

spects to Type FC flat conductor assemblies. Guidelines for the use of this system are given in *NEC Article 328.* Type FCC cable consists of three to five flat conductors placed edge-to-edge, separated, and enclosed in a moisture-resistant and flame-retardant insulating assembly. Accessories include cable connectors, terminators, power source adapters, and receptacles.

This wiring system has been designed to supply floor outlets in office areas and other commercial and institutional interiors. It is meant to be run under carpets so that no floor drilling is required. This system is also suitable for wall mounting. As indicated in *NEC Article 328,* telephone and other communications circuits may share the same enclosure as Type FCC flat cable. The main advantage of the system is its ease of installation. It is the ideal wiring system for use when remodeling or expanding existing office facilities.

2.7.9 Type TC Cable

Guidelines for the use of Type TC (power and control tray) cable are given in *NEC Article 340.* Type TC cable consists of two or more insulated conductors twisted together, with or without associated bare or fully insulated grounding conductors, and covered with a nonmetallic jacket. The cables are rated at 600 volts. The cable is listed in conductor sizes No. 18 AWG to 2,000 kcmil copper or No. 12 AWG to 2,000 kcmil aluminum or copper-clad aluminum (*Figure 13* on page 9–16).

As the *T* in the letter designator indicates, this cable is tray cable. It can be used in cable trays and raceways. It may also be buried directly if the

Discuss Types FC and FCC cable. Pass around examples for the trainees to examine.

Discuss Type TC cable. Pass around an example for the trainees to examine.

- Thermoplastic semiconductive jacket maintains uniform shield impedance.

- Six corrugated copper drain wires embedded in semiconductive jacket provide shielding, instead of tape shield, and allow for ripping back out of jacket to cut off jacket at correct point for stress cone assembly.

Cable insulation of ethylene-propylene rubber

CONDUCTOR

109F10.EPS

Figure 10 ◆ Nonmetallic shielding.

Discuss Type USE cable. Pass around an example for the trainees to examine.

Show Transparency 15 (Figure 14).

Discuss instrumentation control wiring. Pass around examples for the trainees to examine.

Show Transparency 16 (Figure 15A).

sheathing material is suitable for this use. Type TC cable is also good for use in sunlight when indicated by the cable markings.

2.7.10 Type USE Cable

Type USE cable is for underground installation including burial directly in the earth. Type USE cable in sizes No. 4/0 AWG and smaller with all conductors insulated is suitable for all of the underground uses for which Type UF cable is permitted by the NEC.

Type USE cable may consist of either single conductors or a multi-conductor assembly provided with a moisture-resistant covering, but it is not required to have a flame-retardant covering. This type of cable may have a bare copper conductor cabled with the assembly. Furthermore, Type USE single, parallel, or cabled conductor assemblies recognized for underground use may have a bare copper concentric conductor applied. These constructions do not require an outer overall covering. Guidelines for the use of Type USE cable are specified in *NEC Article 338*. See *Figure 14* on page 9–17.

When used as a service-entrance cable, Type USE cable must be installed as specified in *NEC Article*

230. Take the time to read *NEC Article 230* to ensure proper installation. Type USE service-entrance cable may also be used as feeder and branch circuit cable. Guidelines for this use of service-entrance cable are given in *NEC Section 338-3*.

2.8.0 Instrumentation Control Wiring

Instrumentation control wiring links the field-sensing, controlling, printout, and operating devices that form an electronic instrumentation control system. The style and size of instrumentation control wiring must be matched to a specific job.

Instrumentation control wiring usually has two or more insulated conductor wires. These wires may also have a shield and a ground wire. An outer layer called the *jacket* protects the wiring (*Figure 15* on page 9–18). Instrumentation conductor wires come in pairs. The number of pairs in a multi-conductor cable depends on the size of the wire used. A multi-pair cable may have as many as 60 pairs of conductor wires.

2.8.1 Shields

Shields are provided on instrumentation control wiring to protect the electrical signals traveling

ELECTRICAL LEVEL ONE—TRAINEE MODULE 26109

Instructor's Notes:

BASIC COMPONENTS

4-CONDUCTOR
CABLE ASSEMBLY

RACEWAY
(U-CHANNEL)

END VIEW
INSTALLED

TOP (PHASE 1)

INSULATING END CAP   FIXTURE HANGER   TERMINAL BLOCK FOR CHANNEL BOX

ACCESSORIES

109F11.EPS

Figure 11 ◆ Channel wire components and accessories.

through the conductors from electrical interference or noise. Shields are usually constructed of aluminum foil bonded to a plastic film (*Figure 16*). If the wiring is not properly shielded, electrical noise may cause erratic or erroneous control signals, false indications, and improper operation of control devices.

2.8.2 Grounding

A ground wire is a bare copper wire used to provide continuous contact with a specified grounding terminal. A ground wire allows connection of all the instruments within a loop to a common grounding system. In some electronic systems, the grounding wire is called a *drain wire*. Always refer to the loop diagram to determine whether or not the ground wire is to be terminated.

Usually, instrumentation is not grounded at both ends of the wire. This is to prevent unwanted ground loops in the system. If the ground is not to be connected at the end of the wire you are installing, do not remove the ground wire. Fold it back and tape it to the cable. This is called *floating the ground*. This is done in case the ground at the other end ever develops a problem. In that case, that ground can be removed and the wire taped to the cable can be untaped and installed.

2.8.3 Jackets

A plastic jacket covers and protects the components within the wire. Polyethylene (PE) and polyvinyl chloride (PVC) jackets are the most commonly used (*Figure 17* on page 9–19). Some jackets have a nylon rip cord that allows the jacket to be peeled back without the use of a knife or cable cutter. This eliminates nicking of the conductor insulation when preparing for termination.

Discuss shields, ground wires, and jackets.

Show Transparency 17 (Figure 16).

NEUTRAL CONDUCTOR

PHASE CONDUCTOR

CABLE SUPPORT CHANNEL

TYPE FC CABLE 4 CONDUCTORS

COLOR-CODED TAP-IN ADAPTER

SADDLE CLAMP

WELDED LOCKNUTS

BUSHING

LOCKNUT

CORD

4" SQUARE BOX

STRAIN-RELIEF CONNECTOR

GROUNDING CONNECTION

109F12.EPS

Figure 12 ◆ Type FC connection.

Classroom

Discuss types of fish tape and pass around examples for the trainees to examine.

Audiovisual

Show Transparencies 18 and 19 (Figures 18 and 19).

VERONICA WESTFALL   109F13.EPS

Figure 13 ◆ Type TC cable.

3.0.0 ◆ INSTALLING CONDUCTORS IN CONDUIT SYSTEMS

Conductors are installed in all types of conduit by pulling them through the conduit. This is done by using fish tape, pull lines, and pulling equipment.

3.1.0 Fish Tape

Fish tape can be made of flexible steel or nylon and is available in coils of 25 to 200 feet. It should be kept on a reel to avoid twisting. Fish tape has a hook or loop on one end to attach to the conductors to be pulled (*Figure 18* on page 9–19). Broken or damaged fish tape should not be used. To prevent electrical shock, fish tape should not be used near or in live circuits.

Fish tape is fed through the conduit from its reel. The tape usually enters at one outlet or junction box and is fed through to another outlet or junction box (*Figure 19* on page 9–20).

Instructor's Notes:

_____
_____
_____
_____
_____
_____
_____
_____
_____
_____

Figure 14 ◆ Type USE cable.

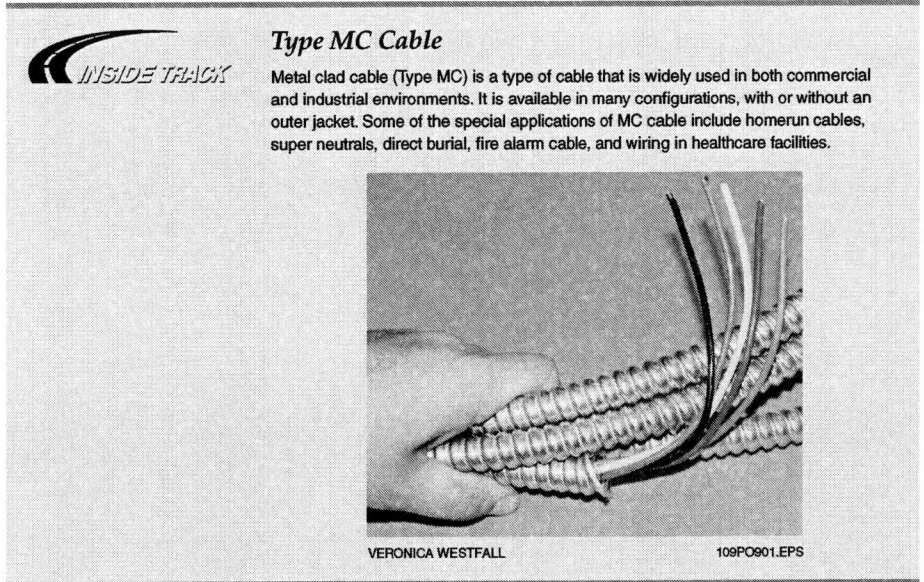

3.1.1 Power Conduit Fishing Systems

String lines can be installed by using different types of power systems. The power system is similar to an industrial vacuum cleaner and pulls a string or rope attached to a piston-like plug (sometimes called a mouse) through the conduit. Once the string emerges at the opposite end, either the conductor or a pull rope is then attached and pulled through the conduit, either manually or with power tools. See *Figure 20* on page 9–21.

The hose connection on these vacuum systems can also be reversed to push the mouse through the conduit as shown in *Figure 21* on page 9–21. In other words, the system can either suck or blow the mouse through the conduit, depending on which method is best in a given situation. In either case, a fish tape is then attached to the string for retrieving through the conduit.

3.1.2 Connecting Wire to a String Line

Once the string is installed in the conduit run, a fish tape is connected to it and pulled back through the conduit. Conductors are then attached to the hooked end of the fish tape or else

Discuss power fishing systems.

Show Transparencies 20 and 21 (Figures 20C and 21).

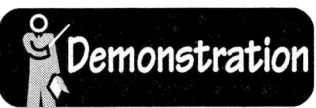

Demonstrate how to connect a wire to a string line.

CONDUCTORS—TRAINEE MODULE 26109

9–17

CONDUCTORS   CONDUCTORS

INSULATION

SHIELD

GROUND

JACKET

109F15A.EPS

VERONICA WESTFALL   109F15B.EPS

Figure 15 ◆ Instrumentation control wiring.

RIP CORD

DRAIN WIRE

PAIR SHIELD

CABLE SHIELD

109F16.EPS

Figure 16 ◆ Multi-conductor instrumentation control wiring with shields.

Instructor's Notes:

_____

_____

_____

_____

_____

_____

_____

_____

_____

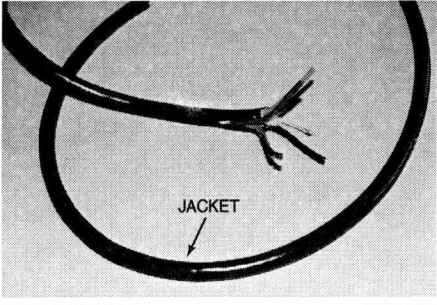

JACKET

VERONICA WESTFALL   109F17.EPS

Figure 17 ◆ Wire jacket.

HOOKED END

109F18.EPS

Figure 18 ◆ Fish tape.

INSIDE TRACK

Fish Tape Selection

Metal fish tape (A) generally comes in longer lengths and is the type used most often. Nylon fish tape (B) generally comes in shorter lengths and is used when fishing in areas where it is impossible to remove the power.

VERONICA WESTFALL   (A)   109PO902.EPS   VERONICA WESTFALL   (B)   109PO903.EPS

---

connected to a basket grip. In most cases, all required conductors are pulled at one time.

3.2.0 Wire Grips

Wire grips are used to attach the cable to the pull tape. One type of wire grip used is a basket grip (sometimes called *Chinese Fingers*). A basket grip is a steel mesh basket that slips over the end of a large wire or cable (*Figure 22* on page 9–22). The

fish tape hooks onto the end and the pull on the fish tape tightens the basket over the conductor.

3.3.0 Pull Lines

After the tape has been inserted into the conduit run, you must determine if the pull on the wire will be easy or difficult. If the pull is going to be difficult because of bends in the conduit or the

Classroom

Discuss the purpose of wire grips and pass around examples for the trainees to examine.

Discuss when a pull line should be used.

OUTLET
BOX

OUTLET
BOX

FISH
TAPE

CONDUCTOR
ATTACHES
TO HOOK

109F19.EPS

Figure 19 ◆ Fish tape installation.

size of the conductors, or if several conductors are to be pulled together, a pull line should be used.

A pull line is usually made of nylon or some other synthetic fiber. It is made with a factory-spliced eye for easy connection to fish tape or conductors.

 WARNING!
When using pull lines, exercise extreme caution and *never* stand in a direct line with the pulling rope. If the rope breaks, the line will whip back with great force. This can result in serious injury or death.

3.4.0 Safety Precautions

The following are several important safety precautions that will help to reduce the chance of being injured while pulling cable.

- To avoid electrical shock, never use fish tape near or in live circuits. If wires must be pulled into boxes that contain live circuits, use rubber blankets over the exposed live circuitry.

- Read and understand both the operating and safety instructions for the pull system before pulling cable.

- When moving reels of cable, avoid back strain by using your legs to lift (rather than your back) and asking for help with heavy loads. Also, when manually pulling wire, spread your legs to maintain your balance and do not stretch.

- Select a rope that has a pulling load rating greater than the estimated forces required for the pull.

- Use only low-stretch rope such as multiplex and double-braided polyester for cable pulling. High-stretch ropes store energy much like a stretched rubber band. If there is a failure of the rope, pulling grip, conductors, or any other component in the pulling system, this potential energy will suddenly be unleashed. The whipping action of a rope can cause considerable damage, serious injury, or death.

- Inspect the rope thoroughly before use. Make sure there are no cuts or frays in the rope. Remember, the rope is only as strong as its weakest point.

- When designing the pull, keep the rope confined in conduit wherever possible. Should the rope break or any other part of the pulling system fail, releasing the stored energy in the rope,

Safety

Review the safety precautions associated with pulling cable.

ELECTRICAL LEVEL ONE—TRAINEE MODULE 26109

Instructor's Notes:

ELECTRICAL LEVEL ONE—INSTRUCTOR'S GUIDE MODULE 26109

VERONICA WESTFALL   (A)   109F20A.EPS   VERONICA WESTFALL   (B)   109F20B.EPS

(C) FISHING SYSTEM IN
VACUUM MODE
                                    (C)   109F20C.EPS

Figure 20 ◆ Power fishing system: (A) vacuum/blower unit; (B) foam plugs; (C) fishing system in vacuum mode.

109F21.EPS

Figure 21 ◆ Power fishing system (blower mode).

Straightening a Bent Fish Tape

To straighten a bent fish tape, drive five 16-penny (16d) nails into a 2 × 4 about 1" apart in a straight line. Then wind the fish tape through the nails in a slalom fashion. This will straighten the tape.

109UA0902.EPS

VERONICA WESTFALL   109F22.EPS

Figure 22 ◆ Basket grip.

the confinement in the conduit will work against the whipping action of the rope by playing out much of this energy within the conduit.

• Do not stand in a direct line with the pulling rope.
• Wrap up the pulling rope after use to prevent others from tripping over it.

3.5.0 Pulling Equipment

Many types of pulling equipment are available to help pull conductors through conduit. Pulling equipment can be operated both manually and electrically. A manually operated puller is used mainly for smaller pulling jobs where hand pulling is not possible or practical (*Figure 23*). It is also used in many locations where hand pulling would put an unnecessary strain on the conductors because of the angle of the pull involved.

Electrically driven power pullers are used where long runs, several bends, or large conductors are involved (*Figure 24*).

The main parts of a power puller are the electric motor, the chain or sprocket drive, the capstan, the sheave, and the pull line.

109F23.EPS

Figure 23 ◆ Manual wire puller.

The pull line is routed over the sheave to ensure a straight pull. The pull line is wrapped around the capstan two or three times to provide a good grip on the capstan. The capstan is driven by the electric motor and does the actual pulling. The pull line is unwound by hand at the same speed at which the capstan is pulling. This eliminates the need for a large spool on the puller to wind the pull line.

Attachments to power pullers, such as special application sheaves and extensions, are available for most pulling jobs (*Figure 25*). Follow the manufacturer's instructions for setup and operation of the puller.

ELECTRICAL LEVEL ONE—TRAINEE MODULE 26109

Instructor's Notes:

_____
_____
_____
_____
_____
_____
_____
_____
_____

PULL LINE

CAPSTAN

CHAIN OR
SPROCKET DRIVE

SHEAVE

ELECTRIC
MOTOR

109F24.EPS

Figure 24 ◆ Power puller.

Replacing the Hook on a Metal Fish Tape

The hook on a fish tape can be replaced using a propane torch. IMPORTANT: Always work in an appropriate environment and wear the necessary protective equipment. Hold the fish tape securely in a pair of pliers, heat the end with a propane torch until it softens, then use a second set of pliers to form a hook in the fish tape. Allow it to air cool.

CAUTION

Before using power pullers, a qualified person must verify the amount of pull or tension that can be withstood by the conductors being pulled.

3.6.0 Feeding Conductors Into Conduit

After the fish tape or pull line is attached to the conductors, they must be pulled back through the conduit. As the fish tape is pulled, the attached conductors must be properly fed into the conduit.

Usually, more than one conductor is fed into the conduit during a wire pull. It is important to keep the conductors straight and parallel, and free from kinks, bends, and crossovers. Conductors that are allowed to cross each other will form a bulge and make pulling difficult. This could also damage the conductors.

Spools and rolls of conductors must be set up so that they unwind easily, without kinks and bends.

When several conductors must be fed into the conduit at the same time, a reel cart is used (*Figure 26*). The reel cart will allow the spools to turn freely and help prevent the wires from tangling.

3.7.0 Conductor Lubrication

When conductors are fed into long runs of conduit or conduit with several bends, the wires are lubricated with a compound designed for wire lubrication.

Several types of formulated compounds designed for wire lubrication are available in either

Emphasize the importance of verifying allowable conductor tension during a pull.

Discuss feeding conductors into conduit. Emphasize a slow and steady approach.

Discuss the use of reel carts.

Discuss conductor lubricants.

UP PULL THROUGH EXPOSED CONDUIT

DOWN PULL THROUGH EXPOSED CONDUIT

HORIZONTAL PULL WITH PIPE ADAPTER

FLUSH-MOUNTED PULL WITH PIPE ADAPTER

109F25.EPS

Figure 25 ◆ Power puller uses.

dry powder, paste, or gel form. These compounds must be noncorrosive to the insulation material of the conductor and to the conduit itself. The compounds are applied by hand to the conductors as they are fed into the conduit. Avoid excess lubrication.

3.8.0 Conductor Termination

The amount of free conductor at each junction or outlet box must meet certain NEC specifications.

For example, there must be sufficient free conductor so that bends or terminations inside the box, cabinet, or enclosure may be made to a radius as specified in the NEC. The NEC specifies a minimum of six inches for connections made to wiring devices or for splices. Where conductors pass through junction or pull boxes, enough slack should be provided for splices at a later date.

When a box is used as a pull box, the conductors are not necessarily spliced. They may merely

ELECTRICAL LEVEL ONE—TRAINEE MODULE 26109

Instructor's Notes:

ELECTRICAL LEVEL ONE—INSTRUCTOR'S GUIDE MODULE 26109

enter the pull box via one conduit run and exit via another conduit run. The purpose of a pull box, as the name suggests, is to facilitate pulling conductors on long runs. A junction box, however, is not only used to facilitate pulling conductors through the raceway system, but it also provides an enclosure for splices in the conductors.

VERONICA WESTFALL   109F26.EPS

Figure 26 ◆ Reel cart.

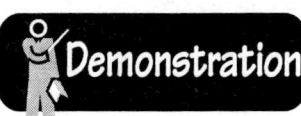

Demonstrate how to pull, feed, lubricate, and terminate a single cable.

Have the trainees practice pulling, feeding, lubricating, and terminating a single cable.

Cutting Cable

Large cable is best cut using cable cutters.

VERONICA WESTFALL   109PO905.EPS

Summary

Knowing the different types of conductors, how they are rated, and what they are used for will help you when selecting conductors for a specific job. Pulling wires and cables through conduit systems is an important part of your job as an electrician. The more you learn about the concepts involved in pulling cable, the safer and more efficient you will be.

Cleanup

A professional always takes the time to clean up the work area, removing excess lubricant from the boxes, bushings, and conductors. Completing a job in a professional manner is not only good business, it is also a NEC requirement (*NEC Section 110-12*).

Discuss the "Think About It."

Putting It All Together

Think about the design of conductor installations. How does the location of pull points affect the ease of the pull?

Classroom

Have the trainees complete the Review Questions and go over the answers prior to administering the Module Examination.

Examination

Administer the Module Examination and Performance Test. Be sure to record the testing results on Craft Training Report Form 200 and submit the results to the Training Program Sponsor.

Review Questions

1. Where can the ampacity ratings of conductors be found in the NEC?
   a. *NEC Chapter 1*
   b. *NEC Articles 348 through 352*
   c. *NEC Tables 310-16 through 310-19*
   d. *NEC Chapter 9*

2. How many wires are contained in the three configurations of stranded cable?
   a. 7, 19, and 37
   b. 5, 7, and 9
   c. 3, 5, and 7
   d. 6, 18, and 36

3. What is the purpose of stranding?
   a. To stiffen the wire
   b. To increase the conductivity of the wire
   c. To decrease the conductivity of the wire
   d. To increase the flexibility of the wire

4. What is the most common conductor material?
   a. Silver
   b. Aluminum
   c. Copper
   d. Platinum

5. Which of the following does *not* help to determine how good a material will be for the construction of wire?
   a. Availability
   b. Cost
   c. Conductivity
   d. Molecular weight

6. What letter must be included in the marking of a conductor if it is to be used in a wet, outdoor application?
   a. D
   b. O
   c. I
   d. W

7. What service conductor may be green or green with a yellow stripe?
   a. The grounding conductor of a multi-conductor cable
   b. The neutral conductor of a multi-conductor cable
   c. The hot conductor of a multi-conductor cable
   d. Both a and b.

8. What are the colors of insulation on the conductors for a three-conductor cable?
   a. One white, one red, and one black
   b. Two white and one red
   c. One white, one red, one black, and a grounding conductor
   d. One green, one white, one blue, and a grounding conductor

9. Type NMC cable is suitable for _____ locations.
   a. dry
   b. damp
   c. corrosive
   d. All of the above.

10. Type USE cable can be used for _____.
    a. above-ground installation only
    b. underground installation within a special PVC pipeline
    c. underground installation including direct burial
    d. None of the above.

Instructor's Notes:

_____

_____

_____

_____

_____

_____

_____

_____

Matthew Fine, Elwood Electric

Matthew Fine twice competed in the National Craft Olympics during his apprentice training, winning first place in 1996. His story shows that youth is no barrier to advancement in the industry.

How did you choose a career in the electrical field?
My father was an electrician and I often helped him during the summer. Through that experience and the courses I had in high school, I became very interested in electrical control systems, especially motor controls.

What was your apprenticeship experience like?
The training I received during my four-year apprenticeship helped me to understand the principles behind the work I was doing on the job. Once you understand the principles, any task becomes easier. I should mention that I still think of myself as an apprentice because I am learning all the time.

The first time I competed in the Craft Olympics in 1995, I didn't place, but I did learn a lot about handling pressure. The next year I was able to apply what I learned and came away with a first-place finish.

Tell us about your present job.
I do residential and commercial installation and repair work, but I also work as a foreman, designer, estimator, and instructor. In a large company, there are different people for all of these jobs, but if you're going to succeed in a small company, you have to be able to wear more than one hat.

What factor contributed the most to your success?
My employer, Brad Elwood, gave me the opportunity to succeed. He has been my mentor in the business since high school. Without his help and direction, I would not be where I am today or have the attitude I have now.

What advice would you give to apprentice trainees?
Never pass up an opportunity to learn something new. The field changes constantly. If you don't keep up with it, you will be left behind. I myself plan to stay as far ahead as I can.

Terms Introduced
in This Module

Ampacity: The current in amperes a conductor can carry continuously under the conditions of use without exceeding its temperature rating.

Capstan: The turning drum of the cable puller on which the rope is wrapped and pulled.

Fish tape: A hand device used to pull a wire through a conduit run.

Mouse: A cylinder of foam rubber that fits inside the conduit and is then propelled by compressed air or vacuumed through the conduit run, pulling a line or tape.

Wire grip: A device used to link pulling rope to cable during a pull.

Instructor's Notes:

_____

_____

_____

_____

_____

_____

_____

_____

_____

Answers to Review Questions

Answer	Section
1. c	2.1.0
2. a	2.3.2
3. d	2.3.2
4. c	2.4.0
5. d	2.4.0
6. d	2.5.2
7. a	2.5.3
8. c	2.5.3
9. d	2.7.2
10. c	2.7.10

Additional Resources

For advanced study of topics covered in this task module, the following book is suggested:

National Electrical Code Handbook, Latest Edition. Quincy, MA: National Fire Protection Association.

Instructor's Notes:

_____

_____

_____

_____

_____

_____

_____

_____

_____

NCCER CRAFT TRAINING USER UPDATES

The NCCER makes every effort to keep these textbooks up-to-date and free of technical errors. We appreciate your help in this process. If you have an idea for improving this textbook, or if you find an error, a typographical mistake, or an inaccuracy in the NCCER's Craft Training textbooks, please write us, using this form or a photocopy. Be sure to include the exact module number, page number, a detailed description, and the correction, if applicable. Your input will be brought to the attention of the Technical Review Committee. Thank you for your assistance.

Instructors – If you found that additional materials were necessary in order to teach this module effectively, please let us know so that we may include them in the Equipment/Materials list in the Instructor's Guide.

Write: Curriculum Revision and Development Department
National Center for Construction Education and Research
P.O. Box 141104, Gainesville, FL 32614-1104

Fax: 352-334-0932

E-mail: curriculum@nccer.org

Craft _____ Module Name _____

Copyright Date _____ Module Number _____ Page Number(s) _____

Description _____

_____

_____

_____

_____

(Optional) Correction _____

_____

_____

(Optional) Your Name and Address _____

_____

_____

_____

TEACHING TIPS

Section 2.1.0 *Think About It—Overheating*

Insulation can become overheated as a result of excessive heat generated in the wire. When overheated, insulation becomes discolored. Eventually, it will melt or otherwise disintegrate and expose the enclosed wire. Factors which affect the ampacity of an insulated conductor are the physical size of the wire, the material the wire is made of, and the ambient temperature surrounding the wire.

Section 2.3.2 *Think About It—Wire Size*

The size of the wires used in a specific circuit depends on how much current the wires must conduct, how far they must conduct it, and in what type of environment or location the wires are to be placed. Use of too small a wire can result in overheating of both the wire and the load device(s) connected to the wire. This can cause poor operation of the load devices. If the overheating is excessive, it can severely damage both the wiring and load devices and may even cause a fire. Use of a wire size larger than needed only adds expense to an installation from which no benefit is derived.

Section 2.5.0 *Think About It—Conductor Insulation*

In addition to protection from shock, insulation on a wire provides corrosion protection and protection for the wires when pulling wires, and the insulation's color code helps in wire identification. Ground wires are often left bare. In addition, *NEC Section 230-41* allows the use of an uninsulated conductor in interior service-entrance cable in certain installations. *NEC Section 230-30* allows a grounded service-lateral conductor to be uninsulated under certain conditions.

Section 2.5.2 *Think About It—Insulation Types*

NEC Table 310-13 shows many cables with insulation suitable for use in wet locations. Two common types are THW and THWN.

Section 2.5.3 *Think About It—Color Coding Ungrounded Conductors*

Color coding of ungrounded conductors provides a method of identifying hot wires in a system and distinguishing between the different hot wires in multi-phase systems. Color coding also provides for standardization in wiring systems and aids when troubleshooting.

Section 3.8.0 *Think About It—Putting It All Together*

There is a direct relationship between the location and number of pull points provided in a conductor installation and how easy (or difficult) it will be to pull the cables. Using a generous number of pull points located in areas where there is adequate space to use pulling equipment and to maneuver while working will help to make the pulling process much easier.

Introduction to Electrical Blueprints

26110-00

MODULE OVERVIEW

This course introduces the electrical trainee to electrical blueprint reading, including the interpretation of various types of scales, drawings, schedules, and specifications.

PREREQUISITES

Please refer to the Course Map in the Trainee Module. Prior to training with this module, it is recommended that the trainee shall have successfully completed the following modules:

Core Curriculum; Electrical Level One, Modules 26101 through 26109

LEARNING OBJECTIVES

Upon completion of this module, the trainee will be able to:

1. Explain the basic layout of a blueprint.
2. Describe the information included in the title block of a blueprint.
3. Identify the types of lines used on blueprints.
4. Identify common symbols used on blueprints.
5. Understand the use of architect's and engineer's scales.
6. Interpret electrical drawings, including site plans, floor plans, and detail drawings.
7. Read equipment schedules found on electrical blueprints.
8. Describe the type of information included in electrical specifications.

PERFORMANCE OBJECTIVES

Under supervision of the instructor, the trainee should be able to:

1. Using Sheet 2 of Transparency 2, make a material takeoff of the lighting fixtures specified in Sheet 3 of Transparency 2. The takeoff requires that all lighting fixtures be counted, and where applicable, the total number of lamps for each fixture type must be calculated.

   Answer:

   Lighting Fixture Takeoff

Lighting Fixture Type	Manufacturer and Catalog Number	Number and Type of Lamps	Total Number of Fixtures	Total Number of Lamps for Fixture Type
FA	Lithonia LB 440	4-F40CS	59	236
FB	Lithonia LB 240	2-F20U	13	26
FC	Lithonia LP/RFB-3	INCL.	13	N/A
A	Hitek TWP 150	1-450HPS	4	4
EX	Lithonia XSIG-EL	INCL.	5	N/A
EM	Lithonia ELU-2	INCL.	2	N/A

2. Demonstrate the use of an architect's scale.

NCCER STANDARDIZED CRAFT TRAINING PROGRAM

The National Center for Construction Education and Research (NCCER) provides a standardized national program of accredited craft training. Key features of the program include instructor certification, competency-based training, and performance testing. The program provides trainees, instructors, and companies with a standard form of recognition through a National Craft Training Registry. The program is described in full in the *Guidelines for Accreditation*, published by the NCCER. For more information on standardized craft training, contact the NCCER at P.O. Box 141104, Gainesville, FL 32614-1104, 352-334-0911, visit our Web site at www.NCCER.org, or e-mail info@NCCER.org.

HOW TO USE THIS ANNOTATED INSTRUCTOR'S GUIDE

Each page presents two sections of information. The larger section displays each page exactly as it appears in the Trainee Module. The narrow column ties suggested trainee and instructor actions to each page and provides icons to call your attention to material, safety, audiovisual, or testing requirements. The bottom of each page includes space for your notes.

 If you see the Teaching Tip icon, that means there is a teaching tip associated with this section. Also refer to the suggested teaching tips at the end of the module.

SAFETY CONSIDERATIONS

Ensure that the trainees are equipped with appropriate personal protective equipment.

PREPARATION

Before teaching this module, you should review the Module Outline, Learning and Performance Objectives, and the Materials and Equipment List. Be sure to allow ample time to prepare your own training or lesson plan and gather all required equipment and materials.

MATERIALS AND EQUIPMENT LIST

Materials:
Transparencies

Markers/chalk

Copy of the latest edition of the *National Electrical Code*®

Module Examinations*

Performance Profile Sheets*

Equipment:
Overhead projector and screen

Whiteboard/chalkboard

Appropriate personal protective equipment

Architect's scales (both flat and triangular)

Engineer's scale

Set of electrical drawings

*Located in the Test Booklet packaged with this Annotated Instructor's Guide.

ADDITIONAL RESOURCES

This module is intended to present thorough resources for task training. The following reference works are suggested for both instructors and motivated trainees interested in further study. These are optional materials for continued education rather than for task training.

American Electrician's Handbook, Latest Edition. New York, NY: McGraw-Hill.

Illustrated Guide to the NEC, Latest Edition. Carlsbad, CA: Craftsman Book Company.

National Electrical Code Handbook, Latest Edition. Quincy, MA: National Fire Protection Association.

NOTES

The designations "National Electrical Code," "NE Code," and "NEC," where used in this document, refer to the *National Electrical Code*®, which is a registered trademark of the National Fire Protection Association, Quincy, MA. All National Electrical Code (NEC) references in this module refer to the 1999 edition of the NEC.

If you feel that additional math instruction would be helpful, Prentice Hall offers a basic math textbook entitled *Fundamentals of Electrical and Mechanical Mathematics*. It covers the basic math requirements for electrical trainees and may be ordered by contacting Prentice Hall Customer Service at 1-800-922-0579.

TEACHING TIME FOR THIS MODULE

An outline for use in developing your lesson plan is presented below. Note that each Roman numeral in the outline equates to one session of instruction. Each session has a suggested time period of 2½ hours. This includes 10 minutes at the beginning of each session for administrative tasks and one 10-minute break during the session. Approximately 7½ hours are suggested to cover *Introduction to Electrical Blueprints*. You will need to adjust the time required for hands-on activity and testing based on your class size and resources.

Topic   **Planned Time**

Session I: Introduction to Blueprint Reading; Blueprint Layout; Drafting Lines; Electrical Symbols; Scale Drawings

    A. Introduction to Blueprint Reading   _____

       1. Site Plan   _____

       2. Floor Plans   _____

       3. Elevations   _____

       4. Sections   _____

       5. Electrical Drawings   _____

    B. Blueprint Layout   _____

       1. Title Block   _____

       2. Approval Block   _____

       3. Revision Block   _____

    C. Drafting Lines   _____

       1. Electrical Drafting Lines   _____

    D. Electrical Symbols   _____

    E. Scale Drawings   _____

       1. Architect's Scale   _____

           a. Types of Architect's Scales   _____

       2. Engineer's Scale   _____

       3. Metric Scale   _____

Session II. Analyzing Electrical Drawings; Power Plans; Lighting Floor Plan; Electrical Details and Diagrams

    A. Analyzing Electrical Drawings   _____

       1. Development of Site Plans   _____

    B. Power Plans   _____

       1. Key Plan   _____

       2. Symbol List   _____

       3. Floor Plan   _____

           a. Notes and Building Symbols   _____

           b. Busways   _____

    C. Lighting Floor Plan   _____

       1. Drawing Schedules   _____

    D. Electrical Details and Diagrams   _____

       1. Power-Riser Diagrams   _____

       2. Schematic Diagrams   _____

       3. Drawing Details   _____

Session III. Written Specifications; Module Examination and Performance Testing

    A. Written Specifications   _____

       1. How Specifications Are Written   _____

       2. Format of Specifications   _____

          a. CSI Format   _____

    B. Module Examination   _____

       1. Trainees must score 70% or higher to receive recognition from the NCCER.

       2. Record the testing results on Craft Training Report Form 200 and submit the results to the Training Program Sponsor.

    C. Performance Testing   _____

       1. Trainees must perform each task to the satisfaction of the instructor to receive recognition from the NCCER.

       2. Record the testing results on Craft Training Report Form 200 and submit the results to the Training Program Sponsor.

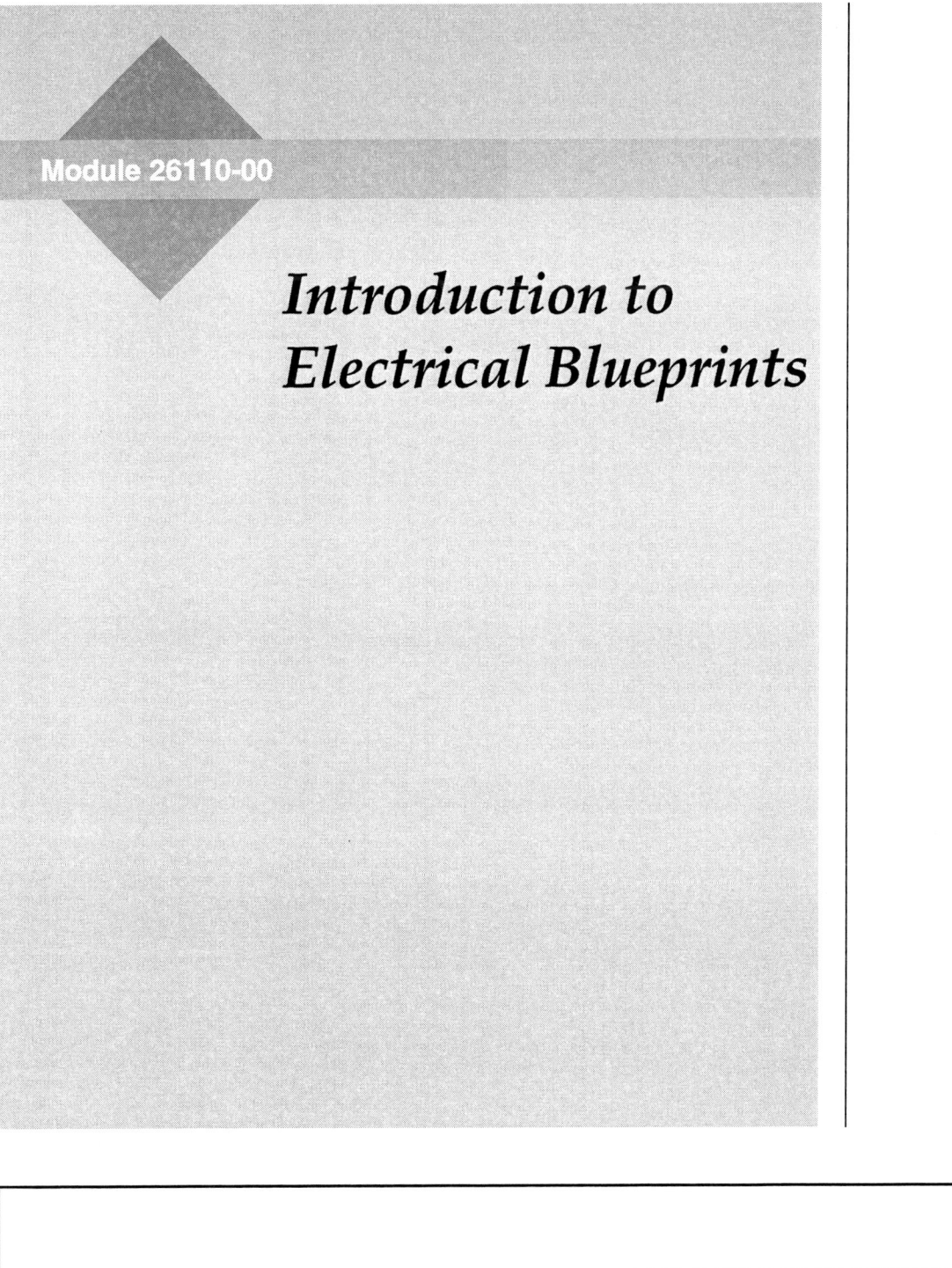

Module 26110-00

Introduction to Electrical Blueprints

Course Map

This course map shows all of the task modules in the first level of the Electrical curricula. The suggested training order begins at the bottom and proceeds up. Skill levels increase as a trainee advances on the course map. The training order may be adjusted by the local Training Program Sponsor.

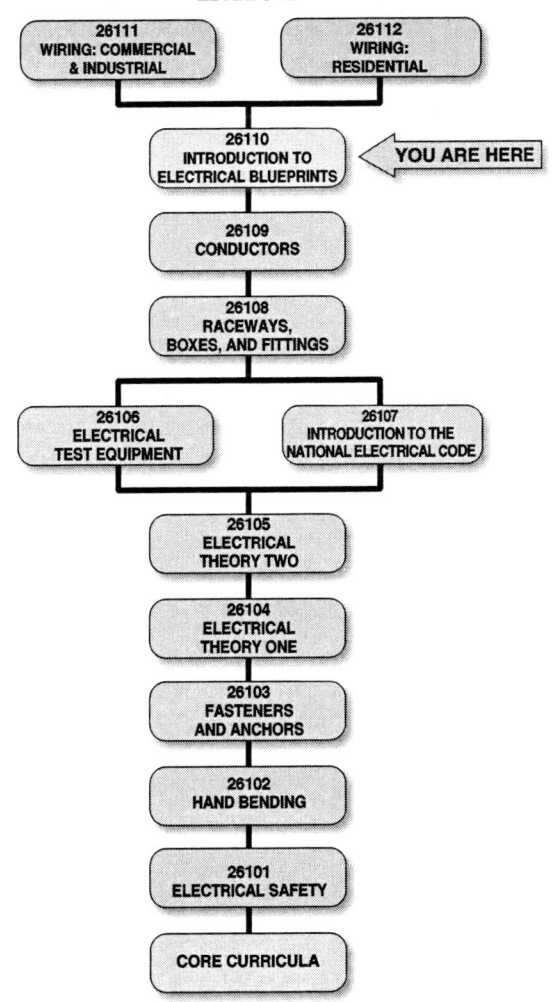

LEVEL ONE

26111
WIRING: COMMERCIAL & INDUSTRIAL

26112
WIRING: RESIDENTIAL

26110
INTRODUCTION TO ELECTRICAL BLUEPRINTS ◁ YOU ARE HERE

26109
CONDUCTORS

26108
RACEWAYS, BOXES, AND FITTINGS

26106
ELECTRICAL TEST EQUIPMENT

26107
INTRODUCTION TO THE NATIONAL ELECTRICAL CODE

26105
ELECTRICAL THEORY TWO

26104
ELECTRICAL THEORY ONE

26103
FASTENERS AND ANCHORS

26102
HAND BENDING

26101
ELECTRICAL SAFETY

CORE CURRICULA

110CMAP.EPS

Assign reading of Module 26110.

Instructor's Notes:

_____

_____

_____

_____

_____

_____

_____

_____

_____

Figures

Instructor's Notes:

_____

_____

_____

_____

_____

_____

_____

_____

_____

Materials

Ensure you have everything required to teach the course. Check the Materials and Equipment List at the front of this Instructor's Guide.

Introduction to Electrical Blueprints

Objectives

Upon completion of this module, the trainee will be able to:

1. Explain the basic layout of a blueprint.
2. Describe the information included in the title block of a blueprint.
3. Identify the types of lines used on blueprints.
4. Identify common symbols used on blueprints.
5. Understand the use of architect's and engineer's scales.
6. Interpret electrical drawings, including site plans, floor plans, and detail drawings.
7. Read equipment schedules found on electrical blueprints.
8. Describe the type of information included in electrical specifications.

Prerequisites

Successful completion of the following Task Modules is recommended before beginning study of this Task Module: Core Curriculum; Electrical Level One, Modules 26101 through 26109.

Required Trainee Materials

1. Trainee Task Module
2. Copy of the latest edition of the *National Electrical Code*
3. Appropriate Personal Protective Equipment

1.0.0 ◆ INTRODUCTION TO BLUEPRINT READING

In all large construction projects and in many of the smaller ones, an architect is commissioned to prepare complete working drawings and specifications for the project. These drawings usually include:

- A site plan indicating the location of the building on the property.
- Floor plans showing the walls and partitions for each floor or level.
- Elevations of all exterior faces of the building.
- Several vertical cross sections to indicate clearly the various floor levels and details of the footings, foundation, walls, floors, ceilings, and roof construction.
- Large-scale detail drawings showing such construction details as may be required.

For projects of any consequence, the architect usually hires consulting engineers to prepare structural, electrical, and mechanical drawings, with the latter encompassing pipe-fitting, instrumentation, plumbing, heating, ventilating, and air conditioning drawings.

1.1.0 Site Plan

This type of plan of the building site looks as if the site is viewed from an airplane and shows the property boundaries, the existing contour lines,

Note: The designations "National Electrical Code," "NE Code," and "NEC," where used in this document, refer to the *National Electrical Code®*, which is a registered trademark of the National Fire Protection Association, Quincy, MA. *All National Electrical Code (NEC) references in this module refer to the 1999 edition of the NEC.*

10–1

Show Transparency 1, Course Objectives.

Show Transparency 2, Performance Profile Tasks.

Explain that items shown in bold (blue) are defined in the Glossary at the back of this module.

Discuss the types of drawings typically included in the blueprint set for a project.

Discuss the types of information conveyed by site plans.

Discuss the types of information conveyed by floor plans.

Show Transparencies 4 through 6 (Figures 2 and 3).

the new contour lines (after grading), the location of the building on the property, new and existing roadways, all utility lines, and other pertinent details. The drawing scale is also shown. Descriptive notes may also be found on the site (plot) plan listing names of adjacent property owners, the land surveyor, and the date of the survey. A legend or symbol list is also included so that anyone who must work with the site plan can readily read the information. See *Figure 1*.

1.2.0 Floor Plans

The plan view of any object is a drawing showing the outline and all details as seen when looking directly down on the object. It shows only two dimensions, length and width. The floor plan of a building is drawn as if a horizontal cut were made through the building—at about window height—and then the top portion removed to reveal the bottom part. See *Figure 2*.

SCALE: 1" = 20'0"

110F01.EPS

Figure 1 ◆ Typical site plan.

Instructor's Notes:

PERSPECTIVE VIEW
SHOWING SECTION CUTS

TOP HALF OF SECTION
REMOVED

RESULTING FLOOR PLAN
IS WHAT THE REMAINING
STRUCTURE LOOKS LIKE
WHEN VIEWED FROM
ABOVE

Figure 2 ◈ Principles of floor plan layout.

110F02.EPS

If a plan view of a home's basement is needed, the part of the house above the middle of the basement windows is imagined to be cut away. By looking down on the uncovered portion, every detail and partition can be seen. Likewise, imagine the part above the middle of the first floor windows being cut away. A drawing that looks straight down at the remaining part would be called the *first floor plan* or *lower level*. A cut through the second floor windows would be called the *second floor plan* or *upper level*. See *Figure 3*.

1.3.0 Elevations

The *elevation* is an outline of an object that shows heights and may show the length or width of a particular side, but not depth. *Figures 4* and *5* show elevation drawings for a building.

> **Note**
> These elevation drawings show the heights of windows, doors, and porches, the pitch of roofs, etc., because all of these measurements cannot be shown conveniently on floor plans.

1.4.0 Sections

A section or sectional view (*Figure 6*) is a cutaway view that allows the viewer to see the inside of a structure. The point on the plan or elevation showing where the imaginary cut has been made is indicated by the section line, which is usually a dashed line. The section line shows the location of the section on the plan or elevation. It is necessary to know which of the cutaway parts is represented in the sectional drawing. To show this, arrow points are placed at the ends of the section lines.

In architectural drawings, it is often necessary to show more than one section on the same

Discuss the types of information conveyed by elevation drawings.

Show Transparencies 7 and 8 (Figures 4 and 5).

Discuss the types of information conveyed by section views.

Show Transparency 9 (Figure 6).

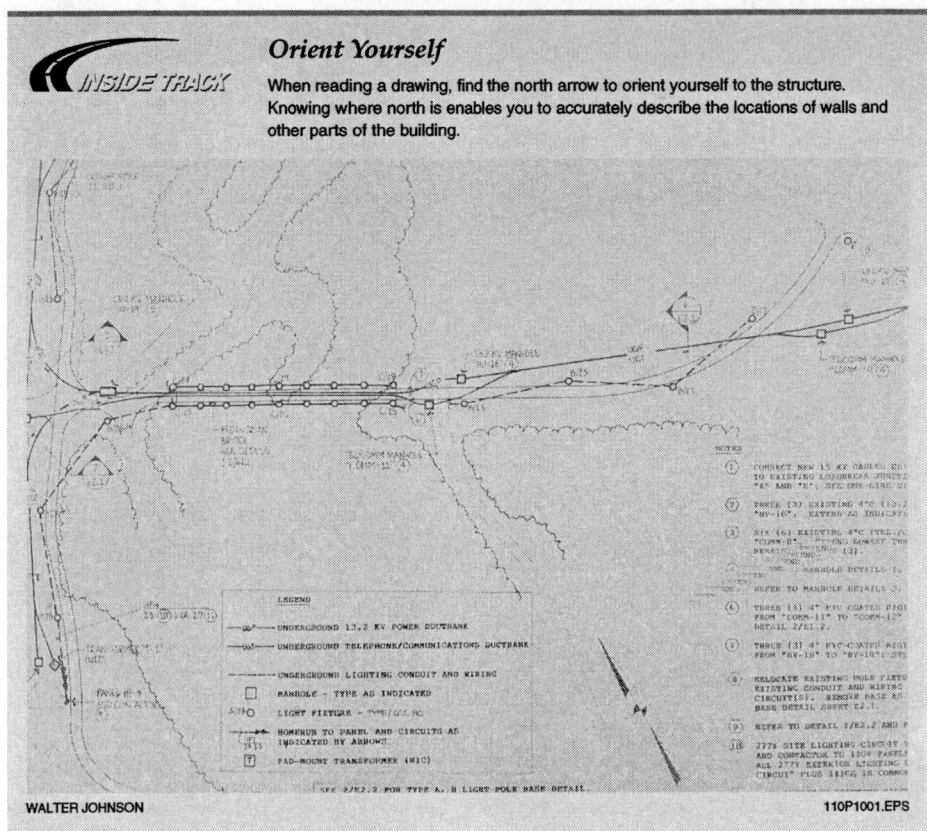

Orient Yourself

INSIDE TRACK When reading a drawing, find the north arrow to orient yourself to the structure. Knowing where north is enables you to accurately describe the locations of walls and other parts of the building.

WALTER JOHNSON   110P1001.EPS

Explain the difference between a longitudinal section and a cross section.

drawing. The different section lines must be distinguished by letters, numbers, or other designations placed at the ends of the lines. These section letters are generally large so as to stand out on the drawings. To further avoid confusion, the same letter is usually placed at each end of the section line. The section is named according to these letters (e.g., Section A-A, Section B-B, and so forth).

A longitudinal section is taken lengthwise while a cross section is usually taken straight across the width of an object. Sometimes, however, a section is not taken along one straight line. It is often taken along a zigzag line to show important parts of the object.

A sectional view, as applied to architectural drawings, is a drawing showing the building, or portion of a building, as though it were cut through on some imaginary line. This line may be either vertical (straight up and down) or horizontal. Wall sections are nearly always made vertically so that the cut edge is exposed from top to bottom. In some ways, the wall section is one of the most important of all the drawings to construction workers, because it answers the questions as to how a structure should be built. The floor plans of a building show how each floor is arranged, but the wall sections tell how each part is constructed and usually indicate the material to be used. The electrician needs to know this information when determining wiring methods that comply with the NEC.

ELECTRICAL LEVEL ONE—TRAINEE MODULE 26110

Instructor's Notes:

FLOOR PLAN

UPPER LEVEL

110F03A.EPS

LOWER LEVEL

110F03B.EPS

Figure 3 ◆ Floor plans of a building.

FRONT ELEVATION

REAR ELEVATION

110F04.EPS

Figure 4 ◆ Front and rear elevations.

Discuss the "Think About It."

Using All of the Drawings

Look back over the information on floor plans, elevations, and sections. What kinds of information would an electrician get from each of these drawings? What could a sectional drawing show that a floor plan could not?

Interpreting Electrical Drawings

A good example of when an electrician must interpret the drawings is when wiring a log cabin. The drawings will show the receptacle and switch locations in branch circuits as usual, but the electrician must figure out how to route wires and install boxes where there is no hollow wall and sometimes no ceiling space.

Instructor's Notes:

_____

_____

_____

_____

_____

_____

_____

_____

LEFT ELEVATION

RIGHT ELEVATION

110F05.EPS

Figure 5 ◆ Left and right elevations.

1.5.0 Electrical Drawings

Electrical drawings show in a clear, concise manner exactly what is required of the electricians. The amount of data shown on such drawings should be sufficient, but not overdone. This means that a complete set of electrical drawings could consist of only one 8½" × 11" sheet, or it could consist of several dozen 24" × 36" (or larger) sheets, depending on the size and complexity of a given project. A shop drawing, for example, may contain details of only one piece of equipment, while a set of working drawings for an industrial installation may contain dozens of

Classroom

Describe the types of information conveyed by electrical drawings.

PLAN

PLAN SECTION C

SECTION A

DETAIL SECTION A

CUTTING PLANE

110F06.EPS

Figure 6 ◆ Sectional drawing.

Instructor's Notes:

_____
_____
_____
_____
_____
_____
_____
_____
_____

drawing sheets detailing the electrical system for lighting and power, along with equipment, motor controls, wiring diagrams, schematic diagrams, equipment schedules, and a host of other pertinent data.

In general, the electrical working drawings for a given project serve three distinct functions:

- They provide electrical contractors with an exact description of the project so that materials and labor may be estimated to project a total cost of the project for bidding purposes.
- They provide workers on the project with instructions as to how the electrical system is to be installed.
- They provide a map of the electrical system once the job is completed to aid in maintenance and troubleshooting for years to come.

Electrical drawings from consulting engineering firms will vary in quality from sketchy, incomplete drawings to neat, precise drawings that are easy to understand. Few, however, will cover every detail of the electrical system. Therefore, a good knowledge of installation practices must go hand-in-hand with interpreting electrical working drawings.

Sometimes electrical contractors will have electrical drafters prepare special supplemental drawings for use by the contractors' employees. On certain projects, these supplemental drawings can save supervision time in the field once the project has begun.

2.0.0 ◆ BLUEPRINT LAYOUT

Although a strong effort has been made to standardize drawing practices in the building construction industry, blueprints prepared by different architectural or engineering firms will rarely be identical. Similarities, however, will exist between most sets of blueprints, and with a little experience, you should have no trouble interpreting any set of drawings that might be encountered.

Most drawings used for building construction projects will be drawn on sheets ranging from 11" × 17" to 24" × 36" in size. Each drawing sheet will have border lines framing the overall drawing and one or more title blocks, as shown in *Figure 7*. The type and size of title blocks varies with each firm preparing the drawings. In addition, some drawing sheets will also contain a revision block near the title block, and perhaps an approval block. This information is normally found on each drawing sheet, regardless of the type of project or the information contained on the sheet.

Discuss the layout of a basic blueprint.

Show Transparency 10 (Figure 7).

Figure 7 ◆ Typical blueprint layout.

Discuss the types of information conveyed by a title block.

Show Transparencies 11 and 12 (Figures 8 and 9).

Discuss the types of information conveyed by approval blocks.

Show Transparencies 13 and 14 (Figures 10 and 11).

Using a Drawing Set

Always treat a drawing set with care. It is best to keep two sets, one for the office and one for field use. After you use a sheet from a set of drawings, be sure to refold the sheet with the title block facing up.

2.1.0 Title Block

The architect's title block for a blueprint is usually boxed in the lower right-hand corner of the drawing sheet; the size of the block varies with the size of the drawing and with the information required. See *Figure 8.*

In general, the title block of an electrical drawing should contain the following information:

- Name of the project
- Address of the project
- Name of the owner or client
- Name of the architectural firm
- Date of completion
- Scale(s)
- Initials of the drafter, checker, and designer, with dates under each
- Job number
- Sheet number
- General description of the drawing

Every architectural firm has its own standard for drawing titles, and they are often preprinted directly on the tracing paper or else printed on a sticker, which is placed on the drawing.

Often, the consulting engineering firm will also be listed, which means that an additional title block will be applied to the drawing, usually next to the architect's title block. *Figure 9* shows completed architectural and engineering title blocks as they appear on an actual drawing.

2.2.0 Approval Block

The approval block, in most cases, will appear on the drawing sheet as shown in *Figure 10.* The various types of approval blocks (drawn, checked, etc.) will be initialed by the appropriate personnel. This type of approval block is usually part of the title block and appears on each drawing sheet.

On some projects, authorized signatures are re-

Figure 8 ◆ Typical architect's title block.

Check the Title Block

Always refer to the title block for scale and be aware that it often changes from drawing to drawing within the same set. Also check for the current revision, and when you replace a drawing with a new revision, be sure to remove and file the older version.

ELECTRICAL LEVEL ONE—TRAINEE MODULE 26110

Instructor's Notes:

_____

_____

_____

_____

_____

_____

_____

_____

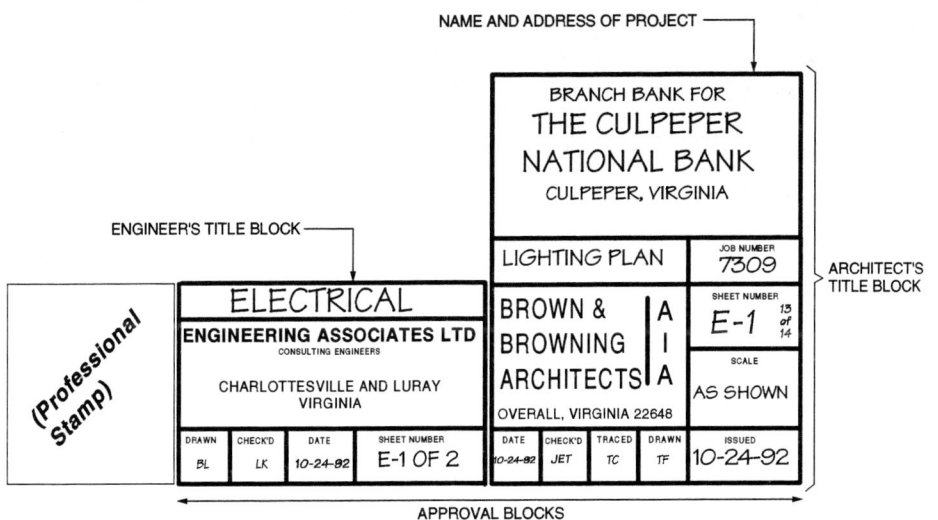

NAME AND ADDRESS OF PROJECT

ENGINEER'S TITLE BLOCK

ARCHITECT'S TITLE BLOCK

APPROVAL BLOCKS

110F09.EPS

Figure 9 ◆ Title blocks.

quired before certain systems may be installed, or even before the project begins. An approval block such as the one shown in *Figure 11* indicates that all required personnel have checked the drawings for accuracy, and that the set meets with everyone's approval. Such an approval block usually appears on the front sheet of the blueprint set and may include:

110F10.EPS

Figure 10 ◆ Typical approval block.

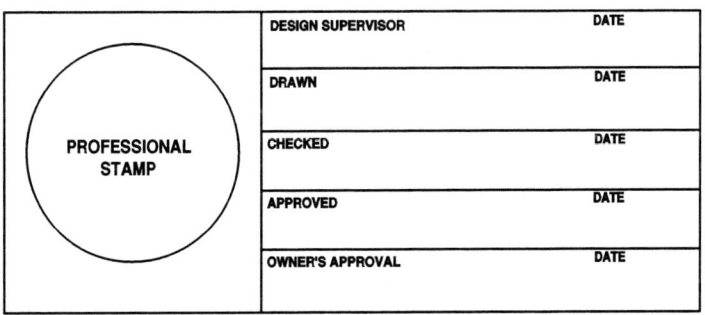

110F11.EPS

Figure 11 ◆ Alternate approval block.

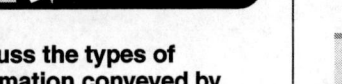

Discuss the types of information conveyed by revision blocks.

Show Transparencies 15 and 16 (Figures 12 and 13).

- *Professional stamp*—Registered seal of approval by the architect or consulting engineer.
- *Design supervisor*—Signature of the person who is overseeing the design.
- *Drawn (by)*—Signature or initials of the person who drafted the drawing and the date it was completed.
- *Checked (by)*—Signature or initials of the person who reviewed the drawing and the date of approval.

CAUTION

When a set of electrical working drawings has been revised, always make certain that the most up-to-date set is used for all future layout work. Either destroy the old, obsolete set of drawings or else clearly mark on the affected sheets, *Obsolete Drawing—Do Not Use.* Also, when working with a set of working drawings and written specifications for the first time, thoroughly check each page to see if any revisions or modifications have been made to the originals. Doing so can save much time and expense to all concerned with the project.

- *Approved*—Signature or initials of the architect/engineer and the date of the approval.
- *Owner's approval*—Signature of the project owner or the owner's representative along with the date signed.

2.3.0 Revision Block

Sometimes electrical drawings will have to be partially redrawn or modified during the construction of a project. It is extremely important that such modifications are noted and dated on the drawings to ensure that the workers have an up-to-date set of drawings to work from. In some situations, sufficient space is left near the title block for dates and descriptions of revisions, as shown in *Figure 12*. In other cases, a revision block is provided (again, near the title block), as shown in *Figure 13*.

Note

Architects, engineers, designers, and drafters have their own methods of showing revisions, so expect to find deviations from those shown here.

Instructor's Notes:

_____

_____

_____

_____

_____

_____

_____

_____

_____

REVISIONS
10/12/92 - REVISED LIGHTING FIXTURE
NO. 3 IN. LIGHTING FIXTURE SCHEDULE

Professional Stamp	ELECTRICAL
	DISTRICT HOME
	LAUNDRY BUILDING
	AUGUSTA COUNTY, VIRGINIA

G. LEWIS CRAIG, ARCHITECT				SHEET NO.
WAYNESBORO, VIRGINIA				**E-1**

COMM. NO. 7215	DATE 9/6/92	DRAWN GK	CHECKED GLC	REVISED TF	

110F12.EPS

Figure 12 ◈ One method of showing revisions on working drawings.

REVISIONS				
REV	**DESCRIPTION**	**DR**	**APPD**	**DATE**
1	FIXTURE NO. 3 IN. LIGHTING-FIXTURE SCHDL	GK	GLC	10/12/92

Professional Stamp	ELECTRICAL
	DISTRICT HOME
	LAUNDRY BUILDING
	AUGUSTA COUNTY, VIRGINIA

G. LEWIS CRAIG, ARCHITECT				SHEET NO.
WAYNESBORO, VIRGINIA				**E-1**

COMM. NO. 7215	DATE 9/6/92	DRAWN GK	CHECKED GLC	REVISED TF	

110F13.EPS

Figure 13 ◈ Alternative method of showing revisions on working drawings.

_____
_____
_____
_____
_____
_____
_____
_____
_____
_____

Discuss the types of
information conveyed by
architectural drafting lines.

Show Transparency 17
(Figure 14).

3.0.0 ◆ DRAFTING LINES

You will encounter many types of drafting lines. To specify the meaning of each type of line, contrasting lines can be made by varying the width of the lines or breaking the lines in a uniform way.

Figure 14 shows common lines used on architectural drawings. However, these lines can vary. Architects and engineers have strived for a common standard for the past century, but unfortunately, their goal has yet to be reached. Therefore, you will find variations in lines and symbols from drawing to drawing, so always consult the legend or symbol list when referring to any drawing. Also, carefully inspect each drawing to ensure that line types are used consistently.

The drafting lines shown in *Figure 14* are used as follows:

- *Light full line*—This line is used for section lines, building background (outlines), and similar uses where the object to be drawn is secondary to the system being shown (e.g., HVAC or electrical).
- *Medium full line*—This type of line is frequently used for hand lettering on drawings. It is further used for some drawing symbols, circuit lines, etc.
- *Heavy full line*—This line is used for borders around title blocks, schedules, and for hand lettering drawing titles. Some types of symbols are frequently drawn with a heavy full line.
- *Extra heavy full line*—This line is used for border lines on architectural/engineering drawings.
- *Centerline*—A centerline is a broken line made up of alternately spaced long and short dashes. It indicates the centers of objects such as holes, pillars, or fixtures. Sometimes, the centerline indicates the dimensions of a finished floor.

LIGHT FULL LINE ─────────────────

MEDIUM FULL LINE ─────────────────

HEAVY FULL LINE ─────────────────

EXTRA HEAVY FULL LINE ━━━━━━━━━━━━━━━

CENTERLINE ─ · ─ · ─ · ─ · ─ · ─ · ─

HIDDEN LINE · · · · · · · · · · · · · · · ·

DIMENSION LINE ◄──────── 3.00" ────────►

SHORT BREAK LINE ══════╗ ╔══════

LONG BREAK LINE ───────╱╲╱───────

MATCH LINE ━ ━ ━ ━ ━ ━ ━ ━ ━

SECONDARY LINE ─ ─ ─ ─ ─ ─ ─ ─ ─

PROPERTY LINE ─ · · ─ · · ─ · · ─ · · ─

Figure 14 ◆ Typical drafting lines.

110F14.EPS

Instructor's Notes:

_____

_____

_____

_____

_____

_____

_____

_____

_____

- *Hidden line*—A hidden line consists of a series of short dashes that are closely and evenly spaced. It shows the edges of objects that are not visible in a particular view. The object outlined by hidden lines in one drawing is often fully pictured in another drawing.
- *Dimension line*—These are thin lines used to show the extent and direction of dimensions. The dimension is usually placed in a break inside the dimension lines. Normal practice is to place the dimension lines outside the object's outline. However, it may sometimes be necessary to draw the dimensions inside the outline.
- *Short break line*—This line is usually drawn freehand and is used for short breaks.
- *Long break line*—This line, which is drawn partly with a straightedge and partly with freehand zigzags, is used for long breaks.
- *Match line*—This line is used to show the position of the cutting plane. Therefore, it is also called the *cutting plane line*. A match or cutting plane line is a heavy line with long dashes alternating with two short dashes. It is used on drawings of large structures to show where one drawing stops and the next drawing starts.
- *Secondary line*—This line is frequently used to outline pieces of equipment or to indicate reference points of a drawing that are secondary to the drawing's purpose.
- *Property line*—This is a light line made up of one long and two short dashes that are alternately spaced. It indicates land boundaries on the site plan.

Other uses of the lines just mentioned include the following:

- *Extension lines*—Extension lines are lightweight lines that start about ⅟₁₆ inch away from the edge of an object and extend out. A common use of extension lines is to create a boundary for dimension lines. Dimension lines meet extension lines with arrowheads, slashes, or dots. Extension lines that point from a note or other reference to a particular feature on a drawing are called *leaders*. They usually end in either an arrowhead or a dot and may include an explanatory note at the end.

- *Section lines*—These are often referred to as *cross-hatch lines*. Drawn at a 45° angle, these lines show where an object has been cut away to reveal the inside.
- *Phantom lines*—Phantom lines are solid, light lines that show where an object will be installed. A future door opening or a future piece of equipment can be shown with phantom lines.

3.1.0 Electrical Drafting Lines

Besides the architectural lines shown in *Figure 14*, consulting electrical engineers, designers, and drafters use additional lines to represent circuits and their related components. Again, these lines may vary from drawing to drawing, so check the symbol list or legend for the exact meaning of lines on the drawing with which you are working. *Figure 15* shows lines used on some electrical drawings.

* Number of arrowheads indicates number of circuits. A number at each arrowhead may be used to identify circuit numbers.

** Half arrowheads are sometimes used for homeruns to avoid confusing them with drawing callouts.

110F15.EPS

Figure 15 ◆ Electrical drafting lines.

Discuss the types of information conveyed by electrical drafting lines.

Show Transparency 18 (Figure 15).

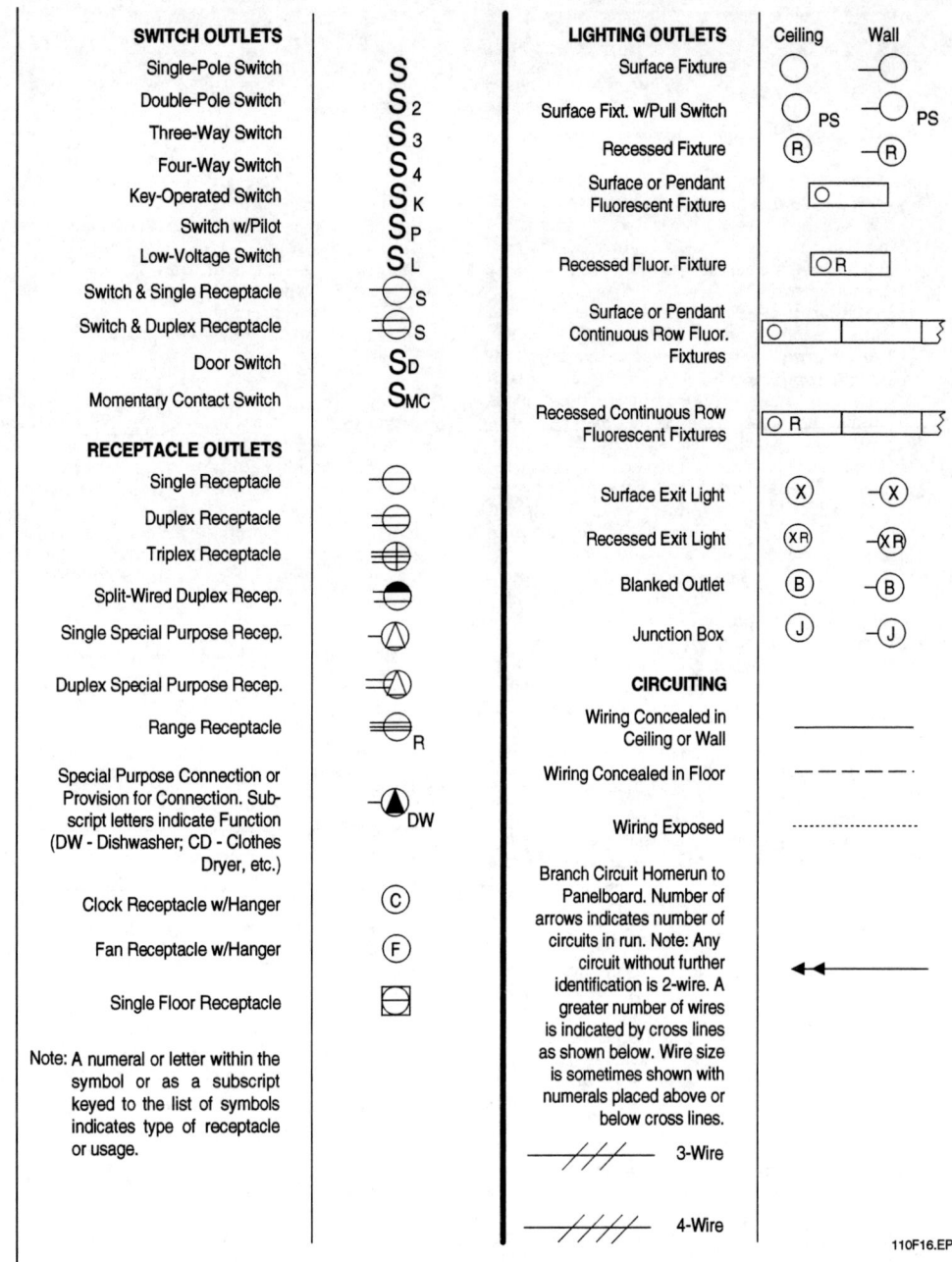

Figure 16 ◆ ANSI electrical symbols.

110F16.EPS

Instructor's Notes:

4.0.0 ◆ ELECTRICAL SYMBOLS

The electrician must be able to correctly read and understand electrical working drawings. This includes a thorough knowledge of electrical symbols and their applications.

An electrical symbol is a figure or mark that stands for a component used in the electrical system. *Figure 16* shows a list of electrical symbols that are currently recommended by the American National Standards Institute (ANSI). It is evident from this list of symbols that many have the same basic form, but, because of some slight difference, their meaning changes. For example, the receptacle symbols in *Figure 17* each have the same basic form (a circle), but the addition of a line or an abbreviation gives each an individual meaning. A good procedure to follow in learning symbols is to first learn the basic form and then apply the variations for obtaining different meanings.

It would be much simpler if all architects, engineers, electrical designers, and drafters used the same symbols; however, this is not the case. Although standardization is getting closer to a reality, existing symbols are still modified, and new symbols are created for almost every new project.

The electrical symbols described in the following paragraphs represent those found on actual electrical working drawings throughout the United States and Canada. Many are similar to those recommended by ANSI and the Consulting Engineers Council/US; others are not. Understanding how these symbols were devised will help you to interpret unknown electrical symbols in the future.

Some of the symbols used on electrical drawings are abbreviations, such as WP for weatherproof and AFF for above finished floor. Others are simplified pictographs, such as (A) in *Figure 18* for a double floodlight fixture or (B) for an infrared electric heater with two quartz lamps.

In some cases, the symbols are combinations of abbreviations and pictographs, such as (C) in *Figure 18* for a fusible safety switch, (D) for a nonfusible safety switch, and (E) for a double-throw safety switch. In each example, a pictograph of a switch enclosure has been combined with an abbreviation: F (fusible), DT (double-throw), and NF (nonfusible), respectively.

Lighting outlet symbols have been devised that represent incandescent, fluorescent, and high-intensity discharge lighting; a circle usually represents an incandescent fixture, and a rectangle is used to represent a fluorescent fixture. These symbols are designed to indicate the physical shape of a particular fixture, and while the circles repre-

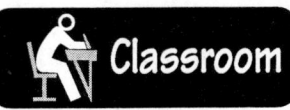

Discuss the types of information conveyed by various electrical symbols.

Show Transparencies 19 through 21 (Figures 16 through 18).

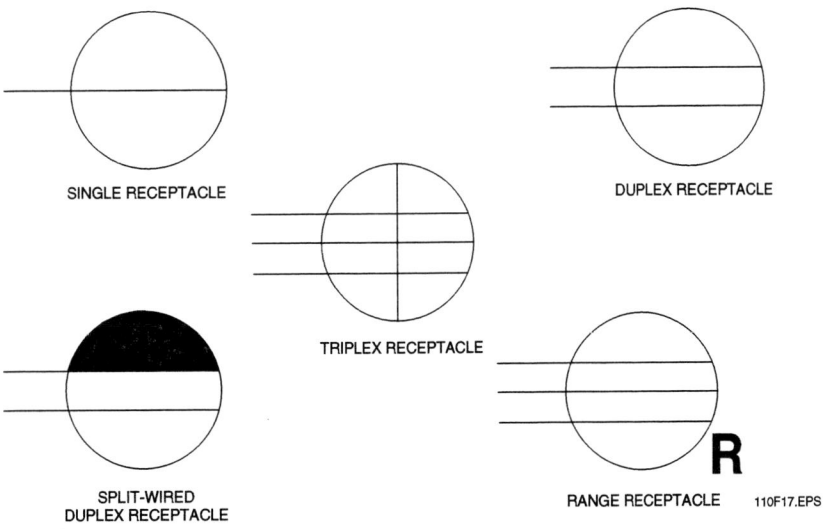

Figure 17 ◆ Various receptacle symbols used on electrical drawings.

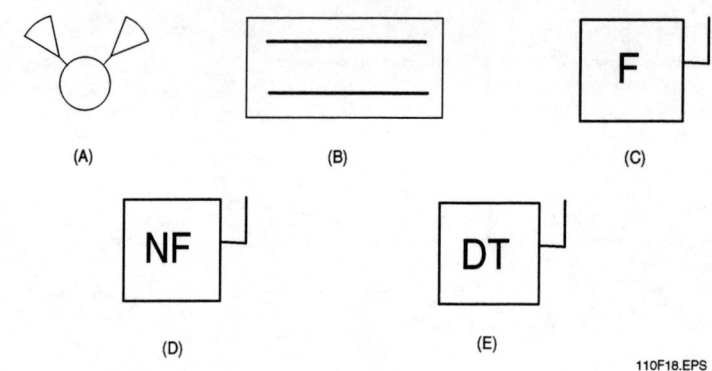

(A) (B) (C)

(D) (E)

110F18.EPS

Figure 18 ◈ General types of symbols used on electrical drawings.

Show Transparencies 22 through 29 (Figures 19 and 20).

Add some fun to the study of symbols by drawing each of the symbols on an index card. Put all of the cards in a hardhat and pass it around the room. Each trainee removes a card and tries to identify it. If they succeed, they keep the card. If not, it goes back into the hardhat. Whoever ends up with the most cards wins.

senting incandescent lamps are frequently enlarged somewhat, symbols for fluorescent fixtures are usually drawn as close to scale as possible. The type of mounting used for all lighting fixtures is usually indicated in a lighting fixture schedule, which is shown on the drawings or in the written specifications.

The type of lighting fixture is identified by a numeral placed inside a triangle or other symbol, and placed near the fixture to be identified. A complete description of the fixtures identified by the symbols must be given in the lighting fixture schedule and should include the manufacturer, catalog number, number and type of lamps, voltage, finish, mounting, and any other information needed for proper installation of the fixture.

Switches used to control lighting fixtures are also indicated by symbols (usually the letter S followed by numerals or letters to define the exact type of switch). For example, S_3 indicates a three-way switch; S_4 identifies a four-way switch; and S_P indicates a single-pole switch with a pilot light.

Main distribution centers, panelboards, transformers, safety switches, and other similar electrical components are indicated by electrical symbols on floor plans and by a combination of symbols and semipictorial drawings in riser diagrams.

A detailed description of the service equipment is usually given in the panelboard schedule or in the written specifications. However, on small projects, the service equipment is sometimes indicated only by notes on the drawings.

Circuit and feeder wiring symbols are getting closer to being standardized. Most circuits concealed in the ceiling or wall are indicated by a solid line; a broken line is used for circuits concealed in the floor or ceiling below; and exposed raceways are indicated by short dashes or else the letter *E* placed in the same plane with the circuit line at various intervals. The number of conductors in a conduit or raceway system may be indicated in the panelboard schedule under the appropriate column, or the information may be shown on the floor plan.

Symbols for communication and signal systems, as well as symbols for light and power, are drawn to an appropriate scale and accurately located with respect to the building. This reduces the number of references made to the architectural drawings. Where extreme accuracy is required in locating outlets and equipment, exact dimensions are given on larger-scale drawings and shown on the plans.

Each different category in an electrical system is usually represented by a basic distinguishing symbol. To further identify items of equipment or outlets in the category, a numeral or other identifying mark is placed within the open basic symbol. In addition, all such individual symbols used on the drawings should be included in the symbol list or legend. The electrical symbols shown in *Figure 19* were modified by a consulting engineering firm for use on a small industrial electrical installation. The symbols shown in *Figure 20* are those recommended by the Consulting Engineers Council/US. You should become familiar with these symbols.

ELECTRICAL LEVEL ONE—TRAINEE MODULE 26110

Instructor's Notes:

Figure 19 ◈ Electrical symbols used by one consulting engineering firm.

The following symbols are shown in the figure:

Symbol	Description
JB (box with three ducts)	UNDERFLOOR DUCT SYSTEM - JUNCTION BOX AND THREE DUCTS (ONE LARGE, TWO STANDARD)
Dashed lines	DOTTED LINES INDICATE BLANK DUCT
Solid black bar	G.E. TYPE LW223 LIGHTING BUSWAY
Hatched bar	G.E. TYPE LW326 LIGHTING BUSWAY
Cross-hatched bar	G.E. TYPE DK-100 BUSWAY
Solid black square	BUSWAY FEED-IN BOX
Panel symbol	PANEL – LIGHTING AND/OR POWER
Solid line	CONDUIT CONCEALED ABOVE CEILING OR WALL
Dashed line	CONDUIT CONCEALED IN FLOOR OR IN WALL
Arrow with crossmarks A-1	HOMERUN TO PANEL; NUMBER OF ARROWS INDICATE NUMBER OF CIRCUITS; LETTER DESIGNATES PANEL; NUMERAL DESIGNATES CIRCUIT NUMBER; CROSSMARKS INDICATE NUMBER OF CONDUCTORS IF MORE THAN TWO
Circle with line	MOTOR CONNECTION
S_T	SWITCH, TOGGLE WITH THERMAL OVERLOAD PROTECTION
Dotted line	CONDUIT EXPOSED
Circle with line through it	DUPLEX RECEPTACLE, GROUNDED
S_K	SWITCH, KEY OPERATED

110F19.EPS

SWITCH OUTLETS		RECEPTACLE OUTLETS	
		Where weatherproof, explosionproof, or other specific types of devices are to be required, use the upper-case subscript letters to specify. For example, weatherproof single or duplex receptacles would have the upper-case WP subscript letters noted alongside the symbol. All outlets must be grounded.	
Single Pole Switch	S		
Double Pole Switch	S_2		
Three-Way Switch	S_3		
Four-Way Switch	S_4	Single Receptacle Outlet	
Key-Operated Switch	S_K	Duplex Receptacle Outlet	
Switch and Fusestat Holder	$S_F H$	Triplex Receptacle Outlet	
Switch and Pilot Lamp	S_P	Quadruplex Receptacle Outlet	
Fan Switch	S_F	Duplex Receptacle Outlet Split Wired	
Switch for Low-Voltage Switching System	S_L	Triplex Receptacle Outlet Split Wired	
Master Switch for Low-Voltage Switching System	S_{LM}	250-Volt Receptacle/Single Phase Use Subscript Letter to Indicate Function (DW - Dishwasher, RA - Range) or Numerals (with explanation in symbols schedule)	
Switch and Single Receptacle	S		
Switch and Duplex Receptacle	S	250-Volt Receptacle/Three Phase	
Door Switch	S_D	Clock Receptacle	C
Time Switch	S_T	Fan Receptacle	F
Momentary Contact Switch	S_{MC}	Floor Single Receptacle Outlet	
Ceiling Pull Switch	S	Floor Duplex Receptacle Outlet	
"Hand-Off-Auto" Control Switch	HOA	Floor Special-Purpose Outlet	*
Multi-Speed Control Switch	M	Floor Telephone Outlet - Public	
Pushbutton	•	Floor Telephone Outlet - Private	
		Use numeral keyed explanation of symbol usage	

110F20A.EPS

Figure 20 ❖ Recommended electrical symbols (1 of 7).

Instructor's Notes:

Example of the use of several floor outlet symbols
to identify a 2, 3, or more gang outlet:

Underfloor duct and junction box
for triple, double, or single
duct system as indicated
by the number of parallel lines

Example of the use of various symbols to identify the
location of different types of outlets or connections for
underfloor duct or cellular floor systems:

Cellular Floor
Heater Duct

CIRCUITING

Wiring Exposed (not in conduit)	—— E ——
Wiring Concealed in Ceiling or Wall	————
Wiring Concealed in Floor	– – – – –
Wiring Existing*	·············
Wiring Turned Up	——○
Wiring Turned Down	——●
Branch Circuit Homerun to Panelboard	2 1 ——➤

Number of arrows indicates number of circuits.
(A number at each arrow may be used to identify
the circuit number.)**

BUS DUCTS AND WIREWAYS

Trolley Duct***	T	T
Busway (Service, Feeder or Plug-in)***	B	B
Cable Trough Ladder or Channel***	C	C
Wireway***	W	W

PANELBOARDS, SWITCHBOARDS AND RELATED EQUIPMENT

Flush Mounted Panelboard
and Cabinet***

Surface Mounted Panelboard
and Cabinet***

Switchboard, Power Control
Center, Unit Substation
(Should be drawn to scale)***

Flush Mounted Terminal Cabinet
(In small scale drawings the TC
may be indicated alongside the
symbol)***

Surface Mounted Terminal Cabinet
(In small scale drawings the TC
may be indicated alongside the
symbol)***

Pull Box (Identify in relation to
Wiring System Section and Size)

Motor or Other Power Controller
May be a starter or contactor***

Externally Operated Disconnection
Switch***

Combination Controller and
Disconnection Means***

*Note: Use heavy-weight line to identify service and feeders.
Indicate empty conduit by notation CO.

**Note: Any circuit without further identification indicates two-wire
circuit. For a greater number of wires, indicate with cross lines,
e.g.:

3 wires ——|||——   4 wires, etc. ——||||——

Neutral wire may be shown longer. Unless indicated otherwise,
the wire size of the circuit is the minimum size required by the
specification. Identify different functions of wiring system (e.g.,
signaling system) by notation or other means.

***Identify by Notation or Schedule

110F20B.EPS

Figure 20 ◆ Recommended electrical symbols (2 of 7).

Figure 20 ◈ Recommended electrical symbols (3 of 7).

Instructor's Notes:

	Ceiling	Wall
Exit Light (Surface or Pendant)	⊗	⊗—
Blanked Outlet	Ⓑ	Ⓑ—
Junction Box	Ⓙ	Ⓙ—
Recessed Incandescent Fixture	▢	
Individual Fluorescent Fixture (Surface or Pendant)	▭	

Continuous Row Fluorescent Fixture (Surface or Pendant)

Letter indicating controlling switch → **A**

Fixture No. ← ①/100 → Wattage

Symbol not needed at each fixture

Bare-Lamp Fluorescent Strip*

ELECTRIC DISTRIBUTION OR LIGHTING SYSTEM, AERIAL

Pole**	◯
Street or Parking Lot Light and Bracket	☼
Transformer**	△
Primary Circuit**	——
Secondary Circuit**	——
Down Guy	→)

Head Guy	—●—
Sidewalk Guy	—○—)
Service Weatherhead**	—◁=

ELECTRIC DISTRIBUTION OR LIGHTING SYSTEM, UNDERGROUND

Manhole	M
Handhole	H
Transformer Manhole or Vault	TM
Transformer Pad	TP

Underground Direct Burial Cable (Indicate type, size, and number of conductors by notation or schedule.) ——

Underground Duct Line (Indicate type, size, and number of ducts by cross-section identification of each run by notation or schedule. Indicate type, size, and number of conductors by notation or schedule.)

Street Light Standard Fed From Underground Circuit** ⊙

*In the case of continuous-row bare-lamp fluorescent strip above an area-wide diffusing means, show each fixture run using the standard symbol; indicate area of diffusing means and type by light shading and/or drawing notation.
**Identify by Notation or Schedule

110F20D.EPS

Figure 20 ◆ Recommended electrical symbols (4 of 7).

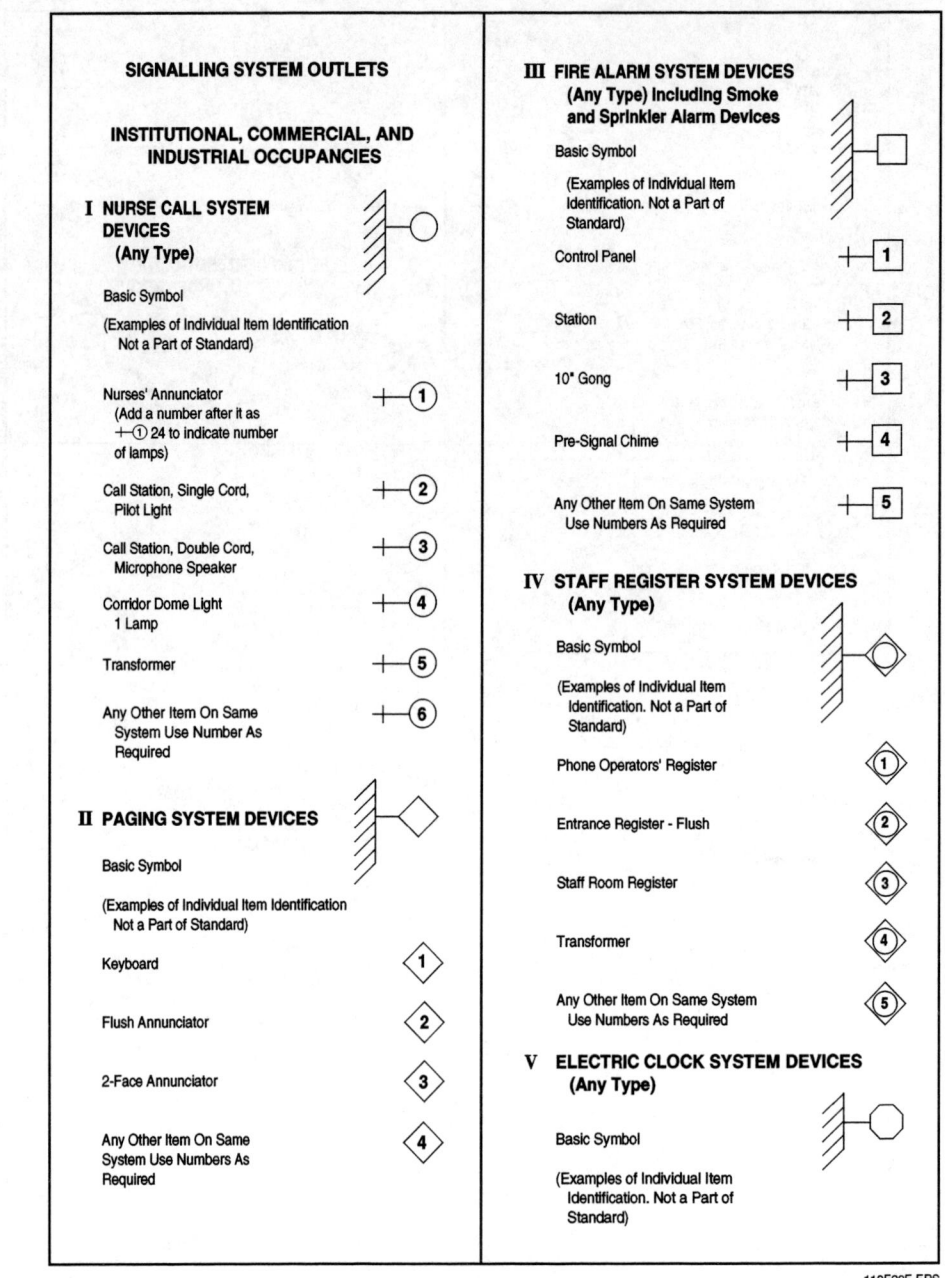

Figure 20 ◈ Recommended electrical symbols (5 of 7).

110F20E.EPS

Instructor's Notes:

Master Clock ⬡ 1

12" Secondary - Flush ⬡ 2

12" Double Dial - Wall Mounted ⬡ 3

18" Skeleton Dial ⬡ 4

Any Other Item On Same System
Use Numbers As Required ⬡ 5

VI PUBLIC TELEPHONE SYSTEM DEVICES

Basic Symbol

(Examples of Individual Item
Identification. Not a Part of
Standard)

Switchboard

Desk Phone

Any Other Item On Same
System Use Numbers As
Required

VII PRIVATE TELEPHONE SYSTEM DEVICES
(Any Type)

Basic Symbol

(Examples of Individual Item
Identification. Not a Part of
Standard)

Switchboard ◁1

Wall Phone ◁2

Any Other Item On Same System
Use Numbers As Required ◁3

VIII WATCHMAN SYSTEM DEVICES
(Any Type)

Basic Symbol

(Examples of Individual Item
Identification. Not a Part of
Standard)

Central Station ⬠ 1

Key Station ⬠ 2

Any Other Item On Same System
Use Numbers As Required ⬠ 3

IX SOUND SYSTEM

Basic Symbol

(Examples of Individual Item
Identification. Not a Part of
Standard)

Amplifier ◁ 1

Microphone ◁ 2

Interior Speaker ◁ 3

Exterior Speaker ◁ 4

Any Other Item On Same System
Use Numbers As Required ◁ 5

X OTHER SIGNAL SYSTEM DEVICES

Basic Symbol

(Examples of Individual Item
Identification. Not a Part of
Standard)

Buzzer ◯1

Bell ◯2

Pushbutton ◯3

Annunciator ◯4

Any Other Item On Same System
Use Numbers As Required ◯5

110F20F.EPS

Figure 20 ◈ Recommended electrical symbols (6 of 7).

10–25

Discuss drawing scales and how to interpret them.

Show Transparency 30 (Figure 21).

RESIDENTIAL OCCUPANCIES

Signalling system symbols for use in identifying standardized residential-type signal system items on residential drawings where a descriptive symbol list is not included on the drawing. When other signal system items are to be identified, use the above basic symbols for such items together with a descriptive symbol list.

Pushbutton	●
Buzzer	
Bell	
Combination Bell - Buzzer	
Chime	CH
Annunciator	
Electric Door Opener	D
Maid's Signal Plug	M
Interconnection Box	
Bell-Ringing Transformer	BT
Outside Telephone	▶
Interconnecting Telephone	▷
Television Outlet	TV

110F20G.EPS

Figure 20 ◆ Recommended electrical symbols (7 of 7).

5.0.0 ◆ SCALE DRAWINGS

In most electrical drawings, the components are so large that it would be impossible to draw them actual size. Consequently, drawings are made to some reduced *scale*; that is, all the distances are drawn smaller than the actual dimensions of the object itself, with all dimensions being reduced in the same proportion. For example, if a floor plan of a building is to be drawn to a scale of ¼" = 1'-0", each ¼" on the drawing would equal 1 foot on the building itself; if the scale is ⅛" = 1'-0", each ⅛" on the drawing equals 1 foot on the building, and so forth.

When architectural and engineering drawings are produced, the selected scale is very important. Where dimensions must be held to extreme accuracy, the scale drawings should be made as large as practical with dimension lines added. Where dimensions require only reasonable accuracy, the object may be drawn to a smaller scale (with dimension lines possibly omitted).

In dimensioning drawings, the dimensions written on the drawing are the actual dimensions of the building, not the distances that are measured on the drawing. To further illustrate this point, look at the floor plan in *Figure 21*; it is drawn to a scale of ½" = 1'-0". One of the walls is drawn to an actual length of 3½" on the drawing paper, but since the scale is ½" = 1'-0" and since 3½" contains 7 halves of an inch (7 × ½ = 3½"), the dimension shown on the drawing will therefore be 7'-0" on the actual building.

As shown in the previous example, the most common method of reducing all the dimensions (in feet and inches) in the same proportion is to choose a certain distance and let that distance represent one foot. This distance can then be divided into twelve parts, each of which represents an inch. If half inches are required, these twelfths are further subdivided into halves, etc. Now the scale represents the common foot rule with its subdivisions into inches and fractions, except that the scaled foot is smaller than the distance known as a foot and, likewise, its subdivisions are proportionately smaller.

When a measurement is made on the drawing, it is made with the reduced foot rule or scale; when a measurement is made on the building, it is made with the standard foot rule. The most common reduced foot rules or scales used in electrical drawings are the architect's scale and the engineer's scale. Drawings may sometimes be encountered that use a metric scale, but using this scale is similar to using the architect's or engineer's scales.

Instructor's Notes:

_____

_____

_____

_____

_____

_____

_____

_____

_____

The distance between the arrowheads to the left measures 3-1/2" on the drawing, but since the drawing is made to a scale of 1/2" = 1'–0", this measurement actually represents 7'–0".

PUMP HOUSE FLOOR PLAN
1/2" = 1'–0"

110F21.EPS

Figure 21 ◆ Typical floor plan showing drawing scale.

Using Electrical Symbols

Although there are many electrical symbols, you must be able to read the common ones at a glance. Looking at the simple pump house drawing in *Figure 21*, see how quickly you can explain the symbols and the circuits that they identify.

5.1.0 Architect's Scale

Figure 22 shows two configurations of architect's scales. The one on the top is designed so that 1" = 1'–0", and the one on the bottom has graduations spaced to represent ⅛" = 1'–0".

Note that on the one-inch scale in *Figure 23*, the longer marks to the right of the zero (with a numeral beneath) represent feet. Therefore, the distance between the zero and the numeral 1 equals one foot. The shorter mark between the zero and 1 represents ½ of a foot, or six inches.

Referring again to *Figure 23*, look at the marks to the left of the zero. The numbered marks are spaced three scaled inches apart and have the numerals 0, 3, 6, and 9 for use as reference points. The other lines of the same length also represent scaled inches, but are not marked with numerals. In use,

you can count the number of long marks to the left of the zero to find the number of inches, but after some practice, you will be able to tell the exact measurement at a glance. For example, the measurement A represents five inches because it is the fifth inch mark to the left of the zero; it is also one inch mark short of the six-inch line on the scale.

The lines that are shorter than the inch line are the half-inch lines. On smaller scales, the basic unit is not divided into as many divisions. For example, the smallest subdivision on some scales represents two inches.

5.1.1 Types of Architect's Scales

Architect's scales are available in several types, but the most common include the triangular

Discuss the "Think About It."

Show an architect's scale and discuss how it is used.

Show Transparencies 31 and 32 (Figures 22 and 23).

Discuss triangular architect's scales and pass around an example for the trainees to examine.

1" = 1'–0"

1/8" = 1'–0"

110F22.EPS

Figure 22 ◆ Two different configurations of architect's scales.

SUBDIVISIONS OF ONE FOOT   ONE FOOT RANGE

HALF-FOOT (6-INCH) MARKS

A

110F23.EPS

Figure 23 ◆ One-inch architect's scale.

Show Transparency 33 (Figure 24).

Discuss the 12 scales typically found on a triangular architect's scale.

scale (*Figure 24*) and the flat scale. The quality of architect's scales also vary from cheap plastic scales (costing a dollar or two) to high-quality wooden-laminated tools that are calibrated to precise standards.

The triangular scale (*Figure 24*) is frequently found in drafting and estimating departments or engineering and electrical contracting firms, while the flat scales are more convenient to carry on the job site.

Triangular architect's scales have 12 different scales—two on each edge—as follows:

- Common foot rule (12 inches)
- ¹⁄₁₆" = 1'–0"
- ³⁄₃₂" = 1'–0"
- ³⁄₁₆" = 1'–0"
- ⅛" = 1'–0"

- ¼" = 1'–0"
- ⅜" = 1'–0"
- ¾" = 1'–0"
- 1" = 1'–0"
- ½" = 1'–0"
- 1½" = 1'–0"
- 3" = 1'–0"

Two separate scales on one face may seem confusing at first, but after some experience, reading these scales becomes second nature.

In all but one of the scales on the triangular architect's scale, each face has one of the scales placed opposite to the other. For example, on the one-inch face, the one-inch scale is read from left to right, starting from the zero mark. The half-inch scale is read from right to left, again starting from the zero mark.

Instructor's Notes:

Figure 24 ◆ Typical triangular architect's scale.

On the remaining foot-rule scale (¹⁄₁₆" = 1'–0"), each ¹⁄₁₆" mark on the scale represents one foot.

Figure 25 shows all the scales found on the triangular architect's scale.

The flat architect's scale shown in Figure 26 is ideal for workers on most projects. It is easily and conveniently carried in the shirt pocket, and the four scales (⅛", ¼", ½", and 1") are adequate for the majority of projects that will be encountered.

The partial floor plan shown in Figure 26 is drawn to a scale of ⅛" = 1'–0". The dimension in question is found by placing the ⅛" architect's scale on the drawing and reading the figures. It can be seen that the dimension reads 24'–6".

Every drawing should have the scale to which it is drawn plainly marked on it as part of the drawing title. However, it is not uncommon to have several different drawings on one blueprint sheet—all with different scales. Therefore, always check the scale of each different view found on a drawing sheet.

5.2.0 Engineer's Scale

The civil engineer's scale is used in basically the same manner as the architect's scale, with the principal difference being that the graduations on the engineer's scale are decimal units rather than feet, as on the architect's scale.

The engineer's scale is used by placing it on the drawing with the working edge away from the user. The scale is then aligned in the direction of the required measurement. Then, by looking down at the scale, the dimension is read.

Civil engineer's scales commonly show the following graduations:

- 1" = 10 units
- 1" = 20 units
- 1" = 30 units
- 1" = 40 units
- 1" = 60 units
- 1" = 80 units
- 1" = 100 units

Show Transparencies 34 and 35 (Figures 25 and 26).

Discuss flat architect's scales and pass around an example for the trainees to examine.

Discuss engineer's scales and explain how they are used. Pass around an example for the trainees to examine.

110F25.EPS

Figure 25 ◆ Various scales on a triangular architect's scale.

INTRODUCTION TO ELECTRICAL BLUEPRINTS—TRAINEE MODULE 26110

10–29

SCALE: 1/8" = 1'0"

110F26.EPS

Figure 26 ◆ Using the ⅛" architect's scale to determine the dimensions on a drawing.

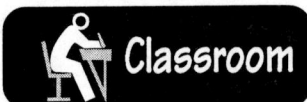
The purpose of this scale is to transfer the relative dimensions of an object to the drawing or vice versa. It is used mainly on site plans to determine distances between property lines, manholes, duct runs, direct-burial cable runs, and the like.

Site plans are drawn to scale using the engineer's scale rather than the architect's scale. On small lots, a scale of 1 inch = 10 feet or 1 inch = 20 feet is used. For a 1:10 scale, this means that one inch (the actual measurement on the drawing) is equal to 10 feet on the land itself.

On larger drawings, where a large area must be covered, the scale could be 1 inch = 100 feet or 1 inch = 1,000 feet, or any other integral power of 10. On drawings with the scale in multiples of 10, the engineering scale marked 10 is used. If the scale is 1 inch = 200 feet, the engineer's scale marked 20 is used, and so on.

Although site plans appear reduced in scale, depending on the size of the object and the size of the drawing sheet to be used, the actual dimensions must be shown on the drawings at all times. When you are reading the drawing plans to scale,

think of each dimension in its full size and not in the reduced scale it happens to be on the drawing (*Figure 27*).

5.3.0 Metric Scale

The metric scale (*Figure 28*) is divided into centimeters (cm), with the centimeters divided into 10 divisions for millimeters (mm) or in 20 divisions for half millimeters. Scales are available with metric divisions on one edge while inch divisions are inscribed on the opposite edge. Many contracting firms that deal in international trade have adopted a dual-dimensioning system expressed in both metric and English symbols. Drawings prepared for government projects may also require metric dimensions.

6.0.0 ◆ ANALYZING ELECTRICAL DRAWINGS

The most practical way to learn how to read electrical construction documents is to analyze an

Instructor's Notes:

Figure 27 ◆ Practical use of the engineer's scale.

110F27.EPS

110F28.EPS

Figure 28 ◆ Typical metric scale.

existing set of drawings prepared by consulting or industrial engineers.

Engineers or electrical designers are responsible for the complete layout of electrical systems for most projects. Electrical drafters then transform the engineer's designs into working drawings, using either manual drafting instruments or computer-aided design (CAD) systems. The following is a brief outline of what usually takes place in the preparation of electrical design and working drawings:

- The engineer meets with the architect and owner to discuss the electrical needs of the building or project and to discuss various recommendations made by all parties.
- After that, an outline of the architect's floor plan is laid out.
- The engineer then calculates the required power and lighting outlets for the project; these are later transferred to the working drawings.
- All communications and alarm systems are located on the floor plan, along with lighting and power panelboards.

- Circuit calculations are made to determine wire size and overcurrent protection.
- The main electric service and related components are determined and shown on the drawings.
- Schedules are then placed on the drawings to identify various pieces of equipment.
- Wiring diagrams are made to show the workers how various electrical components are to be connected.
- A legend or electrical symbol list is drafted and shown on the drawings to identify all symbols used to indicate electrical outlets or equipment.
- Various large-scale electrical details are included, if necessary, to show exactly what is required of the electricians.
- Written specifications are then made to give a description of the materials and installation methods.

6.1.0 Development of Site Plans

In general practice, it is usually the owner's responsibility to furnish the architect/engineer with

Describe the process of drafting electrical drawings.

property and topographic surveys, which are made by a certified land surveyor or civil engineer. These surveys show:

- All property lines
- Existing public utilities and their location on or near the property (e.g., electrical lines, sanitary sewer lines, gas lines, water-supply lines, storm sewers, manholes, telephone lines, etc.)

A land surveyor does the property survey from information obtained from a deed description of the property. A property survey shows only the property lines and their lengths, as if the property were perfectly flat.

The topographic survey shows both the property lines and the physical characteristics of the land by using contour lines, notes, and symbols. The physical characteristics may include:

- The direction of the land slope
- Whether the land is flat, hilly, wooded, swampy, high, or low, and other features of its physical nature

All of this information is necessary so that the architect can properly design a building to fit the property. The electrical engineer also needs this information to locate existing electrical utilities and to route the new service to the building, provide outdoor lighting and circuits, etc.

Electrical site work is sometimes shown on the architect's plot plan. However, when site work involves many trades and several utilities (e.g., gas, telephone, electric, television, water, and sewage), it can become confusing if all details are shown on one drawing sheet. In cases like these, it is best to have a separate drawing devoted entirely to the electrical work, as shown in *Figure 29*. This project is an office/warehouse building for Virginia Electric, Inc. The electrical drawings consist of four 24" × 36" drawing sheets, along with a set of written specifications, which will be discussed later in this module.

The electrical site or plot plan shown in *Figure 29* has the conventional architect's and engineer's title blocks in the lower right-hand corner of the drawing. These blocks identify the project and

project owners, the architect, and the engineer. They also show how this drawing sheet relates to the entire set of drawings. Note the engineer's professional stamp of approval to the left of the engineer's title block. Similar blocks appear on all four of the electrical drawing sheets.

When examining a set of electrical drawings for the first time, always look at the area around the title block. This is where most revision blocks or revision notes are placed. If revisions have been made to the drawings, make certain that you have a clear understanding of what has taken place before proceeding with the work.

Refer again to the drawing in *Figure 29* and note the North Arrow in the upper left corner. A North Arrow shows the direction of true north to help you orient the drawing to the site. Look directly down from the North Arrow to the bottom of the page and notice the drawing title, *Plot Utilities*. Directly beneath the drawing title you can see that the drawing scale of 1" = 30' is shown. This means that each inch on the drawing represents 30 feet on the actual job site. This scale holds true for all drawings on the page unless otherwise noted.

An outline of the proposed building is indicated on the drawing by cross-hatched rectangles along with a callout stating, *Proposed Bldg. Fin. Flr. Elev. 590.0'*. This means that the finished floor level of the building is to be 590 feet above sea level, which in this part of the country will be about two feet above finished grade around the building. This information helps the electrician locate conduit sleeves and stub-ups to the correct height before the finished concrete floor is poured.

The shaded area represents asphalt paving for the access road, drives, and parking lot. Note that the access road leads into a highway, which is designated Route 35. This information further helps workers to orient the drawing to the building site.

Existing manholes are indicated by a solid circle, while an open circle is used to show the position of the five new pole-mounted lighting fixtures that are to be installed around the new building. Existing power lines are shown with a light solid line with the letter *E* placed at intervals along the line. The new underground electric service is

Interpreting Site Plans

THINK ABOUT IT

Study *Figure 29* and explain as many of its features as you can. How much can be understood using common sense? What features require special information?

Instructor's Notes:

WV ROUTE 35

SECTION THROUGH
DUCT BANK

EXISTING MANHOLE
SEE DETAIL THIS
SHEET

NEW UNDERGROUND
ELECTRIC SERVICE
TO BE INSTALLED
IN EXISTING DUCT
SYSTEM

EXISTING POWER
LINE

ACCESS ROAD

A
E-1

E

MANHOLE DETAIL
NO SCALE

NORTH

NEW TELEPHONE
SERVICE

#8 AWG
TO TIME
CLOCK

NEW PADMOUNT
TRANSFORMER

T

PROPOSED BLDG.
FIN. FLR. ELEV. 590.0'

G

GAS MAIN BY
GAS COMPANY

G

TYPICAL - EXTERIOR FIXTURE.
SEE SPECS. AND DETAILS ON
SHEET E-4

PLOT UTILITIES
SCALE: 1" = 30'

110F29.EPS

Figure 29 ◆ Typical electrical site plan.

Discuss the types of information conveyed by power plans.

Show Transparency 39 (Figure 30).

Pass around an electrical power plan from an instructor-supplied set of drawings and have the trainees identify various types of information.

shown in the same way, except the lines are somewhat wider and darker on the drawing. Note that this new high-voltage cable terminates into a pad-mount transformer near the proposed building. New telephone lines are similar except the letter *T* is used to identify the telephone lines.

The direct-burial underground cable supplying the exterior lighting fixtures is indicated with dashed lines on the drawing—shown connecting the open circles. A homerun for this circuit is also shown to a time clock.

The manhole detail shown to the right of the North Arrow may seem to serve very little purpose on this drawing since the manholes have already been installed. However, the dimensions and details of their construction will help the electrical contractor or supervisor to better plan the pulling of the high-voltage cable. The same is true of the cross section shown of the duct bank. The electrical contractor knows that three empty ducts are available if it is discovered that one of them is damaged when the work begins.

Although the electrical work will not involve working with gas, the main gas line is shown on the electrical drawing to let the electrical workers know its approximate location while they are installing the direct-burial conductors for the exterior lighting fixtures.

7.0.0 ◆ POWER PLANS

The electrical power plan (*Figure 30*) shows the complete floor plan of the office/warehouse building with all interior partitions drawn to scale. Sometimes, the physical locations of all wiring and outlets are shown on one drawing; that is, outlets for lighting, power, signal and communications, special electrical systems, and related equipment are shown on the same plan. However, on complex installations, the drawing would become cluttered if both lighting and power were shown on the same floor plan. Therefore, most projects will have a separate drawing for power and another for lighting. Riser diagrams and details may be shown on yet another

Figure 30 ◆ Electrical power plan.

ELECTRICAL LEVEL ONE—TRAINEE MODULE 26110

Instructor's Notes:

ELECTRICAL LEVEL ONE—INSTRUCTOR'S GUIDE MODULE 26110

drawing sheet, or if room permits, they may be shown on the lighting or power floor plan sheets.

A closer look at this drawing reveals the title blocks in the lower right corner of the drawing sheet. These blocks list both the architectural and engineering firms, along with information to identify the project and drawing sheet. Also note that the floor plan is titled *Floor Plan "B"—Power* and is drawn to a scale of ⅛" = 1'–0". There are no revisions shown on this drawing sheet.

7.1.0 Key Plan

A key plan appears on the drawing sheet immediately above the engineer's title block (*Figure 31*). The purpose of this key plan is to identify that part of the project to which this sheet applies. In this case, the project involves two buildings: Building A and Building B. Since the outline of Building B is cross-hatched in the key plan, this is the building to which this drawing applies. Note

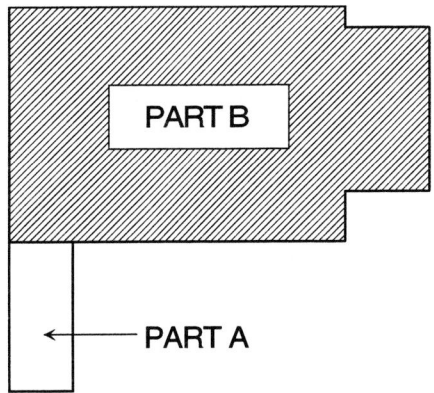

KEY PLAN
NO SCALE

NOTE: NO ELECTRICAL WORK IN PART "A"

110F31.EPS

Figure 31 ◈ Key plan appearing on electrical power plan.

that this key plan is not drawn to scale—only its approximate shape.

Although Building A is also shown on this key plan, a note below the key plan title states that there is no electrical work required in Building A.

On some larger installations, the overall project may involve several buildings requiring appropriate key plans on each drawing to help the workers orient the drawings to the appropriate building. In some cases, separate drawing sheets may be used for each room or area in an industrial project—again requiring key plans on each drawing sheet to identify applicable drawings for each room.

7.2.0 Symbol List

A symbol list appears on the electrical power plan (immediately above the architect's title block) to identify the various symbols used for both power and lighting on this project. In most cases, the only symbols listed are those that apply to the particular project. In other cases, however, a standard list of symbols is used for all projects with the following note:

These are standard symbols and may not all appear on the project drawings; however, wherever the symbol on the project drawings occurs, the item shall be provided and installed.

Only electrical symbols that are actually used for the office/warehouse drawings are shown in the list on the example electrical power plan. A close-up look at these symbols appears in *Figure 32*.

7.3.0 Floor Plan

A somewhat enlarged view of the electrical floor plan drawing is shown in *Figure 33*. However, due to the size of the drawing in comparison with the size of the pages in this module, it is still difficult to see very much detail. This illustration is meant to show the overall layout of the floor plan and how the symbols and notes are arranged.

In general, this plan shows the service equipment (in plan view), receptacles, underfloor duct system, motor connections, motor controllers, electric heat, busways, and similar details. The electric panels and other service equipment are drawn close to scale. The locations of other electrical outlets and similar components are only approximated on the drawings because they have to be exaggerated to show up on the prints. To illustrate, a common duplex receptacle is only about three inches wide. If such a receptacle were to be located on the floor plan of this building (drawn to a scale of ⅛" = 1'–0"), even a small dot on the

Explain the purpose of a key plan.

Show Transparency 40 (Figure 31).

Discuss symbol lists and emphasize that they will vary from drawing to drawing.

Show Transparencies 41 and 42 (Figures 32 and 33).

Discuss the types of information conveyed by floor plans.

Figure 32 ◆ Sample electrical symbols list.

Show Transparency 43 (Figure 34).

Show Transparency 43 (Figure 34).

Pass around a floor plan from an instructor-supplied set of drawings and have the trainees identify switches, outlets, junction boxes, etc.

Discuss notes and building symbols.

drawing would be too large to draw the receptacle exactly to scale. Therefore, the receptacle symbol is exaggerated. When such receptacles are scaled on the drawings to determine the proper location, a measurement is usually taken to the center of the symbol to determine the distance between outlets. Junction boxes, switches, and other electrical connections shown on the floor plan will be exaggerated in a similar manner. The partial floor plan drawing in *Figure 34* on page 10–38 allows a better view of the drawing details.

7.3.1 Notes and Building Symbols

Referring again to *Figure 33*, you will notice numbers placed inside an oval symbol in each room. These numbered ovals represent the room name or type and correspond to a room schedule in the architectural drawings. For example, room number 112 is designated as the lobby in the room schedule (not shown), room number 113 is designated as office No. 1, etc. On some drawings, these

room symbols are omitted and the room names are written out on the drawings.

There are also several notes appearing at various places on the floor plan. These notes offer additional information to clarify certain aspects of the drawing. For example, only one electric heater is to be installed by the electrical contractor; this heater is located in the building's vestibule. Rather than have a symbol in the symbol list for this one heater, a note is used to identify it on the drawing. Other notes on this drawing describe how certain parts of the system are to be installed. For example, in the office area (rooms 112, 113, and 114), you will see the following note: *CONDUIT UP AND STUBBED OUT ABOVE CEILING.* This empty conduit is for telephone/communications cables that will be installed later by the telephone company.

7.3.2 Busways

The office/warehouse project utilizes three types of busways: two types of lighting busways and

ELECTRICAL LEVEL ONE—TRAINEE MODULE 26110

Instructor's Notes:

Figure 33 ◆ Power plan for office/warehouse building.

110F33.EPS

10–37

Discuss the "Think About It."

110F34.EPS

Figure 34 ◆ Partial floor plan for office/warehouse building.

Watch Specified Dimensions

When devices are to be located at heights specified above the finished floor (a.f.f.), be sure to find out the actual height of the flooring to be installed. Some materials, such as ceramic tile, can add significantly to the height of the finished floor.

Instructor's Notes:

Reading Notes

The notes are crucial elements of the drawing set. Receptacles, for example, are hard to position precisely based on a scaled drawing alone, and yet the designer may call for exact locations. For example, the designer may want receptacles exactly 6" above the kitchen counter backsplash and centered on the sink.

one power busway. Only the power busway is shown on the power plan; the lighting busways will appear on the lighting plan.

Figure 33 shows two runs of busways: one running the length of the building on the south end (top wall on drawing), and one running the length of the north wall. The symbol list in *Figure 32* shows this busway to be designated by two parallel lines with a series of X's inside. The symbol list further describes the busway as General Electric Type DK-100. These busways are fed from the main distribution panel (circuits MDP-1 and MDP-2) through GE No. DHIBBC41 tap boxes.

The NEC defines a *busway* as a grounded metal enclosure containing factory-mounted, bare or insulated conductors, which are usually copper or aluminum bars, rods, or tubes.

The relationship of the busway and hangers to the building construction should be checked prior to commencing the installation so that any problems due to space conflicts, inadequate or inappropriate supporting structure, openings through walls, etc. are worked out in advance so as not to incur lost time.

For example, the drawings and specifications may call for the busway to be suspended from brackets clamped or welded to steel columns. However, the spacing of the columns may be such that additional supplementary hanger rods suspended from the ceiling or roof structure may be necessary for the adequate support of the busway. To offer more assistance to workers on the office/warehouse project, the engineer may also provide an additional drawing that shows how the busway is to be mounted.

Other details that appear on the floor plan in *Figure 34* include the general arrangement of the underfloor duct system, junction boxes and feeder conduit for the underfloor duct system, and plan views of the service and telephone equipment, along with duplex receptacle outlets. A note on the drawing requires all receptacles in the toilets to be provided with ground fault circuit interrupter (GFCI) protection. The letters EWC next to the receptacle in the vestibule designates this receptacle for use with an electric water cooler.

8.0.0 ◆ LIGHTING FLOOR PLAN

A skeleton view of a lighting floor plan is shown in *Figure 35*. Again, the architect's/engineer's title blocks appear in the lower right corner of the drawing. A key plan, as discussed previously, appears above the engineer's title block. This plan is drawn to the same scale as the power plan; that is, ⅛" = 1'-0". A lighting fixture schedule appears in the upper right corner of the drawing and some installation notes appear below the schedule.

The lighting outlet symbols found on the drawing for the office/warehouse building represent both incandescent and fluorescent types; a circle on most electrical drawings usually represents an incandescent fixture, and a rectangle represents a fluorescent one. All of these symbols are designed to indicate the physical shape of a particular fixture and are usually drawn to scale.

The type of mounting used for all lighting fixtures is usually indicated in a lighting fixture schedule, which in this case is shown on the drawings. On some projects, the schedule may be found only in the written specifications.

The type of lighting fixture is identified by a numeral placed inside a triangle near each lighting fixture. If one type of fixture is used exclusively in one room or area, the triangular indicator need only appear once with the word *ALL* lettered at the bottom of the triangle.

8.1.0 Drawing Schedules

A schedule is a systematic method of presenting notes or lists of equipment on a drawing in tabular form. When properly organized and thoroughly understood, schedules are powerful timesaving devices for both those preparing the drawings and workers on the job.

For example, the lighting fixture schedule shown in *Figure 36* lists the fixture and identifies each fixture type on the drawing by number. The manufacturer and catalog number of each type are given along with the number, size, and type of lamp for each.

Discuss the types of information conveyed by lighting floor plans.

Show Transparency 44 (Figure 35).

Discuss the types of information conveyed by drawing schedules.

Show Transparency 45 (Figure 36).

Pass around the lighting floor plan and associated schedule from a set of instructor-supplied drawings and have the trainees identify various fixture types and their locations.

Figure 35 ◆ Sample lighting plan.

LIGHTING FIXTURE SCHEDULE

SYMBOL	TYPE	MANUFACTURER AND CATALOG NUMBER	MOUNTING	LAMPS
⊥	A	LIGHTOLIER 10234	WALL	2-40W T-12WWX
▭	B	LIGHTOLIER 10420	SURFACE	2-40W T-12 WWX
✦	C	ALKCO RPC-210-6E	SURFACE	2-8W T-5
⊷O	D	P 7 S AL 2936	WALL	1-100W 'A'
O	E	P 7 S 110	SURFACE	1-100W 'A'

110F36.EPS

Figure 36 ◆ Lighting fixture schedule.

Instructor's Notes:

At times, all of the same information found in schedules will be duplicated in the written specifications, but combing through page after page of written specifications can be time consuming, and workers do not always have access to the specifications while working, whereas they usually do have access to the working drawings. Therefore, the schedule is an excellent means of providing essential information in a clear and accurate manner, allowing the workers to carry out their assignments in the least amount of time.

Other schedules that are frequently found on electrical working drawings include:

• Connected load schedule
• Panelboard schedule
• Electric heat schedule
• Kitchen equipment schedule
• Schedule of receptacle types

There are also other schedules found on electrical drawings, depending upon the type of project. However, most will deal with lists of equipment such as motors, motor controllers, and similar items.

9.0.0 ◆ ELECTRICAL DETAILS AND DIAGRAMS

Electrical diagrams are drawings that are intended to show electrical components and their related connections. They show the electrical association of the different components, but are seldom, if ever, drawn to scale.

9.1.0 Power-Riser Diagrams

One-line (single-line) block diagrams are used extensively to show the arrangement of electric service equipment. The power-riser diagram in *Figure 37*, for example, was used on the office/warehouse building under discussion and is typical of such drawings. The drawing shows all pieces of electrical equipment as well as the connecting lines used to indicate service-entrance conductors and feeders. Notes are used to identify the equipment, indicate the size of conduit necessary for each feeder, and show the number, size, and type of conductors in each conduit.

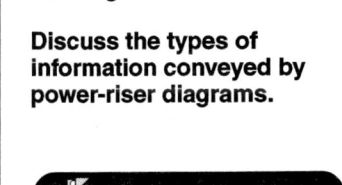

Show the electrical drawings from a set of instructor-supplied drawings.

Discuss the types of information conveyed by power-riser diagrams.

Show Transparency 46 (Figure 37).

Figure 37 ◆ Typical power-riser diagram.

Discuss the types of information conveyed by panelboard schedules.

Show Transparency 47 (Figure 38).

Discuss the types of information conveyed by schematic diagrams.

Show Transparencies 48 and 49 (Figures 39 and 40).

A panelboard schedule (*Figure 38*) is included with the power-riser diagram to indicate the exact components contained in each panelboard. This panelboard schedule is for the main distribution panel. On the actual drawings, schedules would also be shown for the other two panels (PNL A and PNL B).

In general, panelboard schedules usually indicate the panel number, type of cabinet (either flush- or surface-mounted), panel mains (ampere and voltage rating), phase (single- or three-phase), and number of wires. A four-wire panel, for example, indicates that a solid neutral exists in the panel. Branches indicate the type of overcurrent protection; that is, they indicate the number of poles, trip rating, and frame size. The items fed by each overcurrent device are also indicated.

9.2.0 Schematic Diagrams

Complete schematic wiring diagrams are normally used only in complicated electrical systems, such as control circuits. Components are represented by symbols, and every wire is either shown by itself or included in an assembly of several wires, which appear as one line on the drawing. Each wire should be numbered when it enters an assembly and should keep the same number when it comes out again to be connected to some electrical component in the system. *Figure 39* shows a complete schematic wiring diagram for a three-phase, AC magnetic non-reversing motor starter.

Note that this diagram shows the various devices in symbol form and indicates the actual connections of all wires between the devices. The three-wire supply lines are indicated by L_1, L_2, and L_3; the motor terminals of motor M are indicated by T_1, T_2, and T_3. Lines L_1 and L_3 each have a thermal overload protection device (OL) connected in series with normally open line contactors C_1 and C_3, respectively, which are both controlled by the magnetic starter coil, C. Each contactor has a pair of contacts that close or open during operation. The control station, consisting of start pushbutton 1 and stop pushbutton 2, is connected across lines L_1 and L_2. An auxiliary contactor (C_4) is connected in series with the stop pushbutton and in parallel with the start pushbutton. The control circuit also has normally closed overload contactors (OC) connected in series with the magnetic starter coil (C).

Any number of additional pushbutton stations may be added to this control circuit similarly to the way in which three-way and four-way switches are added to control a lighting circuit. When adding pushbutton stations, the stop buttons are always connected in series and the start buttons are always connected in parallel. *Figure 40* shows the same motor starter circuit in *Figure 39*, but this time it is controlled by two sets of start/stop buttons.

Schematic wiring diagrams have only been touched upon in this module; there are many other details that you will need to know to perform your work in a proficient manner. Later modules cover wiring diagrams in more detail.

PANELBOARD SCHEDULE

| PANEL No. | CABINET TYPE | PANEL MAINS | | | BRANCHES | | | | | ITEMS FED OR REMARKS |
		AMPS	VOLTS	PHASE	1P	2P	3P	PROT.	FRAME	
MDP	SURFACE	600A	120/208	3φ,4-W	-	-	1	225A	25,000	PANEL "A"
					-	-	1	100A	18,000	PANEL "B"
					-	-	1	100A		POWER BUSWAY
					-	-	1	60A		LIGHTING BUSWAY
					-	-	1	70A		ROOFTOP UNIT #1
					-	-	1	70A	↓	SPARE
					-	-	1	600A	42,000	MAIN CIRCUIT BRKR

110F38.EPS

Figure 38 ◆ Typical panelboard schedule.

Instructor's Notes:

_____

_____

_____

_____

_____

_____

_____

_____

_____

Figure 39 ◆ Schematic wiring diagram.

Figure 40 ◆ Circuit being controlled by two sets of start/stop buttons.

9.3.0 Drawing Details

A detail drawing is a drawing of a separate item or portion of an electrical system, giving a complete and exact description of its use and all the details needed to show the electrician exactly what is required for its installation. For example, the power plan for the office/warehouse has a sectional cut through the busduct. This is a good example of where an extra, detailed drawing is desirable.

Classroom

Discuss the types of information conveyed by detail drawings.

Using the electrical drawings from an instructor-supplied set of drawings, have the trainees identify the power-riser diagram, panelboard schedule, and any schematic or detail drawings.

Understanding Contact Symbols

When a drawing shows normally open or normally closed contacts, the word *normally* refers to the condition of the contacts in their deenergized or shelf state.

Don't Just Check the Electrical Plan

Always review all of the drawings in a drawing set, not just the electrical plan. Several drawings in the set will have information of relevance to the electrician. For example, you should review:

• Site plans for utility lines and elevation information
• Mechanical drawings for routing, clearances, and HVAC equipment and controls
• Architectural drawings for the type of construction (block, wood, metal stud, etc.), fire ratings, and special details.
• Finish drawings (e.g., reflected ceiling plans) for locations of fixtures, fans, and other devices
• Room finish schedules for ceiling heights and floor and wall finishing details
• Plumbing drawings for pumps, water service, and sprinklers

A set of electrical drawings will sometimes require large-scale drawings of certain areas that are not indicated with sufficient clarity on the small-scale drawings. For example, the site plan may show exterior pole-mounted lighting fixtures that are to be installed by the contractor.

10.0.0 ◆ WRITTEN SPECIFICATIONS

The written specifications for a building or project are the written descriptions of work and duties required of the owner, architect, and consulting engineer. Together with the working drawings, these specifications form the basis of the contract requirements for the construction of the building or project. Those who use the construction drawings and specifications must always be alert to discrepancies between the working drawings and the written specifications. Such discrepancies may occur when:

• Architects or engineers use standard or prototype specifications and attempt to apply them without any modification to specific working drawings.
• Previously prepared standard drawings are changed or amended by reference in the specifications only and the drawings themselves are not changed.
• Items are duplicated in both the drawings and specifications, but an item is subsequently amended in one and overlooked in the other contract document.

In such instances, the person in charge of the project has the responsibility to ascertain whether the drawings or the specifications take precedence. Such questions must be resolved, preferably before the work begins, to avoid added cost to the owner, architect/engineer, or contractor.

10.1.0 How Specifications Are Written

Writing accurate and complete specifications for building construction is a serious responsibility for those who design the buildings because the specifications, combined with the working drawings, govern practically all important decisions made during the construction span of every project. Compiling and writing these specifications is not a simple task, even for those who have had considerable experience in preparing such documents. A set of written specifications for a single project will usually contain thousands of products, parts, and components, and the methods of installing them, all of which must be covered in either the drawings and/or specifications. No one can memorize all of the necessary items required to accurately describe the various areas of construction. One must rely upon reference materials such as manufacturer's data, catalogs, checklists, and, most of all, a high-quality master specification.

Classroom

Discuss the types of information conveyed by written specifications.

ELECTRICAL LEVEL ONE—TRAINEE MODULE 26110

Instructor's Notes:

_____

_____

_____

_____

_____

_____

_____

_____

_____

ELECTRICAL LEVEL ONE—INSTRUCTOR'S GUIDE MODULE 26110

10.2.0 Format of Specifications

For convenience in writing, speed in estimating, and ease of reference, the most suitable organization of the specifications is a series of sections dealing successively with the different trades. All the work of each trade should be incorporated into the section devoted to that trade. Those people who use the specifications must be able to find all information needed without spending too much time looking for it.

10.2.1 CSI Format

The Construction Specification Institute (CSI) has developed the Uniform Construction Index that allows all specifications, product information, and cost data to be arranged into a uniform system. This format is followed on most large construction projects in North America. All construction is divided into 16 divisions, and each division has several sections and subsections. The following outline describes the various divisions normally included in a set of specifications for building construction.

Division 1: General Requirements—This division summarizes the work, alternatives, project meetings, submissions, quality control, temporary facilities and controls, products, and the project closeout. Every responsible person involved with the project should become familiar with this division.

Division 2: Site Work—This division outlines work involving such items as paving, sidewalks, outside utility lines (electrical, plumbing, gas, telephone, etc.), landscaping, grading, and other items pertaining to the outside of the building.

Division 3: Concrete—This division covers work involving footings, concrete formwork, expansion and contraction joints, cast-in-place concrete, specially finished concrete, precast concrete, concrete slabs, and similar items.

Division 4: Masonry—This division covers concrete, mortar, stone, masonry accessories, and similar items.

Division 5: Metals—Metal roofs, structural metal framing, metal joists, metal decking, ornamental metal, and expansion control normally fall under this division.

Division 6: Carpentry—Items falling under this division include rough carpentry, heavy timber construction, trestles, prefabricated structural wood, finish carpentry, wood treatment, architectural woodwork, and the like. Plastic fabrications may also be included in this division.

Division 7: Thermal and Moisture Protection—Waterproofing is the main topic discussed under this division. Other related items such as dampproofing, building insulation, shingles and roofing tiles, preformed roofing and siding, membrane roofing, sheet metal work, wall flashing, roof accessories, and sealants are also included.

Division 8: Doors and Windows—All types of doors and frames are included under this division: metal, plastic, wood, etc. Windows and framing are also included, along with hardware and other window and door accessories.

Division 9: Finishes—Included in this division are the types, quality, and workmanship of lath and plaster, gypsum wallboard, tile, terrazzo, acoustical treatment, ceiling suspension systems, wood flooring, floor treatment, special coatings, painting, and wallcovering.

Division 10: Specialties—Specialty items such as chalkboards and tackboards, compartments and cubicles, louvers and vents that are not connected with the mechanical system, wall and corner guards, access flooring, specialty modules, pest control, fireplaces, flagpoles, identifying devices, lockers, protective covers, postal specialties, partitions, scales, storage shelving, wardrobe specialties, and similar items are covered in this division of the specifications.

Division 11: Equipment—The equipment included in this division could include central vacuum cleaning systems, bank vaults, darkrooms, food service, vending machines, laundry equipment, and many similar items.

Division 12: Furnishing—Items such as cabinets and storage files, fabrics, furniture, rugs and mats, seating, and similar furnishings are included under this division.

Division 13: Special Construction—Such items as air-supported structures, incinerators, and other special items will fall under this division.

Division 14: Conveying Systems—This division covers conveying apparatus such as dumbwaiters, elevators, hoists and cranes, lifts, material-handling systems, turntables, moving stairs and walks, pneumatic tube systems, and power scaffolding.

Division 15: Mechanical—This division includes plumbing, heating, ventilating, and air conditioning and related work. Electric heat is sometimes covered under Division 16, especially if individual baseboard heating units are used in each room or area of the building.

Division 16: Electrical—This division covers all electrical requirements for the building including lighting, power, alarm and communications systems, special electrical systems, and related electrical equipment. This is the division that electricians will use the most.

A sample set of electrical specifications is shown in *Figure 41*.

Classroom

Discuss the 16 divisions in the CSI format for specifications.

Refer the trainees to *Figure 41*. Have them identify various types of information from these sample specifications.

```
 1    SECTION 16011 - ELECTRICAL OUTLINE OF WORK
 2
 3    PART 1 - GENERAL
 4
 5
 6    The work included in Division 16 electrical includes, but is not necessarily limited to the following
 7    items and systems:
 8
 9    .      Two new 480/277 volt, 3 phase underground services from Duke Power Company pad
10           mounted transformer.
11    .      Connection to power company transformers.
12    .      Concrete pad for power company transformers.
13    .      Installation of new power company meter box.
14    .      All coordination, labor and materials required to connect new chiller service to existing
15           Duke Power Company pad mounted transformer.
16    .      Motor control center.
17    .      Lighting and receptacle panelboard.
18    .      Feeder circuits, including conduits, conductors, troughs and fittings.
19    .      Branch circuits, including conduits, conductors, outlets, boxes, receptacles, switches and
20           fittings.
21    .      Lighting fixtures including lamps.
22    .      Equipment tests.
23    .      Wiring devices
24    .      Grounding systems.
25    .      Safety switches.
26    .      Power connection to all equipment requiring power.
27    .      Provide new lay-in ceiling in corridor of Building 'B'.
28    .      Remove existing ceiling in part of Guidance Area of Building 'A' and provide new lay-in
29           ceiling throughout the entire Guidance Area.
30
31
32    END OF SECTION
33
```

16011-1

110F41A.TIF

Figure 41 ◆ Sample electrical specifications (1 of 12).

ELECTRICAL LEVEL ONE—TRAINEE MODULE 26110

Instructor's Notes:

_____

_____

_____

_____

_____

_____

_____

_____

_____

SECTION 16110 - RACEWAYS

PART 1 - GENERAL:

GENERAL

All wiring shall be in rigid metal conduit, 'RMC', except as otherwise noted.

Electric metallic tubing 'EMT' may be used for concealed and exposed work, except as listed below.

        (1) Exposed to the weather, or in damp location.
        (2) In earth or stone.
        (3) In concrete slabs on grade.
        (4) Where obviously subject to mechanical injury.
        (5) Where specified otherwise.

EMT may be used in lieu of 'RMC' in the following sizes, 1/2 inch through 2-inch for power circuits, 1/2 inch through 4 inch for communications circuits and control wiring as applicable, subject to the use limitations specified above. Intermediate metal conduit 'IMC' may be used for 'RMC'.

Rigid non-metallic conduit (RNMC) may be used in in the following applications only. (All stub-ups shall be 'RMC' or 'IMC' elbows.)

        Below slab, encased all around in not less than three inches of
        concrete and not less than 18 inches of ground cover for power service
        and power feeders, under building. Except, that conduit under
        building foundation, or conduit run through footings or foundations,
        shall be rigid steel conduit.

        Underground, buried not less than 30" below grade for area lighting
        circuits.

        In concrete slab on grade for branch circuits utilizing 3/4 inch or
        smaller conduit when encased by at least 2-inches of concrete all around.

Flexible metallic conduit shall not be used as a wiring method, other than when specifically noted to be used, without prior permission of the Architect/Engineer.

Type AC (BX) armored cable is not permitted in this project.

SLEEVES AND PENETRATIONS:

See Section 16115 for required sleeves and method of achieving raceway penetrations.

APPLICABLE SPECIFICATIONS AND STANDARDS:

The materials specified here shall meet the following specifications and standards in their current edition.

16110-1

110F41B.TIF

Figure 41 ◈ Sample electrical specifications (2 of 12).

```
 1        (1)  UL Standards
 2
 3             Electric metallic Tubing
 4             Flexible Metal Conduits-UL-1
 5             Rigid Metal Conduit UL-6
 6             Intermediate Metal Conduit UL-1242
 7
 8        (2)  NEMA Standards
 9
10             Electric plastic conduit-TC-2
11
12        (3)  ANSI Standards
13
14             Specifications for Rigid Steel Conduit, Zinc Coated,
15             ANSI C80.1.
16
17   PART 2 - PRODUCTS:
18
19   RACEWAYS:
20
21   General: Minimum size conduit shall be 1/2 inch.
22
23   Rigid Metal Conduit:
24
25   Rigid metal conduit shall be schedule 40, of the best quality steel.
26
27   The interior and exterior surfaces of the conduit shall be protected with a metallic zinc coating.
28   Rigid steel conduit shall be galvanized by the Hot-Dip process in accordance with ASTM A 123.
29
30   Fittings for 'RMC' shall be threaded UL listed.
31
32   Electric Metallic Tubing:
33
34   Electrical metallic tubing shall be rigid metal conduit of the thin-wall type in straight lengths,
35   elbows or bends for use as raceways for wire or cables in an electrical system.
36
37    Electrical metallic tubing shall utilize hexagonal steel type compression threadless fittings of
38   galvanized steel throughout.  All fittings shall be UL listed for concrete-tight and rain-tight
39   construction.  All EMT entrance fittings shall be provided with insulated throats.
40
41   Flexible Metallic Conduit:
42
43   Flexible metallic conduit shall conform to UL standard 'Flexible Steel Conduit'. All steel used in the
44   fabrication of the conduit shall be zinc coated.
45
46   Liquid-tight flexible steel conduit shall be provided with a protective jacket of polyvinyl chloride
47   extruded over a flexible interlocked galvanized steel core to protect wiring against moisture, oil,
48   chemicals and corrosive fumes.
49
50   Flexible conduit connectors shall be UL listed T & B nylon-insulated "Tite-Bite", or equivalent from
51   "Blackhawk.'
```

16110-2

110F41C.TIF

Figure 41 ◆ Sample electrical specifications (3 of 12).

Instructor's Notes:

1 Rigid Non-Metallic Conduit:
2
3 Schedule 40 (EPC-40), heavy wall polyvinyl chloride plastic conduit and fittings, UL listed, suitable
4 for 90 degree C. conductors.
5
6 Minimum size RNMC conduit shall be 3/4-inch.
7
8 Intermediate Metal Conduit:
9
10 Intermediate metal conduit 'IMC' shall be zinc coated steel, UL listed and labeled.
11
12 Fittings shall be the same type as for 'RMC.'
13
14 Wireways and Troughs:
15
16 Wireways or troughs shall be of the size noted on the drawings or as required by the NEC of code
17 gauge galvanized steel. Sizes 6 inch x 6 inch and smaller wireways and all troughs shall be of the
18 hinged cover type except as otherwise noted on the drawings. Larger sized wireways shall be of
19 the flangeless screw cover lay-in type. All shall be without knockout, and shall be provided with
20 fittings, supports, and apurtenances as required.
21
22 PART 3 - EXECUTION:
23
24 INSTALLATION OF CONDUIT AND TUBING:
25
26 Metallic raceways shall not be stored exposed to the weather.
27
28 Conduits shall be concealed within the walls, ceilings, and floors, where possible, and shall be kept
29 at least 6 inches from parallel runs of flues, steam pipes, or hot water pipes. Exposed runs of
30 conduit or tubing, and conduit or tubing run above suspended ceilings, shall have supports spaced
31 not more than 8 feet apart and shall be installed with runs parallel or perpendicular to walls,
32 structural members, or intersections of vertical planes and ceilings with right-angle turns consisting
33 of cast metal fittings or symmetrical bends. All raceways shall be run in a neat and orderly
34 fashion. Conduits or tubing run in diagonal or disorganized way shall be removed from the
35 premises if so structed by the A/E. Bends and offsets shall be avoided where possible, but where
36 necessary shall be made with an approved hickey or conduit bending machine. Conduit or tubing
37 which has been crushed or deformed in any way shall not be installed.
38
39 Conduit and tubing shall be supported on approved types of galvanized wall brackets, ceiling
40 trapezes, strap hangers, or pipe straps, secured by means of toggle bolts on hollow masonry units,
41 expansion bolts in concrete or brick, machine screws on metal surfaces, and wood screws on
42 wood construction. Nails shall not be used as the means of fastening boxes on conduits. Wooden
43 plugs inserted in masonry or concrete shall not be used as a base to secure conduit supports.
44
45 Conduits shall be installed in such manner as to insure against trouble from the collection of
46 trapped condensation, and all runs of conduit shall be arranged so as to be devoid of traps where
47 feasible. The contractor shall exercise the necessary precautions to prevent the lodgment of dirt,
48 plaster, or trash in conduit, tubing, fittings, and boxes during the course of installation by the use
49 of T & B pushpennies, appleton pennies, or equal closures. A run of conduit or tubing which has
50 become clogged shall be entirely freed of these accumulations or shall be replaced.
51

16110-3   110F41D.TIF

Figure 41 ◆ Sample electrical specifications (4 of 12).

1   Conduit shall be securely fastened to all sheet metal enclosures with double galvanized locknuts
2   and insulated bushings, care being observed to see that full number of threads project through to
3   permit the bushing to be drawn tight against the end of conduit, after which the locknuts shall be
4   made up sufficiently tight to insure positive ground continuity between conduit and box.

6   Double locknuts shall be used on all feeder and motor circuit conduits and where insulated
7   bushings are used. Insulated bushings of molded bakelite shall be used on all conduit entrances,
8   one inch over in size, into junction boxes, panel boxes and motors starters having sheet metal
9   enclosures. Galvanized steel insulated throat fittings shall be used for EMT work.

11 Rigid metal conduit 'RMC' installed underground 5 feet or more from the building shall have a
12 minimum cover of three feet. 'RMC' directly installed underground shall be coated with two coats
13 of bitumastic or asphalt paint. Conduit installed underground or in concrete on ground shall be
14 made watertight by wrapping the joints with 'Teflon' tape.

16 <u>EXPANSION FITTINGS:</u>

18 Conduit crossing expansion joints in concrete slabs shall be provided with suitable expansion
19 fittings, or other suitable means shall be provided to compensate for the building expansion and
20 contraction.

22 <u>PULL WIRE:</u>

24 Nylon pull wire not less than 5/32 inches in diameter shall be installed in all empty conduit longer
25 than 10 feet. Pull wire shall be secured at each end and tagged for identification of the use of the
26 conduit.

29 END OF SECTION

16110-4

110F41E.TIF

Figure 41 ◆ Sample electrical specifications (5 of 12).

Instructor's Notes:

_____

_____

_____

_____

_____

_____

_____

_____

_____

1   SECTION 16120 - CONDUCTORS
2
3   PART 1 - GENERAL
4
5   SCOPE:
6
7   This Section applies to secondary conductors for systems rated 600 volts and below.
8
9   A complete system of conductors shall be installed in the raceway systems as specified here and
10   shown on drawings.
11
12   APPLICABLE SPECIFICATIONS AND STANDARDS:
13
14   Compliance:
15
16   The materials specified here shall meet the following specifications and standards in their current
17   edition.
18
19   (1) UL Standards:
20
21   Insulation tape
22   Wire Connectors and Soldering lugs
23
24   (2) NEMA Standards:
25
26   Thermoplastic - Insulated WC 5 (IPCEA S-61-402)
27
28   PART 2 PRODUCTS
29
30   CONDUCTORS:
31
32   All conductors shall be made of copper.
33
34   Conductors, unless otherwise noted, shall be heat and moisture resistant grade, thermoplastic
35   insulated. Conductors No. 8 AWG and larger shall be stranded copper conductors, dual rated, Type
36   THHN-THWN; Conductors No. 10 and smaller shall be solid copper, Type THHN-THWN (dual
37   rated), unless otherwise required below. Branch circuit conductors for all other lighting fixtures
38   shall have a temperature rating of not less than what is required by the UL listing of the fixture
39   with a minimum rating of 90 degrees C.
40
41   Conductors for signal and control circuits above 50 volts AC may be THWN-THHN as permitted
42   by NEC, No. 14 AWG. Conductors for signal and control circuits below 50 volts AC may be 300-
43   volt, PVC insulated, No. 14 AWG.
44
45   Branch circuit conductors shall be not smaller than No. 12 AWG, except that conductors for branch
46   circuits whose length from panel to center of load exceeds 75 feet for the 280/120 volt system,
47   or 150 feet for 277/480 volt system, shall not be smaller than No. 10 AWG from the panel to the
48   first outlet box in the circuit regardless of what is scheduled on panelboard schedule.
49
50   Conductors being connected to transformers and other equipment shall have a temperature rating
51   as required by the transformer or equipment manufacturer.
52
53   PART 3 - EXECUTION:
54

16120-1   110F41F.TIF

Figure 41 ◆ Sample electrical specifications (6 of 12).

```
1   SPLICES:
2
3   Solid Conductor Splices:
4
5   Solid conductors namely those sized #10, #12, and #14 AWG copper, and smaller, shall be spliced
6   by twisting securely and by means of hot-dipped solder plus gum rubber tape, plus friction tape,
7   or plastic tape approved as a substitute for friction tape. The contractor shall use Ideal "wire-nuts"
8   for recessed lighting fixture lead splices to branch circuit conductors.
9
10  Stranded Conductor Splices:
11
12  Namely #8 AWG and larger, shall be spliced by approved mechanical connectors plus gum tape,
13  plus friction or plastic tape. Solderless mechanical connectors, for splices and taps provided with
14  U.L. approval insulating covers, may be used instead of mechanical connectors plus tape.
15
16  INSTALLATION OF CONDUCTORS:
17
18  Conductors shall be continuous from outlet to outlet, and no splices shall be made except within
19  outlet or junction boxes, troughs and gutters. Junction boxes may be utilized where required. No
20  'condulet' type fitting shall be used on any service conduits. If other than long radius bends are
21  required, pull boxes sized in accordance with the NEC shall be used.
22
23  Home runs may be combined in one conduit, provided all connections are in accordance with
24  National Electrical Code requirements, and the maximum unbalanced current in the neutral does
25  not extend the capacity of the conductor, and the conductors are not required to be derated to
26  below circuit capacity.
27
28  Conductors in vertical runs shall be supported as required by NEC 300-19.
29
30  COLOR CODING:
31
32  Conductors, feeders, and branch circuits shall be color coded by phases as follows:
33
34  480/277-volt systems: Phase A-brown; Phase B-orange; Phase C-yellow; Neutral - white with
35  identifiable color stripe, other than green; grounding wire - green.
36
37  208/120-volt systems: Phase A-black; Phase B-red; Phase C-blue; neutral-white; Grounding wire-
38  green.
39
40  Insulating tape of proper color shall be used to identify the phase conductors No. 6 AWG and
41  largerconductors.
42
43  All feeders, sub-feeds to panels, motors, etc., shall be completely phased outas to sequence
44  and rotation. Phase sequence shall be A-B-C from front to rear, top to bottom, left to right when
45  facing equipment.
46
47
```

16120-2

110F41G.TIF

Figure 41 ◆ Sample electrical specifications (7 of 12).

ELECTRICAL LEVEL ONE—TRAINEE MODULE 26110

Instructor's Notes:

```
 1    SECTION 16136 - WIRING DEVICES, BOXES & ENCLOSURES
 2
 3    PART 1 GENERAL
 4
 5    WORK INCLUDED:
 6
 7    Work under this Section includes but is not necessarily limited to the following.
 8
 9    Receptacles, toggle switches, dimmers and photoelectric switching devices.
10    Outlet Boxes
11    Cabinets and Enclosures
12
13    APPLICABLE SPECIFICATIONS AND STANDARDS:
14
15    Equipment specified in this Section shall meet the following specifications and standards.
16
17        (1)  UL Standards
18
19            Attachment Plug and Receptacles
20            Electric Cabinet and Boxes
21            Outlet Boxes and Fittings
22            Snap Switches
23
24        (2)  NEMA Standards
25
26            Boxes, OS1
27            Wiring Devices, General - Purpose, WD1
28            Wiring Devices, Specific - Purpose, WD5
29
30    PART 2  PRODUCTS
31
32    OUTLET BOXES, PULL BOXES, CABINETS AND ENCLOSURES:
33
34    Boxes:
35
36    Boxes shall have sufficient volume to accomodate the number of conductors entering the box in
37    accordance with the requirements of NFPA 70, Article 370.  Boxes that are exposed to the
38    weather or that are in normally wet locations shall be of the cast-metal type having threaded hubs.
39    Boxes shall be of suitable construction for installation in the environment of their location.  Unless
40    otherwise specifically stated all boxes shall be metallic boxes.
41
42    Zinc-coated or cadium-plated sheet steel boxes, or a class to satisfy the conditions of each outlet,
43    shall be used in concealed work or in exposed work above eight feet from floor.
44
45    Fixture outlet boxes on ceiling shall be not less than 4 inch octagonal.  Fixture outlet boxes in
46    concrete ceiling shall be of the 4 inch octagonal concrete type, set flush with the finished surface.
47    Fixture outlet boxes on plastered ceilings shall be fitted with open covers set to come flush with
48    the finished surface.
49
50    Switch and receptacle outlet boxes in dry walls, plastered walls and pour-in concrete walls shall
51    be not less than 4 inches square cut with appropriate extension to set flush with the finished
52    surface.  One-piece gang or gangable boxes not less than 2 inches deep shall be utilized where the
```

16136-1

110F41H.TIF

Figure 41 ◆ Sample electrical specifications (8 of 12).

1   use of 4-inch square boxes is not feasible.
2
3   Unless otherwise noted on the drawings, outlet, junction or pull boxes not larger than 5 inches
4   square and within eight feet from floor level in exposed work shall be of cast steel or alloy with
5   threaded hubs and appropriate covers.
6
7   Outlet boxes in unplastered masonry and gypsum drywall walls shall be tile type.
8
9   Outlet boxes for use with conduit and tubing systems shall be not less than 1-1/2 inches deep.
10
11   A device plate or cover shall be provided for each outlet to suit the outlet.
12
13   <u>Pull Boxes:</u>
14
15   Pull boxes shall be constructed of code-gauge galvanized sheet metal. Boxes shall be of not less
16   than the minimum size required by the National Electrical Code and shall be furnished with screw
17   fastened covers. When several feeders pass through a common pull box they shall be tagged to
18   indicate clearly their electrical characteristics, circuit numbers and panel designations.
19
20   Pull boxes shall be furnished and installed where necessary in the raceway system to facilitate
21   conductor installation. Except as otherwise noted for telephone raceways, conduit runs longer
22   than 150 feet, or with more than 360 degrees compound angle bends, shall have a pull box
23   insalled at a convenient intermediate location. Normally, when feeder routing is shown on
24   drawings, pull boxes are not acceptable. It is the responsibility of the electrical contractor to
25   provide pull boxes as necessary to meet the stated requirements.
26
27   <u>Cabinets:</u>
28
29   Cabinet boxes shall be constructed of zinc-coated sheet steel and shall conform with the
30   requirements of Underwriters' Laboratories "Standards for Cabinets and Cutout Boxes". Unless
31   otherwise noted, surface mounted cabinet trims shall have a corrosion inhibiting primer and a
32   lacquer finish. Flush mounted cabinets shall be factory primed ready for finish painting by the
33   General Contractor. Cabinets shall be of suitable construction for installation in the environment
34   of their locations.
35
36   Systems cabinet, if shown, shall be provided with interior dimensions not less than those indicated
37   on the drawings. Trims shall provide maximum size openings to the box interiors. Cabinets shall
38   be provided with 5/8 inch fire retardant plywood back-boards having an insulating varnish finish.
39
40   <u>Device Plates:</u>
41
42   A device plate shall be provided for each outlet (including telephone and computer system outlets)
43   to suit the device installed. Screws shall be of metal with countersunk heads, with finish to match
44   the finish of the plate.
45
46   Device plates shall be of the one-piece type, of suitable shape for the devices to be covered. The
47   use of sectional device plates will not be permitted. Plates shall be as follows:
48
49   .   Plates on surface boxes in unfinished areas shall be of galvanized steel with beveled edges.
50   .   All plates on walls with flush switches or receptacles shall be 302 stainless steel, 0.32"
51   nominal thickness.
52

16136-2

110F41I.TIF

Figure 41 ◆ Sample electrical specifications (9 of 12).

Instructor's Notes:

_____

_____

_____

_____

_____

_____

_____

_____

_____

1   Plates on surface boxes in finished areas shall be as for flush outlets.

2

3   <u>Switches and Receptacles:</u>

4

5   Switches and receptacles shall be as shown on the drawings and in the symbol schedule, and shall
6   meet the latest federal specifications W-S 896 or W-C 596 as verified by UL. Devices shall be the
7   product of one of the followingmanufacturers complying with referenced NEMA Standards - Arrow-
8   Hart Electric Company, Bryant Electric Company, General Electric Company, Harvey Hubbell, Inc.,
9   Slater Electric, Inc., Pass & Seymour, Inc., Sierra Electric, or Leviton. 20-ampere and 15 ampere
10   receptacles shall be heavy duty, hospital grade with nylon body. Switches shall be 20-ampere,
11   specification grade.

12

13   Color of switches and receptacles shall be gray.

14

15   Receptacles with ground fault interrupting (GFI) Protection.

16

17   GFI receptacles shall be NEMA 5-15R configuration, UL listed with"noise-suppressed" circuitry to
18   eliminate false trippings. GFI receptacles shall provide protection to all other receptacles installed
19   "downstream" on the same branch circuit. Provide separate neutral conductor for GFI receptacle
20   circuit.

21

22   <u>Wiring Device Schedule</u>

23

24   For clarity and to identify the class and type of devices required, refer to the following schedule
25   of devices by Pass & Seymour (P&S). Device color shall be as previous specified.

26

27   <u>Switches:</u>   <u>Catalogue Number</u>

28

	Switches	Catalogue Number
29	Single Pole	20 AC 1
30	Double Pole	20 AC 2
31	Three Way	20 AC 3

32

33   <u>Receptacles:</u>

34

35   <u>NEMA Configuration</u>   <u>P&S Catalogue Number</u>

36

	NEMA Configuration	P&S Catalogue Number
37	5-15R	5262
38	5-20R	5362
39	5-30R	3802
40	5-50R	3803
41	6-20R	5862
42	6-30R	3801
43	6-50R	3804
44	10-30R	3860
45	10-50R	3890
46	14-30R	3864
47	15-20R	3821
48	15-30R	5740
49	18-20R	3822
50	L5-20R	L520-R
51	L5-30R	L530-R
52	L6-20R	L620-R

16136-3

110F41J.TIF

Figure 41 ◈ Sample electrical specifications (10 of 12).

1	L6-30R	L630-R
2	L15-20R	L1520-R
3	L15-30R	L1530-R

4

5 MULTI-OUTLET ASSEMBLIES AND SURFACE METAL RACEWAYS:

6

7 Surface metal raceways shall be provided as indicated on the drawings, complete with all
8 appropriate fittings as required, to provide a safe and complete installation. All components shall
9 be UL listed. Raceways shall be supported on approximately 18" centers with #8 flat lead
10 fasterners. The entire installation shall meet the requirements of Article 352 of the NEC. All field
11 cuts of the raceway shall be made square and shall have no rough edges.

12

13 PART 3 - EXECUTION:

14

15 INSTALLATION OF OUTLETS:

16

17 Location of outlets shown on drawings, other than those dimensioned, are only approximate, the
18 Owner shall have the right to make slight changes in the position of outlets if the Contractor is
19 notified before roughing-in is done. The Contractor shall study the general building plans in relation
20 to the spaces surrounding each outlet in order that his work may fit the other work required by
21 these specifications. When necessary, the Contractor shall relocate outlets of junction boxes so
22 that, when fixtures or other fittings are installed, they will be symmetrically located according to
23 room layout and will not interfere with other work or equipment. Do not install outlets back to
24 back. Outlets flush mounted in fire or sound wall shall be mounted not closer than 24" apart
25 horizontally when on same face of wall and 12" apart horizontally when mounted on opposite sides
26 of wall. For those outlets in fire or sound rated walls that must be closer than 24" apart, provide
27 Nelson Firestop system 'putty pads' all around outlets to maintain fire rating.

28

29 Minimum length of conduit connecting to adjacent flush in sound rated walls outlets sound rated
30 walls shall be 18".

31

32 Boxes shall be installed in a rigid and satisfactory manner, either by wood screws on wood,
33 expansion shields on masonry, or machine screws on steel work.

34

35 Recessed boxes in dry wall type construction shall be supported from both adjacent studs, or by
36 the use of metal stud brackets as manufactured by E-Z Mount Bracket Co., or equivalent.

37

38 PLATES:

39

40 Plates for receptacles and switches shall be installed with all four edges in continuous contact with
41 the finished wall surfaces without the use of mats or similar devices. Plaster fillings will not be
42 permitted. Plates shall be installed vertically and with an alignment tolerance of 1/16 inch.

43

44 SURFACE METAL RACEWAYS

45

46 Surface metal raceways shall be installed where noted on the drawings. Support on approximately
47 30" centers with #8 flat head screws. In addition, couplings, fittings and boxes shall be supported
48 independent of the raceway.

49

50 INSTALLATION OF PULL BOXES

51

52 Pull boxes shall be installed overhead or on walls at locations free of interference with equipment,

16136-4

110F41K.TIF

Figure 41 ◆ Sample electrical specifications (11 of 12).

Instructor's Notes:

_____

_____

_____

_____

_____

_____

_____

_____

_____

```
1   ducts, piping and activities being carried out at the premises.  Pull boxes with covers larger than
2   4 square feet shall be provided with two handles.
3
4
5   END OF SECTION
6
```

16136-5

110F41L.TIF

Figure 41 ◆ Sample electrical specifications (12 of 12).

Putting It All Together

Study the specifications in *Figure 41*. How would you need to combine information from the drawings and the written specifications to create a finished installation?

Note

The above specifications refer to color coding. Please refer to *NEC Section 210-4(d)* for specific requirements.

Summary

In this module, you learned the symbols and conventions used on architectural and engineering drawings. As an electrician, you need to know how to recognize the basic symbols used on electrical drawings and other drawings used in the building construction industry. You should also know where to find the meaning of symbols that you do not immediately recognize. Schedules, diagrams, and specifications often provide detailed information that is not included on the working drawings.

Reading architectural and engineering drawings takes practice and study. Now that you have the basic skills, take the time to master them.

Discuss the "Think About It."

Have the trainees complete the Review Questions and go over the answers prior to administering the Module Examination.

Administer the Module Examination and Performance Test. Be sure to record the testing results on Craft Training Report Form 200 and submit the results to the Training Program Sponsor.

INTRODUCTION TO ELECTRICAL BLUEPRINTS—TRAINEE MODULE 26110

10–57

Review Questions

Questions 1 through 7 refer to the seven electrical symbols shown below. In the spaces provided, place the letter corresponding to the correct answer found in the list.

1. ──o──)──   _____   a. Single Receptacle Outlet

2. ─⊖─   _____   b. Duplex Receptacle Outlet

3. ○   ○─   _____   c. Triplex Receptacle Outlet
   PC   PC

4. ≡⊕   _____   d. Incandescent Fixture
                                (Surface or Pendant)

5. ─⊖─   _____   e. Incandescent Fixture
                                with Pull Chain
   Ceiling   **Wall**   (Surface or Pendant)

6. ○   ○─   _____   f. Head Guy

7. ──●──   _____   g. Sidewalk Guy

110EX.EPS

Questions 8 through 10 refer to the sample electrical specifications shown in *Figure 41*.

8. Which of the following tasks is *not* covered under Division 16?
   a. Lighting fixtures
   b. New lay-in ceiling in the corridor of Building B
   c. Concrete pad for power company transformers
   d. New chiller

9. What is the minimum size RNMC conduit allowed by the specifications?
   a. ½ inch
   b. ¾ inch
   c. 1 inch
   d. ⁵⁄₁₆ inch

10. The color code specified for a 208/120V system is _____.
    a. phase A—black; phase B—red; phase C—blue
    b. neutral—white; grounding wire—green
    c. phase A—brown; phase B—orange; phase C—yellow
    d. Both a and b.

Instructor's Notes:

_____

_____

_____

_____

_____

_____

_____

_____

_____

Jerry Bass, Watson Electrical Construction Co.

Jerry Bass is a good example of an electrician who climbed the ladder of success one rung at a time. After 33 years in the industry, he finds it as exciting and challenging today as he did as an apprentice.

How did you choose a career in the electrical field?
I applied for a summer job with an electrical contractor right out of high school. I found that I really liked the construction environment, especially the electrical work. I was never bored and I enjoyed the fellowship of other construction workers. I knew after that summer that electrical construction would be my career.

What was your apprenticeship like?
I went through a four-year program that included classroom instruction two nights a week and on-the-job training during the day. The classroom materials we had then were nowhere near as progressive as the NCCER materials we have today. There were no structured materials and overheads like we have in the NCCER program. Sometimes it was pretty boring.

The practical, on-the-job experience made up for those long nights in class. I was fortunate enough to have the same outstanding field trainer for my entire apprenticeship. He acted as a mentor to me, and we became close friends as well as coworkers.

What positions have you held and how did they help you get where you are today?
I have gone from an electrician's helper to an apprentice, journeyman, foreman, superintendent, estimator, project manager, division manager, district manager, and finally vice president. I feel lucky to have learned the business from the bottom up. I know

I'm a better manager and administrator because of my experience at so many levels.

What would you say is the greatest single factor that contributed to your success?
I've always had the desire to be the best that I can possibly be at my chosen career. At each step along the way, I have tried to be the best of the best.

What does your present job entail?
I oversee four construction divisions with annual revenues of $38 million. I'm responsible for ensuring that these divisions meet their operating goals and produce profitable growth for the company. As an executive, I try to exemplify positive traits such as fairness, creativity, and a commitment to getting the job done in a way that produces the maximum benefit for the employees, customers, and communities we serve.

What advice would you give to apprentices entering the field today?
Participate in industry activities. I believe the industry prospers because of programs such as NCCER's Standardized Craft Training, and I am happy to have been able to give something back to the industry through my participation in the development of this program. It has given me another opportunity to pass on my knowledge and experience to the next generation of electricians.

Trade Terms Introduced in This Module

Architectural drawings: Working drawings consisting of plans, elevations, details, and other information necessary for the construction of a building. Architectural drawings usually include:

- A site (plot) plan indicating the location of the building on the property
- Floor plans showing the walls and partitions for each floor or level
- Elevations of all exterior faces of the building
- Several vertical cross sections to indicate clearly the various floor levels and details of the footings, foundations, walls, floors, ceilings, and roof construction
- Large-scale detail drawings showing such construction details as may be required

Block diagram: A single-line diagram used to show electrical equipment and related connections. See *power-riser diagram.*

Blueprint: An exact copy or reproduction of an original drawing.

Detail drawing: An enlarged, detailed view taken from an area of a drawing and shown in a separate view.

Dimensions: Sizes or measurements printed on a drawing.

Electrical drawing: A means of conveying a large amount of exact, detailed information in an abbreviated language. Consists of lines, symbols, dimensions, and notations to accurately convey an engineer's designs to electricians who install the electrical system on a job.

Elevation drawing: An architectural drawing showing height, but not depth; usually the front, rear, and sides of a building or object.

Floor plan: A drawing of a building as if a horizontal cut were made through a building at about window level, and the top portion removed. The floor plan is what would appear if the remaining structure were viewed from above.

One-line diagram: A drawing that shows, by means of lines and symbols, the path of an electrical circuit or system of circuits along with the various circuit components. Also called a *single-line diagram.*

Plan view: A drawing made as though the viewer were looking straight down (from above) on an object.

Power-riser diagram: A single-line block diagram used to indicate the electric service equipment, service conductors and feeders, and subpanels. Notes are used on power-riser diagrams to identify the equipment; indicate the size of conduit; show the number, size, and type of conductors; and list related materials. A panelboard schedule is usually included with power-riser diagrams to indicate the exact components (panel type and size), along with fuses, circuit breakers, etc., contained in each panelboard.

Scale: On a drawing, the size relationship between an object's actual size and the size it is drawn. Scale also refers to the measuring tool used to determine this relationship.

Schedule: A systematic method of presenting equipment lists on a drawing in tabular form.

Schematic drawing: A cutaway drawing that shows the inside of an object or building.

Instructor's Notes:

_____

_____

_____

_____

_____

_____

_____

_____

_____

Shop drawing: A drawing that is usually developed by manufacturers, fabricators, or contractors to show specific dimensions and other pertinent information concerning a particular piece of equipment and its installation methods.

Site plan: A drawing showing the location of a building or buildings on the building site. Such drawings frequently show topographical lines, electrical and communication lines, water and sewer lines, sidewalks, driveways, and similar information.

Written specifications: A written description of what is required by the owner, architect, and engineer in the way of materials and workmanship. Together with working drawings, the specifications form the basis of the contract requirements for construction.

Metric Conversion Chart

Instructor's Notes:

_____

_____

_____

_____

_____

_____

_____

_____

_____

METRIC CONVERSION CHART

inches (fractions)	inches (decimals)	mm
—	.0004	.01
—	.004	.10
—	.01	.25
1/64	.0156	.397
—	.0197	.50
—	.0295	.75
1/32	.03125	.794
—	.0394	1.
3/64	.0469	1.191
—	.059	1.5
1/16	.062	1.588
5/64	.0781	1.984
—	.0787	2.
3/32	.094	2.381
—	.0984	2.5
7/64	.109	2.778
—	.1181	3.
1/8	.125	3.175
—	.1378	3.5
9/64	.141	3.572
5/32	.156	3.969
—	.1575	4.
11/64	.172	4.366
—	.177	4.5
3/16	.1875	4.763
—	.1969	5.
13/64	.203	5.159
—	.2165	5.5
7/32	.219	5.556
15/64	.234	5.953
—	.2362	6.
1/4	.250	6.350
—	.2559	6.5
17/64	.2656	6.747
—	.2756	7.
9/32	.281	7.144
—	.2953	7.5
19/64	.297	7.541
5/16	.312	7.938
—	.315	8.
21/64	.328	8.334
—	.335	8.5
11/32	.344	8.731
—	.3543	9.
23/64	.359	9.128
—	.374	9.5
3/8	.375	9.525
25/64	.391	9.922
—	.3937	10.
13/32	.406	10.319
—	.413	10.5
27/64	.422	10.716
—	.4331	11.
7/16	.438	11.113
—	.453	11.509
29/64	.469	11.906
15/32	.469	11.906
—	.4724	12.
31/64	.484	12.303
—	.492	12.5
1/2	.500	12.700
—	.5118	13.
33/64	.5156	13.097
17/32	.531	13.494
35/64	.547	13.891
—	.5512	14.
9/16	.563	14.288
—	.571	14.5
37/64	.578	14.684
—	.5906	15.
19/32	.594	15.081
39/64	.609	15.478
5/8	.625	15.875
—	.6299	16.
41/64	.6406	16.272
—	.6496	16.5
21/32	.656	16.669
—	.6693	17.
43/64	.672	17.066
11/16	.6875	17.463
45/64	.703	17.859
—	.7087	18.
23/32	.719	18.256
—	.7283	18.5
47/64	.734	18.653
—	.7480	19.
3/4	.750	19.050
49/64	.7656	19.447

inches (fractions)	inches (decimals)	mm
25/32	.781	19.844
—	.7874	20.
51/64	.797	20.241
13/16	.8125	20.638
—	.8268	21.
53/64	.828	21.034
27/32	.844	21.431
55/64	.859	21.828
—	.8661	22.
7/8	.875	22.225
57/64	.8906	22.622
—	.9055	23.
29/32	.9062	23.019
59/64	.922	23.416
15/16	.9375	23.813
—	.9449	24.
61/64	.953	24.209
31/32	.969	24.606
—	.9843	25.
63/64	.9844	25.003
1	1.000	25.400
—	1.0236	26.
1-1/32	1.0312	26.194
1-1/16	1.062	26.988
—	1.063	27.
1-3/32	1.094	27.781
—	1.1024	28.
1-1/8	1.125	28.575
1-5/32	1.156	29.369
—	1.1811	30.
1-3/16	1.1875	30.163
1-7/32	1.219	30.956
—	1.2205	31.
1-1/4	1.250	31.750
—	1.2598	32.
1-9/32	1.281	32.544
—	1.2992	33.
1-5/16	1.312	33.338
—	1.3386	34.
1-11/32	1.344	34.131
1-3/8	1.375	34.925
—	1.3779	35.
1-13/32	1.406	35.719
—	1.4173	36.
1-7/16	1.438	36.513
—	1.4567	37.
1-15/32	1.469	37.306
—	1.4961	38.
1-1/2	1.500	38.100
1-17/32	1.531	38.894
—	1.5354	39.
1-9/16	1.562	39.688
—	1.5748	40.
1-19/32	1.594	40.481
—	1.6142	41.
1-5/8	1.625	41.275
—	1.6535	42.
1-21/32	1.6562	42.069
1-11/16	1.6875	42.863
—	1.6929	43.
1-23/32	1.719	43.656
—	1.7323	44.
1-3/4	1.750	44.450
—	1.7717	45.
1-25/32	1.781	45.244
—	1.8110	46.
1-13/16	1.8125	46.038
1-27/32	1.844	46.831
—	1.8504	47.
1-7/8	1.875	47.625
—	1.8898	48.
1-29/32	1.9062	48.419
—	1.9291	49.
1-15/16	1.9375	49.213
—	1.9685	50.
1-31/32	1.969	50.006
2	2.000	50.800
—	2.0079	51.
2-1/32	2.03125	51.594
—	2.0472	52.
2-1/16	2.062	52.388
—	2.0866	53.
2-3/32	2.094	53.181
2-1/8	2.125	53.975
—	2.126	54.
2-5/32	2.156	54.769

inches (fractions)	inches (decimals)	mm
—	2.165	55.
2-3/16	2.1875	55.563
—	2.2047	56.
2-7/32	2.219	56.356
—	2.244	57.
2-1/4	2.250	57.150
2-9/32	2.281	57.944
—	2.2835	58.
2-5/16	2.312	58.738
—	2.3228	59.
2-11/32	2.344	59.531
—	2.3622	60.
2-3/8	2.375	60.325
—	2.4016	61.
2-13/32	2.406	61.119
2-7/16	2.438	61.913
—	2.4409	62.
2-15/32	2.469	62.706
—	2.4803	63.
2-1/2	2.500	63.500
—	2.5197	64.
2-17/32	2.531	64.294
—	2.559	65.
2-9/16	2.562	65.088
2-19/32	2.594	65.881
—	2.5984	66.
2-5/8	2.625	66.675
—	2.638	67.
2-21/32	2.656	67.469
—	2.6772	68.
2-11/16	2.6875	68.263
—	2.7165	69.
2-23/32	2.719	69.056
2-3/4	2.750	69.850
—	2.7559	70.
2-25/32	2.781	70.6439
—	2.7953	71.
2-13/16	2.8125	71.4376
—	2.8346	72.
2-27/32	2.844	72.2314
—	2.8740	73.
2-7/8	2.875	73.025
2-29/32	2.9062	73.819
—	2.9134	74.
2-15/16	2.9375	74.613
—	2.9527	75.
2-31/32	2.969	75.406
—	2.9921	76.
3	3.000	76.200
3-1/32	3.0312	76.994
—	3.0315	77.
3-1/16	3.062	77.788
—	3.0709	78.
3-3/32	3.094	78.581
—	3.1102	79.
3-1/8	3.125	79.375
—	3.1496	80.
3-5/32	3.156	80.169
3-3/16	3.1875	80.963
—	3.1890	81.
3-7/32	3.219	81.756
—	3.2283	82.
3-1/4	3.250	82.550
—	3.2677	83.
3-9/32	3.281	83.344
—	3.3071	84.
3-5/16	3.312	84.1377
3-11/32	3.344	84.9314
—	3.3464	85.
3-3/8	3.375	85.725
—	3.3858	86.
3-13/32	3.406	86.519
—	3.4252	87.
3-7/16	3.438	87.313
—	3.4646	88.
3-15/32	3.469	88.106
3-1/2	3.500	88.900
—	3.5039	89.
3-17/32	3.531	89.694
—	3.5433	90.
3-9/16	3.562	90.4877
—	3.5827	91.
3-19/32	3.594	91.281
—	3.622	92.
3-5/8	3.625	92.075
3-21/32	3.656	92.869
—	3.6614	93.

inches (fractions)	inches (decimals)	mm
3-11/16	3.6875	93.663
—	3.7008	94.
3-23/32	3.719	94.456
—	3.7401	95.
3-3/4	3.750	95.250
—	3.7795	96.
3-25/32	3.781	96.044
3-13/16	3.8125	96.838
—	3.8189	97.
3-27/32	3.844	97.631
—	3.8583	98.
3-7/8	3.875	98.425
—	3.8976	99.
3-29/32	3.9062	99.219
—	3.9370	100.
3-15/16	3.9375	100.013
3-31/32	3.969	100.806
—	3.9764	101.
4	4.000	101.600
4-1/16	4.062	103.188
4-1/8	4.125	104.775
—	4.1338	105.
4-3/16	4.1875	106.363
4-1/4	4.250	107.950
4-5/16	4.312	109.538
—	4.3307	110.
4-3/8	4.375	111.125
4-7/16	4.438	112.713
4-1/2	4.500	114.300
—	4.5275	115.
4-9/16	4.562	115.888
4-5/8	4.625	117.475
4-11/16	4.6875	119.063
—	4.7244	120.
4-3/4	4.750	120.650
4-13/16	4.8125	122.238
4-7/8	4.875	123.825
—	4.9212	125.
4-15/16	4.9375	125.413
5	5.000	127.000

inches (fractions)	inches (decimals)	mm
—	5.1181	130.
5-1/4	5.250	133.350
5-1/2	5.500	139.700
—	5.5118	140.
5-3/4	5.750	146.050
—	5.9055	150.
6	6.000	152.400
6-1/4	6.250	158.750
—	6.2992	160.
6-1/2	6.500	165.100
—	6.6929	170.
6-3/4	6.750	171.450
7	7.000	177.800
—	7.0866	180.
—	7.4803	190.
7-1/2	7.500	190.500
—	7.8740	200.
8	8.000	203.200
—	8.2677	210.
8-1/2	8.500	215.900
—	8.6614	220.
9	9.000	228.600
—	9.0551	230.
—	9.4488	240.
9-1/2	9.500	241.300
—	9.8425	250.
10	10.000	254.001
—	10.2362	260.
—	10.6299	270.
11	11.000	279.401
—	11.0236	280.
—	11.4173	290.
—	11.8110	300.
12	12.000	304.801
13	13.000	330.201
—	13.7795	350.
14	14.000	355.601
15	15.000	381.001
—	15.7480	400.
16	16.000	406.401
17	17.000	431.801
—	17.7165	450.
18	18.000	457.201
19	19.000	482.601
—	19.6850	500.
20	20.000	508.001

110UA1003.TIF

Answers to Review Questions

Answer	Section
1. g	4.0.0
2. b	4.0.0
3. e	4.0.0
4. c	4.0.0
5. a	4.0.0
6. d	4.0.0
7. f	4.0.0
8. d	10.2.1/Figure 41
9. b	10.2.1/Figure 41
10. d	10.2.1/Figure 41

Instructor's Notes:

_____

_____

_____

_____

_____

_____

_____

_____

_____

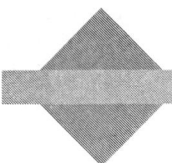

Additional Resources

For advanced study of topics covered in this task module, the following books are suggested:

American Electrician's Handbook, Latest Edition. New York: McGraw-Hill.

Illustrated Guide to the NEC, Latest Edition. Carlsbad, CA: Craftsman Book Company.

National Electrical Code Handbook, Latest Edition. Quincy, MA: National Fire Protection Association.

_____

_____

_____

_____

_____

_____

_____

_____

_____

_____

NCCER CRAFT TRAINING USER UPDATES

The NCCER makes every effort to keep these textbooks up-to-date and free of technical errors. We appreciate your help in this process. If you have an idea for improving this textbook, or if you find an error, a typographical mistake, or an inaccuracy in the NCCER's Craft Training textbooks, please write us, using this form or a photocopy. Be sure to include the exact module number, page number, a detailed description, and the correction, if applicable. Your input will be brought to the attention of the Technical Review Committee. Thank you for your assistance.

Instructors – If you found that additional materials were necessary in order to teach this module effectively, please let us know so that we may include them in the Equipment/Materials list in the Instructor's Guide.

Write: Curriculum Revision and Development Department
National Center for Construction Education and Research
P.O. Box 141104, Gainesville, FL 32614-1104

Fax: 352-334-0932

E-mail: curriculum@nccer.org

Craft _____ Module Name _____

Copyright Date _____ Module Number _____ Page Number(s) _____

Description _____

_____

_____

_____

_____

(Optional) Correction _____

_____

_____

(Optional) Your Name and Address _____

_____

_____

_____

TEACHING TIPS

Section 1.4.0   *Think About It—Using All of the Drawings*

Encourage the trainees to closely examine the floor plan, elevation drawing, and section view to determine the specific types of information shown on each type of drawing. They should discover that each drawing contains a wealth of information for the electrician. Section views provide important information not shown on other types of drawings. For example, section views show the structural members and materials used inside the walls and on the outside wall surfaces. The height, thickness, and shape of the walls are shown, all of which are important in the selection and installation of outlets and fixtures.

Section 5.0.0   *Think About It—Using Electrical Symbols*

This exercise is intended to show the trainees that the symbols convey a great deal of information. Have the trainees read the plan carefully, then describe the details of the pump and receptacle circuits shown on the plan.

Section 6.1.0   *Think About It—Interpreting Site Plans*

The symbols shown on the typical electrical site plan in *Figure 29* are described in detail in the module. The important thing here is to make sure that the trainees fully understand the purpose of this drawing and that they can correctly interpret the information contained in it.

Section 7.3.2   *Think About It—Power Plans*

Inspection of the power plan in *Figure 33* shows that power for the building applied from the underground electric service enters the building at the right end of the south wall (top wall on drawing). This power is applied via a C/T cabinet to the main distribution panel (MDP). From there, the power is further distributed to the building busways and power panels A and B. Some connections are left to the discretion of the electrician.

Section 10.2.1   *Think About It—Putting It All Together*

Written specifications supplement the related working drawings in that they contain details not shown on the drawings. Specifications define and clarify the scope of the job. They describe the specific types and characteristics of the electrical components that are to be used on the job and the methods for installing some of them. Many components are identified specifically by the manufacturer's model and part numbers. This type of information is used to purchase the various items of hardware needed to accomplish the installation in accordance with the contractual requirements. Using the sample electrical specifications given in the module, have the trainees determine the requirements for some specific items of hardware such as conduit, conductors, switches, etc. Have them also determine the requirements (if any) for the installation of these items.

Wiring: Commercial and Industrial

26111-00

MODULE OVERVIEW

This course introduces the electrical trainee to the various types of switches and receptacles used in commercial and industrial wiring, along with their basic installation procedures and NEC requirements.

PREREQUISITES

Please refer to the Course Map in the Trainee Module. Prior to training with this module, it is recommended that the trainee shall have successfully completed the following modules:

Core Curriculum; Electrical Level One, Modules 26101 through 26110

LEARNING OBJECTIVES

Upon completion of this module, the trainee will be able to:

1. Identify and state the functions and ratings of single-pole, double-pole, three-way, four-way, dimmer, special, and safety switches.
2. Explain NEMA classifications as they relate to switches and enclosures.
3. Explain the NEC requirements concerning wiring devices.
4. Identify and state the functions and ratings of straight blade, twist lock, and pin and sleeve receptacles.
5. Identify and define receptacle terminals and disconnects.
6. Identify and define ground fault circuit interrupters.
7. Explain the box mounting requirements in the NEC.
8. Use a wire stripper to strip insulation from a wire.
9. Use a solderless connector to splice wires together.
10. Identify and state the functions of limit switches and relays.
11. Identify and state the function of switchgear.

PERFORMANCE OBJECTIVES

Under supervision of the instructor, the trainee should be able to:

1. Draw single-pole, three-way, and four-way switches as they are connected in typical lighting circuits.
2. Based on a given set of environmental conditions, select the correct NEMA enclosure for use with a motor controller.
3. Use a wire stripper to safely remove insulation from a wire.
4. Use a solderless connector to splice two wires together.

NCCER STANDARDIZED CRAFT TRAINING PROGRAM

The National Center for Construction Education and Research (NCCER) provides a standardized national program of accredited craft training. Key features of the program include instructor certification, competency-based training, and performance testing. The program provides trainees, instructors, and companies with a standard form of recognition through a National Craft Training Registry. The program is described in full in the *Guidelines for Accreditation*, published by the NCCER. For more information on standardized craft training, contact the NCCER at P.O. Box 141104, Gainesville, FL 32614-1104, 352-334-0911, visit our Web site at www.NCCER.org, or e-mail info@NCCER.org.

HOW TO USE THIS ANNOTATED INSTRUCTOR'S GUIDE

Each page presents two sections of information. The larger section displays each page exactly as it appears in the Trainee Module. The narrow column ties suggested trainee and instructor actions to each page and provides icons to call your attention to material, safety, audiovisual, or testing requirements. The bottom of each page includes space for your notes.

If you see the Teaching Tip icon, that means there is a teaching tip associated with this section. Also refer to the suggested teaching tips at the end of the module.

SAFETY CONSIDERATIONS

Ensure that the trainees are equipped with appropriate personal protective equipment. Emphasize the importance of following all safety precautions and procedures when working with electrical circuits.

PREPARATION

Before teaching this module, you should review the Module Outline, Learning and Performance Objectives, and the Materials and Equipment List. Be sure to allow ample time to prepare your own training or lesson plan and gather all required equipment and materials.

MATERIALS AND EQUIPMENT LIST

Materials:
Transparencies

Markers/chalk

Copy of the latest edition of the *National Electrical Code®*

Pencils/scratch paper

Wire (No. 14 or No. 12 AWG)

Assorted solderless connectors

Assorted crimp-on wire lugs

Module Examinations*

Performance Profile Sheets*

Equipment:
Overhead projector and screen

Whiteboard/chalkboard

Appropriate personal protective equipment

Screwdrivers

Wire strippers

Assortment of switches, including:
  Single-pole, single-throw
  Double-pole, single-throw
  Single-pole, double-throw
  Double-pole, double-throw
  Four-way
  Dimmer

Assorted safety/disconnect switches housed in NEMA Type 1 through Type 13 enclosures

Assortment of duplex and other receptacles

Ground fault circuit interrupter (GFCI) receptacles

Relays

Limit switches

*Located in the Test Booklet packaged with this Annotated Instructor's Guide.

ADDITIONAL RESOURCES

This module is intended to present thorough resources for task training. The following reference work is suggested for both instructors and motivated trainees interested in further study. This is optional material for continued education rather than for task training.

National Electrical Code Handbook, Latest Edition. Quincy, MA: National Fire Protection Association.

NOTES

The designations "National Electrical Code," "NE Code," and "NEC," where used in this document, refer to the *National Electrical Code®*, which is a registered trademark of the National Fire Protection Association, Quincy, MA. All National Electrical Code (NEC) references in this module refer to the 1999 edition of the NEC.

If you feel that additional math instruction would be helpful, Prentice Hall offers a basic math textbook entitled *Fundamentals of Electrical and Mechanical Mathematics*. It covers the basic math requirements for electrical trainees and may be ordered by contacting Prentice Hall Customer Service at 1-800-922-0579.

TEACHING TIME FOR THIS MODULE

An outline for use in developing your lesson plan is presented below. Note that each Roman numeral in the outline equates to one session of instruction. Each session has a suggested time period of 2½ hours. This includes 10 minutes at the beginning of each session for administrative tasks and one 10-minute break during the session. Approximately 7½ hours are suggested to cover *Wiring: Commercial and Industrial*. You will need to adjust the time required for hands-on activity and testing based on your class size and resources.

Topic   **Planned Time**

Session I. Switches; Types of Switches; NEMA Classifications

- A. Switches _____
    1. Common Terms _____
    2. Identifying Switches _____
- B. Types of Switches _____
    1. Single-Pole Switches _____
    2. Double-Pole Switches _____
    3. Three-Way Switches _____
    4. Four-Way Switches _____
    5. Special Switches _____
    6. Disconnects _____
        a. Safety Switches _____
- C. NEMA Classifications _____
    1. Switch Classifications _____
        a. General-Duty Switches _____
        b. Heavy-Duty, Motor-Rated Switches _____
    2. Enclosure Classifications _____
        a. General Purpose Indoor (Type 1) _____
        b. Drip-Proof Indoor (Type 2) _____
        c. Dust-Tight, Rain-Tight, and Sleet-Resistant (Type 3) _____
        d. Rainproof and Sleet-Resistant (Type 3R) _____
        e. Dust-Tight, Rain-Tight, and Sleet-Proof (Type 3S) _____
        f. Watertight and Dust-Tight (Type 4) _____
        g. Submersible, Watertight, and Dust-Tight (Type 6) _____
        h. Hazardous Locations, Class I (Type 7) _____
        i. Hazardous Locations, Class I (Type 8) _____
        j. Hazardous Locations, Class II (Type 9) _____
        k. Bureau of Mines (Type 10) _____
        l. Corrosion-Resistant, Drip-Proof Indoor (Type 11) _____
        m. Dust-Tight and Drip-Tight Industrial Use (Type 12) _____
        n. Oil-Tight and Dust-Tight (Type 13) _____
        o. Ingress Protection (IP) Classifications _____

Session II. Receptacles; Wiring Techniques; Laboratory

    A. Receptacles   _____

        1. Identifying Receptacles   _____

        2. Types of Receptacles   _____

            a. Straight Blade Receptacles   _____

            b. Twist Lock Receptacles   _____

            c. Pin and Sleeve Receptacles   _____

        3. Receptacle Terminals   _____

        4. Ground Fault Circuit Interrupters   _____

    B. Wiring Techniques   _____

        1. Bending Wire   _____

        2. Branch Circuit Wiring   _____

            a. Planning   _____

            b. Installation   _____

            c. Splicing   _____

    C. Laboratory   _____

    Under instructor supervision, have the trainees practice stripping and splicing wires.

Session III. Control Devices; Power Distribution Equipment; Module Examination and Performance Testing

    A. Control Devices   _____

        1. Relays   _____

        2. Limit Switches   _____

    B. Power Distribution Equipment   _____

    C. Module Examination   _____

        1. Trainees must score 70% or higher to receive recognition from the NCCER.

        2. Record the testing results on Craft Training Report Form 200 and submit the results to the Training Program Sponsor.

    D. Performance Testing   _____

        1. Trainees must perform each task to the satisfaction of the instructor to receive recognition from the NCCER.

        2. Record the testing results on Craft Training Report Form 200 and submit the results to the Training Program Sponsor.

Module 26111-00

Wiring: Commercial and Industrial

Instructor's Notes:

_____

_____

_____

_____

_____

_____

_____

_____

_____

Course Map

This course map shows all the task modules in the first level of the Electrical curricula. The suggested training order begins at the bottom and proceeds up. Skill levels increase as a trainee advances on the course map. The training order may be adjusted by the local Training Program Sponsor.

Assign reading of Module 26111.

Instructor's Notes:

_____

_____

_____

_____

_____

_____

_____

_____

_____

Figures

Ensure you have everything required to teach the course. Check the Materials and Equipment List at the front of this Instructor's Guide.

Wiring: Commercial and Industrial

Objectives

Upon completion of this module, the trainee will be able to:

1. Identify and state the functions and ratings of single-pole, double-pole, three-way, four-way, dimmer, special, and safety switches.
2. Explain NEMA classifications as they relate to switches and enclosures.
3. Explain the NEC requirements concerning wiring devices.
4. Identify and state the functions and ratings of straight blade, twist lock, and pin and sleeve receptacles.
5. Identify and define receptacle terminals and disconnects.
6. Identify and define ground fault circuit interrupters.
7. Explain the box mounting requirements in the NEC.
8. Use a wire stripper to strip insulation from a wire.
9. Use a solderless connector to splice wires together.
10. Identify and state the functions of limit switches and relays.
11. Identify and state the function of switchgear.

Prerequisites

Successful completion of the following Task Modules is recommended before beginning study of this Task Module: Core Curriculum; Electrical Level One, Modules 26101 through 26110.

Required Trainee Materials

1. Trainee Task Module
2. Copy of the latest edition of the *National Electrical Code*
3. Appropriate Personal Protective Equipment

1.0.0 ◆ INTRODUCTION

As an electrician, a large part of your job will be involved with wiring and wiring devices. There are many different types of wiring devices and methods of installing them. This module will help you become familiar with some common devices and the methods for installing them. As with almost all aspects of an electrician's job, the NEC is an important reference for use when installing wiring devices.

2.0.0 ◆ SWITCHES

The purpose of a switch is to make and break an electrical circuit. In that way, the switch can control the operation of lighting and various kinds of equipment connected in the circuit. *NEC Article 380* is one of the areas in the NEC that covers the installation and use of switches.

In the past, all switches were mechanical devices that were manually operated to open or

Note: The designations "National Electrical Code," "NE Code," and "NEC," where used in this document, refer to the National Electrical Code®, which is a registered trademark of the National Fire Protection Association, Quincy, MA. All National Electrical Code (NEC) references in this module refer to the 1999 edition of the NEC.

11–1

Show Transparency 1, Course Objectives.

Show Transparency 2, Performance Profile Tasks.

Introduce wiring devices.

Explain the purpose of a switch.

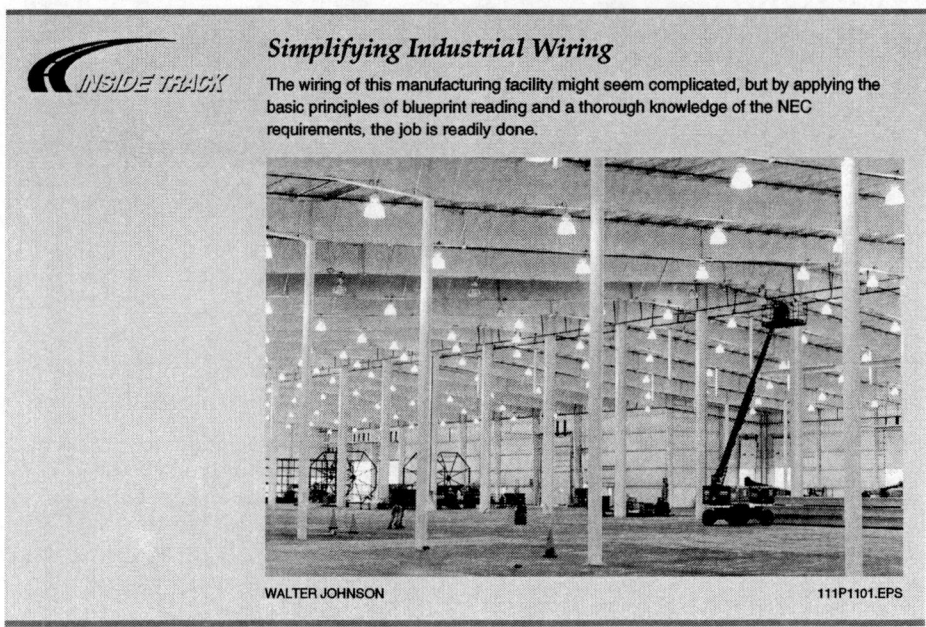

Simplifying Industrial Wiring

INSIDE TRACK

The wiring of this manufacturing facility might seem complicated, but by applying the basic principles of blueprint reading and a thorough knowledge of the NEC requirements, the job is readily done.

WALTER JOHNSON   111P1101.EPS

Define *pole* and *throw.*

Show Transparency 3 (Figure 1).

close a circuit. Modern technology has greatly expanded the role and design of early switches. Today, switches can be activated by light, heat, chemicals, and electrical energy.

2.1.0 Common Terms

Two general terms are used to define switches. The term *pole* refers to the number of conducting paths that the switch will control in the circuit. A single-pole switch breaks the connection on only one conducting path in a circuit. A double-pole switch is used to break the connection on both conducting paths in a two-conductor circuit, usually a 240V, two-wire circuit.

The term *throw* refers to the number of internal operations that a switch can perform. For example, a single-pole, single-throw (SPST) switch will complete a single circuit only when it is thrown in one direction (the ON position). This circuit will be opened when the switch is thrown in the opposite direction (the OFF position). The common ON/OFF snap switch is an SPST switch.

A double-pole, single-throw (DPST) switch opens or closes two conducting paths at the same time. In this case, both circuits are either open or closed. A DPST switch is common to 240V circuits. In these circuits, both conductors are considered to be energized.

The single-pole, double-throw (SPDT) switch, also known as a *three-way switch,* is often used to control a single load, such as a lamp, from two different locations. SPDT switches are discussed in more detail later in this module.

A double-pole, double-throw (DPDT) switch usually directs a 240V, two-wire circuit through one of two different paths. One example of a DPDT switch is an electrical transfer switch used to energize certain circuits from either the main electric service or from an emergency standby generator. The DPDT switch prevents the circuits from being energized from both sources at once.

A four-way switch cannot be classified as an example of a double-pole, double-throw switch. Although a four-way switch does throw the current path in two different directions by moving the switch toggle, only one pole on each side of the switch is allowing current to pass.

Figure 1 summarizes the basic switches and shows their symbols.

Instructor's Notes:

_____

_____

_____

_____

_____

_____

_____

_____

_____

SWITCH TYPE	ABBREVIATION	SYMBOL
SINGLE-POLE SINGLE-THROW	S1	
DOUBLE-POLE SINGLE-THROW	S2	
SINGLE-POLE DOUBLE-THROW (THREE-WAY)	S3	
DOUBLE-POLE DOUBLE-THROW	N/A	
FOUR-WAY	S4	

111F01.EPS

Figure 1 ❖ Switch symbols.

2.2.0 Identifying Switches

Switches vary in their grades, capacities, and purposes. For example, a switch that is appropriate for activating a tungsten-filament lamp would not meet the NEC requirements for controlling motor starting. It is very important that switches be used for their designated purpose.

There is a great deal of information given on the switch itself, which can help you determine its function and rating (see *Figure 2*). Most information related to the switch rating will be found on the front of the switch.

- The testing laboratory label is an indication that the device has undergone extensive testing by a nationally recognized testing laboratory and has met the minimum safety requirements. In *Figure 2*, the switch is marked with a "UL," which stands for Underwriters' Laboratories, which was created by the National Board of Fire Underwriters to test electrical devices and materials.

- The Canadian Standards Association (CSA) label is an indication that the material or device has undergone a similar testing procedure by the Canadian Standards Association and is acceptable for use in Canada.

- Current and voltage ratings are listed by maximum amperage and maximum voltage. In *Figure 2*, the maximum current through the device should not exceed 15 amps and the maximum voltage should not exceed 120V. This switch is to be used on alternating current (AC) only. Failure to follow these recommendations could result in damage to the unit or a possible fire hazard.

- The CO/ALR symbol indicates that a switch may be used with copper, aluminum, or copper-clad aluminum wire. The letters ALR stand for aluminum revised. The CO/ALR mark replaces an earlier mark, CU/AL. It was discovered that the earlier CU/AL-rated switches were not suitable for aluminum wire in the 15- to 20-amp range.

Discuss the types of information given on switches.

Show Transparency 4 (Figure 2).

Identifying Switches

The differences between a light switch and a motor switch will become obvious as you begin to use them. This motor switch has built-in overload protection.

OVERLOAD PROTECTION

VERONICA WESTFALL   111P1102.EPS

COPPER/
ALUMINUM
WIRE

TESTING
LABORATORY
LISTING

STRIP GAUGE

CANADIAN
STANDARDS
ASSOCIATION

MAXIMUM AMPERAGE 15A
MAXIMUM VOLTAGE 120V
RESTRICTED TO AC ONLY

111F02.EPS

Figure 2 ◆ Typical single-pole switch.

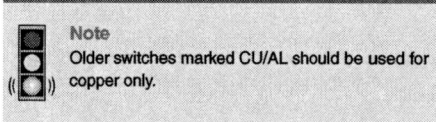

Note
Older switches marked CU/AL should be used for copper only.

- A tungsten (T) rating indicates that the switch can be used with lamps that use tungsten. Tungsten is a metal that is used as the filament in standard incandescent lamps. When it is cold, tungsten has a very low resistance. This low resistance causes the initial current flow to be six to ten times that of the normal current

flow. Once the tungsten heats up, which takes about ½₅₀th of a second, the current flow drops back to normal. This severe overloading can reduce the life expectancy of the switch.

Another way to identify the function and rating of a switch is by color coding. Switches are typically constructed with a ground screw attached to the metallic strap of the switch. The ground screw is usually a hex-head screw and is indicated by the color green.

On a three-way switch, the color black or bronze (brass) is used to show the common pivot point for the switch. Current will always enter or leave through the pivot point.

ELECTRICAL LEVEL ONE—TRAINEE MODULE 26111

Instructor's Notes:

_____

_____

_____

_____

_____

_____

_____

_____

_____

3.0.0 ◆ TYPES OF SWITCHES

There are many types of switches available. This section covers the most common types.

3.1.0 Single-Pole Switches

Single-pole switches are easy to identify because they only have two terminals for conductor connections (*Figure 3*).

Note
Figure 3 shows the white wire to the switch as the hot wire. This allows the use of standard black/white pair cable to be used for switch legs, while retaining a black wire for the hot wire to the load.

Many single-pole switches have a third or ground terminal. This type of switch is probably the most widely used switch in residential appli-

cations. It is used for the on and off control of lights, fans, and numerous other appliances from a single point.

Note
According to *NEC Section 200-7(c)(2)*, single-pole, three-way, and four-way switch loops can use the white wire (usually reserved for the grounded conductor) for supply to the switch, but the conductor must be permanently re-identified for this use by painting or other effective means.

Figure 4 shows how a single-pole switch controlling a single light would be represented on a set of blueprints. *Figure 5* shows a schematic representation of a single-pole switch.

3.2.0 Double-Pole Switches

Double-pole switches are really two switches in one. They are used primarily when it is necessary to completely isolate the load. Double-pole switches are connected so that both circuit conductors can be opened or closed at the same time. The double-pole switch has four terminals for conductor connections. These terminals are identified according to polarity. Like the single-pole switch, the double-pole switch has an identified on or off position.

3.3.0 Three-Way Switches

A three-way switch does not perform three separate operations, as its name might suggest. The three-way switch is a single-pole, double-throw switch without a center off position. A common use for a three-way switch is to control a lamp from two separate locations, such as from the top and bottom of a stairway. In this case, a pair of three-way switches would be used. *Figure 6* shows a pictorial drawing of such a circuit.

RE-IDENTIFIED PER
NEC SECTION 200-7(c)(2)

111F03.EPS

Figure 3 ◆ Single-pole pictorial for cable installation.

111F04.EPS

Figure 4 ◆ Single-pole switch as shown on a blueprint.

Discuss single-pole, double-pole, and three-way switches and their applications. Pass around examples for the trainees to examine.

Show Transparencies 5 through 8 (Figures 3 through 6).

Explain how various switches are wired in a typical lighting circuit.

Figure 5 ◆ Single-pole schematic.

Double-Pole Switches

How would a double-pole switch isolate an electrical load? Why might you want to isolate a device or branch circuit?

Figure 6 ◆ Three-way pictorial for cable installation.

Instructor's Notes:

_____

_____

_____

_____

_____

_____

_____

_____

The three-way switch has three terminal connections (four if a ground or green terminal is present). The single terminal located at one end of the switch is called the *common*. This terminal is darker in color (either black or brass-colored) than the other two terminals. The hot leg is always connected to this darker terminal. The remaining two terminals are lighter in color and are referred to as the *traveler terminals. Figure 7* shows how the actual connection of the three-way switching circuit is made.

With the hot leg connected to the common of the first switch, the current is given a choice of paths to follow depending upon the relative position of the switch. The current then proceeds along one of the travelers (or alternating switch loops) to the second switch. At the second switch, the circuit will either be completed or opened, depending upon the position of this second switch. There are four possible combinations of the switch contact positions, as shown in *Figure 8.*

Of the four possible combinations of switch contact positions, only two will result in a complete circuit. Changing the position of either switch will always change the state of the lamp.

Figure 9 illustrates how a three-way switch is shown on a blueprint.

3.4.0 Four-Way Switches

Four-way switches can be used if it is necessary to control a light or group of lights from more than two locations. The four-way switch (or switches) must always be connected in the traveler conductors between the three-way switches. *Figure 10* shows how a four-way switch could be added to the circuit previously shown in *Figure 6*. The path for the current to follow is relative to the positions of the switch handle.

Note that the only difference between this circuit and the three-way switch circuit is the inclusion of the four-way switch in the traveler conductors between the two three-way switches. If it is necessary to control a lamp (or lamps) from *more* than three positions, more four-way switches can be installed in the travelers or carrier conductors.

3.5.0 Special Switches

Another common switch is the dimmer switch (*Figure 11*). This switch not only turns the light on and off, but also provides a means of controlling the light's brilliance. Dimmer switches are used where full brightness is not always desired. The standard models are available in single-pole and three-way types. They mount in a standard switch box. Single-pole dimmers can replace single-pole switches. However, three-way dimmers must be used with a three-way toggle switch at the other

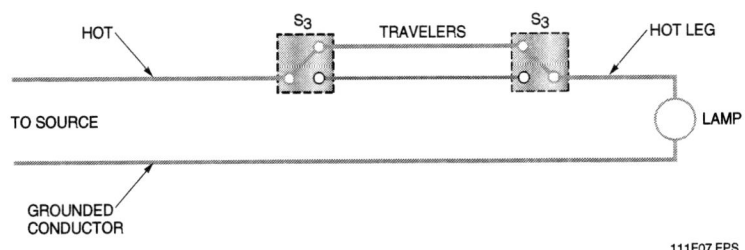

Figure 7 ◆ Three-way schematic.

111F07.EPS

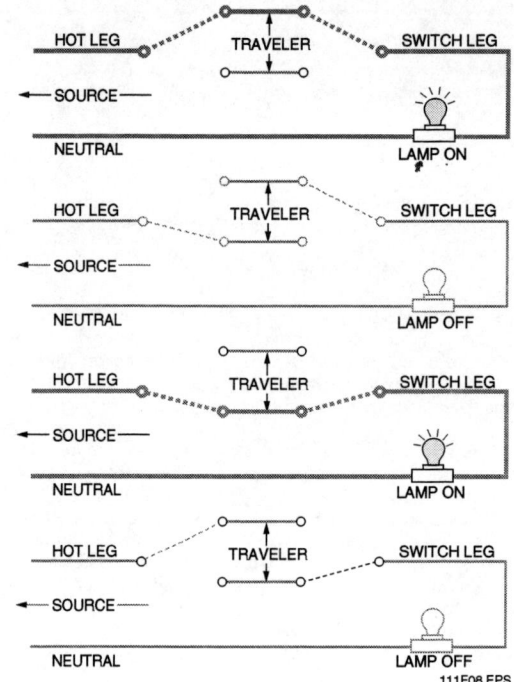

111F08.EPS

Figure 8 ◆ Possible combinations of a three-way switch.

111F09.EPS

Figure 9 ◆ Three-way switch as shown on a blueprint.

switch location. In other words, two three-way dimmers cannot be used to control the same load.

Incandescent dimmers are normally rated at 600, 1,000, 1,500, or 2,000 watts, 120 volts AC. Because incandescent dimmers are rated by wattage, it is easy to select lamp loads that will not exceed the dimmer rating. Specialized dimmer systems are also available. Incandescent dimmers should not be used to control receptacles, appliances, fluorescent lamps, or fluorescent ballasts.

The wiring procedure for dimmer switches is similar to the method used for standard switches. However, it is important to give special attention to the connections and the grounding. Some dimmer switches are solid-state devices, and their electronic circuitry can be damaged if connected improperly.

3.6.0 Disconnects

NEC Article 100 defines a disconnect as a device, group of devices, or other means by which the conductors of a circuit can be disconnected from their source of supply. All disconnect devices (switches or circuit breakers) for load devices and circuits must be clearly and permanently marked to show the purpose of the disconnect. This is a must according to the NEC. Under OSHA, this rule applies to all existing electrical systems, no matter how old, and also to new, modernized, expanded, or altered

Instructor's Notes:

_____

_____

_____

_____

_____

_____

_____

_____

_____

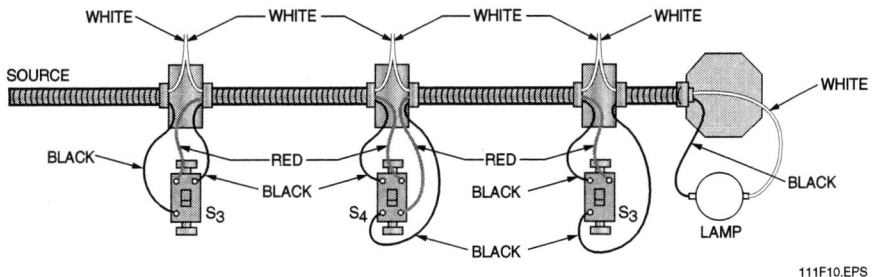

Figure 10 ◆ Four-way pictorial.

VERONICA WESTFALL   111F11.EPS

Figure 11 ◆ Dimmer switches.

CAUTION

Effectively identified disconnect devices are critically important to safety when a switch or circuit breaker has to be opened to quickly deenergize a circuit.

electrical systems. *NEC Section 110-22* covers the identification of disconnecting means.

3.6.1 Safety Switches

In the early 1900s, an open knife switch was commonly used for manual switching and disconnecting. It consisted of a stationary jaw and a movable blade pivoting in a hinge post. These early devices carried current adequately, but were insufficient from a safety standpoint. The load-make and load-break manual switching relied on an operator to open or close the blade quickly, and there were also exposed live parts.

Because of the danger of exposed live parts, the switches were eventually mounted in service boxes to increase operator safety. In the early 1920s, an externally operated handle was added to make the switch safe. A quick-make, quick-break mechanism was connected to the handle and installed inside the enclosure in the 1930s. This made opening and closing the breaker contacts independent of the operator's hand speed. These devices became known as *safety switches*.

Further safety features have been added over the years:

- Arc shields for electrical protection and maximum horsepower ratings
- Interiors that could be removed easily for wire pulling
- Coilproof dual-defeatable interlocks that increase safety and ease of inspection by authorized personnel
- Bold on/off and handle markings that clearly indicate the contact position from a distance
- Faceplates that cover the front of the switch and protect exposed wires

Figure 12 shows a safety switch that is typically used in commercial and industrial applications.

Safety switches are individually enclosed, manually operated disconnects that must be installed and operated in accordance with the NEC. Safety switches are voltage-rated and must be used with the appropriate voltage level. Unless otherwise noted, all safety switches are listed under UL 98 (Enclosed Switches). Safety switches must provide several features that are required by the NEC, including the following:

- Protection from live parts by a metallic enclosure that completely surrounds the disconnecting means

Discuss safety switches and their applications. Pass around examples for the trainees to examine.

VERONICA WESTFALL   111F12.EPS

Figure 12 ◆ Safety switch.

- Highly visible indication of the contact position
- Provisions for padlocking the external, manually operated handle in the OFF position

4.0.0 ◆ NEMA CLASSIFICATIONS

The National Electrical Manufacturers' Association (NEMA) is a nonprofit organization supported by the manufacturers of electrical equipment and supplies. NEMA develops standards that are used when purchasing equipment. The classifications help to match the product to the application.

4.1.0 Switch Classifications

There are two NEMA classifications for enclosed switches: general-duty switches and heavy-duty, motor-rated switches.

4.1.1 General-Duty Switches

A general-duty switch is designed for use in residential and commercial applications. It is used where the service factor is not high (e.g., in lighting, air conditioning, and appliance loads). There are several kinds of general-duty, single-throw switches. They range from a simple two-wire switch to a solid neutral switch with one blade and one fuse, which is rated for 120 volts AC. The other end of the range includes four-wire switches with solid neutral, three-blade, three-fuse construction, which are rated for 240 volts AC. General-duty two- and three-pole switches are designed for industrial use where the service demands are not high. They are rated from 30 to 1,200 amps, up to 600 volts AC or DC, and are available in fusible and nonfusible types. The safety switch shown in *Figure 12* is a general-duty switch.

4.1.2 Heavy-Duty, Motor-Rated Switches

Heavy-duty, motor-rated switches are designed to be used in production industries. Heavy-duty switches are rated from 30 to 1,200 amps, up to 600 volts AC or DC. They are available in both fusible and nonfusible types.

When a heavy-duty switch is in the OFF position, the blades are visible for safety. Most of these switches also feature quick-make, quick-break operating mechanisms and a full-cover interlock. The quick-make, quick-break operation prevents an operator from closing or opening the switch too slowly. Slow operation of the switch can cause the blades and clips to arc when a load is connected. Arcing is more pronounced on larger switches. In time, the blades and clips burn away. *Figure 13* shows a heavy-duty switch.

When the switch is closed, the door is locked. The door can be opened only when the switch handle is in the open position. In addition, the switch cannot be closed with the door open. However, for maintenance purposes, the interlock device can be tripped so that the switch can be closed with the door open. The switch handle can be locked with a padlock to prevent unauthorized operation. When you need to work on the circuit or the switch controls, pull the handle to open the circuit. Next, open the switch door and remove the fuses. Padlock the handle in the open position; that way, the switch cannot be closed by anyone while you are working. Tie a safety tag with your name onto the padlock. When you are finished

 Classroom

Discuss the NEMA switch classifications.

Discuss the differences between general-duty and heavy-duty switches.

 INSIDE TRACK

Check the Specifications Before Wiring

Always refer to the project plans and/or specifications before selecting a wiring method. Often, the specifications will disallow certain techniques. Also, be sure to check that the equipment is suitable for the application. For example, equipment installed in a classified area must meet or exceed the rating for that area.

Instructor's Notes:

_____

_____

_____

_____

_____

_____

_____

_____

_____

VERONICA WESTFALL   111F13.EPS

Figure 13 ◆ Heavy-duty switch.

working on the circuit, replace the fuses and remove your lock and tag. After verifying that it is safe to close the switch, do so.

WARNING!
Make sure to always follow your site-specific lockout/tagout procedure.

The manufacturer's list shows the switch's current rating, voltage rating, and the types of enclosures available. If the switch is intended for interrupting a motor circuit, the switch must also be horsepower-rated.

4.2.0 Enclosure Classifications

NEMA also classifies the enclosures designed for specific conditions. When choosing devices for use in hazardous locations, the proper NEMA-type enclosure must be specified. The NEMA numbers are not marked on the enclosures, but are included in the manufacturers' catalogs. *NEC Table 430-91* provides a complete list of enclosures.

4.2.1 General Purpose Indoor (Type 1)

The Type 1 general purpose indoor enclosure is made of sheet metal. This enclosure is primarily intended to protect people against accidental contact with equipment. It is not dust-tight. This en-

closure often has pre-punched knockouts that are easily removed for connection to a raceway. *Figure 13* shows a NEMA Type 1 enclosure.

4.2.2 Drip-Proof Indoor (Type 2)

The Type 2 general purpose indoor enclosure is similar to Type 1, but it also has drip shields or their equivalent.

4.2.3 Dust-Tight, Rain-Tight, and Sleet-Resistant (Type 3)

The Type 3 enclosure is intended for outdoor use to protect against windblown dust and water.

4.2.4 Rainproof and Sleet-Resistant (Type 3R)

The Type 3R enclosure (*Figure 14*) is intended for outdoor use. This enclosure meets the requirements of UL 508. This enclosure often contains knockouts below the level of the lowest live component.

4.2.5 Dust-Tight, Rain-Tight, and Sleet-Proof (Type 3S)

Similar to Type 3, the Type 3S enclosure is intended for use outdoors to protect equipment from windblown dust and water. However, the enclosure can operate even when covered by external snow and sleet.

VERONICA WESTFALL   111F14.EPS

Figure 14 ◆ NEMA Type 3R enclosure.

Classroom

Discuss NEMA enclosure types and their applications. Show as many examples as possible and allow the trainees to examine them.

Note

Types 3, 3R, and 3S are not considered watertight but are intended to keep water at a safe distance from live internal parts. These enclosures may have weep holes built in to allow any water that may enter to drain quickly.

4.2.6 Watertight and Dust-Tight (Type 4)

The Type 4 enclosure is made of rigid cast metal or sheet metal. It is designed to withstand a standard hose-stream test.

4.2.7 Submersible, Watertight, and Dust-Tight (Type 6)

The Type 6 enclosure can be submerged in water. It is also resistant to dust and sleet (ice), and can be installed either indoors or outdoors.

4.2.8 Hazardous Locations, Class I (Type 7)

The Type 7 enclosure is used for air-break devices. It meets the NEC requirements for Class I, Groups A, B, C, or D hazardous locations. These locations may have flammable gases or vapors present. See *Figure 15* for an example of a Type 7 enclosure.

4.2.9 Hazardous Locations, Class I (Type 8)

A Type 8 enclosure is similar to the Type 7 enclosure. However, the Type 8 enclosure is designed for oil-immersed equipment.

VERONICA WESTFALL   111F15.EPS

Figure 15 ◆ Snap switch in explosion-proof (Type 7) enclosure.

4.2.10 Hazardous Locations, Class II (Type 9)

The Type 9 enclosure meets the NEC requirements for Class II, Groups E, F, and G hazardous locations. These locations have combustible dust in the air.

4.2.11 Bureau of Mines (Type 10)

The Type 10 enclosure meets the requirements of the U.S. Bureau of Mines.

4.2.12 Corrosion-Resistant, Drip-Proof Indoor (Type 11)

The Type 11 enclosure is designed for oil-immersed equipment and is suitable for use in locations subject to corrosive acid or fumes.

4.2.13 Dust-Tight and Drip-Tight Industrial Use (Type 12)

The Type 12 enclosure keeps out dust, lint, fibers, and other airborne particles. A gasket fits between the case and the cover. There are no holes in the enclosure and no conduit knockouts or conduit openings.

4.2.14 Oil-Tight and Dust-Tight (Type 13)

The Type 13 enclosure is used indoors mainly to house pilot devices, including limit switches, foot switches, and pushbuttons. This enclosure protects the devices from lint, dust, and seepage.

Note

Hazardous locations are described in *NEC Article 500.*

4.2.15 Ingress Protection (IP) Classifications

The Ingress Protection (IP) Classification System uses a two-digit number to classify the degree of protection provided by an enclosure (*Table 1*). The first digit (0 through 6) represents the degree of protection against solid objects, while the second digit (0 through 8) represents the degree of protection against water.

Instructor's Notes:

_____

_____

_____

_____

_____

_____

_____

_____

_____

Switch Enclosures

THINK ABOUT IT What type of enclosure would you use for a pool lighting fixture that is 3' from the pool's edge? What type would you use in the paint room of an auto body shop?

Table 1 IP Classification System

FIRST NUMBER Degree of protection against solid objects	SECOND NUMBER Degree of protection against water
0. Not protected. 1. Protected against a solid object greater than 50mm, such as a hand. 2. Protected against a solid object greater than 12mm, such as a finger. 3. Protected against a solid object greater than 2.5mm, such as a wire or tool. 4. Protected against a solid object greater than 1.0mm, such as wire or thin strips of metal. 5. Dust-protected. Prevents ingress of dust sufficient to cause harm. 6. Dust-tight. No dust ingress.	0. Not protected. 1. Protected against water dripping vertically, such as condensation. 2. Protected against dripping water when tilted up to 15°. 3. Protected against water when spraying at an angle of up to 60°. 4. Protected against water splashing from any direction. 5. Protected against jets of water from any direction. 6. Protected against heavy seas or powerful jets of water. Prevents ingress sufficient to cause harm. 7. Protected against harmful ingress of water when immersed between a depth of 150mm to 1m. 8. Protected against submersion. Suitable for continuous immersion in water.

Note
The IP Classification System does not indicate corrosion resistance.

5.0.0 ◆ RECEPTACLES

A receptacle is a contact device that is installed at an outlet so that a single attachment plug can be connected. According to *NEC Article 100,* a receptacle outlet is an outlet where one or more receptacles are installed. Receptacles are rated according to their voltage and amperage capacity. This rating determines the number and configuration of the contacts. *Figure 16* shows a portion of the NEMA configurations for general purpose nonlocking plugs and receptacles, as well as other common receptacles. In most cases, a distinction is made between the system grounded conductor (W) and the grounding conductor (G). You will note that plugs

		15 AMPERE		20 AMPERE		30 AMPERE		50 AMPERE	
		RECEPTACLE	PLUG	RECEPTACLE	PLUG	RECEPTACLE	PLUG	RECEPTACLE	PLUG
2-POLE 3-WIRE GROUNDING	5 125V	5-15R	5-15P	5-20R	5-20P	5-30R	5-30P	5-50R	5-50P
	6 250V	6-15R	6-15P	6-20R	6-20P	6-30R	6-30P	6-50R	6-50P
	7 277V	7-15R	7-15P	7-20R	7-20P	7-30R	7-30P	7-50R	7-50P

111F16.EPS

Figure 16 ◆ Typical NEMA plug configurations.

Teaching Tip

Discuss the "Think About It."

Classroom

Discuss the IP classification system.

Discuss receptacles and explain how they are rated.

Audiovisual

Show Transparency 13 (Figure 16).

are often polarized (one blade is wider than the other). This prevents the user from plugging the device in the wrong way.

Grounded conductors, which are labeled W, may carry current in normal operation. *NEC Article 100* defines a grounded conductor as a system or circuit conductor that is intentionally grounded. Grounded conductors are commonly referred to as *neutrals*.

Grounding conductors (equipment grounds) are labeled G. They carry current only when part of the equipment, which was never intended to be energized, becomes energized through a fault. These are commonly referred to as *grounding conductors*. *NEC Article 100* defines a grounding conductor as a conductor used to connect equipment or the grounded circuit of a wiring system to a grounding electrode or electrodes.

NEC Section 210-7 states that receptacles installed on 15A and 20A branch circuits must be of the grounding type. *Figure 17* shows a grounding-type duplex receptacle. On all new installations, a grounding-type device must be installed.

The receptacle current rating must be at least equal to that of the conductors feeding it. If there is more than one receptacle on the circuit, the total load plugged into these receptacles cannot exceed the rating of the conductors. *NEC Tables 210-21(b)(2) and (3)* list these requirements.

> **Note**
> *NEC Section 210-21* allows the installation of 15A receptacles on a 20A circuit if the circuit feeds more than one receptacle.

5.1.0 Identifying Receptacles

Receptacles, like switches, have various symbols and information printed on them that help to determine their proper use and ratings (*Figure 18*). In addition, the shapes and positions of the openings also determine the use of the receptacle.

LIGHTNING PANEL

15-AMP OR 20-AMP OVERCURRENT DEVICE

NEUTRAL

BOTH TERMINALS ARE TIED TOGETHER

RACEWAY, CABLE ARMOR, OR SEPARATE GROUNDING CONDUCTOR TO GROUND THE GROUNDING TERMINAL AND MOUNTING YOKE

GROUND TERMINAL IS GREEN HEX-HEAD SCREW

MOUNTING YOKE AND GREEN GROUNDING SCREW ARE BONDED TOGETHER AND CONNECTED INTERNALLY TO THE GROUNDING TERMINAL SLOTS

111F17.EPS

Figure 17 ◆ Grounding-type duplex receptacle.

Instructor's Notes:

_____

Identifying Receptacles

The picture shown here will get you started on identifying various receptacles. From left to right, the receptacles pictured are: TOP: Insulated ground receptacle; 220V receptacle; GFCI with ground screw. BOTTOM: Twist lock receptacle; standard 120V receptacle; GFCI with ground tail.

VERONICA WESTFALL   111P1103.EPS

THE HOT SIDE OF THE RECEPTACLE IS INDICATED BY BRASS-COLORED SCREWS

THE NEUTRAL SIDE OF THE RECEPTACLE IS INDICATED BY SILVER-COLORED SCREWS

ON POLARIZED RECEPTACLE LONG SLOT INDICATES NEUTRAL SMALL SLOT INDICATES HOT

CANADIAN STANDARDS ASSOCIATION

UNDERWRITERS' LABORATORIES LISTING

GROUNDING SLOT

GROUNDING CONNECTION HAS GREEN SCREW

111F18.EPS

Figure 18 ◆ Standard duplex receptacle.

The printed information on the receptacle is similar to the information printed on switches. The information on receptacles includes the following:

- The UL label
- The CSA label
- Current and voltage ratings listed by maximum amperage and maximum voltage (may be shown on the front or back, depending on the switch)

5.2.0 Types of Receptacles

There are many different types of receptacles. This section covers the most common types.

5.2.1 Straight Blade Receptacles

Straight blade receptacles, as their name implies, will accept a straight blade connector or plug. These receptacles are the most common type.

5.2.2 Twist Lock Receptacles

Twist lock receptacles accept a curved blade connector or plug. The plug/connector and the receptacle lock together with a slight twist. The locking action prevents accidental unplugging of the equipment.

5.2.3 Pin and Sleeve Receptacles

Pin and sleeve devices have a unique locking feature. These receptacles are made with an extremely heavy-duty plastic housing that makes

them virtually indestructible. They come with long brass pins for long life and are color coded according to voltage for easy identification.

5.3.0 Receptacle Terminals

The terminals on a receptacle are color coded. The green terminal is the equipment ground connection and is associated with the U-shaped slot. The silver-colored terminal is associated with the neutral (white) conductor and the larger of the two vertical slots on the receptacle. The brass-colored terminal is associated with the hot (black) conductor and the smaller vertical slot on the receptacle.

Similar to switches, receptacles have either screw terminals for attaching wires or quick-wire terminals located on the back. Although UL-approved, the push-in quick-wire terminals on receptacles and switches do not provide as good a connection as a wire that is properly terminated on the screw terminal. Use quick-wire terminals only for lightly loaded circuits. Receptacle terminals must be listed for the type of wire that is used in the system. If copper or copper-clad wire is used, all types of terminals are acceptable. If aluminum wire is used, the terminal must be rated for aluminum.

5.4.0 Ground Fault Circuit Interrupters

The NEC defines a ground fault circuit interrupter (GFCI) as a protective device that functions to

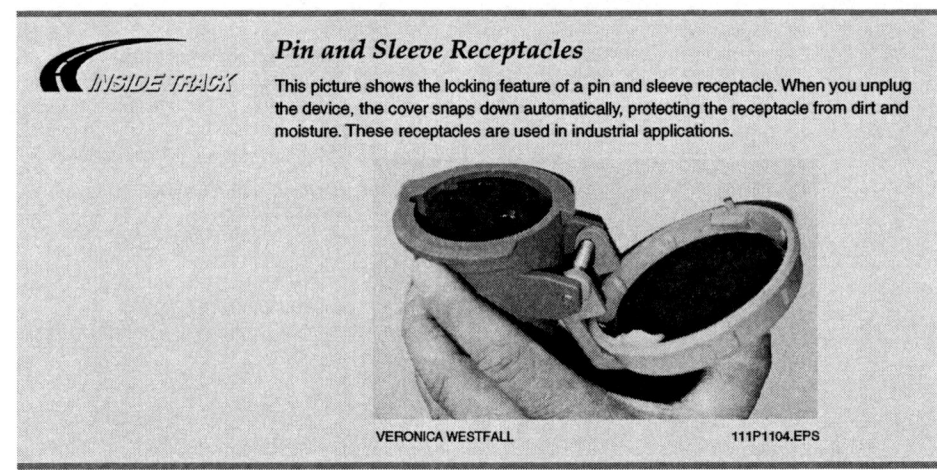

Pin and Sleeve Receptacles

This picture shows the locking feature of a pin and sleeve receptacle. When you unplug the device, the cover snaps down automatically, protecting the receptacle from dirt and moisture. These receptacles are used in industrial applications.

VERONICA WESTFALL   111P1104.EPS

Instructor's Notes:

_____

_____

_____

_____

_____

_____

_____

_____

_____

deenergize a circuit or portion thereof within an established period of time when a current to ground exceeds some predetermined value that is less than that required to operate the overcurrent protective device of the supply circuit. *NEC Section 210-8* defines locations for GFCIs.

Electricity normally flows into a tool or appliance through an energized wire and returns through a neutral wire. A GFCI compares the amount of current going out of the hot wire to the amount of current returning on the neutral wire. Any current that goes out of the hot wire but does not return on the neutral wire must be taking an unintended path or ground fault. Ground faults may be caused by worn or damp insulation or faulty wiring. If someone is in that current path, that person can complete the circuit necessary for the flow of current. Shock, injury, or electrocution can result, depending on the age and health of the person involved, how well grounded the person is, and the path the current takes through the person's body.

When a ground fault leak of some predetermined value (usually 5mA) is detected, the sensitive electronic circuitry within the GFCI will automatically trip a switch. This interrupts the circuit and cuts off the power. The interruption takes place within 1/40th of a second or less and prevents possible injury. *Figure 19* shows a typical GFCI. GFCIs are available as circuit breakers, receptacles, or on extension cords and adapters that plug into standard receptacles.

Note
When installed properly, one GFCI receptacle can protect all of the receptacles on the same circuit, as shown in *Figure 20*.

NEC Section 305-6 covers the rules that concern GFCI protection for all 15A, 20A, and 30A receptacle outlets on construction sites. The NEC states that ground fault circuit interrupters (either GFCI circuit breakers or GFCI receptacles) must be used to provide personnel protection for all receptacles of the designated rating that are in use by employees and are not part of the permanent wiring of the building or structure.

6.0.0 ◆ WIRING TECHNIQUES

When you terminate conductors at devices, there is one simple rule to follow: Use the proper procedure and the correct tool.

VERONICA WESTFALL   111F19.EPS
Figure 19 ◆ Typical GFCI.

6.1.0 Bending Wire

Suppose you need to wrap the stripped end of a conductor around a screw lug. If you used a pair of needle-nose pliers to grip the conductor, you would probably scar the conductor and weaken it in the process. A conductor can overheat if it is scarred or scratched at its point of termination. Overheating creates a very real fire hazard.

Instead, the right tool to use for bending wire is the wire stripper. Many wire strippers have a hole in each of the jaws. When the stripped end of the conductor is inserted into one of these holes, you can safely bend the conductor into the proper shape for terminating it. Care must be taken to wrap the U-shaped conductor around the screw with the short end of the U wrapped around the screw in a clockwise direction. Before the screw is tightened, the stripped end of the conductor should be bent against the screw. To do this, use the screw shank with the wire stripper and press the end of the conductor against the screw. Then, use the jaws of the wire stripper to bend the conductor, as shown in *Figure 21*.

After the conductor is bent, the screw can be tightened to make a good electrical connection. The wire is actually twisted around the screw as it is tightened. For this reason, the wire must be wrapped around the screw in a clockwise direction.

Explain how GFCIs are wired in a circuit.

Show Transparency 16 (Figure 20).

Discuss the importance of proper technique when terminating conductors.

Demonstrate how to properly bend wire.

Show Transparency 17 (Figure 21).

Figure 20 ◆ GFCI protecting several standard receptacles.

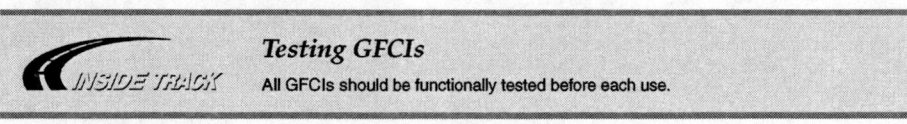

Testing GFCIs

All GFCIs should be functionally tested before each use.

Selecting Wire Strippers

Wire strippers come in various sizes. Always select the correct tool (photo A) for the wire to be stripped and work slowly and carefully (photo B) to avoid damaging the conductor.

(A) CHOOSE THE CORRECT TOOL

VERONICA WESTFALL   111P1105.EPS VERONICA WESTFALL

(B) WORK SLOWLY AND CAREFULLY

111P1106.EPS

111F21.EPS

Figure 21 ◆ Bending the conductor.

Instructor's Notes:

_____

_____

_____

_____

_____

_____

_____

_____

_____

_____

CAUTION

The torque applied to the terminal screw must be just right. Too much torque can strip the terminal. Too little torque can cause the termination to come apart.

Note

Certain receptacles are specially rated and designed for stranded wire.

6.2.0 Branch Circuit Wiring

According to *NEC Article 100,* a branch circuit is the circuit conductors between the final overcurrent device protecting the circuit and the outlet(s). That means that all outlets are supplied by branch circuits.

6.2.1 Planning

The first step in branch circuit wiring is to check the drawings to determine the exact location of all outlet boxes for general purpose receptacles, switches, lighting fixtures, and special purpose receptacles. The NEC requirements for the number of circuits, protective measures, and special circuits are determined at this time.

Note

The drawings may not show the actual conductor runs or homeruns for the wiring installation. Prior to planning the branch circuit wiring for an electrical installation, you should review the applicable articles of the NEC.

Classroom

Explain that careful examination of the drawings is the first step in installing branch wiring.

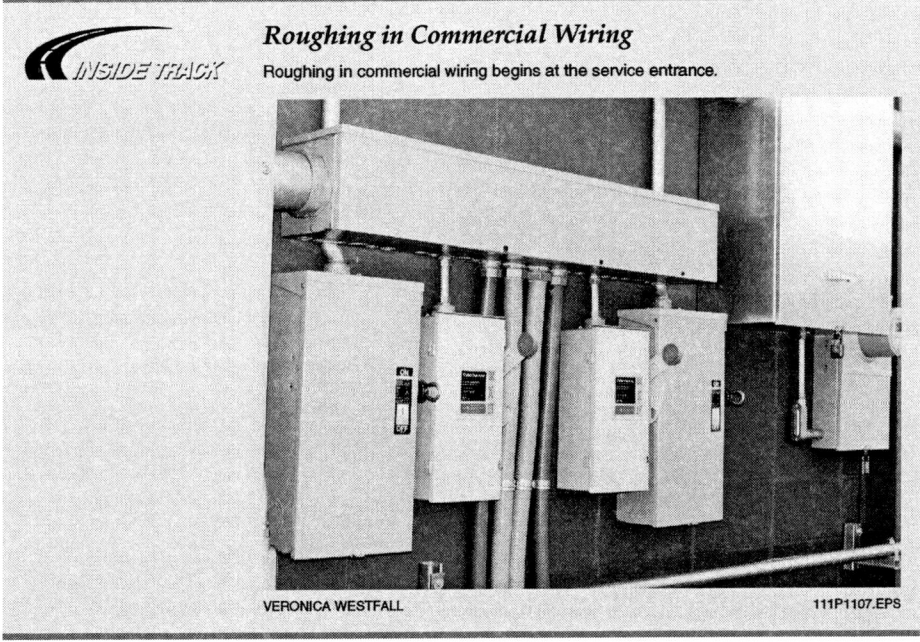

INSIDE TRACK

Roughing in Commercial Wiring

Roughing in commercial wiring begins at the service entrance.

VERONICA WESTFALL   111P1107.EPS

Making Connections

Always get in the habit of making your connections in the following order: first the grounding conductors, then the grounded conductors, and finally, the ungrounded conductors. If you approach each job in a systematic manner, it will soon become second nature to you and you'll be less likely to make mistakes.

Terminations

For each branch circuit, terminate the devices starting from the furthest location in the circuit and working back toward the panel. Terminate the branch circuit breakers last.

Classroom

Discuss the installation process for branch circuit wiring.

6.2.2 Installation

After you have studied the electrical drawings and determined the general layout of the electrical system, the general rough-in stage begins. The rough-in stage involves the following:

- Installing outlet boxes and the service-entrance panel
- Establishing the conductor pathway
- Pulling or installing the cable or raceway

Roughing-in and the following step (installing switches, receptacles, etc.) require that you have an overall knowledge of both tasks. For example, it is difficult to install the proper conductors for a four-way switch in a particular location without knowing the working characteristics of the switch or the various configurations in which the switch can be used.

> **Note**
> Before you begin the roughing-in stage of the wiring, review the applicable articles of the NEC.

Before any actual installation begins, study the electrical floor plan and consult with the builder or architect to ensure that no last-minute changes have been made in the electrical system. At this time, you should select the type of boxes that will be used at particular locations. The sizes of the required boxes must be determined in accordance with *NEC Article 370*. Pay particular attention to *NEC Section 370-16(a) and (b)*.

The size of the box is rated according to its cubic-inch capacity (volume). Although the NEC states what practices must be followed, the NEC does not explain why such a practice is required except that all provisions are designed to promote safety and prevent damage. Minimum volume capacities are required for different reasons. For example, too many wires too close together can cause overheating. In addition, adequate volume allows for greater air circulation and reduces crowding, which could cause damage to the insulation or connections of the wires.

>
> **WARNING!**
> Strict observance of the NEC is mandatory and all required calculations must be taken very seriously.

According to the NEC, an outlet box or junction box has to be installed at every point in the electrical wiring system where the wiring cable is going to be spliced or terminated. Each of these points must also be accessible for future revisions or repairs. Failure to make all boxes accessible is a violation of the NEC.

Many industrial/commercial outlet boxes are surface-mounted and connect directly to the electrical raceway (*Figure 22*). These boxes are often mounted to the building steel or to a support channel or strut that is welded or clamped to the steel. The installation of outlet boxes or raceways should be coordinated so as not to interfere with piping, ductwork, fireproofing, and other items that may or may not already be in place.

ELECTRICAL LEVEL ONE—TRAINEE MODULE 26111

Instructor's Notes:

VERONICA WESTFALL   111F22.EPS

Figure 22 ◆ Outlet box.

Commercial wiring may involve flush wiring similar to residential work. *NEC Sections 370-20 and 370-21* require that recessed switch, receptacle, and fixture boxes be mounted flush with the finished face of the wall surface when combustible material, such as plywood or paneling, is being used. If noncombustible material is used as the finish surface, the box may be recessed by as much as ¼ inch. The distance that the box is extended past the wall stud or building member is referred to as the *mounting depth* or *setout*. Many boxes have several setout measurements inscribed on the box (*Figure 23*).

These marks provide a quick and easy method of measuring the setout required for a particular wall finish. This saves time and effort when installing the box.

> **Note**
> Front-mounting bracket boxes will have the setout determined by the bracket assembly.

Install all boxes in accordance with the electrical drawings. The spacing should be as even as possible. The box center is the midpoint on the vertical dimension of the box. Make sure that you check the door swing direction so that the switches are not installed behind a door. Measure the height of the switch boxes from the floor so that they will be at the proper height when installed.

111F23.EPS

Figure 23 ◆ Box with setout measurements inscribed.

6.2.3 Splicing

Wires are spliced when they are joined or connected. Splices are required when installing switches, lighting, and other electrical equipment. There are three major steps for making a splice: removing the insulation, making the splice, and insulating the splice.

Wire strippers should be used to remove the insulation. Knives should not be used because of the possibility of nicking the wire. Use the following procedure to strip wire using a wire stripper.:

Step 1 Insert the wire into the proper slot on the side of the stripper marked for solid wire.

Step 2 Twist the stripper back and forth a few times to cut through the insulation.

Step 3 Pull the stripper back to pull the insulation off.

Most industrial and commercial lighting and receptacle circuits are spliced using solderless connectors. Solderless connectors, commonly referred to as *Wirenuts*®, are plastic caps that contain a tapered and threaded metal insert, as shown in *Figure 24*.

When using a solderless connector, screw the wire nut over the twisted ends to hold the conductors together. (Some manufacturers recommend first twisting the conductors together to ensure good electrical contact.) Solderless connectors are rated for size, temperature, and number of conductors. Make sure you use a properly rated connector, and always follow the manufacturer's instructions when using any connector.

Most industrial and commercial control circuits and signal wires (for devices such as transmitters, motor stop/start circuits, etc.) are spliced on fixed terminal blocks with crimp-on lugs. Examples of lugs are shown in *Figure 25*. Lugs help to keep the conductors organized and greatly aid in troubleshooting. When installing lugs, make sure your

Show Transparency 18 (Figure 23).

Demonstrate how to strip wire in order to make a splice.

Show Transparency 19 (Figure 24A).

Demonstrate how to splice wires using a solderless connector.

Discuss crimp-on wire lugs and their applications. Pass around examples for the trainees to examine.

Show Transparency 20 (Figure 25A).

WIRING: COMMERCIAL AND INDUSTRIAL—TRAINEE MODULE 26111   11–21

111F24A.EPS

111F25A.EPS

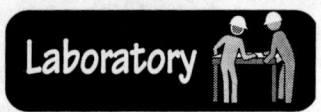

Have the trainees practice stripping and splicing wires.

VERONICA WESTFALL   111F24B.EPS

Figure 24 ◆ Solderless connectors.

VERONICA WESTFALL   111F25B.EPS

Figure 25 ◆ Crimp-on wire lugs.

Discuss relays and their applications. Pass around an example for the trainees to examine.

Show Transparency 21 (Figure 26).

Discuss limit switches and their applications. Pass around an example for the trainees to examine.

Show Transparency 22 (Figure 27).

crimping tool is designed for the particular lugs you are using.

7.0.0 ◆ CONTROL DEVICES

One of the major differences between industrial/commercial wiring and residential wiring is the complexity of controls for electrical machinery. Whereas most loads in a home are turned on manually using a wall switch, industrial machinery is controlled by a variety of devices, including relays and limit switches.

7.1.0 Relays

A relay acts as a remote-controlled switch that opens and closes to supply power to a load such as a motor, heater, or light. The electromagnetic coil of a relay acts as a load for the control circuit. This allows switches and other controlling devices to control a large load while only carrying a very small current. For this reason, these control-

ling devices are often referred to as *pilot devices*. The pilot devices work together in a control circuit to ultimately close a relay that has contacts that are rated for the full load current. These relay contacts are what connects the load to its source voltage. *Figure 26* shows a typical relay along with its internal electrical arrangement.

The pilot devices that control a relay may be activated by a variety of methods, such as air or hydraulic pressure for pressure switches, heat for temperature switches, or light for photoelectric switches, to name a few. One of the most popular industrial pilot devices is the mechanical limit switch.

7.2.0 Limit Switches

Limit switches use mechanical movement to open or close electrical contacts. Limit switches are frequently used on manufacturing equipment, conveyor belts, automatic doors, and many other devices that have repeated motion or travel limits. A typical limit switch is shown in *Figure 27*. Limit switches can be fitted with numerous types of mechanical arms to suit specific equipment configurations.

Instructor's Notes:

_____

_____

_____

_____

_____

_____

_____

_____

Figure 26 ◆ Typical plug-in relay.

ON (RED)
24VAC (RECTIFIED)
OFF (BLACK)
ON COIL   OFF COIL
LOAD
ARMATURE
LINE
LEADS TO LIGHTING FIXTURE
RELAY

111F26.EPS

Relays

Relays are very common switching controls that we take for granted. You often use relays (either directly or indirectly) in your everyday life. What examples can you think of?

THINK ABOUT IT

111F27.EPS

Figure 27 ◆ Limit switch.

8.0.0 ◆ POWER DISTRIBUTION EQUIPMENT

In residential wiring, power distribution equipment is usually limited to a breaker panel or fuse panel. Modern panels contain the main and branch circuit breakers in one relatively small enclosure. Industrial and commercial wiring calls for distribution equipment that is usually much larger.

Due to the high amount of load current and higher voltages found in large facilities, the main circuit breaker itself may be larger than the average refrigerator. This main breaker may feed a step-down transformer, which in turn feeds feeder circuit breakers. Power is distributed from the feeder breakers to the required loads. This configuration (main breaker, transformer, and feeder breakers) may be repeated at more than one location, depending on the size of the facility.

Distribution equipment, commonly referred to as switchgear, may be comprised of individually enclosed and mounted devices that are wired together, or it may all be included in one enclosure, as shown in *Figure 28*. Despite the many configurations of switchgear available, the purpose is always the same: to provide a safe system of controlling a large power source and supplying this power to the remote loads where it is needed.

Discuss the "Think About It."

Discuss power distribution equipment.

VERONICA WESTFALL   111F28.EPS

Figure 28 ◆ Switchgear.

Summary

Wiring and installing wiring devices will make up a large portion of your job as an electrician. Knowing how to relate the rules and regulations in the NEC to your job will improve your skills and ensure that the devices are installed properly.

*Minimizing Conduit for
Large Service Entrances*

The large industrial plant shown here uses busduct at the service entrance rather than conduit. It provides a nicer appearance as well as minimizes the time required for installation and the space needed for the service entrance conductors. If this industrial service used standard conduit, it would take up several times the space of the copper busduct used here.

VERONICA WESTFALL   111P1108.EPS

Teaching Tip

Discuss the "Think About It."

THINK
ABOUT IT

Putting It All Together

Think about the use of various devices in two vastly different applications, such as a hospital and a small office building. How do you imagine the installations will be similar? How will they be different?

Instructor's Notes:

_____

_____

_____

_____

_____

_____

_____

_____

_____

1. The designation for a double-pole, single-throw switch is _____.
   a. S1
   b. S2
   c. S3
   d. S4

2. A three-way switch is a type of _____ switch.
   a. SPST
   b. DPDT
   c. DPST
   d. SPDT

3. Older switches marked CU/AL should be used for _____.
   a. copper wire only
   b. aluminum wire only
   c. either copper or aluminum wire
   d. aluminum-clad copper wire only

4. Four-way switches can be used to control lights from more than _____ location(s).
   a. one
   b. two
   c. three
   d. four

5. Dimmer switches are normally rated at _____ watts.
   a. 600
   b. 1,000
   c. 1,500
   d. All of the above.

6. Which of the following is *not* true regarding modern safety switches?
   a. They include arc shields.
   b. They have removable interiors.
   c. They have coilproof dual-defeatable interlocks.
   d. They are called *open knife switches.*

7. NEMA Type 1 enclosures _____.
   a. can be used in all locations
   b. can only be used indoors
   c. are dust-tight
   d. Both b and c.

8. For a Class II, Group E hazardous location, you would use a _____ enclosure.
   a. Type 7
   b. Type 8
   c. Type 9
   d. Type 10

9. An IP Classification of 6-8 means that _____.
   a. the enclosure offers the highest degree of protection
   b. the enclosure is dust-tight
   c. the enclosure is suitable for continuous immersion
   d. All of the above.

10. Outlet boxes are covered extensively in _____.
    a. *NEC Article 500*
    b. *NEC Article 310*
    c. *NEC Article 370*
    d. *NEC Article 422*

Classroom

Have the trainees complete the Review Questions and go over the answers prior to administering the Module Examination.

Examination

Administer the Module Examination and Performance Test. Be sure to record the testing results on Craft Training Report Form 200 and submit the results to the Training Program Sponsor.

Jeff Eldridge, Ready Electric Co.

Jeff Eldridge is one of the countless Americans who have proven that those who are willing to put in the effort can start at the bottom and work their way to the top. Jeff started with a dream and eventually became part owner of a large electrical contracting company.

How did you choose a career in the electrical field?
My mother suggested it to me when I was 13 years old. From then on, it stuck with me. Whenever someone asked what I was going to be, I told them I was going to be an electrician.

Tell us about your apprenticeship experience.
I took two years of electrical training at a vocational school before I was hired by Ready Electric. Then I started the first year of a four-year apprenticeship program conducted by ABC (Associated Builders and Contractors). When I finished my four-year apprenticeship, I became a foreman on a residential crew.

What positions have you held in the industry?
I started out as a residential foreman and then became residential manager. After that, I spent three years in commercial and industrial work, first as an electrician and then as a foreman. Following that, I was an estimator and project manager for several years, and eventually became part owner of the business.

What would you say is the primary factor in achieving success?
Hard work, pure and simple. Just work harder than everyone else.

What does your current job involve?
I have responsibility for cost estimating, project management, scheduling of personnel and equipment, and direct management of personnel, including hiring and evaluation.

Do you have any advice for someone just entering the trade?
First, treat others as you would like to be treated. It's one of the oldest rules of life. Second, work hard. We live in a world where you reap what you sow. I believe that if you treat people right and work hard, you will eventually reap the rewards.

Instructor's Notes:

_____

_____

_____

_____

_____

_____

_____

_____

_____

Trade Terms Introduced in This Module

Attachment plug (plug cap): A device which, by insertion in a receptacle, establishes a connection between the conductors of the attached flexible cord and the conductors connected permanently to the receptacle.

Branch circuit: The circuit conductors between the final overcurrent device protecting the circuit and the outlet(s).

Disconnecting means: A device, group of devices, or other means by which the conductors of a circuit can be disconnected from their source of supply.

Grounding: The process of connecting noncurrent-carrying metal parts of equipment to earth to provide an intentional path for fault current.

Limit switch: A type of pilot device that uses a mechanical movement to open or close electrical contacts.

Outlet: A point on a wiring system at which current is taken to supply utilization equipment.

Pilot device: Any of several types of electrical switches that are actuated by air pressure, temperature, level, or other mechanical or electrical means. These devices are wired to energize relays, which in turn connect power to larger loads.

Polarized: A physical difference between the grounded and positive and negative poles of a plug or receptacle. The difference is accomplished by making the grounded (neutral) blade of a plug larger than the ungrounded blade. The purpose of polarizing is to ensure that the grounded lead of a plug-and-cord assembly cannot be connected to the ungrounded terminal of a receptacle.

Receptacle: A contact device installed at the outlet for the connection of a single attachment plug.

Receptacle outlet: An outlet in which one or more receptacles are installed.

Relay: An electromagnetic switch for controlling one or more separate circuits remotely by applying voltage to the coil. Contacts attached to a moving plunger are connected to, or disconnected from, stationary contacts as the plunger is pulled by the magnetic field of the energized coil.

Switchgear: A term used for a wide variety of enclosed circuit breakers, disconnects, busways, and associated items that distribute power from a common point.

Answers to Review Questions

Answers	Section
1. b	2.1.0
2. d	2.1.0
3. a	2.2.0
4. b	3.4.0
5. d	3.5.0
6. d	3.6.1
7. b	4.2.1
8. c	4.2.10
9. d	4.2.15
10. c	6.2.2

Instructor's Notes:

_____

_____

_____

_____

_____

_____

_____

_____

_____

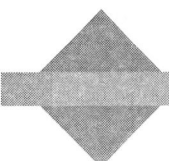

Additional Resources

For advanced study of topics covered in this task module, the following book is suggested:

National Electrical Code Handbook, Latest Edition. Quincy, MA: National Fire Protection Association.

NCCER CRAFT TRAINING USER UPDATES

The NCCER makes every effort to keep these textbooks up-to-date and free of technical errors. We appreciate your help in this process. If you have an idea for improving this textbook, or if you find an error, a typographical mistake, or an inaccuracy in the NCCER's Craft Training textbooks, please write us, using this form or a photocopy. Be sure to include the exact module number, page number, a detailed description, and the correction, if applicable. Your input will be brought to the attention of the Technical Review Committee. Thank you for your assistance.

Instructors – If you found that additional materials were necessary in order to teach this module effectively, please let us know so that we may include them in the Equipment/Materials list in the Instructor's Guide.

Write: Curriculum Revision and Development Department
National Center for Construction Education and Research
P.O. Box 141104, Gainesville, FL 32614-1104

Fax: 352-334-0932

E-mail: curriculum@nccer.org

Craft _____ Module Name _____

Copyright Date _____ Module Number _____ Page Number(s) _____

Description _____
_____
_____
_____
_____

(Optional) Correction _____
_____
_____

(Optional) Your Name and Address _____
_____
_____

TEACHING TIPS

Section 3.2.0   *Think About It—Double-Pole Switches*

Double-pole switches isolate the load because they open both conductors. They are used in commercial/industrial applications for safety purposes. One example is a situation where conductors run through a hazardous environment, such as lighting wires that run though a gasoline pump enclosure. They are also used as safety disconnects for air-conditioning systems.

Section 4.2.15   *Think About It—Switch Enclosures*

The swimming pool would require a NEMA Type 3R enclosure, while the paint shop environment calls for a Class I (Type 7) enclosure. Have the trainees look up these specific requirements in the NEC. Show examples of the two types of enclosures and point out the differences.

Section 7.1.0   *Think About It—Relays*

Explain that relays and contactors are used in lighting control and security systems, as well as furnaces, air conditioners, and other devices that contain electric motors. Explain that because relays usually have more than one set of contacts, they allow several devices to be controlled with only one input signal, such as the closing of a furnace thermostat.

Section 8.0.0   *Think About It—Putting It All Together*

Refer the trainees to *NEC Article 517* to observe the many special requirements that apply to patient-care areas of health care facilities. Point out that requirements for special receptacles are designed to protect patients from shock, and that other special features are needed to protect patients and staff from fire and explosion hazards due to the presence of oxygen and flammable anesthetics. Office areas of health care facilities can be treated in the same way as those in commercial office buildings. Explain that the most important point in this exercise is to recognize that a device, material, or method that can be used in one environment may not be acceptable for another.

Wiring: Residential

26112-00

MODULE OVERVIEW

This course introduces the electrical trainee to the various types of devices and installation procedures used in residential wiring. It also covers service-entrance and branch circuit calculations and NEC requirements.

PREREQUISITES

Please refer to the Course Map in the Trainee Module. Prior to training with this module, it is recommended that the trainee shall have successfully completed the following modules:

Core Curriculum; Electrical Level One, Modules 26101 through 26111

LEARNING OBJECTIVES

Upon completion of this module, the trainee will be able to:

1. Describe how to determine electric service requirements for dwellings.
2. Explain the grounding requirements of a residential electric service.
3. Calculate and select service-entrance equipment.
4. Select the proper wiring methods for various types of residences.
5. Explain the role of the NEC in residential wiring.
6. Compute branch circuit loads and explain their installation requirements.
7. Explain the types and purposes of equipment grounding conductors.
8. Explain the purpose of ground fault circuit interrupters and tell where they must be installed.
9. Size outlet boxes and select the proper type for different wiring methods.
10. Describe rules for installing electric space heating and HVAC equipment.
11. Describe the installation rules for electrical systems around swimming pools, spas, and hot tubs.
12. Explain how wiring devices are selected and installed.
13. Describe the installation and control of lighting fixtures.

PERFORMANCE OBJECTIVES

Under supervision of the instructor, the trainee should be able to:

1. For a residential dwelling of a given size, and equipped with a given list of major appliances, demonstrate or explain how to:
    * Compute the lighting, small appliance, and laundry loads.
    * Compute the loads for large appliances.
    * Determine the number of branch circuits required.
    * Size and select the service-entrance equipment (conductors, panelboard, and protective devices).
2. Using an unlabeled diagram of a panelboard (Sheet 2 of Transparency 2), label the lettered components.
3. Select the proper type and size outlet box needed for a given set of wiring conditions.

NCCER STANDARDIZED CRAFT TRAINING PROGRAM

The National Center for Construction Education and Research (NCCER) provides a standardized national program of accredited craft training. Key features of the program include instructor certification, competency-based training, and performance testing. The program provides trainees, instructors, and companies with a standard form of recognition through a National Craft Training Registry. The program is described in full in the *Guidelines for Accreditation*, published by the NCCER. For more information on standardized craft training, contact the NCCER at P.O. Box 141104, Gainesville, FL 32614-1104, 352-334-0911, visit our Web site at www.NCCER.org, or e-mail info@NCCER.org.

HOW TO USE THIS ANNOTATED INSTRUCTOR'S GUIDE

Each page presents two sections of information. The larger section displays each page exactly as it appears in the Trainee Module. The narrow column ties suggested trainee and instructor actions to each page and provides icons to call your attention to material, safety, audiovisual, or testing requirements. The bottom of each page includes space for your notes.

 If you see the Teaching Tip icon, that means there is a teaching tip associated with this section. Also refer to the suggested teaching tips at the end of the module.

SAFETY CONSIDERATIONS

Ensure that the trainees are equipped with appropriate personal protective equipment. Emphasize the importance of following all safety precautions and procedures when working with electrical circuits.

PREPARATION

Before teaching this module, you should review the Module Outline, Learning and Performance Objectives, and the Materials and Equipment List. Be sure to allow ample time to prepare your own training or lesson plan and gather all required equipment and materials.

MATERIALS AND EQUIPMENT LIST

Materials:
Transparencies

Markers/chalk

Copy of the latest edition of the *National Electrical Code*®

Calculator

Pencils/scratch paper

Examples of made-type grounding electrodes

GFCI circuit breakers and receptacles

Assortment of metallic and plastic outlet boxes

Assorted types of electrical receptacles

Assortment of switches, including:
    Single-pole
    Three-way
    Four-way
    Dimmer

Panelboard

Examples of cable, including:
    Type NM
    Type AC
    Type UF
    Type SE/USE

Examples of raceways, including:
    Rigid
    IMC
    EMT
    Flexible
    PVC

Module Examinations*

Performance Profile Sheets*

Equipment
Overhead projector and screen

Whiteboard/chalkboard

Appropriate personal protective equipment

*Located in the Test Booklet packaged with this Annotated Instructor's Guide.

ADDITIONAL RESOURCES

This module is intended to present thorough resources for task training. The following reference works are suggested for both instructors and motivated trainees interested in further study. These are optional materials for continued education rather than for task training.

Illustrated Dictionary for Electrical Workers, Latest Edition. Albany, NY: Delmar Publishers.

Illustrated Guide to the NEC, Latest Edition. Carlsbad, CA: Craftsman Book Co.

National Electrical Code Handbook, Latest Edition. Quincy, MA: National Fire Protection Association.

NOTES

The designations "National Electrical Code," "NE Code," and "NEC," where used in this document, refer to the *National Electrical Code*®, which is a registered trademark of the National Fire Protection Association, Quincy, MA. All National Electrical Code (NEC) references in this module refer to the 1999 edition of the NEC.

If you feel that additional math instruction would be helpful, Prentice Hall offers a basic math textbook entitled *Fundamentals of Electrical and Mechanical Mathematics*. It covers the basic math requirements for electrical trainees and may be ordered by contacting Prentice Hall Customer Service at 1-800-922-0579.

TEACHING TIME FOR THIS MODULE

An outline for use in developing your lesson plan is presented below. Note that each Roman numeral in the outline equates to one session of instruction. Each session has a suggested time period of 2½ hours. This includes 10 minutes at the beginning of each session for administrative tasks and one 10-minute break during the session. Approximately 15 hours are suggested to cover *Wiring: Residential*. You will need to adjust the time required for hands-on activity and testing based on your class size and resources.

Topic	Planned Time
Session I. Sizing the Electrical Service	
A. Sizing the Electrical Service	_____
1. Floor Plans	_____
2. General Lighting Loads	_____
3. Calculating the Electric Service Load	_____
a. Small Appliance Loads	_____
b. Laundry Circuit	_____
c. Demand Factors	_____
d. Required Service Size	_____
4. Demand Factors	_____
5. General Lighting and Receptacle Load Demand Factors	_____
6. Small Appliance and Laundry Loads for Dwelling Units	_____
a. Laundry Circuit Load	_____
b. Dryer Load	_____
c. Range Load	_____
d. Demand Loads for Electric Ranges	_____
e. Calculating the Demand Load	_____
7. Demand Factors for Neutral Conductors	_____
Session II. Sizing Residential Neutral Conductors; Sizing the Load Center	
A. Sizing Residential Neutral Conductors	_____
B. Sizing the Load Center	_____
1. Ground Fault Circuit Interrupters	_____
a. Single-Pole GFCI Circuit Breakers	_____
b. Two-Pole GFCI Circuit Breakers	_____
c. Direct-Wired GFCI Receptacles	_____

Session III. Grounding Electric Services

   A. Grounding Electric Services   _____

      1. Grounding Methods   _____

         a. Grounding Conductors   _____

      2. Grounding Path to Ground Electrode at Service   _____

         a. Use of a Grounded (Neutral) Circuit Conductor for Grounding Equipment   _____

         b. Method of Bonding Service Equipment   _____

         c. Main and Equipment Bonding Jumpers   _____

         d. Bonding and Grounding Requirements for Other Systems   _____

         e. Grounding Electrode Conductor Connections   _____

         f. Bonding of Piping Systems   _____

         g. Other Metal Piping   _____

         h. Grounding Electrode System   _____

         i. Made Electrodes and Other Electrodes   _____

         j. Resistance of Made Electrodes   _____

         k. Use of Lightning Rods   _____

         l. Grounding Conductor Material   _____

         m. Installation of Grounding Electrode Conductors   _____

         n. Size of Grounding Electrode Conductor   _____

Session IV. Installing the Service Entrance; Panelboard Location; Wiring Methods; Equipment Grounding System; Branch Circuit Layout for Power

   A. Installing the Service Entrance   _____

      1. Service Drop Locations   _____

      2. Vertical Clearances of Service Drop   _____

      3. Service Drop Clearances for Building Openings   _____

   B. Panelboard Location   _____

   C. Wiring Methods   _____

      1. Cable Systems   _____

         a. Nonmetallic-Sheathed Cable   _____

         b. Armored Cable   _____

         c. Underground Feeder Cable   _____

         d. Service-Entrance Cable   _____

      2. Raceways   _____

   D. Equipment Grounding System   _____

   E. Branch Circuit Layout for Power   _____

      1. Branch Circuits and Feeders   _____

      2. Locating Receptacles   _____

      3. Split-Wired Duplex Receptacles   _____

      4. Multiwire Branch Circuits   _____

      5. 240-Volt Circuits   _____

Session V. Branch Circuit Layout for Lighting; Outlet Boxes; Wiring Devices; Lighting Control

   A. Branch Circuit Layout for Lighting   _____

   B. Outlet Boxes   _____

      1. Mounting Outlet Boxes   _____

   C. Wiring Devices   _____

      1. Receptacles   _____

   D. Lighting Control   _____

      1. Three-Way Switches   _____

      2. Four-Way Switches   _____

      3. Photoelectric Switch   _____

      4. Relays   _____

      5. Dimmers   _____

      6. Switch Locations   _____

      7. Low-Voltage Electrical Systems   _____

         a. NEC Requirements for Low-Voltage Systems   _____

         b. Relays   _____

         c. Switches   _____

         d. Wiring Methods   _____

Session VI. Electric Heating; Residential Swimming Pools, Spas, and Hot Tubs; Module Examination and Performance Testing

   A. Electric Heating   _____

   B. Residential Swimming Pools, Spas, and Hot Tubs   _____

   C. Module Examination   _____

      1. Trainees must score 70% or higher to receive recognition from the NCCER.

      2. Record the testing results on Craft Training Report Form 200 and submit the results to the Training Program Sponsor.

   D. Performance Testing   _____

      1. Trainees must perform each task to the satisfaction of the instructor to receive recognition from the NCCER.

      2. Record the testing results on Craft Training Report Form 200 and submit the results to the Training Program Sponsor.

Wiring: Residential

_____

_____

_____

_____

_____

_____

_____

_____

_____

Course Map

This course map shows all of the task modules in the first level of the Electrical curricula. The suggested training order begins at the bottom and proceeds up. Skill levels increase as a trainee advances on the course map. The training order may be adjusted by the local Training Program Sponsor.

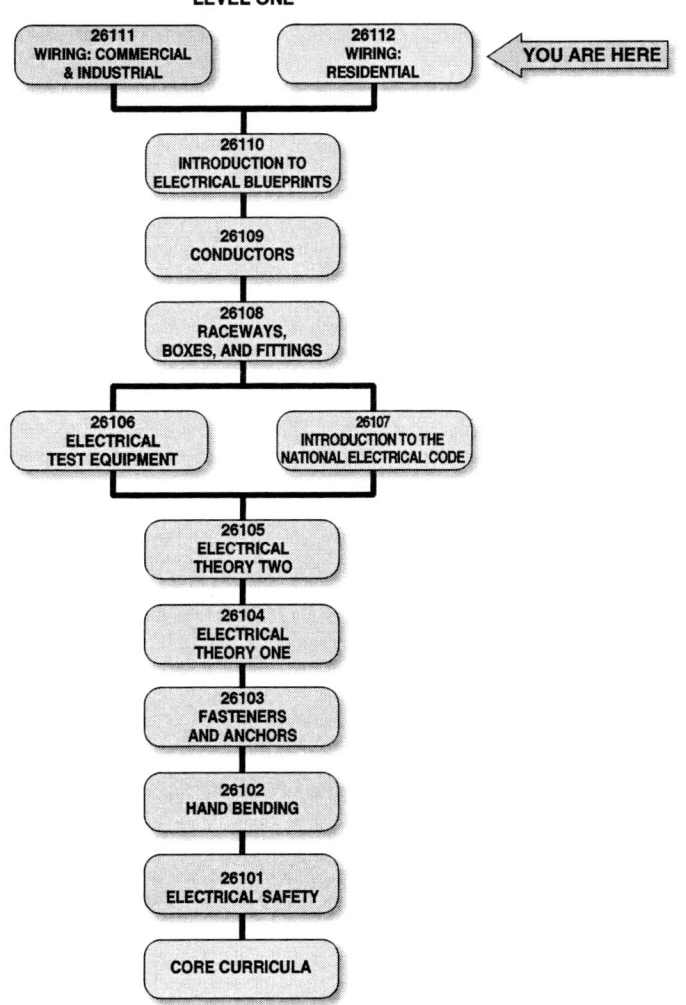

LEVEL ONE

26111 WIRING: COMMERCIAL & INDUSTRIAL

26112 WIRING: RESIDENTIAL ◁ YOU ARE HERE

26110 INTRODUCTION TO ELECTRICAL BLUEPRINTS

26109 CONDUCTORS

26108 RACEWAYS, BOXES, AND FITTINGS

26106 ELECTRICAL TEST EQUIPMENT

26107 INTRODUCTION TO THE NATIONAL ELECTRICAL CODE

26105 ELECTRICAL THEORY TWO

26104 ELECTRICAL THEORY ONE

26103 FASTENERS AND ANCHORS

26102 HAND BENDING

26101 ELECTRICAL SAFETY

CORE CURRICULA

112 CMAP.EPS

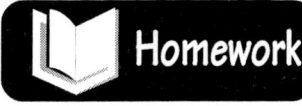

Assign reading of Module 26112.

12–iii

Instructor's Notes:

_____

_____

_____

_____

_____

_____

_____

_____

_____

Figures

Materials

Ensure you have everything required to teach the course. Check the Materials and Equipment List at the front of this Instructor's Guide.

Instructor's Notes:

_____

_____

_____

_____

_____

_____

_____

_____

_____

Wiring: Residential

Objectives

Upon completion of this module, the trainee will be able to:

1. Describe how to determine electric service requirements for dwellings.
2. Explain the grounding requirements of a residential electric service.
3. Calculate and select service-entrance equipment.
4. Select the proper wiring methods for various types of residences.
5. Explain the role of the NEC in residential wiring.
6. Compute branch circuit loads and explain their installation requirements.
7. Explain the types and purposes of equipment grounding conductors.
8. Explain the purpose of ground fault circuit interrupters and tell where they must be installed.
9. Size outlet boxes and select the proper type for different wiring methods.
10. Describe rules for installing electric space heating and HVAC equipment.
11. Describe the installation rules for electrical systems around swimming pools, spas, and hot tubs.
12. Explain how wiring devices are selected and installed.
13. Describe the installation and control of lighting fixtures.

Prerequisites

Successful completion of the following Task Modules is recommended before beginning study of this Task Module: Core Curriculum; Electrical Level One, Task Modules 26101 through 26111.

Required Trainee Materials

1. Trainee Task Module
2. Copy of the latest edition of the *National Electrical Code*
3. Appropriate Personal Protective Equipment

1.0.0 ◆ INTRODUCTION

The use of electricity in houses began shortly after the opening of the California Electric Light Company in 1879 and Thomas Edison's Pearl Street Station in New York City in 1882. These two companies were the first to enter the business of producing and selling electric service to the public. In 1886, the Westinghouse Electric Company secured patents that resulted in the development and introduction of alternating current; this paved the way for rapid acceleration in the use of electricity.

The primary use of early home electrical systems was to provide interior lighting, but today's uses of electricity include:

- Heating and air conditioning
- Electrical appliances

Note: The designations "National Electrical Code," "NE Code," and "NEC," where used in this document, refer to the *National Electrical Code®*, which is a registered trademark of the National Fire Protection Association, Quincy, MA. All *National Electrical Code (NEC) references in this module refer to the 1999 edition of the NEC.*

Show Transparency 1, Course Objectives.

Show Transparency 2, Performance Profile Tasks.

Discuss the history of residential wiring.

Explain that items shown in bold (blue) are defined in the Glossary at the back of this module.

12–1

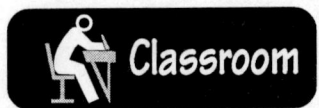

Discuss the factors that must be taken into consideration when designing a residential electrical system.

Explain that the first step in sizing an electrical service is a study of the building's floor plan.

Show Transparencies 3 and 4 (Figures 1 and 2).

Explain that the class will size the service for a sample home.

Discuss codes and standards. Refer the trainees to *Appendix A*.

- Interior and exterior lighting
- Communications systems
- Alarm systems

In planning any electrical system, there are certain general steps to be followed, regardless of the type of construction. In planning a residential electrical system, the electrician must take certain factors into consideration. These include:

- Wiring method
- Overhead or underground electrical service
- Type of building construction
- Type of service entrance and equipment
- Grade of wiring devices and lighting fixtures
- Selection of lighting fixtures
- Type of heating and cooling system
- Control wiring for the heating and cooling system
- Signal and alarm systems

The experienced electrician readily recognizes, within certain limits, the type of system that will be required. However, always check the local code requirements when selecting a wiring method. The NEC provides minimum requirements for the practical safeguarding of persons and property from hazards arising from the use of electricity. These minimum requirements are not necessarily efficient, convenient, or adequate for good service or future expansion of electrical use. Some local building codes require electrical installations that surpass the requirements of the NEC. For example, *NEC Section 230-51(a)* requires that service cable be secured by means of cable straps placed every 30 inches. The electrical inspection department in one area requires these cable straps to be placed at a minimum distance of 18 inches.

If more than one wiring method may be practical, a decision as to which type of system to use should be made prior to beginning the installation.

In a residential occupancy, the electrician should know that a 120/240-volt (V), single-phase *service entrance* will invariably be provided by the utility company. The electrician knows that the service and *feeders* will be three-wire, that the *branch circuits* will be either two- or three-wire, and that the safety *switches*, service equipment, and panelboards will be three-wire, solid neutral.

> **Note**
> See *Appendix A* for a general reference to other codes and electrical standards that apply to residential electrical installations.

On each project, however, the electrician must determine where the point of service-drop attachment to the building will be located and how much of the service is to be provided as part of the electrical contract, with the remaining portion being installed by the utility company.

2.0.0 ◆ SIZING THE ELECTRICAL SERVICE

It may be difficult to decide at times which comes first, the layout of the outlets or the sizing of the electric service. In many cases, the service (main disconnect, panelboard, service conductors, etc.) can be sized using the NEC before the outlets are actually located. In other cases, the outlets will have to be laid out first. However, in either case, the service entrance and panelboard locations will have to be determined before the circuits can be installed—so the electrician will know in which direction (and to what points) the circuit home-runs will terminate. In this module, an actual residence will be used as a model to size the electric service according to the latest edition of the NEC.

2.1.0 Floor Plans

A floor plan is a drawing that shows the length and width of a building and the rooms that it contains. A separate plan is made for each floor.

Figure 1 shows how a floor plan is developed. An imaginary cut is made through the building as shown in the view on the left. The top half of this cut is removed (bottom view), and the resulting floor plan is what the remaining structure looks like when viewed directly from above.

The floor plan for a small residence is shown in *Figure 2*. This building is constructed on a concrete slab with no basement or crawl space. There is an unfinished attic above the living area and an open carport just outside the kitchen entrance. Appliances include a 12 kilovolt-ampere (kVA) electric range and a 4.5kVA water heater.

There is also a washer/dryer (rated at 5.5kVA) in the utility room. Gas heaters are installed in each room with no electrical requirements. In this module, the electrical requirements of this example building will be computed.

2.2.0 General Lighting Loads

General lighting loads are calculated on the basis of *NEC Table 220-3(a)*. For residential occupancies, three volt-amperes (watts) per square foot of living space is the figure to use. This includes non-

---

Instructor's Notes:

_____

_____

_____

_____

_____

_____

_____

_____

_____

PERSPECTIVE VIEW
SHOWING SECTION CUTS

TOP HALF OF SECTION
REMOVED

RESULTING FLOOR PLAN
IS WHAT THE REMAINING
STRUCTURE LOOKS LIKE
WHEN VIEWED FROM
ABOVE

112F01.EPS

Figure 1 ◆ Principles of floor plan layout.

appliance duplex receptacles into which lamps, televisions, etc. may be connected. Therefore, the area of the building must be calculated first. If the building is under construction, the dimensions can be determined by scaling the working drawings used by the builder. If the residence is an existing building, with no drawings, actual measurements will have to be made on the site.

Using the floor plan of the residence in *Figure 2* as a guide, an architect's scale is used to measure the longest width of the building (using outside dimensions). It is determined to be 33 feet. The longest length of the building is 48 feet. These two measurements multiplied together give $33 \times 48 = 1,584$ square feet of living area. However, there is an open carport on the lower left of the drawing. This carport area will have to be calculated and then deducted from 1,584 to give the true amount of living space. This open area (carport) is 12 feet wide by 19.5 feet long: $12 \times 19.5 = 234$ square feet. Subtract the carport area from 1,584 square feet: $1,584 - 234 = 1,350$ square feet of living area.

When using the square-foot method to determine lighting loads for buildings, *NEC Section 220-3(a)* requires the floor area for each floor to be computed from the outside dimensions. When calculating lighting loads for residences, the computed floor area must not include open porches, carports, garages, or unused or unfinished spaces that are not adaptable to future use.

2.3.0 Calculating the Electric Service Load

Figure 3 shows a standard calculation worksheet for a single-family dwelling. This form contains numbered blank spaces to be filled in while making the service calculation. Using this worksheet as a guide, the total area of our sample dwelling has been previously determined to be 1,350 square feet of living space. This figure is entered in the appropriate space (Box 1) on the form and multiplied by 3 volt-amperes (VA) for a total general lighting load of 4,050VA (Box 2).

Explain that general lighting loads are calculated at 3VA/sq. ft.

Explain that a simple worksheet can be used to calculate electric loads.

Show Transparency 5 (Figure 3).

Point out the square footage and lighting load for the sample home on the worksheet.

Figure 2 ◆ Floor plan of a typical residence.

Instructor's Notes:

_____

_____

_____

_____

_____

_____

_____

_____

_____

I. GENERAL LIGHTING LOADS

TYPE OF LOAD	CALCULATION		TOTAL VA		NEC REFERENCE
Lighting load	(1) 1,350	sq. ft. x 3VA =	(2) 4,050	VA	Table 220-3(a)
Small appliance loads	(3)	circuits x 1,500VA=	(4)	VA	Section 220-16(a)
Laundry load	(5)	circuits x 1,500VA=	(6)	VA	Section 220-16(b)
Lighting, small appliance, laundry	Total VA =		(7)	VA	

LOAD	CALCULATION	DEMAND FACTOR	TOTAL VA	
Lighting, small appliance, laundry	First 3,000VA x	100% =	(8) 3,000VA	
Lighting, small appliance, laundry	(9) Remaining VA x	35% =	(10)	VA
Lighting, small appliance, laundry	Add #8 & #10 above	=	(11)	VA

TOTAL CALCULATED LOAD FOR LIGHTING, SMALL APPLIANCES, & LAUNDRY CIRCUIT(S) (Enter Item #11 above in box #12)

(12)

II. LARGE APPLIANCE LOADS

TYPE OF LOAD	NAMEPLATE RATING	DEMAND FACTOR	TOTAL VA		NEC REFERENCE
Electric range	Not over 12kVA	Use 8kVA	(13)	8,000VA	Table 220-19
Clothes dryer	(14) VA	100%	(15)	VA	Table 220-18
Water heater	(16) VA	100%	(17)	VA	
Other appliances	(18) VA	100%	(19)	VA	

TOTAL CALCULATED LOAD FOR LARGE APPLIANCES (Add items # 13, #15, #17, and #19 above). ENTER TOTAL IN BOX #20

(20)

TOTAL CALCULATED LOAD (Add boxes #12 & #20) ENTER TOTAL IN BOX #21

(21)

III. CONVERT VA TO AMPERES

$$\frac{Total\ VA\ in\ Box\ \#21}{240\ (volts)} = amperes \qquad \frac{VA}{240\ (volts)} = \text{(22)} \underline{\hspace{2cm}} amperes$$

IV. FEEDER CONDUCTOR SIZE—NEC SECTION 310-15(b)(6)

FEEDER CONDUCTOR SIZE		
COPPER AWG SIZE	ALUMINUM OR COPPER-CLAD AL	CONDUCTOR SERVICE SIZE
4	2	100
3	1	110
2	1/0	125
1	2/0	150
1/0	3/0	175
2/0	4/0	200

112F03.TIF

Figure 3 ◈ Calculation worksheet for residential requirements.

Discuss the NEC requirements for small appliance loads.

Enter the small appliance load for the sample home on the worksheet.

Discuss the NEC requirements for laundry loads.

Enter the laundry load on the worksheet.

Introduce the concept of demand factors and explain how they apply to the sample home. Note that they will be discussed in more detail in the next section.

Enter the load calculations on the worksheet.

Discuss the NEC requirements for calculating service size.

Enter the service calculation on the worksheet.

2.3.1 Small Appliance Loads

NEC Section 210-11(c)(1) requires at least two 120V, 20A small appliance circuits to be installed for small appliance loads in the kitchen, dining area, breakfast nook, and similar areas where toasters, coffee makers, etc. will be used. *NEC Section 220-16* gives further requirements for residential small appliance circuits; that is, each circuit must be rated at 1,500VA. Since two such circuits are used in the sample residence, the number 2 is entered in Box 3 and then multiplied for a total small appliance load of 3,000VA, which is entered in Box 4.

2.3.2 Laundry Circuit

NEC Section 210-11(c)(2) requires an additional 20A branch circuit to be provided for the exclusive use of the laundry area (Box 5). This circuit must not have any other outlets connected except for the laundry receptacle(s). Therefore, enter 1,500VA in Box 6 on the form.

So far, there is enough information to complete the first portion of the service calculation form:

- General lighting   4,050VA (Box 2)
- Small appliance load   3,000VA (Box 4)
- Laundry load   1,500VA (Box 6)
- Total general lighting
  and appliance loads   8,550VA (Box 7)

2.3.3 Demand Factors

All residential electrical outlets are never used at one time. There may be a rare instance when all the lighting may be on for a short time every night, but even so, all the small appliances, all burners on the electric range, water heater, furnace, dryer, washer, and the numerous receptacles throughout the house will never be used simultaneously. Knowing this, the NEC allows a diversity or *demand factor* to be used when sizing electric services. Our calculation continues as follows:

- The first 3,000VA is rated
  at 100%   3,000VA (Box 8)
- The remaining 5,550VA
  (Box 9) may be rated
  at 35% (the allowable
  demand factor).
  Therefore, 5,550 × 0.35 =   1942.5VA (Box 10)
- Net general lighting and
  small appliance load   4,942.5VA (Box 11)

Also enter this number in Box 12 on the form.

The electric range, water heater, and clothes dryer must now be considered in the service calcu-

lation. Although it was previously learned that the nameplate rating of the electric range is 12kVA, seldom will every burner be on the maximum setting at one time. Nor will the oven remain on all the time during cooking. When the oven reaches the temperature set on the oven controls, the thermostat shuts off the power until it cools down. Again, the NEC allows a diversity or demand factor to be used for this appliance.

When one electric range is installed and the nameplate rating is not over 12kVA, *NEC Table 220-19* allows a demand factor resulting in a total rating of 8kVA. Therefore, 8kVA may be used in the service calculation instead of the nameplate rating of 12kVA. The electric clothes dryer and water heater, however, must be calculated at 100% when using this method to calculate residential electric services. The total large appliance load is entered in Box 20 on the form. Here is the service calculation thus far:

- Net general lighting   4,942.5VA (Box 11)
  and small appliance load   and (Box 12)
- Electric range
  (using demand factor)   8,000VA (Box 13)
- Clothes dryer   5,500VA (Box 14)
                                             and (Box 15)
- Water heater   4,500VA (Box 16)
                                             and (Box 17)
- Total large appliance load   18,000VA (Box 20)
- Total load   22,942.5VA
                                             (Box 21)

2.3.4 Required Service Size

The conventional electric service for residential use is 120/240V, three-wire, single-phase. Services are sized in amperes, and when the volt-amperes are known on single-phase services, amperes may be found by dividing the highest voltage into the total volt-amperes. For example:

$$22,942.5VA \div 240V = 95.6A \text{ (Box 22)}$$

NEC Section 230-79(c) requires a minimum of 100 amps for the service disconnecting means for a one-family dwelling. *NEC Section 230-42(b)* requires ungrounded conductors to have an ampacity of not less than the service disconnecting means specified in *NEC Section 230-79(c)*. Therefore, the minimum service size for a single-family dwelling is 100 amps.

2.4.0 Demand Factors

NEC Section 220 Part B provides the rules regarding the application of demand factors to certain types of loads. Recall that a demand factor is the maximum amount of volt-amp load expected at

Instructor's Notes:

_____

_____

_____

_____

_____

_____

_____

_____

_____

any given time compared to the total connected load of the circuit. The maximum demand of a feeder circuit is equal to the connected load times the demand factor. The loads to which demand factors apply can be found in the NEC as listed below:

- Lighting loads — *NEC Table 220-11*
- Receptacle loads — *NEC Table 220-13*
- Dryer loads — *NEC Table 220-18*
- Range loads — *NEC Table 220-19*
- Kitchen equipment loads — *NEC Table 220-20*
- Dwelling unit loads — *NEC Table 220-30*
- Existing dwelling unit loads — *NEC Section 220-31*
- Multi-family dwelling unit loads — *NEC Section 220-32*

2.5.0 General Lighting and Receptacle Load Demand Factors

NEC Table 220-11 provides the demand factors allowed for various types of lighting situations.

2.6.0 Small Appliance and Laundry Loads for Dwelling Units

Each two-wire small appliance branch circuit required by *NEC Section 210-11(c)* for small appliances supplied by 15A or 20A receptacles on 20A branch circuits in the kitchen, pantry, dining room, and breakfast room is calculated at 1,500VA.

These loads should be included with the general lighting load and subjected to the same demand factors permitted in *NEC Table 220-11*.

2.6.1 Laundry Circuit Load

A 1,500VA feeder load is added to load calculations for each two-wire laundry branch circuit installed in a home. The branch circuit is required by *NEC Section 210-11(c)*. This load may also be added to the general lighting load and subjected to the same demand factors provided in *NEC Section 220-11*.

2.6.2 Dryer Load

The dryer load for each electric clothes dryer is 5,000VA or the actual nameplate value of the dryer, whichever is larger. Demand factors listed in *NEC Table 220-18* may be applied for more than one dryer in the same dwelling.

2.6.3 Range Load

Range loads and other cooking appliances are covered under *NEC Section 210-19(c)*. The feeder demand loads for household electric ranges, wall-mounted ovens, countertop cooking units, and other similar household appliances individually rated above 1¾kW are permitted to be computed in accordance with *NEC Table 220-19*. If two or more single-phase ranges are supplied by a three-phase, four-wire feeder, the total load is computed by using twice the maximum number connected between any two phases.

2.6.4 Demand Loads for Electric Ranges

- Dwelling unit loads — *NEC Section 220-30*
- Existing dwelling unit loads — *NEC Section 220-31*
- Multi-family dwelling unit loads — *NEC Table 220-32*
- School loads — *NEC Table 220-34*
- Restaurant loads — *NEC Table 220-36*
- Farm loads (individual loads) — *NEC Table 220-40*
- Farm load (total of loads) — *NEC Table 220-41*

2.6.5 Calculating the Demand Load

To illustrate the use of demand factors, consider the following example: The total connected load of a noncontinuous receptacle load in an office building is 26,000VA.

Refer to *NEC Table 220-13* and find that the first 10kVA or less is calculated at 100%:

$$10\text{kVA} \times 100\% = 10,000\text{VA}$$

$$26,000\text{VA} - 10,000\text{VA} = 16,000\text{VA}$$

Demand Factors

Examine *NEC Table 220-19.* Why does the demand factor decrease as the number of appliances increases? Why does the demand factor decrease more for larger ranges than it does for smaller ones?

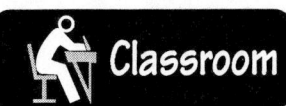

Review the NEC tables and sections applying to demand factors. Cover:
- **General lighting and receptacle loads**
- **Small appliance and laundry loads**

Discuss the "Think About It."

The remainder over 10kVA is calculated at 50%:

$$16,000VA \times 50\% = 8,000VA$$
$$10,000VA + 8,000VA = 18,000VA$$

The calculated receptacle load in this example would be 18,000VA.

2.7.0 Demand Factors for Neutral Conductors

The neutral conductor of electrical systems generally carries only the current imbalance of the phase conductors. For example, in a single-phase feeder circuit with one phase conductor carrying 50 amps and the other carrying 40 amps, the neutral conductor would carry 10 amps. Since the neutral in many cases will never be required to carry as much current as the phase conductors, the NEC allows us to apply a demand factor. (See *NEC Section 220-22* for details.) Note that in certain circumstances such as electrical discharge lighting, data processing equipment, and other similar equipment, a demand factor cannot be applied to the neutral conductors because these types of equipment produce harmonic currents that increase the heating effect in the neutral conductor.

3.0.0 ◆ SIZING RESIDENTIAL NEUTRAL CONDUCTORS

The neutral conductor in a three-wire, single-phase service carries only the unbalanced load between the two ungrounded (hot) wires or legs. Since there are several 240V loads in the above calculations, these 240V loads will be balanced and therefore reduce the load on the service neutral conductor. Consequently, in most cases, the service neutral does not have to be as large as the ungrounded (hot) conductors.

In the previous example, the water heater does not have to be included in the neutral conductor calculation, since it is strictly 240V with no 120V loads. The clothes dryer and electric range, however, have 120V lights that will cause a current imbalance between phases. The NEC allows a demand factor of 70% for these two appliances. Using this information, the neutral conductor may be sized accordingly:

- General lighting and
  appliance load   4,942.5VA
- Electric range (8,000VA × 0.70)   5,600VA
- Clothes dryer (5,500VA × 0.70)   3,850VA
- Total   14,392.5VA

To find the total phase-to-phase amperes, divide the total volt-amperes by the voltage between phases:

$$14,392.5VA \div 240V = 59.96A \text{ or } 60A$$

The service-entrance conductors have now been calculated and must be rated at 100A with a neutral conductor rated for at least 60A. See *Figure 4* for a completed calculation form for the residence in question.

In *NEC Section 310-15(b)(6)*, special consideration is given to 120/240V, single-phase residential services. Conductor sizes are shown in *NEC Table 310-15(b)(6)*. Reference to this table shows that the NEC allows a No. 4 AWG copper or No. 2 AWG aluminum or copper-clad aluminum conductor for a 100A service.

When sizing the grounded conductor for services, the provisions stated in *NEC Sections 215-2, 220-22, and 230-42* must be met, along with other applicable sections.

4.0.0 ◆ SIZING THE LOAD CENTER

Each ungrounded conductor in all circuits must be provided with overcurrent protection in the form of either fuses or circuit breakers. If more than six such devices are used, a means of disconnecting the entire service must be provided using either a main disconnect switch or a main circuit breaker.

To calculate the number of fuse holders or circuit breakers required in the sample residence, look at the general lighting load first. The total general lighting load of 4,050VA can be divided by 120V to find the amperage:

$$4,050VA \div 120V = 33.75A$$

Balanced Phase Conductors

The word *phase* is used in these modules to refer to a hot wire rather than a neutral one. Some electricians call these *legs* rather than phases. Why must the two phase conductors be balanced?

Instructor's Notes:

_____

_____

_____

_____

_____

_____

_____

_____

I. GENERAL LIGHTING LOADS

TYPE OF LOAD	CALCULATION	TOTAL VA	NEC REFERENCE
Lighting load	(1) 1,350 sq. ft. x 3VA =	(2) 4,050 VA	Table 220-3(a)
Small appliance loads	(3) 2 circuits x 1,500VA=	(4) 3,000 VA	Section 220-16(a)
Laundry load	(5) 1 circuits x 1,500VA=	(6) 1,500 VA	Section 220-16(b)
Lighting, small appliance, laundry	Total VA =	(7) 8,550 VA	

LOAD	CALCULATION	DEMAND FACTOR	TOTAL VA
Lighting, small appliance, laundry	First 3,000VA x	100% =	(8) 3,000VA
Lighting, small appliance, laundry	(9) Remaining VA 5,550 x	35% =	(10) 1,942.5 VA
Lighting, small appliance, laundry	Add #8 & #10 above	=	(11) 4,942.5 VA

TOTAL CALCULATED LOAD FOR LIGHTING, SMALL APPLIANCES, & LAUNDRY CIRCUIT(S) (Enter Item #11 above in box #12)

(12) 4,942.5VA

II. LARGE APPLIANCE LOADS

TYPE OF LOAD	NAMEPLATE RATING	DEMAND FACTOR	TOTAL VA	NEC REFERENCE
Electric range	Not over 12kVA	Use 8kVA	(13) 8,000VA	Table 220-19
Clothes dryer	(14) 5,500 VA	100%	(15) 5,500 VA	Table 220-18
Water heater	(16) 4,500 VA	100%	(17) 4,500 VA	
Other appliances	(18) VA	100%	(19) VA	

TOTAL CALCULATED LOAD FOR LARGE APPLIANCES (Add items #15, #17, and #19 above).
ENTER TOTAL IN BOX #20

(20) 18,000VA

TOTAL CALCULATED LOAD (Add boxes #12 & #20)
ENTER TOTAL IN BOX #21

(21) 22,942.5VA

III. CONVERT VA TO AMPERES

$$\frac{Total\ VA\ in\ Box\ \#21}{240\ (volts)} = amperes \qquad \frac{22,942.5VA}{240\ (volts)} = \boxed{(22)\ 95.6\ amperes}$$

112F04.TIF

Figure 4 ◆ Completed calculation form.

Classroom

Have the trainees check these appliances at home or at local stores to determine which one uses the most power. (The wattage will be listed on the device.)

THINK ABOUT IT

Common Loads

Which of the following devices uses the most power?

- Giant-screen television
- Typical hair dryer
- Curling iron
- Crockpot

Demonstrate how to make load service calculations.

Discuss the NEC requirements for the installation of GFCIs.

Either 15A or 20A circuits may be used for the lighting load. Two 20A circuits (2 × 20) equal 40A, so two 20A circuits would be adequate for the lighting. However, two 15A circuits total only 30A and 33.75A are needed. Therefore, if 15A circuits are used, three will be required for the total lighting load. In this example, three 15A circuits will be used.

In addition to the lighting circuits, the sample residence will require a minimum of two 20A for the small appliance load and one 20A circuit for the laundry. So far, the following branch circuits can be counted:

- General lighting load   Three 15A circuits
- Small appliance load   Two 20A circuits
- Laundry load   One 20A circuit
- Total   Six branch circuits

Most load centers and panelboards are provided with an even number of circuit breaker spaces or fuse holders (e.g., four, six, eight, ten, etc.). But before the panelboard can be selected, space must be provided for the 240V loads—each requiring a two-pole circuit breaker or fuse holder. The ratings of overcurrent protection devices are sized by dividing the demand volt-amperes by the voltage. For example, since the demand load of the electric range is 8kVA, 8,000VA divided by 240V equals 33.33A. The closest standard overcurrent protection device is 40A; this will be the size used—a 40A, two-pole circuit breaker. In some existing installations, you might find a two-pole fuse block containing two 40A cartridge fuses being used to feed a residential electric range. The remaining 240V circuits are calculated in a similar fashion, which results in the following:

- Electric range   One 40A, two-pole circuit breaker
- Clothes dryer   One 30A, two-pole circuit breaker
- Water heater   One 30A, two-pole circuit breaker

These three appliances will therefore require an additional six spaces (two poles × three appliances) in the load center or panelboard. Adding these six spaces to the six required for the general lighting and small appliance loads requires at least a 12-space load center to handle the circuits in the sample residence.

4.1.0 Ground Fault Circuit Interrupters

Under certain conditions, the time it takes to open an overcurrent protection device can be critical.

Because of this fact, the NEC requires ground fault circuit interrupters (GFCIs) to be installed on the following:

- Circuits feeding impedance heating units operating at voltages greater than 30V; such units are frequently used for de-icing and snow-melting equipment
- Circuits for electrically operated pool covers
- Power or lighting circuits for swimming pools, fountains, and similar locations
- Receptacles in both commercial and residential garages
- Receptacles in residential bathrooms
- Receptacles installed outdoors where there is direct grade-level access
- Receptacles installed in residential crawl spaces or unfinished basements
- Countertop receptacles mounted within six feet of a wet bar sink
- Receptacles installed in bathhouses
- Receptacles installed in bathrooms of commercial, industrial, or any other buildings
- Receptacles installed on roofs of any building except dwellings
- Kitchen countertop receptacles
- Branch circuits derived from autotransformers

GFCI circuit breakers require the same mounting space as standard single-pole circuit breakers and provide the same branch circuit wiring protection as standard circuit breakers. They also provide Class A ground fault protection.

GFCI breakers that are UL-listed are available in single- and two-pole construction; 15A, 20A, 25A, and 30A ratings; and have a 10,000A interrupting capacity. Single-pole units are rated at 120VAC; two-pole units are rated at 120/240VAC.

GFCI breakers can be used not only in load centers and panelboards, but they are also available factory-installed in meter pedestals and power outlet panels for recreational vehicle (RV) parks and construction sites.

The GFCI sensor continuously monitors the current balance in the ungrounded or energized (hot) load conductor and the neutral load conductor. If the current in the neutral load wire becomes less than the current in the hot load wire, then a ground fault exists, since a portion of the current is returning to the source by some means other than the neutral load wire. When a current imbalance occurs, the sensor, which is a differential current transformer, sends a signal to the solid-state circuit, which activates the ground trip solenoid mechanism and breaks the hot load connection (*Figure 5*). A current imbalance as low as six milliamps (6mA) will cause the circuit

Instructor's Notes:

_____

_____

_____

_____

_____

_____

_____

_____

breaker to interrupt the circuit. This is indicated by the trip indicator on the front of the device.

The two-pole GFCI breaker (*Figure 6*) continuously monitors the current balance between the two hot conductors and the neutral conductor. As long as the sum of these three currents is zero, the device will not trip; that is, if there are ten amps of current in the A load wire, five amps in the neutral, and five amps in the B load wire, then the sensor is balanced and will not produce a signal. A current imbalance from a ground fault condition as low as 6mA will cause the sensor to produce a signal of sufficient magnitude to trip the device.

4.1.1 Single-Pole GFCI Circuit Breakers

The single-pole GFCI breaker has two load lugs and a white wire pigtail in addition to the line side

plug-on or bolt-on connector. The line side hot connection is made by installing the GFCI breaker in the panel just as any other circuit breaker is installed. The white wire pigtail is attached to the panel neutral (S/N) assembly. Both the neutral and hot wires of the branch circuit being protected are terminated in the GFCI breaker. These two load lugs are clearly marked LOAD POWER and LOAD NEUTRAL in the breaker case. Also in the case is the identifying marking for the pigtail, PANEL NEUTRAL.

Single-pole GFCI circuit breakers must be installed on independent circuits. Circuits that employ a neutral common to more than one hot conductor cannot be protected against ground faults by a single-pole breaker because a common neutral cannot be split and retain the necessary hot wire to neutral wire balance under normal use to prevent the GFCI breaker from tripping.

Explain the operation of single-pole and two-pole GFCIs. Pass around examples for the trainees to examine.

Audiovisual

Show Transparencies 7 and 8 (Figures 5 and 6).

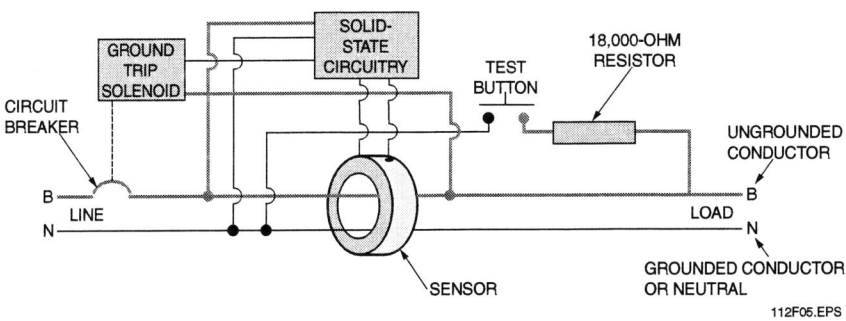

Figure 5 ◆ Operating circuitry of a typical GFCI.

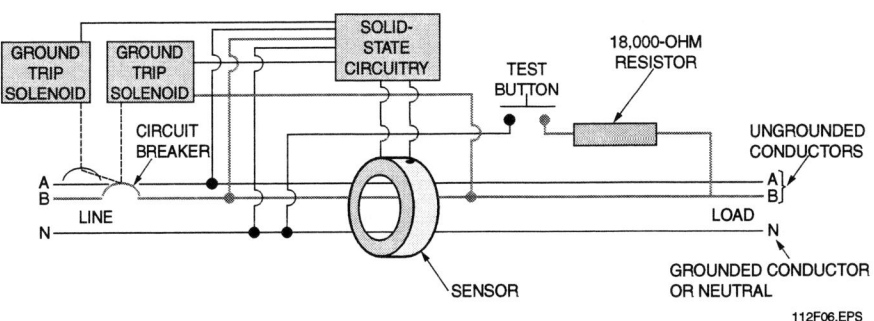

Figure 6 ◆ Operating characteristics of a two-pole GFCI.

Care should be exercised when installing GFCI breakers in existing panels. Be sure that the neutral wire for the branch circuit corresponds with the hot wire of the same circuit. Always remember that unless the current in the neutral wire is equal to that in the hot wire (within 6mA), the GFCI breaker senses this as being a possible ground fault (see *Figure 7*).

4.1.2 Two-Pole GFCI Circuit Breakers

A two-pole GFCI circuit breaker can be installed on a 120/240VAC single-phase, three-wire system; the 120/240VAC portion of a 120/240VAC three-phase, four-wire system; or the two phases and neutral of a 120/208VAC three-phase, four-wire system. Regardless of the application, the installation of the breaker is the same—connections are made to two hot buses and the panel neutral assembly. When installed on these systems, protection is provided for two-wire 240VAC or 208VAC circuits, three-wire 120/240VAC or 120/208VAC circuits, and 120VAC multiwire circuits.

The circuit in *Figure 8* illustrates the problems that are encountered when a common load neutral is used for two single-pole GFCI breakers. Ei-

112F07.EPS

Figure 7 ◆ Operating characteristics of a single-pole circuit breaker with a GFCI.

Instructor's Notes:

ther or both breakers will trip when a load is applied at the #2 duplex receptacle. The neutral current from the #2 duplex receptacle flows through breaker #1; this increase in neutral current through breaker #1 causes an imbalance in its sensor, thus causing it to produce a fault signal. At the same time, there is no neutral current flowing through breaker #2; therefore, it also senses a current imbalance. If a load is applied at the #1 duplex receptacle, and there is no load at the #2 duplex receptacle, then neither breaker will trip because neither breaker will sense a current imbalance.

Junction boxes can also present problems when they are used to provide taps for more than one branch circuit. Even though the circuits are not wired using a common neutral, sometimes all neutral conductors are connected together. Thus, parallel neutral paths are established, producing an imbalance in each GFCI breaker sensor, causing them to trip.

The two-pole GFCI breaker eliminates the problems encountered when trying to use two single-pole GFCI breakers with a common neutral. Because both hot currents and the neutral current pass through the same sensor, no imbalance occurs between the three currents, and the breaker will not trip.

Figure 8 ◆ Circuit depicting the common load neutral.

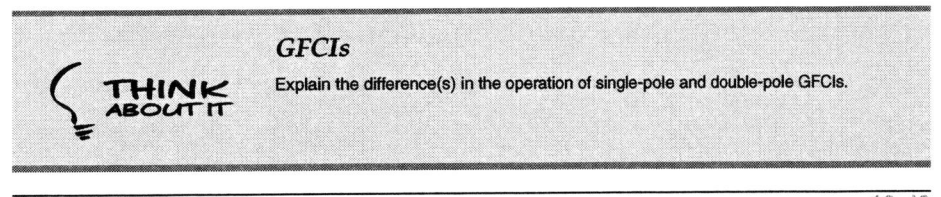

GFCIs

Explain the difference(s) in the operation of single-pole and double-pole GFCIs.

Teaching Tip

Discuss the "Think About It."

**Discuss direct-wired GFCIs.
Pass around examples for
the trainees to examine.**

**Show Transparency 11
(Figure 9).**

**Review the terms
associated with grounding.**

4.1.3 Direct-Wired GFCI Receptacles

Direct-wired GFCI receptacles provide Class A ground fault protection on 120VAC circuits.

They are available in both 15A and 20A arrangements. The 15A unit has a NEMA 5-15R receptacle configuration for use with 15A plugs only. The 20A device has a NEMA 5-20R receptacle configuration for use with 15A or 20A plugs. Both 15A and 20A units have a 120VAC, 20A circuit rating. This is to comply with *NEC Table 210-24,* which requires that 15A circuits use 15A receptacles but permits the use of either 15A or 20A receptacles on 20A circuits. Therefore, GFCI receptacle units that contain a 15A receptacle may be used on 20A circuits.

These receptacles have line terminals for the hot, neutral, and ground wires. In addition, they have load terminals that can be used to provide ground fault protection for other receptacles electrically downstream on the same branch circuit (*Figure 9*). All terminals will accept No. 14 to No. 10 AWG copper wire.

GFCI receptacles have a two-pole tripping mechanism that breaks both the hot and the neutral load connections.

When tripped, the RESET button pops out. The unit is reset by pushing the button back in.

GFCI receptacles have the additional benefit of noise suppression. Noise suppression minimizes false tripping due to spurious line voltages or radio frequency (RF) signals between 10 and 500 megahertz (Mhz).

GFCI receptacles can be mounted without adapters in wall outlet boxes that are at least 1.5 inches deep.

5.0.0 ◆ GROUNDING ELECTRIC SERVICES

The grounding system is a major part of the electrical system. Its purpose is to protect people and equipment against the various electrical faults that can occur. It is sometimes possible for higher-than-normal voltages to appear at certain points in an electrical system or in the electrical equipment connected to the system. Proper grounding ensures that the electrical charges that cause these high voltages are channeled to the earth or ground before damaging equipment or endangering people.

When referring to *ground,* we are talking about ground potential or earth ground. If a conductor is connected to the earth or to some conducting body that serves in place of the earth, such as a driven ground rod (electrode) or cold-water pipe, the conductor is said to be *grounded.* The neutral conductor in a three- or four-wire service, for example, is intentionally grounded and therefore becomes a *grounded conductor.* However, a wire used to connect this neutral conductor to a grounding electrode or electrodes is referred to as a *grounding electrode conductor.* Note the difference in the two meanings; one is grounded, while the other provides a means of grounding.

112F09.EPS

Figure 9 ◆ GFCI receptacle used to protect other outlets on the same circuit.

Instructor's Notes:

_____

_____

_____

_____

_____

_____

_____

_____

_____

Grounding

Systematic grounding wasn't required by the NEC until the mid-1950s; even then, electricians commonly grounded an outlet by wrapping an uninsulated wire around a cold-water pipe and taping it. Three-hole receptacles with grounding terminals became common in the 1960s, but in many older houses, you cannot assume that receptacles are grounded, even when you see a three-hole receptacle. Sometimes, new receptacles have simply been screwed onto old boxes where there is no equipment grounding conductor.

There are two general classifications of protective grounding:

- System grounding
- Equipment grounding

The system ground relates to the service-entrance equipment and its interrelated and bonded components. That is, system and circuit conductors are grounded to limit voltages due to lightning, line surges, or unintentional contact with higher voltage lines, and to stabilize the voltage to ground during normal operation.

Equipment grounding conductors are used to connect the noncurrent-carrying metal parts of equipment, conduit, outlet boxes, and other enclosures to the system grounded conductor, the grounding electrode conductor, or both, at the service equipment or at the source of a separately derived system. Equipment grounding conductors are bonded to the system grounded electrode conductor to provide a low-impedance path for fault current that will facilitate the operation of overcurrent devices under ground fault conditions.

Equipment grounding is discussed later in this module. *NEC Article 250* covers general requirements for grounding and bonding. Read this article carefully.

To better understand a complete grounding system, a conventional residential system will be examined, beginning at the power company's high-voltage lines and transformer, as shown in *Figure 10*. The pole-mounted transformer is fed with a two-wire, single-phase 7,200V system, which is transformed and stepped down to a three-wire, 120/240V, single-phase electric service suitable for residential use. Note that the voltage between line A and line B is 240V. However, by connecting a third wire (neutral) on the secondary winding of the transformer—between the other two—the 240V is split in half, providing 120V between either line A or line B and the neutral conductor. Consequently, 240V is available for household appliances such as

ranges, hot water heaters, clothes dryers, etc., while 120V is available for lights, small appliances, televisions, etc.

Referring again to *Figure 10,* conductors A and B are ungrounded conductors, while the neutral is a grounded conductor. If only 240V loads were connected, the neutral (grounded conductor) would carry no current. However, since 120V loads are present, the neutral will carry the unbalanced load and become a current-carrying conductor. For example, if line A carries 60A and line B carries 50A, the neutral conductor would carry only 10A (60 − 50 = 10). This is why the NEC allows the neutral conductor in an electric service to be smaller than the ungrounded conductors. The typical pole-mounted service entrance is normally routed by messenger cable from a point on the pole to a point on the building being served, terminating in a meter housing. Another service conductor is installed between the meter housing and the main service switch or #2 panelboard. This is the point where most systems are grounded—the neutral bus in the main panelboard. See *Figure 11*.

5.1.0 Grounding Methods

Methods of grounding an electric service are covered in *NEC Section 250-50*. In general, all of the following (if available) and any made electrodes must be bonded together to form the grounding electrode system:

- An underground water pipe may be used if it is in direct contact with the earth for not less than 10 feet.
- The metal frame of a building may be used if it is effectively grounded.
- An electrode encased by at least two inches of concrete may be used if it is located within and near the bottom of a concrete foundation or footing that is in direct contact with the earth.

Explain the difference between system grounding and equipment grounding.

Show Transparencies 12 and 13 (Figures 10 and 11).

Identify the components in an actual panelboard.

Discuss the NEC requirements for grounding electric services.

Figure 10 ◆ Wiring diagram of a 7,200V to 120/240V, single-phase transformer connection.

Show Transparencies 14 and 15 (Figures 12 and 13).

The electrode must also be at least 20 feet long and must be made of electrically conductive coated steel reinforcing bars or rods of not less than ½ inch in diameter, or consisting of at least 20 feet of bare copper conductor not smaller than No. 4 AWG wire size.

- A ground ring encircling the building or structure may be used if it is in direct contact with the earth at a depth below grade of not less than 2½ feet. This ring must consist of at least 20 feet of bare copper conductor not smaller than No. 2 AWG wire size.

In most residential structures, only the water pipe will be available. This water pipe must be supplemented by an additional electrode, as specified in *NEC Sections 250-50(a) and 250-52*. *Figure 12* shows a typical residential electric service and the available grounding electrodes.

The sample residence has a metal underground water pipe that is in direct contact with the earth for more than 10 feet, so this is one valid grounding source. The house also has a metal underground gas piping system, but this may not be used as a grounding electrode [*NEC Section 250-52(a)*]. *NEC*

Section 250-50(a) further states that the underground water pipe must be supplemented by an additional electrode of a type specified in *NEC Section 250-50* or in *NEC Section 250-52*. Since the other grounding methods (a grounded metal building frame, concrete-encased electrode, or a ground ring) are not normally available for most residential applications, *NEC Section 250-52* must be used to determine the supplemental electrode.

In most cases, this supplemental electrode will consist of either a driven rod or pipe electrode, the specifications for which are shown in *Figure 13*.

An alternate method to a pipe or rod is a plate electrode. Each plate electrode must expose not less than two square feet of surface to the surrounding earth. Plates made of iron or steel must be at least ¼-inch thick, while plates of nonferrous metal such as copper need only be .06-inch thick.

Either type of electrode must have a resistance to ground of 25 ohms or less. If not, they must be augmented by an additional electrode, with the two electrodes spaced not less than six feet from each other. In fact, many locations require two electrodes regardless of the resistance to ground. This is not an NEC requirement, but is required by

Instructor's Notes:

SERVICE-ENTRANCE CABLE

NEUTRAL (GROUNDED) CONDUCTOR

UNGROUNDED CONDUCTORS SERVICE 120/240-VOLT HOT WIRES

PANELBOARD

NEUTRAL/GROUND BUS BAR

MAIN DISCONNECT SWITCH

GROUNDING CONDUCTOR

BONDING SCREW

CABLES TO HOUSE CIRCUITS

CABLES TO HOUSE CIRCUITS

THREE-WIRE CABLE TO 240-VOLT APPLIANCES (CONNECT TO DOUBLE BREAKER)

THREE-WIRE CABLE TO 240-VOLT APPLIANCES (CONNECT TO DOUBLE BREAKER)

CIRCUIT BREAKERS

112F11.EPS

Figure 11 ❖ Interior view of panelboard showing connections.

certain power companies and local ordinances in some areas. Always check with the local inspection authority for such rules that surpass the requirements of the NEC.

5.1.1 Grounding Conductors

The grounding conductor connecting the panelboard neutral bus to the water pipe and grounding electrodes must be made of copper, aluminum, or copper-clad aluminum. The material selected must be resistant to any corrosive condition existing at the installation, or it must be suitably protected against corrosion. The grounding conductor may be either solid or stranded, covered or bare, but it must be in one continuous length without a splice or joint, except for the following conditions:

- Splices in busbars are permitted.
- Where a service consists of more than a single enclosure, it is permissible to connect taps to the grounding electrode conductor, provided the taps are made within the enclosures.
- Grounding electrode conductors may also be spliced at any location by means of irreversible compression-type connectors listed for the purpose or by exowelding.

The size of grounding conductors depends on service-entrance size, that is, the size of the largest service-entrance conductor or equivalent for parallel conductors. *NEC Table 250-66* lists the proper sizes of grounding conductors for various sizes of electric services. Equipment grounding is covered in detail later in this module.

Discuss the NEC requirements for grounding conductors.

(IN THIS CONFIGURATION
THE GROUNDED CONDUCTOR
IS TIED TO THE NEUTRAL BUS)

MAIN PANELBOARD

NEUTRAL BUS

GROUNDING
ELECTRODE
CONDUCTOR

UPPER END OF ROD MUST BE
FLUSH WITH OR BELOW GRADE.

GROUNDING ELECTRODE
DRIVEN IN GROUND AT
LEAST BEYOND THE ROOF
OVERHANG.

NOTE: SOME AREAS REQUIRE
TWO GROUNDING ELECTRODES,
SPACED AT LEAST 6 FEET
APART.

GROUNDING CLAMP (CONNECTION MUST
BE MADE WITHIN 5 FEET FROM THE POINT WHERE
THE PIPE ENTERS THE BUILDING)

METAL COLD WATER PIPE

METAL UNDERGROUND GAS PIPING SYSTEM
MAY NOT BE USED AS A GROUNDING ELECTRODE

112F12.EPS

Figure 12 ◆ Components of a residential grounding system.

**Discuss the NEC
requirements for grounding
electrodes.**

**Discuss the use of
grounded conductors for
grounding equipment.**

5.2.0 Grounding Path to Ground Electrode at Service

A grounding electrode conductor is required to be used to connect the equipment grounding conductors, the service-entrance enclosures, and the grounded (neutral) service conductor to the grounding electrode. See *NEC Section 250-24(a)*.

An unspliced main bonding jumper is required to connect the equipment grounding conductor and the service-equipment enclosure to the system grounded (neutral) conductor of the electric utility system inside the service-equipment or service-conductor enclosure.

Where an alternating current system is connected to a grounding electrode in or at a building as specified in *NEC Section 250-24*, the same common grounding electrode must be used to ground conductor enclosures and equipment in or on that building. See *NEC Section 250-58*.

Where separate services supply a building and are required to be connected to a grounding electrode, the same grounding electrode is required to be used. See *NEC Section 250-58*. Two or more grounding electrodes that are effectively bonded together are considered to be a single grounding electrode system.

5.2.1 Use of a Grounded (Neutral) Circuit Conductor for Grounding Equipment

For supply side equipment, a grounded (neutral) circuit conductor is permitted to be used to ground noncurrent-carrying metal parts of equipment, raceways, and other enclosures on the supply (line) side of the service disconnecting means. See *NEC Section 250-58* for additional information on this subject.

ELECTRICAL LEVEL ONE—TRAINEE MODULE 26112

Instructor's Notes:

Grounding Conductors

This residential application has a grounding electrode connected to a rod that travels through the floor and into the ground at least 8'. In addition, it is also connected to a metal cold-water pipe (not shown). What type of grounding system is provided at your home? Does it meet NEC requirements?

VERONICA WESTFALL   112P1202.EPS

5.2.2 Method of Bonding Service Equipment

Electrical continuity at service equipment is required to be assured by one of the methods specified below:

• The service equipment must be bonded to the grounded (neutral) service conductor using a listed pressure connector, listed clamp, or other listed means.

Note
Connection devices or fittings that depend solely upon solder must not be used for this connection.

• Threaded couplings and connectors on enclosures with joints are required to be made up wrench-tight where rigid metal conduit and intermediate metal conduit are installed.

• Threadless couplings and connectors are required to be made up tight for rigid metal conduit, intermediate metal conduit, and electrical metallic tubing. Standard locknuts or bushings cannot be used for the bonding required by this action.
• Bonding jumpers must be used around concentric or eccentric knockouts in enclosures that are punched or otherwise formed so as to impair the electrical connection to ground.
• Other approved devices may be used, such as bonding-type locknuts and bonding bushings.

5.2.3 Main and Equipment Bonding Jumpers

The material of the main and equipment bonding jumpers must be of copper or other corrosion-resistant material. A main bonding jumper may be in the form of a wire, bus, screw, or similar suitable conductor.

Where a main bonding jumper is in the form of a screw, it is required to be identified with a green

Discuss the methods of bonding service equipment.

Discuss bonding jumpers.

GROUNDING ELECTRODES
NEC SECTION 250-52

1/2" MIN.

STAINLESS STEEL, NONFERROUS RODS, OR EQUIVALENT

5/8" MIN.

IRON OR STEEL RODS

3/4" MIN.

PIPE OR CONDUIT OUTER SURFACE MUST BE GALVANIZED OR OTHERWISE METAL-COATED FOR CORROSION PROTECTION

ALUMINUM

ALUMINUM ELECTRODES MUST NOT BE USED

8' MINIMUM

RESISTANCE TO GROUND, 25 OHMS OR LESS

112F13.EPS

Figure 13 ◆ Specifications for rod and pipe grounding electrodes.

Grounding Electrode Conductors

Use *NEC Section 250-24* to find the table that you need to size the grounding electrode conductor for a typical residential service.

finish so that the head of the screw is visible for inspection. A main and equipment bonding jumper must be attached using a listed pressure connector, listed clamp, or other listed means.

The bonding jumper cannot be smaller than the sizes given in *NEC Table 250-66* for grounding electrode conductors. A properly sized main bonding jumper will usually be included as a part of a listed panelboard.

5.2.4 Bonding and Grounding Requirements for Other Systems

An accessible means external to the service-equipment enclosure is required for connecting intersystem conducting and grounding conductors and connections for the communications, radio and television (TV), and community antenna television (CATV) circuits. This can be accomplished using at least one of the following:

- An exposed metallic service raceway
- An exposed grounding electrode conductor
- A No. 6 copper conductor that is not shorter than six inches and is bonded to the metallic service raceway or service-equipment enclosure using a listed and identified fitting

If a copper conductor is used, the other end of this conductor must be accessible and must be located on the exterior outside wall so that an external

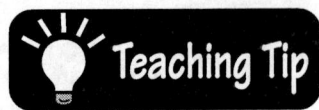

Discuss the "Think About It."

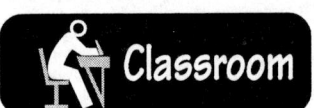

Discuss the bonding/grounding requirements for TV/CATV and other communications systems.

Instructor's Notes:

connection of the intersystem bonding or grounding conductor can be made, as required by *NEC Sections 800-40, 810-21, and 820-40* for communications, radio and TV equipment, and CATV circuits.

> **Note**
> All of the above requirements preclude the use of screw(s) to secure the equipment cover(s), screw on connectors, etc.

Provide a permanent sign at this intersystem connection that reads as follows: *CONNECTION FOR GROUNDING OR BONDING OF COMMUNICATIONS, TV, AND CATV SYSTEMS.*

5.2.5 Grounding Electrode Conductor Connections

The grounding electrode conductor is required to be connected to the grounding fitting using listed pressure connectors, listed clamps, or other listed means. Connections that depend on solder must never be used. To prevent corrosion, the ground clamp must be listed for the material of the grounding electrode and the grounding electrode conductor.

Where used on a pipe, ground rod, or other buried electrodes, the ground clamp must be listed for direct soil burial. More than one conductor is not permitted to be connected to the grounding electrode using a single clamp or fitting unless the clamp or fitting is specifically listed for the connection of more than one conductor.

For the connection to an electrode, you must use one of the following:

- A listed, bolted clamp of cast bronze or brass, or plain or malleable iron
- A pipe fitting, pipe plug, or other approved device that is screwed into a pipe or pipe fitting
- A listed sheet metal strap-type ground clamp with a rigid metal base that seats on the electrode with a strap that will not stretch during or after installation

> **Note**
> The UL listing states that "strap-type ground clamps are not suitable for attachment of the grounding electrode conductor of an interior wiring system to a grounding electrode."

5.2.6 Bonding of Piping Systems

The interior metal water piping system must be bonded to the service-equipment enclosure, the grounded (neutral) conductor at the service, the grounding electrode conductor where of sufficient size, or to the grounding electrode(s) used.

The bonding jumper must be sized in accordance with *NEC Table 250-66* and installed in accordance with *NEC Sections 250-64(b) and (e)*. The points of attachment of the bonding jumper to the metal piping must always be accessible per *NEC Section 250-104*.

An exception for two-family dwellings permits the interior metal water piping system for each occupancy to be bonded to the panelboard enclosure (other than at the service equipment) supplying that occupancy where the interior metal water piping system for each unit is metallically isolated by the use of a nonmetallic water piping system. The size of this bonding jumper must be in accordance with *NEC Table 250-122*.

5.2.7 Other Metal Piping

Interior metal piping that may become energized is required to be bonded to the service-equipment enclosure, the grounded conductor at the service, the grounding electrode conductor where of sufficient size, or the grounding electrode(s) used. This bonding jumper must be sized in accordance with *NEC Table 250-122* using the rating of the circuit that may energize the piping. The equipment grounding conductor for the circuit that may energize the piping is permitted to serve as the bonding means.

> **Note**
> Bonding all piping and metal air ducts within the premises will provide additional safety.

5.2.8 Grounding Electrode System

A bonding jumper is required to be installed in accordance with *NEC Section 250-64(b)*. The bonding jumper must be sized in accordance with *NEC Table 250-66* and be connected using a listed lug, listed pressure connector, or listed clamp. See *NEC Section 250-70*.

The unspliced grounding electrode conductor is permitted to be run to any convenient grounding

Discuss the allowable methods for making grounding electrode conductor connections.

Discuss the NEC requirements for bonding metal piping systems.

Discuss the NEC requirements for the grounding electrode system.

electrode that is available in the grounding electrode system. It will be sized for the largest grounding electrode conductor required among all the available electrodes.

If available on the premises at each building or structure served, each item below, and any made electrodes used in accordance with *NEC Sections 250-52(c) and (d),* are required to be bonded together to form the grounding electrode system.

- A metal underground water pipe may be used, provided that it is in direct contact with the earth for 10 feet or more (including any metal well casing effectively bonded to the pipe) and electrically continuous (or made electrically continuous by bonding around insulating joints or sections or insulating pipe) to the points of connection of the grounding electrode conductor and the bonding conductors. Continuity of the grounding path or the bonding connection to interior piping must never rely upon water meters. A metal underground water pipe must also be supplemented by an additional electrode of a type specified in *NEC Sections 250-50 or 250-52.* The supplemental electrode is permitted to be bonded to the grounding electrode conductor, grounded service-entrance conductor, grounded service raceway, and grounded service enclosure, or the interior metal water piping at any convenient point. Where the supplemental electrode is a made electrode (such as a rod, pipe, or plate), that portion of the bonding jumper which is the sole connection to the supplemental grounding electrode is not required to be larger than No. 6 copper wire or No. 4 aluminum wire. Where used outside, aluminum or copper-clad aluminum grounding electrode conductors are not permitted to be installed within 18 inches of the earth [*NEC Section 250-64(a)*].
- The metal frame of the building may be used, provided that it is effectively grounded.

Note
Effectively grounded means intentionally connected to earth through one or more ground connection(s) of sufficiently low impedance and having sufficient current-carrying capacity to prevent the buildup of voltages that may result in a hazard to people or connected equipment.

- Concrete-encased electrodes may be used when they are encased by at least two inches of concrete. This type of electrode is located within and near the bottom of a concrete foundation or footing that is in direct contact with the earth. It consists of a total of at least 20 feet of one or more lengths of steel reinforcing bars or rods, with each being not less than ½ inch in diameter. It may also consist of a continuous unbroken length of at least 20 feet of No. 4 or larger bare copper conductor.
- A ground ring encircling the building or structure may be used if it is in direct contact with the earth at a depth below the surface of not less than 1½ feet. It must consist of an unbroken length of at least 20 feet of a No. 2 bare copper conductor.

5.2.9 Made Electrodes and Other Electrodes

Where none of the electrodes specified above is available, one or more of the electrodes specified below must be used.

Where practical, made electrodes are required to be embedded below the permanent moisture level. Any nonconductive coatings, such as paint or enamel, must be removed. Where more than one electrode is used, each electrode of one grounding system (including that used for lightning rods) must be at least six feet from any other electrode of another grounding system.

CAUTION
Metal underground gas piping systems must never be used as a grounding electrode.

Where two or more electrodes are effectively bonded together, they are treated as a single electrode system.

Permissible made electrodes are:

- Other local metal underground systems or structures, such as piping systems and underground tanks
- Rod and pipe electrodes

Rod and pipe electrodes are required to be not less than eight feet in length and must consist of the following material:

- Electrodes of pipe or conduit cannot be smaller than ¾ inch in diameter. Pipe or conduit electrodes of iron or steel must have their outer surface galvanized or be otherwise coated for corrosion protection.

Discuss the NEC requirements for made electrodes. Pass around examples of sections of suitable grounding materials.

Instructor's Notes:

- Rods of iron or steel must be at least ⅝ inch in diameter. Nonferrous or stainless steel rods or their equivalent that are less than ⅝ inch in diameter are required to be listed and cannot be less than ½ inch in diameter.

Rod and pipe electrodes must be installed in the following manner:

- The electrode must be driven to a depth of not less than eight feet, unless rock bottom is encountered. In this situation, the electrode is permitted to be driven at an angle not to exceed 45° from the vertical, or it must be buried in a trench that is at least 2½ feet deep.
- The upper end of the electrode is required to be flush with or below ground level unless the above-ground end and the grounding electrode conductor attachment are protected against physical damage. See *NEC Section 250-10.*
- Plate electrodes must expose not less than two square feet of surface to exterior soil. Electrodes of iron or steel plates must be at least ¼ inch thick. Electrodes of nonferrous metal must be at least 0.06 inch thick.
- Aluminum electrodes are not permitted.

5.2.10 Resistance of Made Electrodes

An electrode consisting of a single rod, pipe, or plate that does not register a resistance to ground of 25 ohms or less must be augmented by one additional electrode of any of the types specified in *NEC Sections 250-50 or 250-52.* Where multiple rod, pipe, or plate electrodes are installed to meet the requirements of *NEC Section 250-56,* they must be at least six feet apart. The paralleling efficiency of rods longer than eight feet is improved by spacing greater than six feet.

5.2.11 Use of Lightning Rods

Lightning rod conductors and driven pipes, rods, or other made electrodes used for grounding lightning rods are not to be used in lieu of the made grounding electrodes required by *NEC Sec-*

tion 250-52 for the grounding of wiring systems and equipment.

See *NEC Section 250-106* for more information on lightning protection systems.

5.2.12 Grounding Conductor Material

The materials for grounding conductors must meet the following specifications:

- The grounding electrode conductor must be of copper, aluminum, or copper-clad aluminum. The material selected must be resistant to any corrosive condition existing at the installation or must otherwise be suitably protected against corrosion.
- The grounding electrode conductor may be solid or stranded, and it may be insulated, covered, or bare.
- The grounding electrode conductor must be installed in one continuous length without a splice or joint. An exception permits the connection of taps to the grounding electrode conductor. Each such tap conductor will be required to extend to the inside of each such enclosure. The size for the grounding electrode conductor must be in accordance with *NEC Section 250-66.* Tap conductors are permitted to be sized for the largest conductor serving the respective enclosure(s). Tap conductors must be connected to the grounding electrode conductor in such a manner that the grounding electrode conductor remains without any splice or joint.

5.2.13 Installation of Grounding Electrode Conductors

A grounding electrode conductor or its enclosure is required to be securely fastened to the surface on which it is carried. A No. 4 or larger copper or aluminum grounding electrode conductor is required to be protected where it will be exposed to severe physical damage. A No. 6 or larger grounding electrode conductor that is free from exposure to physical damage is permitted to be run along

What Does a Lightning Rod Do?

An interesting fact about grounding is that a lightning rod (air terminal) isn't meant to bring a bolt of lightning to ground. To do this, its conductors would have to be several feet in diameter. The purpose of the rod is to dissipate the negative static charge that would cause the positive lightning charge to strike the house.

Discuss the NEC requirements for electrode resistance.

Discuss lightning rods (air terminals).

Discuss the allowable materials for use in grounding conductors.

Discuss the installation requirements for grounding electrode conductors.

the surface of the building without metal covering or protection where it is securely fastened to the building. Otherwise, it must be installed in rigid metal conduit, intermediate metal conduit, rigid nonmetallic conduit, electrical metallic tubing, or cable armor.

Grounding electrode conductors smaller than No. 6 must be protected by rigid metal conduit, intermediate metal conduit, rigid nonmetallic conduit, electrical metallic tubing, or cable armor.

Aluminum or copper-clad aluminum grounding electrode conductors cannot be used where they are in direct contact with masonry or the earth or where subject to corrosive conditions. Where used outside, aluminum or copper-clad aluminum grounding electrode conductors must be installed at least 18 inches from the earth.

Metal enclosures for grounding electrode conductors are required to be electrically continuous from the point of attachment to metal cabinets or metallic equipment enclosures to the grounding electrode. They must also be securely fastened to the ground clamp or fitting.

A metal enclosure that is not physically continuous from a metal cabinet or metallic equipment enclosure to the grounding electrode must be made electrically continuous by bonding each end to the enclosed grounding electrode conductor.

5.2.14 Size of Grounding Electrode Conductor

The size of the grounding electrode conductor for an alternating current system cannot be less than that given in *NEC Table 250-66.*

The exceptions to this rule are as follows:

* Where connected to made electrodes as in *NEC Sections 250-52(c) or (d),* that portion of the grounding electrode conductor that is the sole connection to the grounding electrode is not required to be larger than No. 6 copper wire or No. 4 aluminum wire.
* Where connected to a concrete-encased electrode as in *NEC Section 250-50(c),* that portion of the grounding electrode conductor that is the sole connection to the grounding electrode is not required to be larger than No. 4 copper wire.
* Where connected to a ground ring as in *NEC Section 250-50(a)(2),* that portion of the grounding electrode conductor that is the sole connection to the grounding electrode is not required to be larger than No. 6 copper.

Where multiple sets of service-entrance conductors are used as permitted in *NEC Section*

230-40, Exception No. 2, the equivalent size of the largest service-entrance conductor is required to be determined by the largest sum of the areas of the corresponding conductors of each set.

Where there are no service-entrance conductors, the grounding electrode conductor size is required to be determined by the equivalent size of the largest service-entrance conductor required for the load to be served.

6.0.0 ◆ INSTALLING THE SERVICE ENTRANCE

In practical applications, the electric service is normally one of the last components of an electrical system to be installed. However, it is one of the first considerations when laying out a residential electrical system. For instance:

* The electrician must know in which direction and to what location to route the circuit home-runs while roughing in the electrical wiring.
* Provisions must be made for sleeves through footings and foundations in cases where underground systems (service laterals) are used.
* The local power company must be notified as to the approximate size of service required so they may plan the best way to furnish a service drop to the property.

6.1.0 Service Drop Locations

The location of the service drop, electric meter, and load center should be considered first. It is always wise to consult the local power company to obtain their recommendations; where you want the service drop and where they want it may not coincide. A brief meeting with the power company about the location of the service drop can prevent problems later on.

The service drop must be routed so that the service drop conductors have a clearance of not less than three feet from windows, doors, porches, fire escapes, or similar locations. Furthermore, when service drop conductors pass over rooftops, driveways, yards, etc., they must have clearances as specified in *NEC Section 230-24.*

A plot plan (also called a *site plan*) is often available for new construction. The plot plan shows the entire property, with the building or buildings drawn in their proper location on the plot of land. It also shows sidewalks, driveways, streets, and existing utilities—both overhead and underground.

A plot plan of the sample residence is shown in *Figure 14.* In reviewing this drawing, you can see

Explain how grounding electrode conductors are sized.

Discuss the preliminary steps in installing the service entrance.

Show Transparency 16 (Figure 14).

Instructor's Notes:

Figure 14 ◈ Plot plan of the sample residence.

that the closest power pole is located across a public street from the house. By consulting with the local power company, it is learned that the service will be brought to the house from this pole by triplex cable, which will hit the residence at a point on its left (west) end. The steel uninsulated conductor of triplex cable acts as both the grounded conductor (neutral) and as a support for the insulated (ungrounded) conductors. It is also suitable for overhead use.

When Type SE cable is used, it will run directly from the point of attachment and service head to the meter base. However, since the carport is located on the west side of the building, a service mast (*Figure 15*) will have to be installed.

Roof clearances are specified in *NEC Section 230-24(a)*. The NEC requires a clearance of not less than eight feet above the roof surface that is maintained for a distance of not less than three feet in all directions from the roof edge.

6.2.0 Vertical Clearances of Service Drop

NEC Section 225-18 specifies the distances by which service drop conductors must clear the ground. These distances vary according to the surrounding conditions.

In general, the NEC states that the vertical clearances of all service drop conductors that carry 600V or under are based on a conductor temperature of 60°F (15°C) with no wind and with the final unloaded sag in the wire, conductor, or cable. Service drop conductors must be at least 10 feet above the ground or other accessible surfaces at all times. More distance is required under most conditions. For example, if the service conductors pass over residential property and driveways or commercial property that is not subject to truck traffic, the conductors must be at least 15 feet above the ground. However, this distance may be reduced to 12 feet when the voltage is limited to 300V to ground.

In other areas, such as public streets, alleys, roads, parking areas subject to truck traffic, driveways on other-than-residential property, etc., the minimum vertical distance is 18 feet. The conditions of the sample residence are shown in *Figure 16*.

6.3.0 Service Drop Clearances for Building Openings

Service conductors that are installed as open conductors or multiconductor cable without an overall outer jacket must have a clearance of not less

Discuss roof clearances and vertical clearances.

Show Transparencies 17 and 18 (Figures 15 and 16A).

Discuss service drop clearances for building openings.

Voltage between conductors must not exceed 300 volts.
NEC Section 230-24

Where raceway-type service masts are used, all raceway fittings must be identified for use with service masts.
NEC Section 230-28

ROOF CLEARANCES
NEC Section 230-24(a) plus Exceptions

ELECTRIC METER
NEC Section 90-2(b)(5)

Max. and min. distances per local utility standards.

TO MAIN SERVICE PANEL

112F15.EPS

Figure 15 ◆ NEC sections governing service mast installations.

Discuss location considerations for main service disconnects (panelboards).

than three feet from windows that are designed to be opened, doors, porches, balconies, ladders, stairs, fire escapes, or similar locations (*NEC Section 230-9*). However, conductors run above the top level of a window are permitted to be less than three feet from the window opening.

The three-foot clearance is not applicable to raceways or cable assemblies that have an overall outer jacket approved for use as a service conductor. The intention of this requirement is to protect the conductors from physical damage and/or physical contact with unprotected personnel. The exception allows service conductors, including drip loops and service drop conductors, to be located just above window openings because they are considered to be out of reach.

7.0.0 ◆ PANELBOARD LOCATION

The main service disconnect or panelboard is normally located in a portion of an unfinished basement or utility room on an outside wall so that the service cable coming from the electric meter can terminate immediately into the switch or panelboard when the cable enters the building. In the example home, however, there is no basement and the utility room is located in the center of the house with no outside walls. Consequently, a somewhat different arrangement will have to be used. A *load center* is a type of panelboard that is normally located at the service entrance of a residential installation. The load center usually contains a main circuit breaker, which is the main disconnect. Circuit breakers are provided for equipment such as

12–26

ELECTRICAL LEVEL ONE—TRAINEE MODULE 26112

Instructor's Notes:

NEC Section 225-18

(A)

112F16A.EPS

VERONICA WESTFALL   (B)   112F16B.EPS

Figure 16 ◆ Vertical clearances for service drop conductors.

Failure to Deenergize Panelboard

A 31-year-old electrician was finishing the installation of an outdoor floodlight on a new home. He borrowed an aluminum ladder from another contractor and then proceeded with his task. He did not verify that power was removed at the panelboard, and when he used his wire strippers to remove the conductor insulation, his right thumb and index finger contacted a 110V circuit. He received a fatal shock.

The Bottom Line: Always deenergize circuits at the panelboard before beginning any electrical task, and never use aluminum ladders while working in or near electrical devices. In this case, the ladder provided a path to ground.

Discuss the "Case History." Emphasize the deadly consequences of failing to ensure that the power has been removed prior to working on electrical circuits.

electric water heaters, ranges, dryers, air conditioning and heating units, and breakers that feed subpanels such as lighting panels.

NEC Section 230-70 requires that the service disconnecting means be installed in a readily accessible location—either outside or inside the building. If located inside the building, it must be located nearest the point of entrance of the service conductors. In the sample home, there are at least two methods of installing the panelboard in the utility room that will comply with this NEC regulation.

The first method utilizes a weatherproof 100A disconnect (safety switch or circuit breaker enclosure) mounted next to the meter base on the outside of the building. With this method, service

conductors are provided with overcurrent protection; the neutral conductor is also grounded at this point, as this becomes the main disconnect switch. Three-wire cable with an additional grounding wire is then routed from this main disconnect to the panelboard in the utility room. All three current-carrying conductors (two ungrounded and one neutral) must be insulated with this arrangement; the equipment ground, however, may be bare. The panelboard containing overcurrent protection devices for the branch circuits, which is located in the utility room, now becomes a subpanel. See *Figure 17.*

An alternate method utilizes conduit from the meter base that is routed under the concrete slab

Show Transparency 19 (Figure 17).

Figure 17 ◆ One method of wiring a panelboard for the sample residence.

Labels within figure:
- 112F17.EPS
- EQUIPMENT GROUNDING CONDUCTORS
- Neutral bus and equipment grounding bus must be isolated from each other in subpanel.
- GROUNDING CONDUCTOR
- INSUL. NEUTRAL COND.
- UTILITY ROOM
- 100-AMPERE MAIN LUGS-ONLY PANEL
- LIVING ROOM
- 100-amp, three-wire conductor with ground cable is run in finished partitions and in unfinished attic to subpanel in utility room.
- SERVICE MAST
- ELECTRIC METER
- 100A, WP DISCONNECT
- Service ground must be connected at this point.

Instructor's Notes:

Note
Local ordinances in some areas may require a disconnect at the meter base, making the panel in the utility room a subpanel.

and then up to a main panelboard located in the utility room. *NEC Section 230-6* considers conductors installed under at least two inches of concrete to be outside the building. The sample residence has a four-inch-thick reinforced concrete slab—well within the NEC regulations. Therefore, the service conductors from the meter base that are installed under the concrete slab in conduit are considered to be outside the house, and no disconnect is required at the meter base. When this conduit emerges in the utility room, it will run straight up into the bottom of the panelboard, again meeting the NEC requirement that the panel be located nearest the point of entrance of the service conductors. Details of this service arrangement are shown in *Figure 18*.

8.0.0 ◆ WIRING METHODS

Branch circuits and feeders are used in residential construction to provide power wiring to operate components and equipment, and control wiring

to regulate the equipment. Wiring may be further subdivided into either *open* or *concealed* wiring.

In open wiring systems, the cable and/or raceways are installed on the surface of the walls, ceilings, columns, and other areas where they are in view and are readily accessible. Open wiring is often used in areas where appearance is not important, such as in unfinished basements, attics, and garages.

Concealed wiring systems are installed inside walls, partitions, ceilings, columns, and behind baseboards or moldings where they are out of view and are not readily accessible. This type of wiring is generally used in all new construction with finished interior walls, ceilings, and floors, and it is the preferred type of wiring where appearance is important.

In general, there are two basic wiring methods used in the majority of modern residential electrical systems. They are:

• Sheathed cables of two or more conductors
• Raceway (conduit) systems

The method used on a given job is determined by the requirements of the NEC, the type of building construction, the location of the wiring in the building, and the relative costs of different wiring methods.

In most applications, either of the two methods may be used, and both methods are frequently used in combination.

Show Transparency 20 (Figure 18).

Discuss the two basic methods used in residential wiring.

Figure 18 ◆ Alternate method of service installation for the sample residence.

Discuss cable systems:
- **Type NM**
- **Type AC**
- **Type UF**

Pass around examples for the trainees to examine.

Show Transparencies 21 through 24 (Figures 19 through 22).

8.1.0 Cable Systems

Several types of cable are used in wiring systems to feed or supply power to equipment. These include nonmetallic-sheathed cable, armored cable, underground feeder cable, and service-entrance cable.

8.1.1 Nonmetallic-Sheathed Cable

Nonmetallic-sheathed (Type NM) cable is manufactured in two- or three-wire configurations with varying sizes of conductors. In both two- and three-wire cables, conductors are color-coded: one conductor is black while the other is white in two-wire cable; in three-wire cable, the additional conductor is red. Both types also have a grounding conductor, which is usually bare, but it is sometimes covered with green plastic insulation, depending upon the manufacturer. The jacket or covering consists of rubber, plastic, or fiber. Most also have markings on this jacket giving the manufacturer's name or trademark, wire size, and number of conductors (see *Figure 19*). For example, NM 12-2 W/GRD indicates that the jacket contains two No. 12 AWG conductors along with a grounding wire; NM 12-3 W/GRD indicates three conductors plus a grounding wire. This type of cable may be concealed in the framework of buildings, or in some instances, may be run exposed on the building surfaces. It may not be used in any building exceeding three floors above grade, as a service-entrance cable, in commercial garages having hazardous locations, in theaters and similar locations, in places of assembly, in mo-

tion picture studios, in storage battery rooms, in hoistways, embedded in poured concrete or aggregate, or in any hazardous location except as otherwise permitted by the NEC. Type NM cable is frequently referred to as Romex®.

Since Type NM cable is the least expensive type for residential use, this wiring method will be the type encountered most often. *Figure 20* shows additional NEC regulations pertaining to the installation of Type NM cable.

8.1.2 Armored Cable

Armored (Type AC) cable, commonly called BX®, is manufactured in two-, three-, and four-wire assemblies with varying sizes of conductors, and it is used in locations similar to those where Type NM cable is allowed. The metallic spiral covering on Type AC cable offers a greater degree of mechanical protection than with Type NM cable and also provides a continuous grounding bond without the need for additional grounding conductors.

Type AC cable may be used for under-plaster extensions, as provided in the NEC, and embedded in plaster finish, brick, or other masonry, except in damp or wet locations. It may also be run in the air voids of masonry block or tile walls, except where such walls are exposed or subject to excessive moisture or dampness. See *Figures 21 and 22*.

8.1.3 Underground Feeder Cable

Underground feeder (Type UF) cable (*NEC Article 339*) may be used underground, including direct

Type NM cable must be secured by approved staples or straps at intervals not exceeding 4.5 feet, and within 12 inches from every outlet box, junction box, cabinet, or fitting.
NEC Section 336-18

WHITE CONDUCTOR

NM 12-2 W/GRD

NUMBER AND SIZE OF CONDUCTORS

GROUNDING CONDUCTOR

BLACK CONDUCTOR

112F19.EPS

Figure 19 ◆ Characteristics of Type NM cable.

Instructor's Notes:

_____

_____

_____

_____

_____

_____

_____

_____

_____

Where Type NM cable is run through wood joists where the edges of the bored hole is less than 1-1/4" from the nearest edge of the stud, or where studs are notched, a steel plate, not less than 1/16" must be used to protect the cables as shown.
NEC Sections 336-9 and 300-4(a)(1)

Where run across top of floor joists, front edges of rafters or studs, NM cable must be protected by guard strips which are at least as high as the cable.
NEC Sections 336-6(b) & (d) and 333-12

Where the attic space is not accessible by permanent stairs or ladders, guard strips are required only within 6 feet of the nearest edge of the attic entrance.
NEC Sections 336-6(b) & (d) and 333-12

Where cable is carried along the sides of rafters, studs, or floor joists, neither guard strips nor running boards are required.
NEC Sections 336-9 & 300-4(d)

Cables run through holes in wooden joists, rafters, or studs are considered to be supported without additional clamps or straps.
NEC Section 336-18

Cable must be secured within 12" of every cabinet, box, or fitting.
NEC Section 336-18

NM cable must be secured in place at intervals not exceeding 4.5 feet.
NEC Section 336-18

Cables not smaller than two No. 6 or three No. 8 may be secured directly to the lower edges of joists in unfinished basements.
NEC Section 336-6(c)

Where run parallel to the floor joists, cable may be secured to the sides of the joists.
NEC Sections 336-9 & 300-4(d)

Cables smaller than two No. 6 that run on the bottom edge of floor joists in unfinished basements, must be provided with a "running board," and cable must be secured to it.
NEC Section 336-6(c)

Bends must not exceed five times the diameter of the cable.
NEC Section 336-16

Type NM cable may be installed in air voids in masonry block where such walls are not subject to excessive moisture or dampness.
NEC Section 336-4(a)

112F20.EPS

Figure 20 ◆ NEC sections governing the installation of Type NM cable.

burial in the earth, as a feeder or branch circuit cable when provided with overcurrent protection at the rated ampacity as required by the NEC. When Type UF cable is used above grade where it will come in direct contact with the rays of the sun, its outer covering must be sun-resistant. Furthermore, where Type UF cable emerges from the ground, some means of mechanical protection must be provided. This protection may be in the form of conduit or guard strips. Type UF cable resembles Type NM cable; however, the jacket is constructed of weather-resistant material to provide the required protection for direct-burial wiring installations.

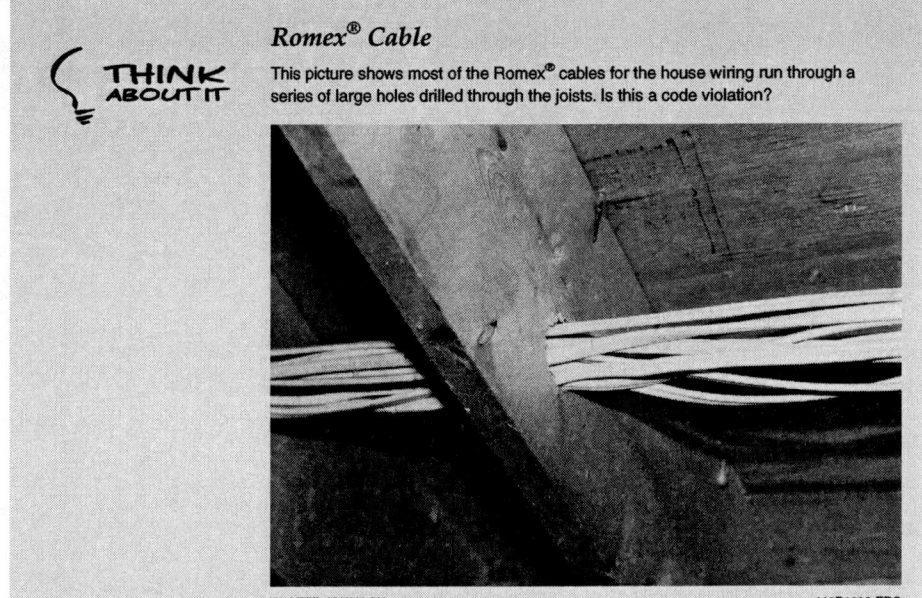

THINK ABOUT IT

Romex® Cable

This picture shows most of the Romex® cables for the house wiring run through a series of large holes drilled through the joists. Is this a code violation?

WALTER JOHNSON
112P1203.EPS

Type AC cable must be secured by approved staples or straps at intervals not exceeding 4.5 feet, and within 12 inches from every outlet box, junction box, cabinet, or fitting.
NEC Section 333-7

An approved insulating bushing or its equivalent must be provided between the conductors and the armor at all points where the armor of the cable terminates.
NEC Section 333-9

At all points where the armor of AC cable terminates, a fitting must be provided to protect wires from abrasion, unless the design of the outlet boxes or fittings provides equivalent protection.
NEC Section 333-9

LOCKNUT

CONNECTOR

112F21.EPS

Figure 21 ◆ Characteristics of Type AC cable.

Instructor's Notes:

_____

_____

_____

_____

_____

_____

_____

_____

_____

Guard strips at least as high as the cable must be provided.
NEC Section 333-12(a)

ATTIC FLOOR JOISTS

BX CABLE

RADIUS

D

5 x D = ALLOWABLE RADIUS
NEC Section 333-8

112F22.EPS

Figure 22 ◆ NEC sections governing the installation of Type AC cable.

8.1.4 Service-Entrance Cable

Service-entrance (Type SE) and underground service-entrance (Type USE) cable, when used for electrical services, must be installed as specified in *NEC Articles 230 and 338*. Service-entrance cable is available with the grounded conductor bare for outside service conductors, and also with an insulated grounded conductor for interior wiring systems.

Type SE cable is permitted for use on branch circuits or feeders provided that all current-carrying conductors are insulated; this includes the grounded or neutral conductor. When Type SE cable is used for interior wiring, all NEC regulations governing the installation of Type NM cable also apply to Type SE cable; however, there are some exceptions. Type SE cable with an uninsulated, covered conductor may be used as a branch circuit or feeder where the fully insulated conductors are used for circuit wiring and the uninsulated covered conductor is used only for equipment grounding purposes. Bends of Type SE and USE cable must be made so that the protective covering of the cable is not damaged, and the radius of the bend to the inner edge of the cable is not less than five times the diameter of the cable.

SE Style R (SER) cable is used in residential applications for subfeeds for ranges, and it is also used for service laterals in multi-family dwellings.

Figure 23 summarizes the installation rules for Type SE cable for both exterior and interior wiring.

8.2.0 Raceways

A raceway is any channel that is designed and used solely for the purpose of holding wires, cables, or busbars. Types of raceways include rigid metal conduit, intermediate metal conduit, rigid non-metallic conduit, flexible metallic conduit, electrical metallic tubing, and auxiliary gutters. Raceways are constructed of either metal or insulating material such as polyvinyl chloride or PVC (plastic). Metal raceways are joined using threaded, compression, or setscrew couplings; nonmetallic raceways are joined using cement-coated couplings. Where a raceway terminates in an outlet box, junction box, or other enclosure, an approved connector must be used.

Raceways provide mechanical protection for the conductors that run in them and also prevent accidental damage to insulation and the conducting material. They also protect conductors from corrosive atmospheres and prevent fire hazards to life and property by confining arcs and flames that may occur due to faults in the wiring system. Conduits or raceways are used in residential applications for service masts, underground wiring embedded in concrete, and sometimes in unfinished basements, shops, or garage areas.

Another function of metal raceways is to provide a continuous equipment grounding system throughout the electrical system. To maintain this feature, it is extremely important that all raceway systems be securely bonded together into a continuous conductive path and properly connected

Discuss service-entrance cable and pass around an example for the trainees to examine.

Show Transparency 25 (Figure 23).

Discuss various types of raceways and their applications. Pass around examples for the trainees to examine.

WEATHERHEAD

30"MAX.

Type SE cable when used as a
service-entrance conductor must be
secured every 30 inches, and within
12 inches of each termination point.
NEC Section 230-51(a)

Cable bends shall be made so that the
cable covering is not damaged and the
radius of the bend is not less than five
times the diameter of the cable.
NEC Section 338-6

CABLE STRAPS

METER BASE

Use Type SE
cable spaces
for interior wiring.
NEC Section 338-4

CLOTHES DRYER

WALL OVEN

These appliances may
be fed by Type SE cable
with a bare neutral; all
other branch circuits
and feeders must have
an insulated neutral.
NEC Section 338-3(b)

ELECTRIC RANGE

RANGE TOP

PANELBOARD

112F23.EPS

Figure 23 ◆ NEC sections governing Type SE cable.

to the system ground. The following section explains how this is accomplished.

9.0.0 ◆ EQUIPMENT GROUNDING SYSTEM

NEC Article 250 covers grounding requirements. *Figure 24* summarizes the equipment grounding rules for most types of equipment that are required to be grounded. These general guidelines apply to all installations except for specific equipment (special applications) as indicated in the NEC. The NEC also lists specific equipment that is to be grounded regardless of voltage.

In all occupancies, major appliances and many hand-held appliances are required to be grounded. Some of these appliances are refrigerators, freezers, air conditioners, clothes dryers, washing machines, dishwashing machines, sump pumps, and electrical aquarium equipment. Tools

that are likely to be used outdoors and in wet or damp locations must also be grounded or have a system of double insulation.

Although most appliance circuits require an equipment grounding conductor, the frames of electric ranges, clothes dryers, and similar appliances that utilize both 120V and 240V may be grounded via the grounded circuit conductor (neutral) under most conditions. In addition, the grounding contacts of any receptacles on the equipment must be bonded to the equipment. If these specified conditions are met, it is not necessary to provide a separate equipment grounding conductor, either for the frames or any outlet or junction boxes that are part of the circuit for these applications.

A bonding jumper is sometimes used to ensure electrical conductivity between metal parts. When the jumper is installed to connect two or more portions of the equipment grounding conductor, the

**Discuss the NEC
requirements for equipment
grounding systems.**

**Show Transparency 26
(Figure 24).**

Instructor's Notes:

_____

_____

_____

_____

_____

_____

_____

_____

GROUNDING SCREW

GROUNDING WEDGE

GROUNDING CLAMP

GROUNDING CLIP

GROUNDING U-BOLT

BONDING BUSHING

Where splices are made in
a junction box, the grounding
conductors must be spliced to
the metal junction box.

GROUNDING CLIP

GROUNDING
SCREW

GROUNDING CONDUCTOR

GROUNDING SCREW

112F24.EPS

Figure 24 ◆ Equipment grounding.

Grounding Devices

Identify each of the devices pictured here and explain their function(s). Which of these devices would *not* be used for equipment grounding?

WALTER JOHNSON

112P1204.EPS

Discuss branch circuits and feeders.

Show Transparencies 27 through 29 (Figures 25 through 27).

jumper is referred to as an *equipment bonding jumper.* Some specific cases in which a bonding jumper is required are as follows:

- Metal raceways, cable armor, and other metal noncurrent-carrying parts that serve as grounding conductors must be bonded whenever necessary to ensure electrical continuity.
- When flexible metal conduit is used for equipment grounding, an equipment bonding jumper is required if the length of the ground return path exceeds six feet or the circuit enclosed is rated over 20A. When the path exceeds six feet, the circuit must contain an equipment grounding conductor, and the bonding may be accomplished using approved fittings.
- A short length of flexible metal conduit that contains a circuit rated over 20A may not serve as a grounding conductor itself, but a separate bonding jumper can be provided in place of a separate equipment grounding conductor for the circuit. This bonding jumper may be installed inside or outside the conduit, but an outside jumper cannot exceed six feet in length.

10.0.0 ◆ BRANCH CIRCUIT LAYOUT FOR POWER

The point at which electrical equipment is connected to the wiring system is commonly called an *outlet.* There are many classifications of outlets: lighting, receptacle, motor, appliance, etc. This section, however, deals with the power outlets normally found in residential electrical wiring systems.

When viewing an electrical drawing, outlets are indicated by symbols (usually a small circle with appropriate markings to indicate the type of outlet). The most common symbols for receptacles are shown in *Figure 25.*

10.1.0 Branch Circuits and Feeders

The conductors that extend from the panelboard to the various outlets are called *branch circuits* and are defined by the NEC as the point of a wiring system that extends beyond the final overcurrent device protecting the circuit. See *Figure 26.*

A feeder consists of all conductors between the service equipment and the final overcurrent device. See *Figure 27.*

12–36

Instructor's Notes:

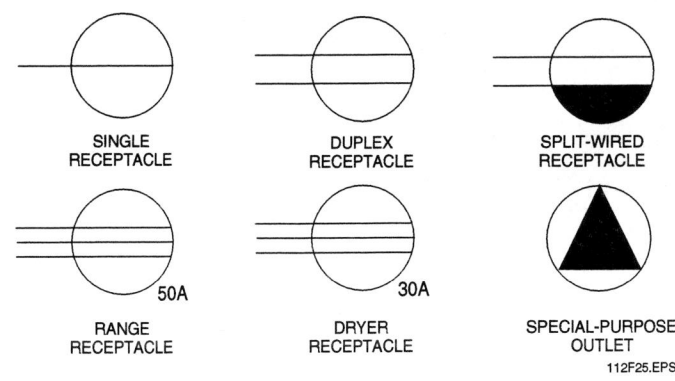

SINGLE
RECEPTACLE

DUPLEX
RECEPTACLE

SPLIT-WIRED
RECEPTACLE

50A

RANGE
RECEPTACLE

30A

DRYER
RECEPTACLE

SPECIAL-PURPOSE
OUTLET

112F25.EPS

Figure 25 ◆ Typical outlet symbols appearing in electrical drawings.

In general, the size of the branch circuit conductors varies depending upon the load requirements of the electrically operated equipment connected to the outlet. For residential use, most branch circuits consist of either No. 14 AWG, No. 12 AWG, No. 10 AWG, or No. 8 AWG conductors.

The basic branch circuit requires two wires or conductors to provide a continuous path for the flow of electric current, plus a third wire for equip-

ment grounding. The usual receptacle branch circuit operates at 120V.

Fractional horsepower motors and small electric heaters usually operate at 120V and are connected to 120V branch circuits by means of a receptacle, junction box, or direct connection.

With the exception of very large residences and tract-development houses, the size of the average residential electrical system of the past has not

PANELBOARD

NEUTRAL BUS

(In this configuration, the grounding conductor is tied to the neutral bus.)

BRANCH
CIRCUIT

DUPLEX
RECEPTACLE

LAST CIRCUIT BREAKER
BEFORE OUTLET

GROUNDING CONDUCTOR

112F26.EPS

Figure 26 ◆ Components of a duplex receptacle branch circuit.

Figure 27 ◆ A feeder being used to feed a subpanel from the main panelboard.

been large enough to justify the expense of preparing complete electrical working drawings and specifications. Such electrical systems were usually laid out by the architect in the form of a sketchy outlet arrangement or else laid out by the electrician on the job, often only as the work progressed. However, many technical developments in residential electrical use—such as electric heat with sophisticated control wiring, increased use of electrical appliances, various electronic alarm systems, new lighting techniques, and the need for energy conservation techniques—have greatly expanded the demand and extended the complexity of today's residential electrical systems.

Each year, the number of homes with electrical systems designed by consulting engineering firms increases. Such homes are provided with complete electrical working drawings and speci-

fications, similar to those frequently provided for commercial and industrial projects. Still, these are more the exception than the rule. Most residential projects will not have a complete set of drawings.

Circuit layout is provided on the drawings to follow for several reasons:

• They provide a visual layout of house wiring circuitry.
• They provide a sample of electrical residential drawings that are prepared by consulting engineering firms, although the number may still be limited.
• They introduce the method of showing electrical systems on working drawings so that you will have a better foundation for tackling advanced electrical systems.

Instructor's Notes:

_____

_____

_____

_____

_____

_____

_____

_____

_____

Branch circuits are shown on electrical drawings by means of a single line drawn from the panelboard (or by homerun arrowheads indicating that the circuit goes to the panelboard) to the outlet or from outlet to outlet where there is more than one outlet on the circuit.

The lines indicating branch circuits can be solid to show that the conductors are to be run concealed in the ceiling or wall, dashed to show that the conductors are to be run in the floor or ceiling below, or dotted to show that the wiring is to be run exposed. *Figure 28* shows examples of these three types of branch circuit lines.

In *Figure 28*, No. 12 indicates the wire size. The slash marks shown through the circuits in *Figure 28* indicate the number of current-carrying conductors in the circuit. Although two slash marks are shown, in actual practice, a branch circuit containing only two conductors usually contains no slash marks; that is, any circuit with no slash marks is assumed to have two conductors. However, three or more conductors are always indicated on electrical working drawings—either by slash marks for each conductor, or else by a note.

Never assume that you know the meaning of any electrical symbol. Although great efforts have been made in recent years to standardize drawing symbols, architects, consulting engineers, and electrical drafters still modify existing symbols or devise new ones to meet their own needs. Always consult the symbol list or legend on electrical working drawings for an exact interpretation of the symbols used.

10.2.0 Locating Receptacles

NEC Section 210-52 states the minimum requirements for the location of receptacles in dwelling units. It specifies that in each kitchen, family room, and dining room, receptacle outlets shall be installed so that no point along the floor line in any wall space is more than six feet (1.83m), measured horizontally, from an outlet in that space, including any wall space two feet (610mm) or more in width and the wall space occupied by fixed panels in exterior walls, but excluding sliding panels. Receptacle outlets shall, insofar as practicable, be spaced equal distances apart. Receptacle outlets in floors shall not be counted as part of the required number of receptacle outlets unless located within 18 inches (457mm) of the wall.

The NEC defines wall space as a wall that is unbroken along the floor line by doorways, fireplaces, or similar openings. Each wall space that is two feet or more in width must be treated individually and separately from other wall spaces within the room.

The purpose of the above NEC requirement is to minimize the use of cords across doorways, fireplaces, and similar openings.

Discuss the NEC requirements for locating receptacles.

Show Transparency 30 (Figure 28).

Figure 28 ◆ Types of branch circuit lines shown on electrical working drawings.

Show Transparency 31 (Figure 29).

Discuss split-wired receptacles.

Show Transparencies 32 and 33 (Figures 30 and 31).

Discuss multiwire branch circuits.

With this NEC requirement in mind, outlets for our sample residence will be laid out (see *Figure 29*). In laying out these receptacle outlets, the floor line of the wall is measured (also around corners), but not across doorways, fireplaces, passageways, or other spaces where a flexible cord extended across the space would be unsuitable.

In general, duplex receptacle outlets must be no more than 12 feet apart. When spaced in this manner, a six-foot extension cord will reach a receptacle from any point along the wall line.

Note that at no point along the wall line are any receptacles more than 12 feet apart or more than six feet from any door or room opening. Where practical, no more than eight receptacles are connected to one circuit. However, *NEC Section 220-3(b)* specifies a demand factor of 180VA per receptacle or group of receptacles. Since a 15A branch circuit is rated at 1,800VA (15A × 120V = 1,800VA), 10 general purpose duplex receptacles may be connected to one circuit.

The utility room has at least one receptacle for the laundry on a separate circuit in order to comply with *NEC Section 210-52(f)*.

One duplex receptacle is located in the vestibule for cleaning purposes, such as feeding a portable vacuum cleaner or similar appliance. It is connected to the living room circuit. An additional duplex receptacle is required per *NEC Section 210-52(h)* in hallways of 10 feet or more.

Although this is not shown in the figure, the living room outlets could be split-wired (the lower half of each duplex receptacle is energized all the time, while the upper half can be switched on or off). The reason for this is that a great deal of the illumination for this area will be provided by portable table lamps, and the split-wired receptacles provide a means to control these lamps from several locations, such as at each entry to the living room, if desired. Split receptacles are discussed in more detail in the next section.

To comply with *NEC Section 210-52(b)*, the kitchen receptacles are laid out as follows. In addition to the number of branch circuits determined previously, two or more 20A small appliance branch circuits must be provided to serve all receptacle outlets (including refrigeration equipment) in the kitchen, pantry, breakfast room, dining room, or similar area of the house. Such circuits, whether two or more are used, must have no other outlets connected to them. All receptacles serving a kitchen countertop require GFCI protection. No small appliance branch circuit shall serve more than one kitchen.

To comply with *NEC Section 210-11(c)(3)*, bathroom receptacle(s) must be on a separate branch circuit supplying only bathroom receptacles or on a circuit supplying a single bathroom with no loads other than that bathroom. All receptacles located within a bathroom require GFCI protection. GFCI protection is also required on garage and exterior receptacles.

Effective January 1, 2002, all bedroom receptacles will be required to have arc-fault circuit interrupter protection to comply with *NEC Section 210-12(b)*.

10.3.0 Split-Wired Duplex Receptacles

In modern residential construction, it is common to have duplex wall receptacles that have one of the outlets wired as a standard duplex outlet (hot all the time) and the other half controlled by a wall switch. This allows table or floor lamps to be controlled by a wall switch and leaves the other outlet available for items that are not to be switched. This wiring method is commonly referred to as a *split receptacle*.

Most duplex 15A and 20A receptacles are provided with a breakoff tab that permits each of the two receptacle outlets to be supplied from a different source or polarity. For example, one outlet would be supplied from the hot leg of a series of outlets and the other outlet supplied from the switch leg of a light switch. A diagram of this arrangement is shown in *Figure 30*.

Another application of split receptacles is shown in *Figure 31*. In this example, one outlet connected from a double-pole circuit breaker supplies 240V for an appliance such as a window air conditioning unit, while the other outlet is connected from one pole of the double-pole circuit breaker and the other side is connected to the neutral or grounded conductor to supply 120V for an appliance such as a lamp. This circuit and the split receptacle mentioned above are both considered multiwire branch circuits.

10.4.0 Multiwire Branch Circuits

NEC Article 100 defines a multiwire branch circuit as "two or more ungrounded conductors having a potential difference between them, and a grounded conductor having equal potential difference between it and each ungrounded conductor of the circuit and that is connected to the neutral conductor of the system."

ELECTRICAL LEVEL ONE—TRAINEE MODULE 26112

Instructor's Notes:

_____
_____
_____
_____
_____
_____
_____
_____
_____
_____

Bathroom receptacles must be on a separate GFCI-protected circuit. *NEC Sections 210-52(d) and 210-8(a)(1)*

Bedroom 1

Bath

Bedroom 2

Utility

(A)

Bedroom 3

Living

No. 12

Kitchen/Dining

To GFCI in panel A

Carport

Receptacles located above countertops in kitchens must be mounted so that no point of the countertop is more than 24" from a receptacle. *NEC Section 210-52(c)*

Receptacles installed to serve the countertop area must be GFCI-protected. *NEC Section 210-8(a)(6)*

112F29.EPS

Figure 29 ◆ Floor plan of the sample residence.

12–41

Figure 30 ◆ Two 120V receptacle outlets supplied from different sources.

Figure 31 ◆ Combination receptacle.

Multiwire branch circuits have many advantages, such as three wires doing the work of four (in place of two two-wire branch circuits), less raceway fill, easier balancing and phasing of a system, and less voltage drop.

10.5.0 240-Volt Circuits

The electric range, clothes dryer, and water heater in the sample residence all operate at 240VAC. Each will be fed by a separate circuit and connected to a two-pole circuit breaker of the appropriate rating in the panelboard. To determine the conductor size and overcurrent protection for the range, proceed as follows:

Step 1 Find the nameplate rating of the electric range. This has previously been determined to be 12kVA.

Step 2 Refer to *NEC Table 220-19*. Since Column A of this table applies to ranges rated at 12kVA (12kW) and under, this will be the column to use in this example.

Step 3 Under the Number of Appliances column, locate the appropriate number of appliances (one in this case), and find the maximum demand given for it in Column A. Column A states that the circuit should be sized for 8kVA (not the nameplate rating of 12kVA).

Step 4 Calculate the required conductor ampacity as follows:

$$\frac{8,000VA}{240V} = 33.33A$$

The minimum branch circuit must be rated at 40A since common residential circuit breakers are

---

Instructor's Notes:

┌─────────────┐
│ THINK   │ **240V Circuits**
│ ABOUT IT   │ Calculate the ampacity required for a kitchen range with an 8kW rating. Now design the
└─────────────┘ practical wiring in a labeled diagram. How will the wires be connected at the service
                 panel and at the appliance? How will the cable be installed?

rated in steps of 15A, 20A, 30A, 40A, etc. A 30A circuit breaker is too small, so a 40A circuit breaker is selected. The conductors must have a current-carrying capacity that is equal to or greater than the overcurrent protection. Therefore, No. 8 AWG conductors will be used.

If a cooktop and wall oven were used instead of the electric range, the circuit would be sized similarly. The NEC specifies that a branch circuit for a counter-mounted cooking unit and not more than two wall-mounted ovens, all supplied from a single branch circuit and located in the same room, is computed by adding the nameplate ratings of the individual appliances and treating this total as equivalent to one range. Therefore, two appliances of 6kVA each may be treated as a single range with a 12kVA nameplate rating.

Figure 32 shows how the electric range circuit may appear on an electrical drawing. The connection may be made directly to the range junction box, but more often a 50A range receptacle is mounted at the range location and a range cord-and-plug set is used to make the connection. This facilitates moving the appliance later for maintenance or cleaning.

Figure 33 shows several types of receptacle configurations used in residential wiring applications. You will eventually recognize these configurations at a glance.

The branch circuit for the water heater in the sample residence must be sized for its full capacity because there is no diversity or demand factor

Figure 32 ◆ Range circuit shown on an electrical drawing.

for this appliance. Since the nameplate rating on the water heater indicates two heating elements of 4,500 watts each, the first inclination would be to size the circuit for a total load of 9,000 watts (volt-amperes). However, only one of the two elements

15 Amp, 125 Volts		30 Amp, 250 Volts
20 Amp, 125 Volts		30 Amp, 125/250 Volts
20 Amp, 250 Volts		50 Amp, 250 Volts
30 Amp, 125 Volts		50 Amp, 125/250 Volts

112F33.EPS

Figure 33 ◆ Residential receptacle configurations.

WIRING: RESIDENTIAL—TRAINEE MODULE 26112

12–43

Figure 34 ◆ Wiring diagram of water heater controls.

Show Transparency 36 (Figure 34).

operates at a time. See *Figure 34*. Note that each element is controlled by a separate thermostat. The lower element becomes energized when the thermostat calls for heat, and at the same time, the thermostat opens a set of contacts to prevent the upper element from operating. When the lower element's thermostat is satisfied, the lower contacts open, and at the same time, the thermostat closes the contacts for the upper element to become energized to maintain the water temperature.

With this information in hand, the circuit for the water heater may be sized as follows:

$$\frac{4{,}500VA}{240V} = 18.75A \times 1.25 = 23.44A$$

Since a water heater will more than likely fall under the continuous load category, the conductor and overcurrent protection should be rated at 125%. Therefore, No. 10 AWG wire should be used and protected with a 30A circuit breaker. A direct

connection is made to the water heater at the integral junction box on top of the heater.

The NEC specifies that electric clothes dryers must be rated at 5kVA or the nameplate rating, whichever is greater. In this case, the dryer is rated at 5.5kVA, and the conductor current-carrying capacity is calculated as follows:

$$\frac{5{,}500VA}{240V} = 22.92A$$

A three-wire, 30A circuit will be provided (No. 10 AWG wire). It is protected by a 30A circuit breaker. The dryer may be connected directly, but a 30A dryer receptacle is normally provided for the same reasons as mentioned for the electric range.

Large appliance outlets rated at 240V are frequently shown on electrical drawings using lines and symbols to indicate the outlets and circuits. In some cases, no drawings are provided.

Instructor's Notes:

_____

_____

_____

_____

_____

_____

_____

_____

_____

11.0.0 ◆ BRANCH CIRCUIT LAYOUT FOR LIGHTING

A simple lighting branch circuit requires two conductors to provide a continuous path for current flow. The usual lighting branch circuit operates at 120V; the white (grounded) circuit conductor is therefore connected to the neutral bus in the panelboard, while the black (ungrounded) circuit conductor is connected to an overcurrent protection device.

Lighting branch circuits and outlets are shown on electrical drawings by means of lines and symbols; that is, a single line is drawn from outlet to outlet and then terminated with an arrowhead to indicate a homerun to the panelboard. Several methods are used to indicate the number and size of conductors, but the most common is to indicate the number of conductors in the circuit by using slash marks through the circuit lines and then indicate the wire size by a notation adjacent to these slash marks. For example, two slash marks indicate two conductors; three slash marks indicate three conductors; etc. Some electrical designers omit slash marks for two-conductor circuits. In this case, the conductor size is usually indicated in the symbol list or legend.

The circuits used to feed residential lighting must conform to standards established by the NEC as well as by local and state ordinances. Most of the lighting circuits should be calculated to include the total load, although at times this is not possible because the electrician cannot be certain of the exact wattage that might be used by the homeowner. For example, an electrician may install four porcelain lampholders for the unfinished basement area, each to contain one 100-watt (100W) incandescent lamp. However, the homeowners may eventually replace the original lamps with others rated at 150W or even 200W. Thus, if the electrician initially loads the lighting circuit to full capacity, the circuit will probably become overloaded in the future.

It is recommended that no residential branch circuit be loaded to more than 80% of its rated capacity. Since most circuits used for lighting are rated at 15A, the total ampacity (in volt-amperes) for the circuit is as follows:

$$15A \times 120V = 1,800VA$$

Therefore, if the circuit is to be loaded to only 80% of its rated capacity, the maximum initial connected load should be no more than 1,440VA.

Figure 35 shows one possible lighting arrangement for the sample residence. All lighting fixtures are shown in their approximate physical location as they should be installed in the residence.

Electrical symbols are used to show the fixture types. Switches and lighting branch circuits are also shown by appropriate lines and symbols. The meanings of the symbols used on this drawing are explained in the symbol list in *Figure 36*.

In actual practice, the location of lighting fixtures and their related switches will probably be the extent of the information shown on working drawings. The circuits shown in *Figure 35* are meant to illustrate how lighting circuits are routed, not to imply that such drawings are typical for residential construction.

12.0.0 ◆ OUTLET BOXES

Electricians installing residential electrical systems must be familiar with outlet box capacities, means of supporting outlet boxes, and other requirements of the NEC. Boxes were discussed in detail in an earlier module, but a general review of the rules and necessary calculations is provided here.

The maximum numbers of conductors of the same size permitted in standard outlet boxes are listed in *NEC Table 370-16(a)*. These figures apply where no fittings or devices such as fixture studs, cable clamps, switches, or receptacles are contained in the box and where no grounding conductors are part of the wiring within the box. Obviously, in all modern residential wiring systems there will be one or more of these items contained in every outlet box installed. Therefore, where one or more of the above-mentioned items are present, the total number of conductors will be one less than that shown in the table.

For example, a deduction of two conductors must be made for each strap containing a wiring device entering the box (based on the largest size conductor connected to the device) such as a switch or duplex receptacle; a further deduction of one conductor must be made for one or more equipment grounding conductors entering the box (based on the largest size grounding conductor). For instance, a 3-inch × 2-inch × 2¾-inch box is listed in the table as containing a maximum of six No. 12 wires. If the box contains cable clamps and a duplex receptacle, three wires will have to be deducted from the total of six, providing for only three No. 12 wires. If a ground wire is used, which is always the case in residential wiring, only two No. 12 wires may be used.

Discuss lighting loads and layouts.

Show Transparencies 37 and 38 (Figures 35 and 36).

Discuss the NEC requirements for sizing outlet boxes.

Figure 35 ◆ Lighting layout of the sample residence.

112F35.EPS

Instructor's Notes:

○ SURFACE-MOUNTED CEILING LIGHTING FIXTURE WITH INCANDESCENT LAMP

○⊣ SURFACE-MOUNTED WALL LIGHTING FIXTURE WITH INCANDESCENT LAMP

◎ RECESSED CEILING LIGHTING FIXTURE WITH INCANDESCENT LAMP

◎→ DIRECTIONAL RECESSED CEILING LIGHTING FIXTURE WITH INCANDESCENT LAMP
ARROW INDICATES DIRECTION THAT LAMP IS POINTED

⊢○⊣ SURFACE-MOUNTED CEILING LIGHTING FIXTURE WITH FLUORESCENT LAMP

S SINGLE-POLE SWITCH

S₃ THREE-WAY SWITCH

⊡DS DOOR-ACTUATED SWITCH

112F36.EPS

Figure 36 ◆ Symbols.

For example, to size a metallic outlet box for two No. 12 AWG conductors with a ground wire, cable clamp, and receptacle, proceed as follows:

Step 1 Calculate the total number of conductors and equivalents [*NEC Section 370-16(b)*]. One ground wire plus one cable clamp plus one receptacle (two wires) plus two No. 12 conductors equals a total of six No. 12 conductors.

Step 2 Determine the amount of space required for each conductor. *NEC Table 370-16(b)* gives the box volume required for each conductor. No. 12 AWG equals 2.25 cubic inches.

Step 3 Calculate the outlet box space required by multiplying the number of cubic inches required for each conductor by the total number of conductors:

$$6 \times 2.25 = 13.5 \text{ cubic inches}$$

Step 4 Once you have determined the required box capacity, again refer to *NEC Table 370-16(a)* and note that a 3-inch × 2-inch × 2¾-inch box comes closest to our requirements. This box is rated for 14 cubic inches.

Now, size the box for two additional conductors. Where four No. 12 conductors enter the box with two ground wires, only the two additional No. 12 conductors must be added to our previous count for a total of 8 conductors (6 + 2 = 8). Remember, any number of ground wires in a box counts as only one conductor; any number of cable clamps also counts as only one conductor. Therefore, the box size required for use with two additional No. 12 conductors may be calculated as follows:

$$8 \times 2.25 = 18 \text{ cubic inches}$$

Again, refer to *NEC Table 370-16(a)* and note that a 3-inch × 2-inch × 3½-inch device box with a rated capacity of 18.0 cubic inches is the closest device box that meets NEC requirements. An alternative is to use a 4-inch × 1¼-inch square box with a single-gang plaster ring, as shown in *Figure 37*. This box also has a capacity of 18.0 cubic inches.

Other box sizes are calculated in a similar fashion. When sizing boxes for different size conductors, remember that the box capacity varies as shown in *NEC Table 370-16(b)*.

Classroom

Demonstrate how to size an outlet box and have the trainees practice sizing outlet boxes.

Audiovisual

Show Transparency 39 (Figure 37A).

Calculating Conductors

In a 4" × 4" × 1½" metal box, one 14/3 cable with ground feeds three 14/2 cables with ground wires. The red wire of the 14/3 cable feeds a receptacle, and the black wire feeds the 14/2 black wires. All of the white wires are spliced together, with one brought out to the receptacle terminal. The ground wires are all spliced, with one brought out to the grounding terminal on the receptacle and one to the ground clip on the box. All four cables are connected with box connectors rather than internal clamps. Using *NEC Section 370-16*, decide whether this wiring violates the code.

WALTER JOHNSON   112P1205.EPS

Discuss the NEC requirements for mounting outlet boxes.

Show Transparency 40 (Figure 38).

12.1.0 Mounting Outlet Boxes

Outlet box configurations are almost endless, and if you research the various methods of mounting these boxes, you will be astonished. In this section, some common outlet boxes and their mounting considerations will be reviewed.

The conventional metallic device box, which is used for residential duplex receptacles and switches for lighting control, may be mounted to wall studs using 16d (penny) nails placed through the round mounting holes passing through the interior of the box. The nails are then driven into the wall stud. When nails are used for mounting outlet boxes in this manner, the nails must be located within ¼ inch of the back or ends of the enclosure.

Nonmetallic boxes normally have mounting nails fitted to the box for mounting. Other boxes

have mounting brackets. When mounting outlet boxes with brackets, use either wide-head roofing nails or box nails about 1¼ inches in length. *Figure 38* shows various methods of mounting outlet boxes.

Before mounting any boxes during the rough wiring process, first find out what type and thickness of finish will be used on the walls. This will dictate the depth to which the boxes must be mounted to comply with NEC regulations. For example, the finish on plastered walls or ceilings is normally ½-inch thick; gypsum board or drywall is either ½-inch or ⅜-inch thick; and wood paneling is normally only ¼-inch thick. (Some tongue-and-groove wood paneling is ½-inch to ⅜-inch thick.)

The NEC specifies the amount of space permitted from the edge of the outlet box to the finished

ELECTRICAL LEVEL ONE—TRAINEE MODULE 26112

Instructor's Notes:

_____

_____

_____

_____

_____

_____

_____

_____

_____

SQUARE BOX

DEVICE BOX

CABLE CLAMP

OCTAGONAL BOX

GROUNDING CLIP, USED TO SECURE EQUIPMENT
GROUNDING CONDUCTOR TO OUTLET BOX

PLASTER RING FOR
WIRING DEVICE

FOUR-INCH BOX COVER
FOR DUPLEX RECEPTACLE
AND SWITCH

PLASTER RING FOR
LIGHTING FIXTURE

112F37A.EPS

WALTER JOHNSON

112F37B.EPS

Figure 37 ◆ Typical metallic outlet boxes with extension (plaster) rings.

Mounting Boxes

To quickly mount each box at the same height from the floor, make a simple height template (story pole) and mark it with the receptacle and switch heights. The story pole consists of an L-shaped jig made out of 2 × 2s or 2 × 4s. After installing the boxes, make sure to push the wires well back into the box so that the sheetrock installers will not damage the wires when they rout out a hole for the receptacle.

wall. When a noncombustible wall finish (such as plaster, masonry, or tile) is used, the box may be recessed ¼ inch. However, when combustible finishes are used (such as wood paneling), the box must be flush (even) with the finished wall or ceiling. See *Figure 39* and *NEC Section 370-20.*

When Type NM cable is used in either metallic or nonmetallic outlet boxes, the cable assembly, including the sheath, must extend into the box by not less than ¼ inch [*NEC Section 370-17(c)*]. In all instances, all permitted wiring methods must be secured to the boxes by means of either cable

**Show Transparency 41
(Figure 39).**

NONMETALLIC DEVICE BOX WITH INTEGRAL NAILS FOR MOUNTING DIRECTLY TO WALL STUD

NONMETALLIC FIXTURE BOX WITH ADJUSTABLE MOUNTING BRACKET FOR MOUNTING BETWEEN CEILING JOISTS

NONMETALLIC DEVICE BOX WITH SIDE BRACKET FOR MOUNTING TO FACE OF WALL STUD

NAIL HOLE

DEPTH GAUGE ON SIDE OF BOX

METALLIC DEVICE BOX

NAIL HOLE

112F38.EPS

Figure 38 ◆ Several methods of mounting outlet boxes.

MOUNTING SCREW

EDGE OF FINISHED WALL

FINISHED WALL OF CONCRETE, TILE OR OTHER NONCOMBUSTIBLE MATERIAL

Front edge of outlet box must not be set back from the finished surface more than $\frac{1}{4}$". In walls (and ceilings) constructed of wood paneling or other combustible material, the front edge of outlet boxes must be **FLUSH.**

MOUNTING SCREW

MAX. 1/4"

112F39.EPS

Figure 39 ◆ Outlet box installation.

Discuss the "Think About It."

Flush Mounting

Why is the NEC concerned with flush mounting of boxes? Is there an apparent danger from boxes that are not flush with the wall or is this just a question of good workmanship?

Instructor's Notes:

clamps or approved connectors. The one exception to this rule is where Type NM cable is used with 2¼-inch × 4-inch (or smaller) nonmetallic boxes where the cable is fastened within eight inches of the box. In this case, the cable does not have to be secured to the box. See *NEC Section 370-17(c), Exception.*

13.0.0 ◆ WIRING DEVICES

Wiring devices include various types of receptacles and switches, the latter being used for lighting control. Switches are covered in *NEC Article 380,* while regulations for receptacles may be found in *NEC Section 210-7* and *NEC Article 410, Part L.*

13.1.0 Receptacles

Receptacles are rated by voltage and amperage capacity. *NEC Section 210-7* requires that receptacles connected to a 15A or 20A circuit have the correct voltage and current rating for the application, and be of the grounding type.

Where there is only one outlet on a circuit, the receptacle's rating must be equal to or greater than the capacity of the conductors feeding it. For example, if one receptacle is connected to a 20A residential laundry circuit, the receptacle must be rated at 20A or more. When more than one outlet is on a circuit, the total connected load must be equal to or less than the capacity of the branch circuit conductors feeding the receptacles.

Refer to *Figure 40* for some of the characteristics of a standard 125V, 15A duplex receptacle. Note that the terminals are color coded as follows:

- *Green*—Connection for the equipment grounding conductor
- *Silver*—Connection for the neutral or grounded conductor
- *Brass*—Connection for the ungrounded conductor

A standard 125V, 15A receptacle is also typically imprinted with the following symbols:

- *UL*—Underwriters' Laboratories listing
- *CSA*—Canadian Standards Association
- *CO/ALR*—Designed for use with both copper and aluminum wire
- *15A*—Receptacle rated for a maximum of 15A
- *125V*—Receptacle rated for a maximum of 125V

The UL label means that the receptacle has undergone testing by Underwriters' Laboratories, Inc., and meets minimum safety requirements. Underwriters' Laboratories was created by the National Board of Fire Underwriters to test electrical devices and materials. The UL label is a safety rating only and does not mean that the device or equipment meets any type of quality standard. The CSA label means that the receptacle is approved by the Canadian Standards Association, the Canadian equivalent to Underwriters' Laboratories.

CO/ALR DESIGNATION INDICATES THAT THE SWITCH IS DESIGNED FOR USE WITH BOTH COPPER AND ALUMINUM WIRE

PLASTER EARS

LONGER SLOT INDICATES NEUTRAL OR GROUNDED CONDUCTOR

SLOT FOR MOUNTING SCREWS

SMALL SLOT INDICATES UNGROUNDED CONDUCTOR

NEUTRAL OR GROUNDED TERMINALS ARE INDICATED BY SILVER-COLORED SCREWS

CANADIAN STANDARDS ASSOC.

AMPERAGE/VOLTAGE RATING

UNGROUNDED CONDUCTORS ARE CONNECTED TO THE BRASS SCREWS ON OPPOSITE SIDE FROM GROUNDED CONDUCTOR SCREWS

UNDERWRITERS' LABORATORIES LISTING

GROUNDING CONNECTION HAS GREEN SCREW HEAD

GROUNDING SLOT

112F40.EPS

Figure 40 ◆ Standard 125V, 15A duplex receptacle.

Discuss lighting control devices.

Show Transparency 43 (Figure 41).

The CSA label means that the receptacle is acceptable for use in Canada.

The CO/ALR symbol means that the device is suitable for use with copper, aluminum, or copper-clad aluminum wire. The CO in the symbol stands for *copper* while ALR stands for *aluminum revised*. The CO/ALR symbol replaces the earlier CU/AL mark, which appeared on wiring devices that were later found to be inadequate for use with aluminum wire in the 15A to 20A range. Therefore, any receptacle or wall switch marked with the CU/AL configuration or anything other than CO/ALR should be used only for copper wire.

These same configurations also apply to wall switches used for lighting control. These will be discussed next.

14.0.0 ◆ LIGHTING CONTROL

There are many types of lighting control devices. These devices have been designed to make the best use of the lighting equipment provided by the lighting industry. They include:

- Automatic timing devices for outdoor lighting
- Dimmers for residential lighting
- Common single-pole, three-way, and four-way switches

For the purposes of this module, a switch is defined as a device that is used on branch circuits to control lighting. Switches fall into the following basic categories:

- Snap-action switches
- Quiet switches

Snap-action switches—A single-pole snap-action switch consists of a device containing two stationary current-carrying elements, a moving current-carrying element, a toggle handle, a spring, and a housing. When the contacts are open, as shown in *Figure 41*, the circuit is broken and no current flows. When the moving element is closed by

Figure 41 ◆ Switch operation, contacts open.

112F41.EPS

Instructor's Notes:

manually flipping the toggle handle, the contacts complete the circuit and the lamp is energized. See *Figure 42*.

Quiet switches—The quiet switch (*Figure 43*) is the most common switch for use in lighting applications. Its operation is much quieter than the snap-action switch.

The quiet switch consists of a stationary contact and a moving contact that are close together when the switch is open. Only a short, gentle movement is required to open and close the switch, producing very little noise. This type of switch may be used only on alternating current.

Quiet switches are common for loads from 10A to 20A, and are available in single-pole, three-way, and four-way configurations.

Many other types of switches are available for lighting control. One type of switch used mainly in residential occupancies is the door-actuated switch. It is generally installed in the door jamb of a closet to control a light inside the closet. When the door is open, the light comes on; when the door is closed, the light goes out. Most refrigerator and oven lights are also controlled by door-actuated switches.

Combination switch/indicator light assemblies are available for use where the light cannot be seen from the switch location, such as an attic or garage. Switches are also made with small neon lamps in the handle that light when the switch is off. These low-current-consuming lamps make the switches easy to find in the dark.

14.1.0 Three-Way Switches

Three-way switches are used to control one or more lamps from two different locations, such as at the top and bottom of stairways, in a room that has two entrances, etc. A typical three-way switch is shown in *Figure 44*.

A three-way switch has three terminals. The single terminal at one end of the switch is called

Show Transparencies 44 and 45 (Figures 42 and 43).

Discuss three-way switches and their applications. Pass around an example for the trainees to examine.

Show Transparency 46 (Figure 44).

Figure 42 ◆ Switch operation, contacts closed.

Figure 43 ◆ Characteristics of a single-pole quiet switch.

Figure 44 ◆ Typical three-way switch.

Discuss the operation of three-way switches.

Show Transparency 47 (Figure 45).

the *common* or *hinge point*. This terminal is easily identified because it is darker than the other two terminals. The feeder (hot wire) or switch leg is always connected to the common dark or black terminal. The two remaining terminals are called *traveler terminals*. These terminals are used to connect three-way switches together.

The connection of two three-way switches is shown in *Figure 45*. By means of the two switches, it is possible to control the lamp from two locations. By tracing the circuit, it may be seen how these three-way switches operate.

A 120V circuit emerges from the left side of the drawing. The white or neutral wire connects di-

rectly to the neutral terminal of the lamp. The hot wire carries current, in the direction of the arrows, to the common terminal of the three-way switch on the left. Since the handle is in the up position, the current continues to the top traveler terminal and is carried by this traveler to the other three-way switch. Note that the handle is also in the up position on this switch; this picks up the current flow and carries it to the common point, which continues to the ungrounded terminal of the lamp to make a complete circuit. The lamp is energized.

Moving the handle to a different position on either three-way switch will break the circuit, which in turn deenergizes the lamp. For example, let's

ELECTRICAL LEVEL ONE—TRAINEE MODULE 26112

Instructor's Notes:

ELECTRICAL LEVEL ONE—INSTRUCTOR'S GUIDE MODULE 26112

WHITE WIRE (NEUTRAL)

120-VOLT SUPPLY CIRCUIT

BLACK WIRE (HOT)

UP   UP

DOWN   DOWN

THREE-WAY SWITCH   THREE-WAY SWITCH

112F45.EPS

Figure 45 ◆ Three-way switches in the ON position; both handles are up.

WHITE WIRE (NEUTRAL)

120-VOLT SUPPLY CIRCUIT

BLACK WIRE (HOT)

UP   UP

DOWN   DOWN

THREE-WAY SWITCH   THREE-WAY SWITCH

112F46.EPS

Figure 46 ◆ Three-way switches in the OFF position; one handle is down, one handle is up.

say a person leaves the room at the point of the three-way switch on the left, and the switch handle is flipped down, as shown in *Figure 46*. Note that the current flow is now directed to the bottom traveler terminal, but since the handle of the three-way switch on the right is still in the up position, no current will flow to the lamp.

If another person enters the room at the location of the three-way switch on the right, and the handle is flipped downward, as shown in *Figure 47*, this change provides a complete circuit to the lamp, which causes it to be energized. In this ex-

ample, current flow is on the bottom traveler. Again, changing the position of the switch handle (pivot point) on either three-way switch will deenergize the lamp.

In actual practice, the exact wiring of the two three-way switches to control the operation of a lamp will be slightly different from the routing shown in these three diagrams. There are several ways that two three-way switches may be connected. One solution is shown in *Figure 48*. In this case, two-wire, Type NM cable is fed to the three-way switch on the left.

Audiovisual

Show Transparencies 48 through 50 (Figures 46 through 48).

WHITE WIRE (NEUTRAL)

120-VOLT SUPPLY CIRCUIT

BLACK WIRE (HOT)

UP   UP

DOWN   DOWN

THREE-WAY SWITCH   THREE-WAY SWITCH

112F47.EPS

Figure 47 ◆ Three-way switches with both handles down; the light is energized.

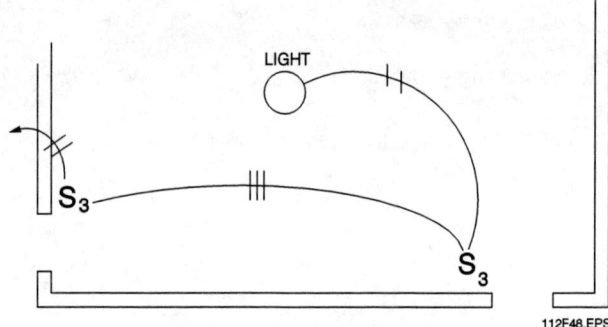

LIGHT

S₃

S₃

112F48.EPS

Figure 48 ◆ Method of showing the wiring arrangement on a floor plan.

The black or hot conductor is connected to the common terminal on the switch, while the white or neutral conductor is spliced to the white conductor of the three-wire, Type NM cable leaving the switch. This three-wire cable is necessary to carry the two travelers plus the neutral to the three-way switch on the right. At this point, the black and red wires connect to the two traveler terminals, respectively. The white or neutral wire is again spliced—this time to the white wire of another two-wire, Type NM cable. The neutral wire is never connected to the switch itself. The black wire of the two-wire, Type NM cable connects to the common terminal on the three-way switch.

This cable, carrying the hot and neutral conductors, is routed to the lighting fixture outlet for connection to the fixture.

Another solution is to feed the lighting fixture outlet with two-wire cable. Run another two-wire cable carrying the hot and neutral conductors to one of the three-way switches. A three-wire cable is pulled between the two three-way switches, and then another two-wire cable is routed from the other three-way switch to the lighting fixture outlet.

Some electricians use a shortcut method that eliminates one of the two-wire cables in the preceding method. In this case, a two-wire cable is

ELECTRICAL LEVEL ONE—TRAINEE MODULE 26112

Instructor's Notes:

ELECTRICAL LEVEL ONE—INSTRUCTOR'S GUIDE MODULE 26112

Wiring Three-Way Switches

THINK ABOUT IT

Using a schematic drawing, explain the actual wiring of two different three-way switches, one in which the load and supply come in from different boxes, and the other in which the load and supply come in from the same box. Be specific about which wires connect to which terminals.

Discuss the "Think About It."

Show Transparency 51 (Figure 49).

Discuss four-way switches and their applications.

Show Transparencies 52 and 53 (Figures 50 and 51).

run from the lighting fixture outlet to one three-way switch. Three-wire cable is pulled between the two three-way switches—two of the wires for travelers and the third for the common point return. This method is shown in *Figure 49*.

14.2.0 Four-Way Switches

Two three-way switches may be used in conjunction with any number of four-way switches to control a lamp, or a series of lamps, from any number of positions. When connected correctly, the actuation of any one of these switches will change the operating condition of the lamp (i.e., turn the lamp either on or off).

Figure 50 shows how a four-way switch may be used in combination with two three-way switches to control a device from three locations. In this ex-

ample, note that the hot wire is connected to the common terminal on the three-way switch on the left. Current then travels to the top traveler terminal and continues on the top traveler conductor to the four-way switch. Since the handle is up on the four-way switch, current flows through the top terminals of the switch and into the traveler conductor going to the other three-way switch.

Again, the switch is in the up position. Therefore, current is carried from the top traveler terminal to the common terminal and then to the lighting fixture to energize it.

If the position of any one of the three switch handles is changed, the circuit will be broken and no current will flow to the lamp. For example, assume that the four-way switch handle is flipped downward. The circuit will now appear as shown in *Figure 51*, and the light will be out.

Figure 49 ◆ One way to connect a pair of three-way switches to control one lighting fixture.

WIRING: RESIDENTIAL—TRAINEE MODULE 26112 12–57

Figure 50 ◆ Three- and four-way switches used in combination; the light is on.

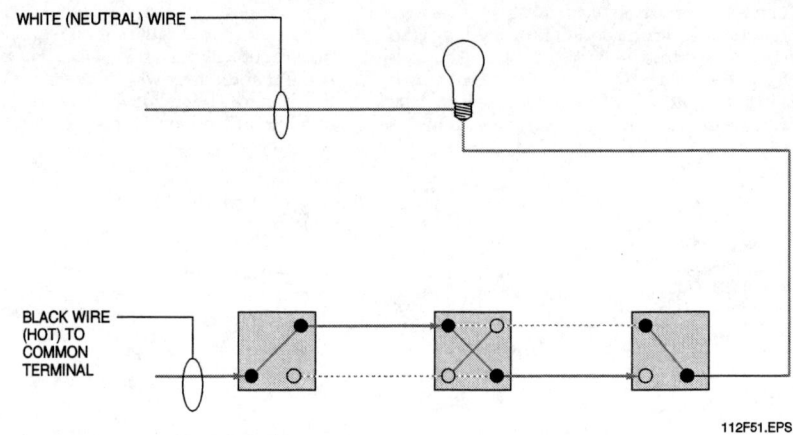

Figure 51 ◆ Three- and four-way switches used in combination; the light is off.

Discuss photoelectric switches and their applications.

Discuss relays and their applications.

Remember, any number of four-way switches may be used in combination with two three-way switches, but two three-way switches are always necessary for the correct operation of one or more four-way switches.

14.3.0 Photoelectric Switch

The chief application of the photoelectric switch is to control outdoor lighting, especially the dusk-to-dawn lights found in suburban areas. This switch has an endless number of possible uses and is a great tool for electricians dealing with outdoor lighting situations.

14.4.0 Relays

Next to switches, relays play the most important part in the control of light. However, the design and application of relays is a study in itself, and

12–58

Instructor's Notes:

---

they are far beyond the scope of this module. Still, a brief mention of relays is necessary to round out your knowledge of lighting controls.

An electric relay is a device whereby an electric current causes the opening or closing of one or more pairs of contacts. These contacts are usually capable of controlling much more power than is necessary to operate the relay itself. This is one of the main advantages of relays.

One popular use of the relay in residential lighting systems is that of remote control lighting. In this type of system, all relays are designed to operate on a 24V circuit and are used to control 120V lighting circuits. They are rated at 20A, which is sufficient to control the full load of a normal lighting branch circuit, if desired.

Remote control switching makes it possible to install a switch wherever it is convenient and practical to do so or wherever there is an obvious need for a switch, no matter how remote it is from the lamp or lamps it is to control. This method enables lighting designs to achieve new advances in lighting control convenience at a reasonable cost. Remote control switching is also ideal for rewiring existing homes with finished walls and ceilings.

One relay is required for each fixture or each group of fixtures that are controlled together. Switch locations for remote control follow the same rules as for conventional direct switching. However, since it is easy to add switches to control a given relay, no opportunities should be overlooked for adding a switch to improve the convenience of control.

Remote control lighting also has the advantage of using selector switches at central locations. For example, selector switches located in the master bedroom or in the kitchen of a home enable the owner to control every lighting fixture on the property from this location. For example, the selector switch may be used to control outside or basement lights that might otherwise be left on inadvertently.

14.5.0 Dimmers

Dimming a lighting system provides control of the quantity of illumination. It may be done to create certain moods or to blend the lighting from different sources for various lighting effects.

For example, in homes with formal dining rooms, a chandelier mounted directly above the dining table and controlled by a dimmer switch becomes the centerpiece of the room while providing general illumination. The dimmer adds versatility since it can set the mood for the activity—low brilliance (candlelight effect) for formal dining or bright for an evening of playing cards. When chandeliers with exposed lamps are used, the dimmer is essential to avoid a garish and uncomfortable atmosphere. The chandelier should be sized in proportion to the dining area.

Note
It is very important that dimmers be matched to the wattage of the application. Check the manufacturer's data.

14.6.0 Switch Locations

Although the location of wall switches is usually provided for convenience, the NEC also stipulates certain mandatory locations for lighting fixtures and wall switches. These locations are deemed necessary for added safety in the home for both the occupants and service personnel.

For example, the NEC requires adequate light in areas where HVAC equipment is placed. Furthermore, these lights must be conveniently controlled so that homeowners and service personnel do not have to enter a dark area where they might come in contact with dangerous equipment. Three-way switches are required under certain conditions. The NEC also specifies regulations governing lighting fixtures in clothes closets, along with those governing lighting fixtures that may be mounted directly to the outlet box without further support. *Figure 52* summarizes some of the NEC requirements for light and switch placement in the home. For further details, refer to the appropriate sections in the NEC.

14.7.0 Low-Voltage Electrical Systems

Conventional lighting systems operate and are controlled by the same system voltage, generally 120V in residential lighting circuits. The NEC permits the use of low-voltage systems to control lighting circuits. There are some advantages to low-voltage systems. One advantage is that the control of lighting from several different locations is more easily accomplished, such as with the remote control system discussed earlier. For example, outside flood lighting can be controlled from several different rooms in a house. The cost of the control wiring is less in that it is rated for a lower voltage and only carries a minimum amount of

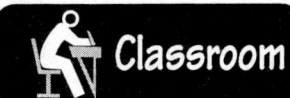

Discuss dimmers and their applications. Pass around an example for the trainees to examine.

Discuss the NEC requirements for switch locations.

Show Transparency 54 (Figure 52).

Discuss the advantages of low-voltage control systems.

THREE-WAY SWITCH
AT EACH FLOOR LEVEL

Where six or more steps separate floor levels, a wall switch is required at each level to control stairway lighting.
NEC Section 210-70(a)

SIX OR MORE STEPS

Lighting fixtures weighing over 50 pounds must be supported independently of the outlet box unless the box is listed for the weight to be supported.
NEC Section 410-16

Install at least one switch-controlled lighting outlet at or near equipment requiring servicing.
NEC Section 210-70(c)

Pendant-mounted lighting fixtures cannot be used in a residential clothes closet.
NEC Section 410-8(c)

HVAC EQUIPMENT

ATTIC

WALL SWITCH

WALL SWITCH
NEC Sections 210-70(a) and (c)

FURNACE ROOM

WALL SWITCH

UNFINISHED BASEMENT

CRAWL SPACE

WALL SWITCH

ENTRANCE

HVAC EQUIPMENT

HVAC EQUIPMENT

112F52.EPS

Figure 52 ◆ NEC requirements for light and switch placement.

Instructor's Notes:

_____

_____

_____

_____

_____

_____

_____

_____

_____

current compared to a standard lighting system. When extensive or complex lighting control is required, low-voltage systems are preferred. Also, since these circuits are low-energy circuits, circuit protection is not required.

14.7.1 NEC Requirements for Low-Voltage Systems

NEC Article 725 governs the installation of low-voltage system wiring. These provisions apply to remote control circuits, low-voltage relay switching, low-energy power circuits, and low-voltage circuits. The NEC divides these circuits into three categories:

• Remote control
• Signaling
• Power-limited circuits

As mentioned earlier, circuit protection of the low-voltage circuit is not required; however, the high-voltage side of the transformer that supplies the low-voltage system must be protected. *NEC Chapter 9, Tables 11(a) and 11(b)* cover circuits that are inherently limited in power output and therefore require no overcurrent protection or are limited by a combination of power source and overcurrent protection.

There are a number of requirements of the power systems described in *NEC Chapter 9, Tables 11(a) and 11(b)* and the notes preceding the tables. You should read and study all applicable portions of the NEC before installing low-voltage power systems.

14.7.2 Relays

A low-voltage remote control wiring system is relay operated. The relay is controlled by a low-voltage switching system and in turn it controls the power circuit that is connected to it (*Figure 53*). The low-voltage, split coil relay is the heart of the low-voltage remote control system (*Figure 54*). When the ON coil of the relay is energized, the solenoid mechanism causes the contacts to move into the ON position to complete the power circuit. The contacts stay in this position until the OFF coil is energized. When this occurs, the contacts are withdrawn and the power circuit is opened. The red wire is the ON wire; the black wire is the OFF wire, and the blue wire is common to the transformer.

The low-voltage relay is available in two mounting styles. One type is designed to mount through a ½-inch knockout opening (*Figure 55*). For a ¾-inch knockout, a rubber grommet is inserted to isolate the relay from the metal. This practice should ensure quieter relay operation. The second relay mounting style is the plug-in relay. This type of relay is used in an installation where several relays are mounted in one enclosure. The advantage of the plug-in relay is that it plugs directly into a busbar. As a result, it is not necessary to splice the line voltage leads.

14.7.3 Switches

The switch used in the low-voltage remote control system is a normally open, single-pole, double-throw, momentary contact switch, as shown in

Discuss the NEC requirements for low-voltage systems.

Discuss the relays used in low-voltage systems.

Show Transparencies 55 through 57 (Figures 53 through 55).

Discuss the switches used in low-voltage systems.

Figure 53 ◆ Low-voltage remote control system.

LINE
VOLTAGE
LEADS

BLACK (OFF)

BLUE (COMMON)

RED (ON)

112F54.EPS

Figure 54 ◆ Low-voltage relay diagram.

GROMMET

112F55.EPS

Figure 55 ◆ Low-voltage relay.

**Show Transparency 58
(Figure 56).**

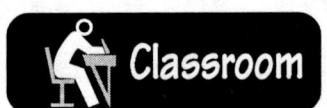

**Discuss the NEC
requirements for wiring
low-voltage systems.**

**Discuss the advantages of
electric heating systems.**

Figure 56. This switch is approximately one-third the size of a standard single-pole switch.

In a low-voltage switch, the white wire is common and is connected to the 24V transformer source. The red wire connects to the ON circuit, and the black wire connects to the OFF circuit.

14.7.4 Wiring Methods

NEC Article 725 governs the installation of low-voltage systems. This article covers remote control circuits, low-voltage relay switching, low-energy power circuits, and low-voltage circuits.

15.0.0 ◆ ELECTRIC HEATING

The use of electric heating in residential occupancies has risen tremendously over the past decade or so, and the practice will no doubt continue. This is due to the following advantages of electric heat over most other heating systems:

- Electric heat is noncombustible and is therefore safer than combustible fuels.
- It requires no storage space, fuel tanks, or chimneys.
- It requires little maintenance.

12–62

ELECTRICAL LEVEL ONE—TRAINEE MODULE 26112

Instructor's Notes:

_____

_____

_____

_____

_____

_____

_____

_____

_____

BLACK

WHITE

RED

112F56.EPS

Figure 56 ◆ Single-pole, double-throw switch.

- The initial installation cost is relatively inexpensive when compared to other types of heating systems.
- The comfort level may be improved since each room may be controlled separately by its own thermostat.

There are also some disadvantages to using electric baseboard heat, especially in northern climates. Some of these disadvantages include:

- Electric heat is often more expensive to operate than other types of fuels.
- Receptacles must not be installed above electric baseboard heaters.
- Electric baseboard heaters tend to discolor the wall area immediately above the heater, especially if there are smokers in the home.

> **Note**
>
> It is very important to calculate the extra electric load of an electric heater installation (especially in an add-on situation). Ensure that the extra load does not exceed the maximum amperage draw of either the circuit or the panel.

The type of electric heating system used for a given residence will usually depend on the structural conditions, the kind of room, and the activities for which the room will be used. The homeowner's preference will also enter into the final decision.

Electric heating equipment is available in baseboard, wall, ceiling, kick space, and floor units; in resistance cable embedded in the ceiling or concrete floor; in forced-air duct systems similar to conventional oil- or gas-fired hot air systems; and in electric boilers for hot water baseboard heat.

Electric heat pumps have also become popular for residential heating, ventilating, and air conditioning (HVAC) systems in certain parts of the country. The term *heat pump,* as applied to a year-round air conditioning system, commonly denotes a system in which refrigeration equipment is used in such a manner that heat is taken from a heat source and transferred to the conditioned space when heating is desired; heat is removed from the space and discharged to a heat sink when cooling and dehumidification are desired.

A heat pump has the unique ability to furnish more energy than it consumes. This is due to the fact that under certain outdoor conditions, electrical energy is required only to move the refrigerant and run the fan; thus, a heat pump can attain a heating efficiency of two or more to one; that is, it will put out an equivalent of two or three watts of heat for every watt consumed. For this reason, its use is highly desirable for the conservation of energy.

In general, electric baseboard heating equipment should be located on the outside wall near the areas where the greatest heat loss will occur, such as under windows, etc. The controls for wall-mounted thermostats should be located on an interior wall, about 50 inches above the floor to sense the average room temperature. *Figure 57* shows an electric heating arrangement for the sample residence. NEC regulations governing the installation of these units are also noted.

16.0.0 ◆ RESIDENTIAL SWIMMING POOLS, SPAS, AND HOT TUBS

The NEC recognizes the potential danger of electric shock to persons in swimming pools, wading pools, and therapeutic pools, or near decorative pools or fountains. This shock could occur from electric potential in the water itself or as a result of a person in the water or a wet area touching an enclosure that is not at ground potential.

Show Transparency 59 (Figure 57).

Discuss the NEC requirements for pools, spas, and hot tubs.

750VA

Bedroom No. 2

2,000VA

Bedroom No. 1

2,000VA

750VA

650VA

Bedroom No. 3

2,000VA

Conductors must be sized for 125% of the heater's nameplate rating.
NEC Section 424-3(b)

No. 10 AWG

Single-pole thermostats that do not break all ungrounded conductors may not be used as a disconnecting means.
NEC Section 424-20(b)

If a thermostat is to serve as both controller and disconnecting means, it must be provided with a marked "OFF" position.
NEC Section 424-20(a)

Branch circuits supplying electric space heating may be rated for 15, 20, or 30 amperes, depending upon the load.
NEC Section 424-3(a)

Permanently-installed baseboard heaters with a factory-installed receptacle outlet is permitted in lieu of a receptacle outlet.
NEC Section 210-52

2,000VA

2,000VA

Living Room

1,500VA

Kitchen

2,000VA kickspace heater w/ integral thermostat

112F57.EPS

Figure 57 ◆ Electric heating arrangement for the sample residence.

Instructor's Notes:

_____

_____

_____

_____

_____

_____

_____

_____

_____

_____

Pools

Pools are a common site for homeowners to add lights, receptacles, or heaters, and code violations are common. If you are called in for a service problem, ask the homeowner about any do-it-yourself wiring.

Residential Wiring

Make a mental wiring tour of your home. Picture several rooms, including the kitchen and utility/laundry room. How is each device connected to the power source, and what is the probable amperage and overcurrent protection? What other devices might or might not be included in the circuit? How many branch circuits serve each room? Later, examine the panelboard and exposed wiring to see how accurately you identified the branch circuits.

Discuss the importance of correct wiring practices in pools, spas, fountains, etc.

Discuss the "Think About It."

Show Transparencies 60 and 61 (Figures 58 and 59).

Accordingly, the NEC provides rules for the safe installation of electrical equipment and wiring in or adjacent to swimming pools and similar locations. *NEC Article 680* covers the specific rules governing the installation and maintenance of swimming pools, spas, and hot tubs.

The electrical installation procedures for hot tubs and swimming pools are too vast to be covered in detail in this module. However, the general requirements for the installation of outlets, overhead fans and lighting fixtures, and other items are summarized in *Figure 58*.

Besides *NEC Article 680,* another good source for learning more about electrical installations in and around swimming pools is from manufacturers of swimming pool equipment, including those who manufacture and distribute underwater lighting fixtures. Many of these manufacturers offer pamphlets detailing the installation of their equipment with helpful illustrations, code explanations, and similar details. This literature is usually available at little or no cost to qualified personnel. You can write directly to manufacturers to request information about available literature, or contact your local electrical supplier or contractor who specializes in installing residential swimming pools. See *Figure 59.*

Figure 58 ◆ Summary of NEC requirements for packaged indoor hot tubs.

All 125-volt receptacles located within 20 feet of the inside walls of the pool must be protected by a ground fault circuit interrupter.

Junction boxes and enclosures for transformers or ground fault circuit interrupters that are connected directly to a forming shell must be equipped with threaded hubs or bosses. They must be constructed of brass, copper, or suitable plastic, or other approved corrosion-resistant material.

Receptacles on the pool property must be located a minimum distance of 10 feet from the inside walls of the pool.

Wall switches must be located at least 5 feet away from the inside wall of the pool.

At least one 125-volt convenience receptacle must be installed and located at a minimum of 10 feet and not more than 20 feet from the inside wall of the pool.

Pool area

Fence

Overhead electrical conductors must not be installed above the pool area extending 10 feet horizontally from inside of pool wall, over a diving structure, or over observation stands, towers, or platforms.
NEC Section 680-8

Receptacles on the pool property must be located a minimum distance of 10 feet from the inside walls of the pool.

18 inches minimum from water level to top of fixture lens.

Normal water level

Utility owned, operated, and maintained communication conductors, cable TV, etc. complying with *NEC Article 820* are permitted provided the cable is at least 10 feet above pool, observation stands, diving boards, etc.
NEC Section 680-8

Transformer must be a two-winding type having a grounded metal barrier between the primary and secondary windings.

Primary winding

Requires GFCI and grounded conductor

Secondary winding

Low-voltage line to supply underwater lighting fixtures.

High-voltage line to power supply

NEC Section 680-5(a)

Transformer, together with the transformer enclosure, must be identified for the purpose of providing power to underwater lighting fixtures.

112F59.EPS

Figure 59 ◆ Summary of NEC regulations for typical swimming pool installations.

Instructor's Notes:

Summary

This module covered the basics of residential wiring, including switches, outlets, conductor sizes, and service sizes. A thorough knowledge of the NEC is essential to the successful installation of residential wiring systems.

Review Questions

1. When sizing electrical services, at what percentage is the first 3,000VA rated?
   a. 20%
   b. 45%
   c. 80%
   d. 100%

2. What section of the NEC requires that fittings be identified for use with service masts?
   a. *NEC Section 230-28*
   b. *NEC Section 230-40*
   c. *NEC Section 250-46*
   d. *NEC Section 250-83*

3. A service conductor without an outer jacket must be _____ feet above a window that can be opened.
   a. two feet
   b. three feet
   c. eight feet
   d. ten feet

4. *NEC Section 230-6* considers conductors installed under at least _____ inch(es) of concrete to be outside the building.
   a. one
   b. two
   c. four
   d. five

5. Type NM cable may not be used in _____ .
   a. buildings over three stories above grade
   b. the framework of a building
   c. protective strips
   d. attic spaces

6. Type AC cable may not be used in _____ .
   a. concrete or plaster where dry
   b. dry masonry
   c. attic spaces
   d. wet locations

7. Type SE cable is available with _____ for interior wiring systems.
   a. a non-insulated ground or neutral conductor
   b. an insulated grounded conductor
   c. no ground conductor
   d. none of the above

8. Type SER cable may be used _____ .
   a. in overhead applications
   b. underground
   c. as a subfeed under certain conditions
   d. none of the above

9. Identify the following receptacles by number.
   a. _____ single receptacle
   b. _____ split-wired receptacle
   c. _____ dryer receptacle
   d. _____ range receptacle
   e. _____ duplex receptacle
   f. _____ special-purpose outlet

112-RQ.EPS

Have the trainees complete the Review Questions and go over the answers prior to administering the Module Examination.

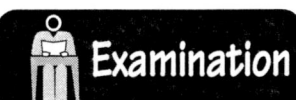

Administer the Module Examination and Performance Test. Be sure to record the testing results on Craft Training Report Form 200 and submit the results to the Training Program Sponsor.

10. Calculate the cubic inches required for the receptacle outlets shown in the table below.

Number and Size of Conductors in Box	Free Space Within Box for Each Conductor	Total Cubic Inches of Box Space Required	What Size Metallic Box May Be Used?
Six No. 12 conductors and three ground wires	2.25	_____	_____
Seven No. 12 conductors and three ground wires with one receptacle	2.25	_____	_____
Two No. 14 conductors and one ground wire	2.0	_____	_____
Four No. 14 conductors and two ground wires	2.0	_____	_____
Six No. 14 conductors and three ground wires with one receptacle	2.0	_____	_____

Instructor's Notes:

_____

_____

_____

_____

_____

_____

_____

_____

_____

Dan Lamphear, Associated Builders and Contractors, Inc.

Like many of other people, Dan Lamphear just fell into his career as an electrician. But once he discovered the electrical trade, he knew he had found a home. Since then, he has progressed from a helper to an apprentice, journeyman, independent contractor, inventor, and finally, a teacher.

It was as much luck as anything else that led Dan Lamphear toward a career as a professional electrician more than two decades ago. Lamphear wasn't sure what he wanted to do with his life after graduating from a suburban Milwaukee high school. But after watching an electrician perform a commercial wiring job at a friend's business—and providing a helping hand—he was hooked.

"It seemed like a challenging career, and I was curious to learn more about how electricity works," he recalls.

Lamphear was hired by that same electrician, under whom he apprenticed for several years before hearing about the NCCER program. He jumped at the chance to further his skills through the program. "Like they say, knowledge is money," he smiles.

After graduating from the program, Dan struck out on his own as an independent electrician, specializing in plant maintenance, industrial, and commercial work. His ability to diagnose and repair electrical problems in factory machinery soon made him a valuable contractor in Milwaukee's industrial sector.

He also discovered his knack for invention, and he has designed and built specialized machinery for a company that hired him as its full-time electrical maintenance supervisor. "Knowing the electrical side of machinery allowed me to understand how they operate mechanically," he says of his work as an inventor.

Dan later returned to the Associated Builders and Contractors (ABC) chapter, which trains out of a local community college, to repay the favor that helped him embark on his career. He teaches Electrical Level 2 courses for Standardized Craft Training students who represent the next generation of professional electricians.

"Knowing how to use test instruments is perhaps the most important aspect of the job," he notes. "I still have some of the same meters I started out with—they're like old friends."

Trade Terms Introduced in This Module

Appliance: Equipment designed for a particular purpose (i.e., using electricity to produce heat, light, mechanical motion, etc.). Appliances are usually self-contained, are generally available for applications other than industrial use, and are normally produced in standard sizes or types.

Armored (Type AC) cable: Cable that consists of wires with a spiral-wound, flexible steel outer jacketing. See *BX®*.

Bonding bushing: A special conduit bushing equipped with a conductor terminal to take a bonding jumper. It also has a screw or other sharp device to bite into the enclosure wall to bond the conduit to the enclosure without a jumper when there are no concentric knockouts left in the wall of the enclosure.

Bonding jumper: A bare or green insulated conductor used to ensure the required electrical conductivity between metal parts required to be electrically connected. Bonding jumpers are frequently used from a bonding bushing to the service-equipment enclosure to provide a path around concentric knockouts in an enclosure wall, and they may also be used to bond one raceway to another.

Branch circuit: The portion of a wiring system extending beyond the final overcurrent device protecting a circuit.

BX®: A name for armored cable; although used generically, BX® is a registered trademark of the General Electric Company.

Feeder: A circuit, such as conductors in conduit or a cable run, that carries current from the service equipment to a subpanel or a branch circuit panel or to some point in the wiring system.

Load center: A type of panelboard that is normally located at the service entrance of a residential installation. It usually contains the main disconnect.

Nonmetallic-sheathed (Type NM) cable: A type of cable that is popular for use in residential and small commercial wiring systems. In general, it may be used for both exposed and concealed work in normally dry locations. See *Romex®*.

Romex®: General Cable's trade name for Type NM cable; however, it is often used generically to refer to any nonmetallic-sheathed cable.

Roughing in: The first stage of an electrical installation, when the raceway, cable, wires, boxes, and other equipment are installed. This is the electrical work that must be done before any finishing work can be done.

Service drop: The overhead conductors, through which electrical service is supplied, between the last power company pole and the point of their connection to the service facilities located at the building.

Service entrance: The point where power is supplied to a building (including the equipment used for this purpose). The service entrance includes the service main switch or panelboard, metering devices, overcurrent protective devices, and conductors/raceways for connecting to the power company's conductors.

Instructor's Notes:

_____

_____

_____

_____

_____

_____

_____

_____

_____

Service-entrance conductors: The conductors between the point of termination of the overhead service drop or underground service lateral and the main disconnecting device in the building.

Service-entrance equipment: Equipment that provides overcurrent protection to the feeder and service conductors, a means of disconnecting the feeders from energized service conductors, and a means of measuring the energy used.

Service lateral: The underground conductors through which service is supplied between the power company's distribution facilities and the first point of their connection to the building or area service facilities located at the building.

Switch: A mechanical device used for turning an electrical circuit on and off.

Switch leg: A circuit routed to a switch box for controlling electric lights.

Reference Information—
Other Codes and Electrical
Standards That Apply to
Electrical Installations

For your reference, information is included here on several other codes and standards that apply to electrical installations in residential structures. It is suggested that to be thoroughly competent in the electrical trade, you should become familiar with the contents of these codes and the terminology used in them. Listed below are the building codes referred to, as well as their addresses. Also included is a list of definitions and phrases that are of particular interest in residential electrical construction.

> International Conference of Building
> Officials (ICBO)
>   5360 South Workman Mill Road
>   Whittier, California 90601
>   (562) 699-0541
> Building Official & Code Administrators
> International, Inc. (BOCA)
>   4051 West Flossmoor Road
>   Country Club Hills, Illinois 60478-5795
>   (708) 799-2300
> Southern Building Code Congress
> International, Inc. (SBCCI)
>   900 Montclair Road
>   Birmingham, Alabama 35213-1206
>   (205) 591-1853
> International Association of Plumbing
> and Mechanical Officials (IAPMO)
>   2001 Walnut Drive South
>   Walnut, California 91789-2825
>   (909) 595-8449

Note
This information is provided as a general reference only. Refer to operative local codes and the latest editions of codes in effect in your area.

◆ UMC—UNIFORM MECHANICAL CODE

Installation of Appliances

Appliances installed in garages, warehouses, or other areas where they may be subjected to mechanical damage, are required to be suitably guarded against such damage by being installed behind protective barriers or by being elevated or located out of the normal path of vehicles. Heating and cooling equipment that generates a glow, spark, or flame that is capable of igniting flammable vapors located in a garage are required to be installed with the pilots and burners or heating elements and switches at least 18 inches above the floor level.

Where such appliances are installed within a garage and are enclosed in a separate approved compartment, having access only from outside the garage, they are permitted to be installed at floor level only where the required combustion air is taken from and discharged to the exterior of the garage. See UMC Section 508.

Instructor's Notes:

_____
_____
_____
_____
_____
_____
_____
_____
_____
_____

Disconnecting Means, Receptacle and Lighting Outlets

Equipment requiring electrical connections of more than 50 volts must have a positive means of disconnect. This disconnect must be adjacent to, and in sight from, the equipment served. A 120-volt receptacle is required to be located within 25 feet of the equipment for service and maintenance purposes. The receptacle is not required to be located on the same level as the equipment.

Where installed within a structure, low-voltage wiring of 50 volts or less is required to be installed in a manner to prevent physical damage. See UMC Section 509.

Where a furnace is installed in an attic, a permanent electric outlet and switch-controlled lighting fixture is required to be located at the required passageway opening. This lighting fixture is required to be installed at or near the furnace. See UMC Section 708.

Where a warm air furnace is located in an under-floor space, a permanent electric outlet and switch-controlled lighting fixture is required to be located at the required passageway opening. This lighting fixture is required to be installed at or near the furnace. See UMC Section 709(5).

Cooling Systems Lighting in Concealed Spaces

Where access is required to equipment located in an under-floor space, attic, or furred space, a permanent electric light outlet and lighting fixture must be installed at or near the equipment. The light must be controlled by a switch located at the required passageway opening.

Exception: Lighting fixtures are not required to be installed where the fixed lighting in the building will provide sufficient light for safe servicing of the equipment. See UMC Section 1203.

Where cooling system equipment requires access for service or maintenance, it must be provided with an unobstructed working space on the control or servicing side of the equipment. This space cannot be less than 30 inches in depth and not less than 6 feet 6 inches in height. Note: Also see the Electrical Code for working space requirements about electrical equipment. See UMC Sections 1202 and Section 1204, Exception.

A cooling system includes all equipment intended or installed for the purpose of cooling air by mechanical means and discharging such air into any room or space. This definition does not include any evaporative cooler. See UMC—91.

◆ BOCA—NATIONAL MECHANICAL CODE

Convenience Outlet

A 125-volt AC grounding-type outlet must be available for all appliances (see definition later in this chapter.) The convenience outlet must be located on the same level within 75 feet of the appliance. The outlet must not be connected to the load side of the appliance disconnect switch. See BOCA Section M-407.2.

Lighting

Permanent lighting must be provided to illuminate the area in which an appliance is located. For remote locations, the light switch must be located near the access opening leading to the appliance. See BOCA M-407.3.

◆ SBCCI—STANDARD MECHANICAL CODE

Air Conditioning, Heating, and Ventilation Equipment

Every appliance must be located with respect to building construction and other equipment so as to permit access for service of the appliance. Sufficient clearance must be maintained to permit cleaning of heating surfaces, the replacement of filters, blowers, motors, burners, controls, and vent connections; the lubrication of moving parts where required; and the adjustment and cleaning of burners and pilots.

SBCCI Mechanical Code Section 303.2.1—Outdoor Installations

Appliances listed for outdoor installations are permitted to be installed without protection in accordance with the provisions of their listing and must be accessible for servicing. Appliances listed for outdoor installation only must not be installed inside a building. See SBCCI Mechanical Code Section 303.2.2.

Attic Installations

Every attic or furred space in which mechanical equipment is installed must be accessible by an opening and passageway as large as the largest piece of the equipment and in no case less than 22 × 36 inches continuous from the opening to the equipment and its controls. The opening to the passageway must be located not more than 20 ft. from the equipment measured along the center line of such passageway. Every passageway must be unobstructed and shall have solid continuous flooring not less than 24 inches wide from the entrance opening to the equipment. On the control side and other sides where access is necessary for servicing the equipment, a level working platform extending a minimum of 30 inches from the edge of the equipment with a 36-inch high clear working space must be provided. Top or bottom service equipment must have a full clearance above or below the unit for component removal. See SBCCI Mechanical Code Section 303.4.1.

A permanent electric light outlet and lighting fixture must be provided at, or near, the mechanical equipment and controlled by a switch located at the required passageway opening. See SBCCI Mechanical Code Section 303.4.5.

Roof or Exterior Wall Installation

Mechanical equipment installations on roofs or exterior walls of buildings must comply with the requirements for roof and wall structures as specified in the Standard Building Code, and must be listed and approved for such use. See SBCCI Mechanical Code Section 303.6.1. Each appliance must have an accessible weatherproof disconnect switch and a 110- to 120-volt AC grounding-type convenience outlet on the roof adjacent to the appliance. The convenience outlet must be on the supply side of the disconnect switch. See SBCCI Mechanical Code Section 303.6.2.

Nameplate Marking

Every electric comfort heating appliance must bear a permanent and legible factory-applied nameplate that must include the following:

- Name and trademark of the manufacturer
- The catalog (model) number or equivalent
- The electric rating in volts, ampacity, and phase
- BTU output rating
- Individual marking for each electrical component in amperes or watts, volts and phase
- Required clearances from combustibles
- A seal indicating approval of the appliance by an approved testing agency (see SBCCI Mechanical Code Section 303.11.2)

Panel Heating Systems

Panel heating is a method of radiant space heating in which heat is supplied by large heated areas of room surfaces. The heating element usually consist of warm water piping, warm air ducts, or electrical resistance elements embedded in or located behind ceiling, wall, or floor surfaces. See SBCCI Mechanical Code Section 303.12.1.

The installation of panel heating systems must be designed and installed in strict accordance with the accepted engineering practices and the requirements of the SBCCI Mechanical Code Section 303.12.2.

◆ SBCCI—STANDARD GAS CODE

Electrical Connections for Electrical Ignition and Control Devices

Devices employing or depending upon an electrical current must not be used to control or ignite a gas supply if of such character that failure of the electrical current could result in the escape of unburned gas or in failure to reduce the supply of gas under conditions which would normally result in its reduction unless other means are provided to prevent the development of dangerous temperatures, pressures, or the escape of gas. See SBCCI Standard Gas Code Section 404.2.

Switches in Air Conditioning Electrical Supply Line

Means for interrupting the electrical supply to the air conditioning appliance and to its associated cooling tower (if supplied and installed in a location remote from the air conditioner) must be provided within sight of and not over 50 ft. from the air conditioner and cooling tower. See Section 508.5 SBCCI Standard Gas Code.

Electric Ignition Systems

Electric ignition systems must ignite only a pilot. The input to the pilot shall not exceed 3% of the maximum input to the main burner as fired. If ignition of the pilot is not obtained within 15 seconds, the gas must shut off automatically. See Section 807.3 SBCCI Standard Gas Code.

◆ SBCCI—STANDARD PLUMBING CODE

Existing Metallic Water Service Piping

Existing metallic water service piping used for electrical grounding must not be replaced with nonmetallic pipe or tubing until other grounding

ELECTRICAL LEVEL ONE—TRAINEE MODULE 26112

Instructor's Notes:

_____

_____

_____

_____

_____

_____

_____

_____

_____

means are provided which are satisfactory to the proper administrative authority having jurisdiction. See Section 504.3.3 SBCCI Standard Plumbing Code.

◆ IAPMO—UNIFORM PLUMBING CODE

Water Heater Locations

Attic and under-floor water heater locations must be provided with an electric light at or near the water heater. Such light must be controlled by a switch located adjacent to the opening or trap door. See Section 1311(e) IAPMO—UPC—91.

The International Association of Plumbing and Mechanical Officials include their Installation Standards in IAPMO—UPC—91. These standards are not considered as part of the Uniform Plumbing Code (UPC) unless they are formally adopted as such. Some sections include the following.

Recessed light fixtures: Polybutylene (PB) tubing shall not be installed within 12 inches of a recessed light fixture.

Identification: A permanent sign with the legible words THIS BUILDING HAS NONMETALLIC INTERIOR WATER PIPING must be fastened on or inside the main electric service panel.

◆ UBC—UNIFORM BUILDING CODE

Requirements for Group R Occupancies

A Group R occupancy includes Division 3—Dwellings and Lodging Houses. A complete code for construction of detached one- and two-family dwellings is located in Appendix Chapter 12 of the UBC. See UBC Section 1201.

Smoke Detectors

Smoke detectors are required in dwelling units that are used for sleeping purposes. They must be installed in accordance with the manufacturer's instructions.

The power source used for a smoke detector must be by a permanently connected wiring method and must be provided with a battery backup where installed in new construction. This supply cannot be operated by any type of switching device or other disconnecting means. This restriction does not include branch-circuit overcurrent protection.

A smoke detector must be installed in all sleeping rooms. Additional smoke detectors must be installed so they are located centrally in the corridor or area giving access to each separate sleeping area. See UBC Section 1210.

Heating

Dwelling units must be provided with heating facilities capable of maintaining a room temperature of 70°F at a point three feet above the floor in all habitable rooms. See UBC Section 1212.

Exterior openings exposed to the weather must be protected and must be flashed in such a manner as to make them weatherproof. See UBC Section 1708(b).

Cutting and Notching Wood

In exterior walls and bearing partitions, any wood stud may be cut or notched to a depth not exceeding 25 percent of its width. Cutting or notching of studs to a depth not greater than 40 percent of the width of the stud is permitted in nonbearing partitions supporting no loads other than the weight of the partition. See UBC Section 2517(g)(8).

Bored Holes in Wooden Studs

A hole not greater in diameter than 40 percent of the stud width may be bored in any wood stud. Bored holes not greater than 60 percent of the width of the stud are permitted in nonbearing partitions or in any wall where each bored stud is doubled, provided not more than two such successive doubled studs are so bored.

In no case shall the edge of the bored hole be nearer than ⅝ inch to the edge of the stud. Bored holes must not be located at the same section of stud as a cut or notch. See UBC Section 2517(g)(9).

Wall and Ceiling Coverings

Where electrical radiant heat cables are installed on ceilings, the stripping, if conductive, may be omitted a distance not to exceed 12 inches from the wall. See UBC Section 4705(a).

◆ BOCA—NATIONAL BUILDING CODE

Bathroom and Toilet Room Lighting

Every bathroom and toilet room must be provided with artificial light. The illumination must have an average intensity of three footcandles measured at a level of 30 inches above the floor. See BOCA Section 703.1.1.

Grounding Metal Veneers

Grounding of metal veneers on all buildings must comply with the requirements of BOCA Article 27 and NFPA 70. See BOCA Section 2103.4.3. See *NEC Section 250-116 (FPN).*

Note
This requirement must be coordinated with the electrical inspector as well. See *NEC Section 250-116 (FPN).*

SBCCI Standard Building Code

Underground inspection: To be made after trenches or ditches are excavated, conduit or cable installed, and before any backfill is put in place.

Rough-in inspection: To be made after the roof, framing, fireblocking, and bracing is in place and prior to the installation of wall or ceiling membranes.

Final inspection: To be made after the building is complete, all required electrical fixtures are in place and properly connected or protected, and the structure is ready for occupancy. See SBCCI Section 103.8.6.

Separation Between Townhouses

Where not more than three stories in height, townhouses (which are defined as a single family dwelling constructed in a series or group of attached units with property lines separating each unit) may be separated by a single wall meeting the following requirements.

The wall is required to provide not less than a two-hour fire resistance rating. Plumbing, piping, ducts, and electrical or other building services are not permitted to be installed within or through the two-hour wall, unless such materials and methods of penetration have been tested in accordance with SBCCI Section 1001.1. See SBCCI Section 403.5.2.

Metal Veneers

Metal veneers fastened to supporting elements which are not a part of the grounded metal framing of a building must be made electrically continuous by contact or interconnection of individual units and must be effectively grounded. The conductor used to ground the veneer must have no greater resistance than the conductor used to ground the electrical system within the building. Where a metal veneer is applied to a building with no electrical wiring system, grounding is required only if determined to be necessary by the Building Official. See SBCCI Section 811.4.6.

Walls, Floors, and Partitions

When walls, floors, and partitions are required to have a minimum of one hour or greater fire resis-

tance rating, cabinets, bathroom components, lighting, and other fixtures must be so installed such that the required fire resistance will not be reduced.

Exception: Fixtures which are listed for such installation are permitted. See SBCCI Section 1001.3.7.

◆ SUMMARY OF APPLICABLE DEFINITIONS FROM BUILDING CODES ABOVE

The following definitions are selected from the codes referenced above and that relate to electrical installations in residential structures. Where the code definitions differ, each definition is included.

ACCESSIBLE—Having access to, but which first may require the removal of, a panel, door, or similar obstruction. BOCA—MECH—90; SBCCI—MECH—91; SBCCI—91; SBCCI—91—GAS; UBC/UMC—91

AIR CONDITIONING—The treatment of air so as to control simultaneously its temperature, humidity, cleanliness, and distribution to meet the requirements of a conditioned space. BOCA—MECH—90; SBCCI—MECH—91; SBCCI—91—GAS

AIR CONDITIONING SYSTEM—An air conditioning system consists of heat exchangers, blowers, filters, supply, exhaust, and return ducts, and shall include any apparatus installed in connection therewith. BOCA—MECH—90; SBCCI—CI—MECH—91

AIR-HANDLING UNIT—A blower or fan used for the purpose of distributing supply air to a room, space, or area. UBC/UMC—91

ALLEY—Any public way or thoroughfare less than 16 feet but not less than 10 feet in width which has been dedicated or deeded to the public for public use. UBC—91

ALTERATION—A change in an air conditioning, heating, ventilating, or refrigeration system that involves an extension, addition, or change to the arrangement, type, or purpose of the original installation. SBCCI—MECH—91

ELECTRICAL LEVEL ONE—TRAINEE MODULE 26112

Instructor's Notes:

_____

_____

_____

_____

_____

_____

_____

_____

_____

AND/OR—In a choice of two code provisions, signifies that use of both provisions will satisfy the code requirement and use of either provision is acceptable also. SBCCI 91—GAS

APPLIANCE—A device which utilizes fuel or other forms of energy to produce light, heat, power, refrigeration, or air conditioning. This definition also shall include a vented decorative appliance. UBC/UMC—91

APPLIANCE—Utilization equipment normally built in standardized sizes which is installed or connected as a unit to perform one or more functions. SBCCI—MECH—GAS—91

APPLIANCE, AUTOMATICALLY CONTROLLED—Appliances equipped with an automatic burner which accomplish complete turn-on and shut-off of the gas to the main burner or burners, and graduate the gas supply to the burner or burners (it does not affect complete shut-off of the gas). SBCCI—91—GAS

APPLIANCE (MECHANICAL)—A device or apparatus—including any attachments or apparatus designed to be attached—which is manufactured and designed to use electricity, natural gas, manufactured gas, mixed gas, liquefied petroleum products, solid fuel, oil, or any gas as a fuel for heating, cooling, or developing light or power. BOCA—MECH—90

ATTIC—The space between the ceiling beams of the top story and the roof rafters. BOCA—90

ATTIC, HABITABLE—A habitable attic is an attic which has a stairway as a means of access and egress and in which the ceiling area at a height of 7⅓ feet above the attic floor is not more than one third of the area of the next floor below. BOCA—90

BASEMENT—Any building story having a floor below grade. SBCCI—91

BASEMENT—Any floor level below the first story in a building, except that a floor level in a building having only one floor level must be classified as a basement unless such floor level qualifies as a first story as defined herein. UBC—91

BASEMENT—That portion of a building which is partly or completely below grade (see Story above grade). BOCA—90

BATHROOM—A room containing a bathtub or shower for use by a person to bathe or cleanse one's self, located in or adjacent to a residence, apartment, hotel, motel, or similar type building. SBCCI—91—GAS

BATHROOM—A room equipped with a shower or bathtub. IAPMO—UPC—91

BUILDING—A building is a structure built, erected, and framed of component structural parts designed for the housing, shelter, enclosure, or support of persons, animals, or property of any kind. IAPMO—UPC—91

BUILDING—Any structure that encloses a space used for sheltering any occupancy. Each portion of a building separated from other portions by a fire wall must be considered as a separate building. SBCCI—91

BUILDING—Any structure used or intended for supporting or sheltering any use or occupancy. UBC—91

BUILDING—Any structure used or intended for supporting or sheltering any use or occupancy. For application of this code, each portion of a building which is completely separated from other portions by fire wall complying with Section 907.0 must be considered as a separate building. BOCA—90

BUILDING CODE—The Uniform Building Code promulgated by the International Conference of Building Officials, as adopted by this jurisdiction. ICBO ELECTRICAL CODE

BUILDING, EXISTING—Any structure erected prior to the adoption of the appropriate code, or one for which a legal building permit has been issued. BOCA—90; SBCCI—91; UBC—91

BUILDING SERVICE EQUIPMENT—The mechanical, electrical, and elevator equipment including piping, wiring, fixtures, and other accessories, which provides sanitation, lighting, heating, ventilation, fire-fighting, and transportation facilities essential for the habitable occupancy of the building or structure for its designated use and occupancy. BOCA—90

COMBUSTION—The rapid oxidation of fuel gases accompanied by the production of heat, or heat and light. SBCCI—91—GAS

COMBUSTIBLE CONSTRUCTION—A wall or surface constructed of wood, composition, or of wooden studding and lath and plaster. SBCCI—91—GAS

DWELLING—A building occupied exclusively for residential purposes by not more than two families, unless qualified otherwise in code text. SBCCI—91

DWELLING—Any building or portion thereof which contains not more than two dwelling units. UBC—91

DWELLING UNIT—Any building or portion thereof which contains living facilities, including provisions for sleeping, eating, cooking, and sanitation, as required by this code, for not more than one family, or a congregate residence for 10 or fewer persons. UBC—91

DWELLING UNIT—A single unit providing complete, independent living facilities for one or more persons, including permanent provisions for living, sleeping, eating, cooking, and sanitation. BOCA—90; SBCCI—91

DWELLING, ONE-FAMILY—A building containing one dwelling unit with not more than five lodgers or boarders. BOCA—90

DWELLING, TWO-FAMILY—A building containing two dwelling units with not more than five lodgers or boarders per family. BOCA—90

EFFICIENCY DWELLING UNIT—A dwelling unit containing only one habitable room. UBC—91

ELECTRIC HEATING APPLIANCE—A device which produces heat energy to create a warm environment by the application of electric power to resistance elements, refrigerant compressors, or dissimilar material junctions. Section 407—UBC/UMC—91

ELECTRIC SPACE HEATERS, PORTABLE—Heaters not intended for permanent connection to a structure or to electric wiring. SBCCI—MECH—91

ELECTRIC SPACE HEATERS, STATIONARY—Heaters permanently mounted in a structure (air duct system) and which are permanently connected to electric wiring. SBCCI—MECH—91

EQUIPMENT, EXISTING—Any equipment covered by this code which was installed prior to the effective date of this code, or for which an application for permit to install was filed with the code official prior thereto. BOCA—90

EXISTING WORK—A mechanical system, or any part thereof, installed prior to the effective date of this code. BOCA—MECH—90

EXIT—See Section 3301(b) Definitions of: Balcony, Exterior Exit; Exit; Exit Court; Exit Passageway; and Horizontal Exit. UBC—91

FIRE PROTECTION—The provision of construction safeguards and exit facilities; and the installation of fire alarm, fire detecting, and fire extinguishing service equipment, to reduce the fire risk and the conflagration hazard. BOCA—90

FLOOR AREA—The area included within the surrounding exterior walls of a building or portion thereof, exclusive of vent shafts and courts. The floor area of a building, or portion thereof, not provided with surrounding exterior walls must be the usable area under the horizontal projection of the roof or floor above. UBC—91

FURNACE—A completely self-contained unit that produces heat by utilizing electric energy or by burning fuel. SBCCI—MECH—91

FURNACE, CENTRAL—A self-contained indirect fired or electrically heated appliance designed to supply heated air through ducts to spaces remote from or adjacent to the appliance. SBCCI—MECH—91

GARAGE, PRIVATE—A building or a portion of a building, not more than 1,000 square feet in area, in which only motor vehicles used by the tenants of the building or buildings on the premises are stored or kept (see UBC Chapter 11). UBC—91

GARAGE, PRIVATE—A garage for four or less passenger motor vehicles without provision for repairing or servicing such vehicles for profit (see Section 608.0). BOCA—90

GRADE—A reference plane representing the average of finished ground level adjoining the building at all exterior walls. When the finished ground level slopes away from the exterior walls, the reference plane must be established by the lowest points within the area between the building and the lot line or between the building and a point six feet from the building, whichever is closer to the building. SBCCI—91

HABITABLE SPACE—A space in a structure for living, sleeping, eating, or cooking. Bathrooms, toilet compartments, closets, halls, storage or utility spaces, and similar areas are not considered habitable spaces. BOCA—90; SBCCI—91

HABITABLE SPACE (ROOM)—Space in a structure for living, sleeping, eating, or cooking. Bathrooms, toilet compartments, closets, halls, storage or utility spaces, and similar areas, are not considered habitable spaces. UBC—91; BOCA—90

Instructor's Notes:

_____

_____

_____

_____

_____

_____

_____

_____

_____

HEAT PUMP—An appliance having heating or heating/cooling capability and which uses refrigerants to extract heat from air, liquid, or earth sources. UBC/UMC—91

HEATING EQUIPMENT—Includes all warm-air furnaces, warm-air heaters, combustion products vents, heating air distribution ducts and fans, all steam and hot water piping, together with all control devices and accessories installed as part of, or in connection with, any environmental heating system or appliance regulated by this code. See UBC/UMC Section 410.

HEATING SYSTEM—A warm-air heating plant consisting of a heat exchanger enclosed in a casing, from which the heated air is distributed through ducts to various rooms and areas. A heating system includes the outside air, return air, and supply air system, and all accessory apparatus and equipment installed in connection therewith. UBC/UMC—91

HEATING SYSTEM, CENTRAL WARM AIR—A heating system consisting of an air heating appliance from which the heated air is distributed by means of ducts, pipes, or plenums including any accessory apparatus and equipment installed in connection therewith. SBCCI—MECH—91

LISTED and LISTING—Terms referring to equipment and materials which are shown in a list published by an approved testing agency, qualified and equipped for experimental testing and maintaining an adequate periodic inspection of current productions and which listing states that the material or equipment complies with accepted national standards which are approved or standards which have been evaluated for conformity with approved standards.

MASONRY—That form of construction composed of stone, brick, concrete, gypsum, hollow clay tile, concrete block or tile, glass block or other similar building units or materials, or a combination of these materials laid up by unit and set in mortar. UBC—91

READILY ACCESSIBLE—Having direct access without the need of removing or moving any panel, door or similar obstruction. BOCA—MECH—90

READILY ACCESSIBLE—Having direct access without the need of removing any panel, door, or similar covering of the item described, and without requiring the use of portable ladders, chairs, etc. SBCCI—MECH—GAS—91

SINGLE-FAMILY DWELLING—A single-family dwelling shall mean a building designed to be used as a home by the owner of such building, and must be the only dwelling located on a parcel of ground with the usual accessory buildings. IAPMO—UPC—91

SMOKE DETECTOR—An approved device that senses visible or invisible particles of combustion. UBC—91

SMOKE DETECTOR—An approved listed detector sensing either visible or invisible particles of combustion. SBCCI—91

STAIRWAY—One or more flights of stairs, either exterior or interior, with the necessary landings and platforms connecting them, to form a continuous and uninterrupted passage from one level to another in a building or structure. SBCCI—91

STRUCTURE—That which is built or constructed, an edifice or building of any kind, or any piece of work artificially built up or composed of parts joined together in some definite manner. UBC—91

TOWNHOUSE—A single-family dwelling constructed in a series or group of attached units with property lines separating each unit. SBCCI—91

USABLE CRAWL SPACE—A crawl space designed to be used for equipment or storage. SBCCI—91

WATER HEATER—An appliance designed primarily to supply hot water and is equipped with automatic controls limiting water temperature to a maximum of 210°F. UBC—91

Answers to Review Questions

Answer	Section
1. d	2.3.3
2. a	6.1.0/Figure 16
3. b	6.3.0
4. b	7.0.0
5. a	8.1.1
6. d	8.1.2
7. b	8.1.4
8. c	8.1.4
9. a. 1	10.0.0
b. 5	
c. 6	
d. 2	
e. 4	
f. 3	
10. See table below	12.0.0

Number and Size of Conductors in Box	Free Space Within Box for Each Conductor	Total Cubic Inches of Box Space Required	What Size Metallic Box May Be Used?
Six No. 12 conductors and three ground wires	2.25	15.75	4 × 1¼ square box
Seven No. 12 conductors and three ground wires with one receptacle	2.25	22.5	4¹¹⁄₁₆ × 1¼ square box
Two No. 14 conductors and one ground wire	2.0	6	3 × 2 × 1½ device box
Four No. 14 conductors and two ground wires	2.0	10	3 × 2 × 2 device box
Six No. 14 conductors and three ground wires with one receptacle	2.0	18	3 × 2 × 3½ device box

Instructor's Notes:

_____

_____

_____

_____

_____

_____

_____

_____

_____

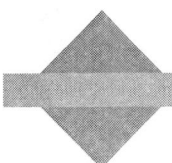

Additional Resources

For advanced study of topics covered in this task module, the following books are suggested:

Illustrated Dictionary for Electrical Workers, Latest Edition. Albany, NY: Delmar Publishers.

Illustrated Guide to the NEC, Latest Edition. Carlsbad, CA: Craftsman Book Co.

National Electrical Code Handbook, Latest Edition. Quincy, MA: National Fire Protection Association.

ELECTRICAL LEVEL ONE—INSTRUCTOR'S GUIDE MODULE 26112

NCCER CRAFT TRAINING USER UPDATES

The NCCER makes every effort to keep these textbooks up-to-date and free of technical errors. We appreciate your help in this process. If you have an idea for improving this textbook, or if you find an error, a typographical mistake, or an inaccuracy in the NCCER's Craft Training textbooks, please write us, using this form or a photocopy. Be sure to include the exact module number, page number, a detailed description, and the correction, if applicable. Your input will be brought to the attention of the Technical Review Committee. Thank you for your assistance.

Instructors – If you found that additional materials were necessary in order to teach this module effectively, please let us know so that we may include them in the Equipment/Materials list in the Instructor's Guide.

Write: Curriculum Revision and Development Department
National Center for Construction Education and Research
P.O. Box 141104, Gainesville, FL 32614-1104

Fax: 352-334-0932

E-mail: curriculum@nccer.org

Craft _____   Module Name _____

Copyright Date _____   Module Number _____   Page Number(s) _____

Description _____

_____

_____

_____

_____

(Optional) Correction _____

_____

_____

(Optional) Your Name and Address _____

_____

_____

_____

TEACHING TIPS

Section 2.6.3   *Think About It—Demand Factors*

Because we are dealing with residential appliances, there is no likelihood that all the appliances will be operated simultaneously at their maximum rating. The demand factor decreases more for larger appliances because the demand factor is a function of usage, not the size of the appliance.

Section 3.0.0   *Think About It—Balanced Phase Conductors*

Phase imbalance causes voltage fluctuations within the phases, which might be seen in flaring and dimming of lights when the neutral connection at the service entrance is poor. A three-phase electric motor can be damaged if the voltage between the legs is out of balance by more than 2% or the current between the legs is out of balance by more than 10%.

Section 4.1.2   *Think About It—GFCIs*

The main difference between single-pole and two-pole GFCI breakers is that the single-pole device must be installed on independent circuits. Circuits that use a neutral common to more than one hot conductor cannot be protected by a single-pole GFCI breaker.

Section 5.2.3   *Think About It—Grounding Electrode Conductors*

NEC Section 250-24 states that the requirements for the grounding electrode conductor are covered by *NEC Article 250, Part C.* There, *NEC Table 250-66* provides the information needed to find the size of the grounding electrode conductor. Its size is based on the size of the largest service conductor and the material the grounding electrode conductor is made from (copper, aluminum, or copper-clad aluminum).

Section 8.1.1   *Think About It—Romex® Cable*

Running several nonmetallic-sheathed cables bundled together through holes drilled in joists is not a code violation. However, the following points should be considered. Drilling large holes in the center of a joist structurally weakens the joist. It is better to drill holes near the top or bottom part of a joist. If the edge of the bored hole is less than 1¼" from the joist edge, a steel plate must be used to protect the cables. Since most joists are spaced no more than 24" apart, derating of the cable ampacities because of bundling is not required.

Section 9.0.0   *Think About It—Grounding Devices*

The items shown from left to right are an equipment grounding conductor (pigtail), grounding clip, crimp lug, and a bonding bushing. The bonding bushing is *not* used for equipment grounding.

Section 10.5.0 *Think About It—240V Circuits*

In accordance with *NEC Table 220-19,* the required ampacity for the freestanding kitchen range is 33.33 amps (8,000VA ÷ 240V). This would require the use of a 40A circuit breaker and No. 8 AWG conductors. Have the trainees sketch and describe the different ways the installation can be accomplished. Typically, the circuit consists of a three-wire cable run from the circuit breaker in the main panel to a heavy-duty range receptacle located adjacent to the range.

Section 12.1.0 *Think About It—Flush Mounting*

The NEC requirements pertaining to the location of the front edge of boxes relative to the finished surface in noncombustible and combustible walls (and ceilings) are meant to prevent the spread of fire if a short circuit occurs within the box.

Section 14.1.0 *Think About It—Wiring Three-Way Switches*

The figures below show two arrangements for connecting three-way switches. One has the load and supply coming in from different boxes and the other has the load and supply coming in from the same box.

LOAD AND SUPPLY FROM DIFFERENT BOXES

LOAD AND SUPPLY FROM SAME BOX

112-IG.EPS

Section 16.0.0 *Think About It—Residential Wiring*

Discuss the results of the trainee surveys. Since you are likely to get a variety of results because of differing ages and types of housing, it is a good opportunity to discuss how code requirements, materials, and installation practices have evolved over the years.